Consumers' Surplus and Louisiana Fish Farming

What happens to a market when an alternative production technique is introduced. One way is to measure the gain to consumers in price reduction, called consumers' surplus. It is used to measure the effect on the Louisiana crawfish industry when fish farms using private ponds compete with standard fisheries using lakes and rivers.

Demography and the Social Security System

Many people mistakenly believe that the taxes they pay to the social security system are put into a savings account to be held until their retirement. Not so! The demographics show that either the system will prove to be a great burden on the work force or the benefits paid to retirees will be greatly reduced.

Minimum Cost Production in the Chemical Industry

Jojoba oil is used for producing amides which, in turn, are an essential ingredient in many chemical processes. Researchers have tried various methods to maximize the amide yield while conserving cost. The chemical industry uses the cost function, which is a function of the two variables temperature and time, to minimize the cost of producing amides.

Volumes: Old and New

The history of ways to calculate volume includes pyramids in ancient Egypt, other solids with Archimedes about 220 B.C., wine casks from the astronomer Kepler about 1600, and a general approach by Newton and Leibniz about 1700. The Case Study applies these methods to famous domes throughout the world and to water in the Great Salt Lake.

Melbourne's Underground Railway: Shaft Design

In the early 1980s the Melbourne, Australia, government started an extensive, controversial plan to build a complex underground system of tunnels that required a new rock-boring machine using rotating cutter heads. Trigonometric functions were used to solve several design problems.

Text Typeface: Times Roman
Compositor: General Graphic Services
Acquisitions Editor: Robert Stern
Developmental Editor: Richard Koreto
Managing Editor: Carol Field
Project Editor: Laura Maier
Copy Editor: Merry Post
Manager of Art and Design: Carol Bleistine
Associate Art Director: Doris Bruey
Text Designer: William Boehme
Cover Designer: Lawrence R. Didona
Text Artwork: Vantage Art, Inc.
Director of EDP: Tim Frelick
Production Manager: Robert Butler
Marketing Manager: Monica Wilson

Cover Credit: Cover illustration produced by James T. Hoffman of the Center for Geometry Analysis, Numerics, and Graphics, University of Massachusetts at Amherst.

Printed in the United States of America

BRIEF CALCULUS WITH APPLICATIONS, Second edition

0-03-072979-3

Library of Congress Catalog Card Number: 92-050298

2345   049   987654321

THIS BOOK IS PRINTED ON **ACID-FREE, RECYCLED** PAPER

Brief Calculus with Applications

Second Edition

Raymond F. Coughlin
David E. Zitarelli

Temple University

$(800)\ 544-6678$

Saunders College Publishing

Harcourt Brace Jovanovich College Publishers

Fort Worth   Philadelphia   San Diego   New York
Orlando   Austin   San Antonio   Toronto
Montreal   London   Sydney   Tokyo

Preface

In the summer of 1991 a committee of the National Research Council, which is part of the National Academy of Sciences, issued a report called *Moving Beyond Myths*.* The report describes many serious problems in undergraduate mathematics education and provides an action plan for attacking them. One of the myths that this committee identified (see page 12) is

Myth: *Only scientists and engineers need to study mathematics.*

Reality: Mathematics is a science of patterns that is useful in many areas. Indeed, the most rapid areas of growth in applications of mathematics have been in the social, biological, and behavioral sciences. Financial analysts, legal scholars, political pollsters, and sales managers all rely on sophisticated mathematical models to analyze data and make projections. Even artists and musicians use mathematically based computer programs to aid in their work. No longer just a tool for the physical sciences, mathematics is a language for all disciplines.

The goal of this book is to reverse this myth. Although the table of contents is standard, the means of achieving the goal differs from other texts in two notable ways. First, in every section the Referenced Exercises refer to areas outside mathematics and are accompanied by footnotes so the student can pursue them in greater depth, or the instructor can assign them as part of special projects. Because such applications deal with real-life situations, calculators are frequently necessary to carry out the computations.

We hope to show students through these exercises that mathematics is everywhere. No matter what major a student chooses, mathematics can play a key role in solving interesting problems in that discipline. Here is a representative list of selected Referenced Exercises, along with the section where each can be found.

Application	Source	Section
stocks and bonds	*Forbes Magazine*	1–3
income tax	*Wall Street Journal*	2–3
telecommunications	*Forecasting Public Utilities*	3–2
regeneration of trees	*Ecology*	4–5
learning curves	*Decision Sciences*	5–2

*William E. Kirwan (Editor), *Moving Beyond Myths: Revitalizing Undergraduate Mathematics*, Washington D.C., National Academy Press, 1991.

Application	Source	Section
pollution control	*American Journal of Agricultural Economics*	6–4
gifted students	*Journal of Mathematical Psychology*	7–1
cancer treatment	*Physics in Medicine and Biology*	8–2
sound transmission	(text) *Musical Acoustics: Piano and Wind Instruments*	9–4

In addition, almost every chapter has at least one Case Study that treats a particular application in greater detail than the Referenced Exercises. The topics have been chosen not only to illustrate areas where mathematics has played a crucial role in solving an important problem, but to highlight areas of particular human interest.

Even if the instructor does not have time to cover all the Referenced Exercises or Case Studies, it is hoped that their relevance and wealth (in both diversity and number) will make a lasting impression.

Changes in the Second Edition

This second edition of the book differs from the first edition in several ways.

1. The exercises are divided into three sets. The first set consists of the standard assortment of problems, usually numbering between 50 and 70. The first 20 or so generally reflect the examples in the text and are routine, while the later problems are a bit more challenging. The second set, titled ''Referenced Exercises,'' contains problems from areas outside mathematics; these exercises are accompanied by complete references to the literature. The third set, ''Cumulative Exercises,'' contains problems whose solutions call on material from the preceding sections in that chapter. These problems often require different skills than do the usual problems. The authors have found this feature very useful in their own classes as a way to continually help students review old material as they study the current section. In total, there are over 4000 exercises in this book, about 20% of them new to this edition. Although the book continues to contain application exercises associated with many subjects, many simple drill problems have been added as well to help students master concepts before applying them.

2. There are many new and updated Referenced Exercises.

3. There is a new Case Study, ''Volumes: Old and New.''

4. The normal exercise sets and the cumulative exercise sets include two types of problems that several users of the first edition requested: those that are stated in words and require the translation from ordinary language into mathematical terms, and those that require a geometric interpretation.

5. The new, inexpensive graphics calculators promise to revolutionize the way we learn mathematics. They are especially helpful when learning calculus because of their amazing capability to sketch the graph of any function

almost instantaneously. We have included two ways to help students incorporate graphics calculators into the study of calculus. One is an appendix for beginners. The other consists of Graphics Calculator Explorations, which appear at the end of every chapter and center on some geometric ideas that lend themselves to this new technology. These Explorations are open-ended in the sense that the student can extend the ideas presented in the text. Found in the Explorations are instructions on the use of Casio and Texas Instruments graphing calculators.

6. Many instructors who used the first edition suggested that two important topics should have their own sections. Section 2-4 is now exclusively devoted to continuous and differentiable functions and Section 8-6 is devoted to volume.

Organization

The text can be divided into three parts: differential calculus (Chapters 1 through 5), integral calculus (Chapters 6 and 7), and special topics (Chapters 8 and 9).

Differential Calculus

Chapter 1 covers the algebraic techniques needed for the subsequent material. Functions are defined and studied, and various properties of polynomial functions and rational functions are illustrated. Chapter 2 explains the limit of a function by discussing velocity, rate of change of a moving object, and the tangent to a graph. The definition of the derivative is used to calculate the derivatives of several functions.

Chapter 3 presents various techniques for computing the derivatives of functions. Chapter 4 applies these techniques in a wide variety of problems.

Chapter 5 introduces exponential and logarithmic functions. Their derivatives are obtained by calculator experiments aimed at suggesting general rules. Those rules are then proved rigorously.

Integral Calculus

Chapter 6 introduces integration via antidifferentiation. After indefinite integrals have been defined, the Fundamental Theorem of Calculus relates the derivative to the integral. Then the integral is used to compute areas bounded by curves.

Chapter 7 presents several techniques of integration, including the use of tables of integrals and numerical integration.

Special Topics

Chapter 8 extends the definition of the derivative and the definition of the integral to functions of more than one variable. Partial derivatives are defined and applied to the sketching of surfaces, while double integrals are evaluated for functions of two variables.

The trigonometric functions, studied in Chapter 9, allow us to create models in such diverse disciplines as astronomy, ecology, navigation, and music, when we must consider "periodic" functions, which are functions that repeat their functional values periodically.

Format

We have tried to keep the length of each section to what can be covered in a typical 50-minute class. Sometimes, however, the topic has dictated more extensive coverage. Each numbered section has been partitioned into subsections to help the instructor prepare lectures and to help the student organize the material.

There are two kinds of examples. One explains a new skill that is being encountered for the first time. It is labeled simply EXAMPLE. The other illustrates a skill that was explained beforehand. It uses the following format:

E X A M P L E

Problem

Solution

The student should be able to make a good attempt at solving the problem independently before reading its solution.

Each chapter ends with a review of the terms, notation, and formulas that were introduced in the chapter. It also contains review problems that can be used to review for tests.

Supplements

A number of supplements are available for use with this text.

The **Instructor's Manual**, by Richard Shores, Lynchburg College, contains the partially worked-out solutions to all exercises in the text. (Answers to the odd-numbered exercises are also in the back of the text.) Also contained are additional teaching hints and aids for the instructor.

A **Student Solutions Manual and Study Guide**, also by Richard Shores, contains fully worked-out solutions to every other odd-numbered problem. This manual is designed to help students with their problem-solving skills: These solutions can be used as models in solving similar problems.

The software package **Graph 2D/3D** by George Bergeman, Northern Virginia Community College, is available free to users of this text. It supplies graphical and computational support for many of the important topics in each chapter. The software requires an IBM or IBM-compatible computer with at least 512K.

The test bank by Robert Kurtz and Pao-sheng Hsu of the University of Maine at Orono, contains about 1000 questions, all new to this edition. "Writing Across the Curriculum" problems, which apply mathematics to different fields, make this manual unique. Even instructors who use their own test banks will find this one useful.

The test bank is also available in a computerized format for Macintosh, IBM, and IBM-compatible computers. This computerized version allows an instructor to custom-design tests and to sort the questions by several different categories. The IBM version requires 256K and two disk drives or 384K and a hard drive, graphics card and monitor, and a printer capable of handling graphs. Full instructions are included.

Raymond F. Coughlin
David E. Zitarelli
Philadelphia, PA
December 1992

Acknowledgments

It is a pleasure to acknowledge the help we received from many people. Our biggest support has continued to come from our wives, Anita and Judy, and children, Paul and Nicole, and Virginia, Christina, and Sara.

Several reviewers deserve special thanks for their constructive criticism of the entire manuscript.

Second Edition Reviewers

Carol Benson, Illinois State University
Stephen R. Bernfeld, University of Texas at Arlington
Edward A. Boyno, Montclair State College
James W. Brewer, Florida Atlantic University
Priscilla Chaffe-Stengel, California State University, Fresno
Charles C. Clever, South Dakota State University
Lynne Doty, Marist College
Keith Ferland, Plymouth State College
Nancy Fisher, University of Alabama
Anne M. Fitzmaurice, University of Hartford
Jerry Goldman, DePaul University
Frances Gulick, University of Maryland, College Park
Denise Hennicke, Collin County Community College
David Jones, Glendale Community College
Robert Kurtz, University of Maine, Orono
Sarah L. Mabrouk, Boston University
M.N. Manougian, University of South Florida
Claire C. McAndrew, Fitchburg State College
Richard J. McGovern, Marist College
Richard Nadel, Florida International University
James Osterburg, University of Cincinnati
Jack R. Porter, The University of Kansas
H. Suey Quan, Golden West College
Richard J. Shores, Lynchburg College
Lowell Stultz, Kalamazoo Valley Community College
William R. Trott, University of Mississippi
Melvin F. Tuscher, West Valley College
P.L. Waterman, Northern Illinois University

S.K. Wyckoff, Brigham Young University
Arnold R. Vobach, University of Houston

We would like to thank again the very helpful reviewers who read the manuscript of the first edition at various stages of development.

First Edition Reviewers

Gail Broome, Providence College
Garret Etgen, University of Houston
Lou Hoelzle, Bucks County Community College
Ron Jeppson, Moorhead State University
Lawrence Maher, North Texas State University
Giles Maloof, Boise State University
Gordon Schilling, University of Texas at Arlington
Daniel Symancyk, Anne Arundel Junior College

A very important part of the reviewing process is the checking of errors. For their aid in this process, we would like to thank Bob Martin, Tarrant County Junior College, and Paul Allen, University of Alabama, who performed the arduous task of checking galleys and page proofs for mathematical accuracy.

We extend special gratitude to our friend Vincent Damiano, who is a Senior Project Engineer for Chevron USA, Inc. He read the entire manuscript and suggested several realistic linear programming problems.

The staff at Saunders has been a pleasure to work with. We especially thank our editors, Robert Stern and Richard Koreto, for the interest, encouragement, guidance, and persistence they have bestowed upon this project. We also thank the project editor, Laura Maier, and the art directors, Carol Bleistine and Doris Bruey, not only for their professional expertise but for their patience and humor as well. Our appreciation also goes to Merry Post for her excellent copyediting.

Contents

5 Exponential and Logarithmic Functions / 236

6 The Integral / 295

7 Techniques of Integration / 346

8 Functions of Several Variables / 379

Index of Applications

Ecology

Social Sciences

General Topics

Brief Calculus
with Applications

Functions

Powerlifting and polynomials: perfect together. (Walt Evans)

Chapter Overview

There are two parts of calculus, differentiation and integration, both of which are based on the concepts of functions and limits. This chapter presents the relevant material on functions, and the next chapter introduces limits.

Section 1–1 discusses graphs of equations, and Section 1–2 studies straight lines and their graphs. In Sections 1–3 and 1–4 the equations are regarded as functions. Since not all equations are functions, a criterion for distinguishing between them is elaborated. The material is applied to an area called cost analysis.

Section 1–4 introduces power functions, polynomial functions, and rational functions. Two additional kinds of functions are described: those that are defined by different formulas over different intervals and those that are defined implicitly by a graph.

Section 1–5 discusses five operations on functions: the four rational operations and composition. It also introduces quality control. Section 1–6 presents some methods for solving polynomial equations, including a brief review of factoring and the quadratic formula. This section describes how to find the points of intersection of the graphs of functions.

CASE STUDY PREVIEW

"Which one is stronger, an ant or an elephant?" The Case Study will not answer this proverbial question, but it will provide some food for thought by examining an analogous question about human beings.

Powerlifters can lift vast amounts of weight. Their meets are divided into weight classes, and each weight class has a champion. But powerlifting meets also feature a "Champion of Champions" award, and the person who lifts the largest weight may not necessarily win it. The reason is the so-called "Schwartz Formula," a handicapping scheme that distributes points according to a person's bodyweight. The Case Study describes how it is defined in terms of a polynomial function.

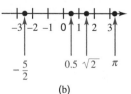

Unit distance

(a)

(b)

FIGURE 1

1–1 The Cartesian Coordinate System

About 1630 two Frenchmen, René Descartes and Pierre Fermat, independently revolutionized the way of expressing scientific laws. The idea they hit upon has come to be called "analytic geometry" or "algebraic geometry," because it is a blend of algebra and geometry. Descartes is usually given credit for being the founder of this field, and the term "Cartesian coordinate system" reflects humankind's debt to him, but Fermat deserves at least as much credit for his pioneering work.* Sections 1–1 and 1–2 are devoted to their discovery.

*For the historical details, see Section 17.11 (pp. 386–388) of Carl B. Boyer and Uta C. Merzbach, *A History of Mathematics* (second edition), New York, John Wiley & Sons, 1989.

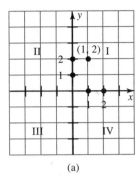

(a)

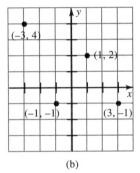

(b)

FIGURE 2

The Graph of an Equation

Consider the formula $y = 3x + 1$. It is an algebraic expression that describes the relationship between the variables x and y, but it can also be viewed as a geometric object, namely, a line in the plane. To accomplish this both Descartes and Fermat introduced a *coordinate system,* which we will describe briefly.

We start with the **real line.** It is a straight line with one point marked as the **origin,** corresponding to the number 0 (see Figure 1a). A point to the right of 0 is marked and it corresponds to the number 1. The distance from 0 to 1 is called a **unit distance.** Points that are located at one-unit intervals to the right of 1 are marked 2, 3, and so on. Negative numbers are located to the left of 0. In this way, each real number a corresponds to the point that is $|a|$ units to the right of 0 if a is positive or to the left if a is negative. Additional points, corresponding to some rational and irrational numbers, are marked off in Figure 1b.

Next construct a vertical line perpendicular to the horizontal real line at the origin, and mark off a unit length on it. The positive direction is upward. Usually the horizontal line is called the **x-axis** and the vertical line the **y-axis.** Every point in the plane is represented by an ordered pair of numbers. The ordered pair (1, 2) corresponds to the point that is one unit to the right of the y-axis and two units above the x-axis. (See Figure 2a.) In the ordered pair (1, 2), 1 is called the **x-coordinate** and 2 the **y-coordinate.**

In Figure 2b additional points are plotted in the coordinate system. The numbers in the ordered pairs are also called the first and second **coordinates.** This coordinate system is called the **Cartesian coordinate system** after Descartes. The choice of the letters x and y is arbitrary; we will change them when a particular problem dictates. The Cartesian coordinate system divides the plane into four distinct **quadrants** denoted by the Roman numerals shown in Figure 2a.

Consider the equation $y = 3x + 1$. If $x = 1$, then $y = 4$. These values are treated as the point (1, 4) in the Cartesian coordinate system. The point (1, 4) is said to "satisfy" the equation $y = 3x + 1$. The set of all points that satisfy an equation is called the **graph** of the equation. The graph is obtained from a table of values of x and the corresponding values of y. Example 1 illustrates the procedure for drawing the graph of an equation.

EXAMPLE 1

Consider the equation $y = 3x + 1$. To find its graph, make a table of some values of x and the corresponding values of y. If $x = -2$, then $y = 3x + 1 = 3(-2) + 1 = -5$. This means that the point $(-2, -5)$ lies on the graph. The table lists four additional points corresponding to $x = -1, 0, 1,$ and 2.

x	-2	-1	0	1	2
y	-5	-2	1	4	7
point	$(-2, -5)$	$(-1, -2)$	$(0, 1)$	$(1, 4)$	$(2, 7)$

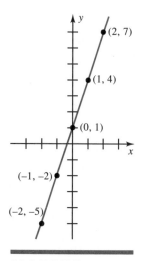

FIGURE 3

Next plot the five points $(-2, -5)$, $(-1, -2)$, $(0, 1)$, $(1, 4)$, and $(2, 7)$ on a Cartesian coordinate system. When these points are joined the resulting graph appears to be the line drawn in Figure 3.

How can we be sure that the graph of $y = 3x + 1$ is a line? One way is to plot additional points. By letting x be *any* real number in the equation $y = 3x + 1$ there will be an infinite number of points of the form $(x, 3x + 1)$ on the graph. These points constitute the graph of $y = 3x + 1$ shown in Figure 3.

The five values that we chose for x in the table were selected arbitrarily. The graph of $y = 3x + 1$ will be the same line even if other values of x are selected.

Example 1 illustrates how algebra and geometry were molded into one mathematical discipline called analytic geometry. The essence is that equations in two variables correspond to curves in two dimensions. The procedure for sketching the graph of an equation is called the **plotting method.** It consists of three steps:

1. Make a table of values of x and y.
2. Treat the values as ordered pairs on a coordinate system and plot them.
3. Connect the points with a smooth curve.

The graph of the equation is the curve that results from step 3. In Section 1–2 we will see how to proceed in the opposite direction, from the graph to the equation.

This simple idea allows algebraic equations to be pictured geometrically, reflecting the old adage that "a picture is worth a thousand words." The next example illustrates it.

EXAMPLE 2

Problem

Sketch the graph of the equation $4x + 2y = 1$ by the plotting method.

Solution Construct a table for the equation $4x + 2y = 1$ by choosing a value for x and substituting it into the equation to get the corresponding value of y. If $x = 1$, then $4(1) + 2y = 1$, so $y = -\frac{3}{2}$. Therefore $(1, -\frac{3}{2})$ is a point on the graph of $4x + 2y = 1$. The table provides four additional points.

x	-2	-1	0	1	2
y	$\frac{9}{2}$	$\frac{5}{2}$	$\frac{1}{2}$	$-\frac{3}{2}$	$-\frac{7}{2}$
point	$(-2, \frac{9}{2})$	$(-1, \frac{5}{2})$	$(0, \frac{1}{2})$	$(1, -\frac{3}{2})$	$(2, -\frac{7}{2})$

Plot these points and draw the curve through them. The graph is shown in Figure 4.

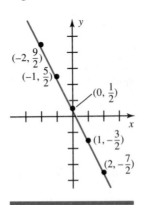

$$\text{IF } x = 1, \quad 4(1) + 2y = 1$$
$$4 + 2y = 1$$
$$2y = -3$$
$$y = -\frac{3}{2}$$

FIGURE 4

Notice that the graph of the equation $4x + 2y = 1$ shown in Figure 4 is a line. Plotting additional points will confirm this fact. It is possible to prove the following statement.

> The graph of an equation of the form $Ax + By = C$ is a line when A, B, and C are real numbers such that A and B are not both zero.

An equation in two variables whose graph is a line is called a **linear equation.** Examples 1 and 2 show that $y = 3x + 1$ and $4x + 2y = 1$ are linear equations.

Applications

For many businesses the total cost C (in dollars) of producing a commodity and the number x of units produced can be written as a linear equation. For instance, let

$$C = 500 + 4x$$

If no units are produced, then $x = 0$, so $C = \$500$. This amount is called the fixed cost. If $x = 25$ units are produced, then the total cost is $C = \$600$.

Often the total revenue R (in dollars) derived from selling the commodity can be written as a linear equation. For instance, let

$$R = 8x$$

If 25 units are produced, then $R = \$200$, in which case the revenue is $400 less than the cost. The business will have to produce more units in order to turn a profit. In Section 1–3 we will explain how to compute the number of units needed to derive a profit from the sales, but for now we lay the foundation for the solution.

EXAMPLE 3

Problem

Sketch the graphs of the cost equation and the revenue equation on the same coordinate axis.

Solution Use the plotting method to sketch the graph of the cost $C = 500 + 4x$. A calculator can be quite helpful in making the table. For instance, if $x = 150$, then the value of C can be obtained from the following sequence of key strokes.

$$4 \boxed{\times} 150 \boxed{=} \boxed{+} 500 \boxed{=}$$

x	0	50	100	150
C	500	700	900	1100
point	(0, 500)	(50, 700)	(100, 900)	(150, 1100)

Sketch the graph of $R = 8x$ for the same values of x.

x	0	50	100	150
R	0	400	800	1200
point	(0, 0)	(50, 400)	(100, 800)	(150, 1200)

The graphs of the two equations are shown in Figure 5.

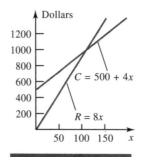

FIGURE 5

Most large firms have a planning group which constructs sales forecasts. These forecasts are critical for production scheduling, plant design, and financial planning. Graphs are often used for viewing the relationship between sales growth and external financial requirements. Example 4 examines the Arinson Products Company, whose owner, John Arinson, sought funds to maintain exclusive control of the company.†

EXAMPLE 4

EXTERNAL FUNDS REQUIRED

Let x represent the sales growth rate of the Arinson Products Company and let y represent the external funds required (in thousands of dollars), where

SALES GROWTH

$$y = 494x - 16$$

Problem

(a) Draw the graph of the equation. (b) If the sales growth rate is 5%, what external funding is required? (c) If $107,500 is available for external funding, what is the sales growth rate?

Solution (a) If $x = 0.1$ (corresponding to a 10% sales growth rate), then $y = 494(0.1) - 16 = 33.4$. The table lists several other values of x and the corresponding values of y.

x	-0.2	-0.1	0	0.1	0.2
y	-114.8	-65.4	-16	33.4	82.8

The graph is sketched in Figure 6.

(b) For a sales growth rate of 5% set $x = 0.05$. Then $y = 494(0.05) - 16 = 8.7$. Since the units for y represent thousands of dollars, the amount of external funds required is $8700.

(c) If $107,500 is available for external funding, set $y = 107.5$. The sales growth rate is obtained by substituting $y = 107.5$ into the equation and solving for x.

$$494x - 16 = y$$

$$494x - 16 = 107.5$$

$$x = \frac{107.5 + 16}{494} = 0.25$$

The sales growth rate is $x = 25\%$.

†B. Campsey and E. Brigham, *Introduction to Financial Management*, Hinsdale, Ill., The Dryden Press, 1985, pp. 190–204.

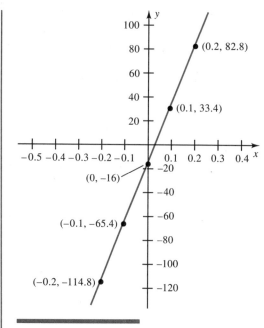

FIGURE 6

EXAMPLE 5

Based on figures supplied by the U.S. Census Bureau, the population of California in the decade 1980–1990 can be described by the equation $C = 617,134.8x + 23,667,902$, where x is the number of years after 1980.

Problem

(a) What was the population of California in 1980? (b) What was the population of California in 1990? (c) What was the increase in California's population in the 1980s?

Solution (a) Since x is the number of years after 1980, the population in 1980 can be found by setting $x = 0$. Then $C = 23,667,902$, which was California's population in 1980.
(b) For the year 1990 set $x = 10$. Then $C = 6,171,348 + 23,667,902 = 29,839,250$. This was California's population in 1990.
(c) The increase in population in the 1980s is found by subtracting the population in 1980 from the population in 1990. Hence the increase was $29,839,250 - 23,667,902 = 6,171,348$.

Notice that the answer to Example 5(c) is ten times the coefficient of x in the equation $C = 617,134.8x + 23,667,902$. This is not an accident. The exercises in this section and the next two sections will explore the matter further.

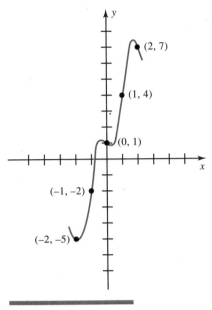

FIGURE 7

What's in a Name?

All of the graphs in this section have been lines. Of course, these are not the only graphs of algebraic equations according to the method of Descartes and Fermat; there are also circles, parabolas, and many others. It is useful to be able to picture a graph when its equation is given.

So far, graphs have been drawn by the plotting method, but this method has its shortcomings. Given the table of five points for the equation $y = 3x + 1$ in Example 1, it is conceivable that its graph could be the curve in Figure 7. In Section 1–2 we will see why it must be a straight line.

EXERCISE SET 1–1

In Problems 1 to 4 fill in the table for the given values.
Supply your own values of x in Problems 3 and 4.

1. $y = 2x - 3$

x	-2	-1	0	1	2	3
y	-7	-5	-3	-1	1	3
point						

2. $y = x - 2$

x	-2	-1	0	1	2	3
y						
point						

3. $y = 4 - x$

x	-2	-1	0	1	2	3
y	6	5	4	3	2	1
point						

4. $3x - y = 2$

x					
y					
point					

In Problems 5 to 10 use the plotting method to sketch the graph of the equation.

5. $y = 2x - 3$

6. $y = x - 2$

7. $y = 4 - x$

8. $3x - y = 2$

9. $2x + y = 6$

10. $4x + y = 7$

In Problems 11 to 14 the graphs of the given equations are straight lines. Plot two points and then sketch the graph.

11. $x - y = 3$

12. $2x + y = 3$

13. $3x - y = -1$

14. $3x + y = 1$

In Problems 15 to 18 sketch the graphs of the total cost C and the total revenue R on the same coordinate axis.

15. $C = 25 + 3x$
    $R = 8x$

16. $C = 40 + x$
    $R = 5x$

17. $C = 100 + 3x$
    $R = 13x$

18. $C = 300 + x$
    $R = 31x$

Problems 19 and 20 refer to the equation $y = 494x - 16$ given in Example 4.

19. (a) If the sales growth rate is 15%, what external funding is required? (b) If $33,400 is available for external funding, what is the sales growth rate?

20. (a) If the sales growth rate is -5%, what external funding is required? (b) If $82,800 is available for external funding, what is the sales growth rate?

In Problems 21 to 24 use the plotting method to sketch the graph of the equation.

21. $x + 4y = 7$

22. $x + 3y = 3$

23. $3x + 4y = -12$

24. $3x - 4y = 6$

In Problems 25 to 34 the equations are nonlinear. Sketch the graph using the plotting method.

25. $y = x^2 - 3$

26. $y = x^2 + 1$

27. $y = -x^2$

28. $y = -2x^2$

29. $y = 1 - 2x^2$

30. $y = 1 - x^2$

31. $y = x^3$

32. $y = x^3 - 8$

33. $y = \sqrt{x}$

34. $y = \sqrt{x} - 1$

In Problems 35 and 36 sketch the graphs of the two equations on the same coordinate axis.

35. $\quad x + \quad y = \quad 120$
    $100x + 40y = 6000$

36. $45x + 9y = 45$
    $10x + 6y = 20$

Problems 37 to 40 refer to the equation $5F - 9C = 160$, where F is degrees Fahrenheit and C is degrees Celsius.

37. Sketch the graph of the equation.

38. Find the degrees Celsius when $F = 68$ degrees.

39. Find the degrees Celsius when $F = 98.6$ degrees.

40. Find the degrees Fahrenheit when $C = 20$ degrees.

Problems 41 and 42 refer to a company whose total cost is $C = 0.5x + 20,000$, where x is the amount of sales and the fixed cost is $20,000.

41. If the company sells $60,000 worth of goods, what is the total cost?

42. If the company sells $30,000 worth of goods, what is the total cost?

43. A person's height can be estimated by the length of the tibia bone, which runs from the knee to the ankle. The formula for the height y [in centimeters (cm)] in terms of the length x (in cm) of the tibia is

$$3y - 8x = 211$$

(a) Draw the graph of the equation.
(b) How tall is a person whose tibia is 37 cm long?
(c) How long is the tibia if the person is 185 cm tall?

44. A person's height can be estimated by the length of the radius bone, which runs from the wrist to the elbow. The formula for the height y (in cm) in terms of the length x (in cm) of the radius is

$$2y - 7x = 166$$

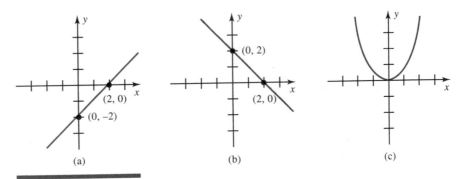

FIGURE 8 Problems 51 and 52.

(a) Draw the graph of the equation.
(b) How tall is a person whose radius bone is 24 cm long?
(c) How long is the radius bone if the person is 174 cm tall?

Problems 45 to 48 refer to Figure 5 in the text.

45. What is the meaning of the value of y at which the cost line and the revenue line meet?

46. What is the meaning of the value of x at which the cost line and the revenue line meet?

47. For what numbers of items produced is there a profit?

48. How many items produce a revenue of $800,000?

Problems 49 and 50 are based on figures supplied by the U.S. Census Bureau.

49. The population of Michigan in the decade 1980–1990 is given by the equation $M = 6,670.6x + 9,262,078$, where x is the number of years after 1980. What was the total increase in Michigan's population in the 1980s?

50. The population of Illinois in the decade 1980–1990 is given by the equation $I = 4,016.4x + 11,426,518$, where x is the number of years after 1980. What was the total increase in Illinois's population in the 1980s?

In Problems 51 and 52 match the graph in Figure 8 to the given equation.

51. $y = x - 2$

52. $y = x^2$

Referenced Exercise Set 1–1

1. The percentage share of the national radio audience for FM radio stations (contrasted with AM radio stations) in any year since 1972 can be approximated by the formula

$$S = 3.514286Y - 226.2953$$

where S is a percent and Y is the last two digits in the year* (for instance, for the year 1989 $Y = 89$).
(a) What was the approximate percentage share in 1985?
(b) In what year was the percentage share about 50%?

2. The amplifier in a stereo system must be strong enough to produce loud peak volume levels (106 decibels) in the speakers. The strength needed depends on the character of the room. For instance, a "live" room is one in which a hand clap will cause a fluttering echo. For Acoustic Research loudspeakers,† the linear relationship between the power y (in watts) and the volume x of a "live" room (in cubic feet) is given by

$$y = \frac{x}{180} + \frac{80}{9}$$

(a) What power is needed for a room measuring 2000 cubic feet?
(b) What power is needed for a room measuring 6500 cubic feet?
(c) If a system produces 40 watts, what size room will allow it to produce loud peak levels?

VCRs have invaded American homes in the last few years. Most VHS systems have a speed switch that allows the

---

*This equation was derived from data supplied by Statistical Research, Incorporated.

†Source: Teledyne Acoustic Research of Canton, Massachusetts. This company manufactures the Acoustic Research loudspeakers.

viewer to record in three different speeds: 2-hour standard play (SP), 4-hour long play (LP), and 6-hour super long standard play (SLP). Problems 3 and 4 involve equations that allow a viewer to get the highest quality from a recording.‡

3. Suppose a viewer wishes to record a program that lasts between 2 and 4 hours. Let x be the length of the program in hours. Then the highest quality recording using the entire T-120 videocassette can be obtained by recording in LP mode for $y = 2x - 4$ hours and in SP mode for $4 - y$ hours. How long should the viewer record in LP mode if the program lasts (a) 3 hours? (b) $3\frac{1}{2}$ hours? (c) 2 hours and 40 minutes?

4. Many people see little difference in quality between LP and SLP modes. Suppose a viewer wishes to record a program that lasts between 2 and 6 hours using LP and SLP modes. Let x be the length of the program in hours. Then the highest quality recording using the entire T-120 videocassette can be obtained by recording in SLP mode for $y = 1.5x - 3$ hours and in SP mode for $4 - y$ hours. How long should the viewer record in SLP mode if the program lasts (a) 4 hours? (b) $3\frac{1}{2}$ hours? (c) 4 hours and 40 minutes?

5. The first federal income tax went into effect in 1862, one year after President Abraham Lincoln signed it into

legislation. Phil Lapsansky, the curator of the Afro-American collection at the Library Company of Philadelphia, recently discovered an original tax form in the stacks of that library, which Benjamin Franklin founded in 1731.§ Computing taxes was a lot easier then; the tax rate was 3% on income from $600 to $10,000 and 5% beyond that amount.
   (a) Write an equation for the tax T of a person whose income in 1862 was x dollars, where x is between $600 and $10,000.
   (b) Write an equation for the tax T of a person whose income in 1862 was x dollars, where x is more than $10,000.

6. Consider the statement: "There are six times as many students as professors at this university." Let S stand for the number of students and P stand for the number of professors.
   (a) If $P = 1200$, what is S?
   (b) If $S = 1200$, what is P?
   (c) Write an equation using the variables S and P to represent the given statement.‖
   (d) Substitute the values you obtained in parts (a) and (b) into the equation in part (c). If they do not satisfy the equation, revise the equation and repeat this part.

---

‡G. N. Fiore, "An Application of Linear Equations to the VCR," *Mathematics Teacher*, October 1988, pp. 570–572.

§Leonard W. Boasberg, "The Tax That Was Due for 1862," *The Philadelphia Inquirer*, April 12, 1990, pp. 1-D, 4-D.

‖It has been reported that only 63% of engineering students and 43% of social science students get this equation correct. Sources for the study addressing this issue can be found in the article by Annie A. Selden and John Selden, Jr., entitled "Do You Know the Students-and-Professors Problem?" which appeared in a joint newsletter of the MAA, AMS, and SIAM, called *UME Trends*, May 1989, p. 6.

1–2 Lines

There are two basic aspects to the Descartes–Fermat invention of analytic geometry. One is a fixed frame of reference, called the Cartesian coordinate system. The other is the relationship between equations and curves. Section 1–1 showed how to begin with a linear equation in two variables and to draw its graph. This section illustrates the opposite direction: given the graph of a line in some form, construct its equation.

General Form

In Section 1–1 we stated that a linear equation in two variables is an equation whose graph is a line. We examine three forms of linear equations.

Recall from Section 1–1 that an equation in two variables x and y is linear if and only if it can be written in the form $Ax + By = C$. This form of a linear equation is called the *general form*. Therefore the equation $4x + 3y = 12$ is linear, but the equations $x^2 - y = 0$ and $\sqrt{x} - y = 0$ are nonlinear.

D E F I N I T I O N

The **general form** of a linear equation is

   $Ax + By = C$

where A, B, and C are real numbers, with A and B not both equal to 0.

Example 1 illustrates a method of graphing linear equations in general form that is more effective than the plotting method shown in Example 2 of Section 1–1. It makes use of the intercepts of a line. The **x-intercept** of a line is the point where the line crosses the x-axis, and the **y-intercept** is the point where it crosses the y-axis. Thus an x-intercept must have a y-coordinate equal to 0 and a y-intercept must have an x-coordinate equal to 0.

EXAMPLE 1

Problem

Sketch the graph of the linear equation

   $3x + 2y = 4$

Solution Find the y-intercept by setting $x = 0$. Then $y = 2$, so the y-intercept is $(0, 2)$. Find the x-intercept by setting $y = 0$. Then $x = \frac{4}{3}$, so the x-intercept is $(\frac{4}{3}, 0)$. Draw the line through the two intercepts. This is shown in Figure 1.

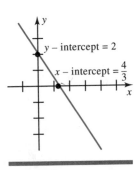

$$2y = 4$$
$$y = 2$$
$$(0, 2)$$

$$3x = 4$$
$$x = \frac{4}{3}$$
$$\left(\frac{4}{3}, 0\right)$$

FIGURE 1   Graph of $3x + 2y = 4$.

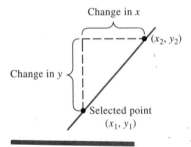

FIGURE 2 Slope.

Slope-Intercept Form

Every equation of the form $y = mx + b$, where m and b are numbers, is linear since it can be written in the general form $mx - y = -b$. Therefore the graph of the equation $y = 3x + 1$ is a line with $m = 3$ and $b = 1$; its graph cannot be the curve shown in Figure 7 of Section 1–1. The expression "the line $y = mx + b$" refers to the graph of the linear equation $y = mx + b$.

The numbers m and b in the linear equation $y = mx + b$ play an important role. The property measured by m is intuitively described as "steepness." It is called the *slope* of the line. To measure the slope, start at some point (x_1, y_1) on the line, proceed in the vertical direction some distance, then proceed in the horizontal direction until the line is intersected at a point (x_2, y_2). (See Figure 2.) The vertical distance is called the "change in y" (or the "rise") and the horizontal distance is called the "change in x" (or the "run"). The slope is then defined to be the ratio of the change in y to the change in x (or the ratio of the rise to the run).

DEFINITION

Let (x_1, y_1) and (x_2, y_2) be distinct points on a line with $x_1 \neq x_2$. The **slope** of the line is the number

$$m = \frac{\text{change in } y}{\text{change in } x} = \frac{y_2 - y_1}{x_2 - x_1} \quad \frac{\text{rise}}{\text{run}}$$

The points $(1, -1)$ and $(4, 5)$ lie on the line $y = 2x - 3$. (See Figure 3.) The slope of the line is

$$m = \frac{\text{change in } y}{\text{change in } x} = \frac{5 - (-1)}{4 - 1} = \frac{6}{3} = 2$$

Figure 4 shows the graphs of several lines. As the values of x increase from left to right, those with positive slope tend upward (or increase) while those with negative slope tend downward (or decrease).

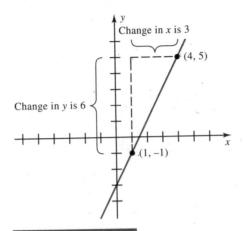

FIGURE 3 Slope of line $y = 2x - 3$.

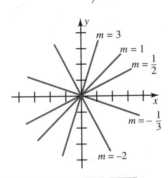

FIGURE 4 Lines with various slopes.

Example 2 investigates the slope determined by two distinct pairs of points on the same line.

EXAMPLE 2

Problem

Find the slope of the line $y = 3x + 1$ through these pairs of points.
(a) $(0, 1)$ and $(3, 10)$
(b) $(-1, -2)$ and $(4, 13)$

Solution (a) The slope of the line through the points $(0, 1)$ and $(3, 10)$ is

$$m = \frac{10 - 1}{3 - 0} = \frac{9}{3} = 3$$

(b) The slope of the line through the points $(-1, -2)$ and $(4, 13)$ is

$$m = \frac{13 - (-2)}{4 - (-1)} = \frac{13 + 2}{4 + 1} = \frac{15}{5} = 3$$

In general the slope of a line is independent of the two points chosen on the line. Example 2 illustrates this fact. Another way of viewing the slope can be seen from a table of points.

x	-2	-1	0	1	2
y	-5	-2	1	4	7

Each time the value of x increases by 1, the value of y increases by the slope 3.

The number b in the linear equation $y = mx + b$ can be obtained by setting $x = 0$. Then $y = m \cdot 0 + b = b$, so the line passes through the point $(0, b)$. The number b is referred to as the y-intercept of the line $y = mx + b$ even though formally the y-intercept is the point $(0, b)$.

The linear equation $y = mx + b$ is called the *slope-intercept form* of a line because m is the slope and b is the y-intercept.

DEFINITION

The **slope-intercept form** of a line is

$$y = mx + b$$

where m is the slope and b is the y-intercept.

EXAMPLE 3

Problem

$y = mx + b$

Sketch the graph of the equation $y = -2x + 3$.

Solution The y-intercept is $b = 3$, so the line passes through $(0, 3)$. The slope is $m = -2$. Since m is negative, the line slopes downward (from left to right). The graph is sketched in Figure 5.

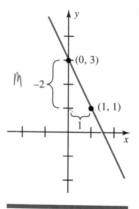

FIGURE 5 Graph of $y = -2x + 3$.

Parallel Lines

Two lines are **parallel** if they do not meet in either direction when extended indefinitely. Since parallel lines have the same steepness, it follows that two lines are parallel if and only if they have the same slope.

EXAMPLE 4

Problem

Write the equation of the line that is parallel to the line $y = -2x + 2$ and passes through the origin.

$y = nx + b$

Solution The graph of the linear equation $y = -2x + 2$ is a line with slope $m = -2$. The new line will have the same slope $m = -2$. Since the new line passes through $(0, 0)$, its y-intercept is $b = 0$. Substituting the values $m = -2$ and $b = 0$ into the slope-intercept form $y = mx + b$ yields $y = -2x$. This is the equation of the line parallel to $y = -2x + 2$ and passing through the origin. The lines are sketched in Figure 6.

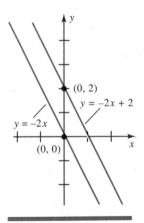

FIGURE 6   Parallel lines.

There are two special kinds of lines. **Horizontal lines** are parallel to the x-axis and **vertical lines** are parallel to the y-axis. The slope of a horizontal line can be determined by choosing any two points $(x_1, 0)$ and $(x_2, 0)$ on the x-axis. By definition the slope of the line through these points is

$$m = \frac{0 - 0}{x_2 - x_1} = 0$$

Thus the slope of the x-axis is 0. Since parallel lines have the same slope, it follows that *the slope of any horizontal line is zero*.

Next consider vertical lines. If $(0, y_1)$ and $(0, y_2)$ are two points on the y-axis, the slope is

$$m = \frac{y_2 - y_1}{0 - 0}$$ NOT DEFINED

Since division by 0 is not permitted, the slope of the y-axis is not defined. It follows that *the slope of any vertical line is not defined.*

EXAMPLE 5

Problem

Write the equations of the horizontal and vertical lines that pass through the point $(2, 3)$.

Solution The slope of the horizontal line is $m = 0$. Since the horizontal line passing through $(2, 3)$ is parallel to the x-axis, its y-intercept is $b = 3$. Substituting $m = 0$ and $b = 3$ into the equation $y = mx + b$ yields $y = 3$.

The equations of vertical lines cannot be obtained from the linear equation $y = mx + b$ because m is not defined. The vertical line through $(2, 3)$ is located two units to the right of the y-axis. Some additional points on it are $(2, 2)$, $(2, 1)$, $(2, 0)$, and $(2, -1)$. The value of x is always 2. Hence the equation of the line is $x = 2$.

Both lines are sketched in Figure 7.

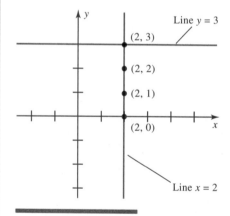

FIGURE 7 Horizontal and vertical lines.

Table 1

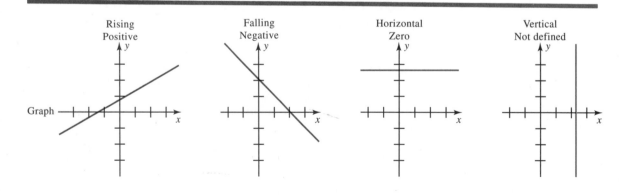

FIGURE 8 Mathematicians always run from left to right.

Example 5 suggests the following standard forms for all horizontal and vertical lines:

horizontal lines: $y = c$

vertical lines: $x = k$

where c and k can be any fixed number. Table 1 summarizes the results on the slopes of lines, while Figure 8 presents the same concepts from a roadrunner's perspective.

Point-Slope Form

The equation of a line that is neither vertical nor horizontal can be expressed in several forms. We have already indicated that the slope-intercept form gives the equation of a line when its slope and y-intercept are known. But what if the slope m and another point (x_1, y_1) on the line are known? Then an arbitrary point (x, y)

on the line can be used with (x_1, y_1) to compute the slope $m = (y - y_1)/(x - x_1)$. So $y - y_1 = m(x - x_1)$. This form is aptly named the *point-slope form*.

DEFINITION

The **point-slope form** of a line is

$$y - y_1 = m(x - x_1)$$

where m is the slope and (x_1, y_1) is the given point on the line.

Example 6 shows how to convert from the point-slope form to the slope-intercept form.

EXAMPLE 6

Problem

Find the point-slope form and the slope-intercept form of the line that passes through the point (3, 2) and has slope 4.

Solution We have $x_1 = 3$, $y_1 = 2$, and $m = 4$. The equation of the line in point-slope form is

$$y - 2 = 4(x - 3)$$

Use the properties of algebra indicated on the right to write the equation of the line in slope-intercept form.

$y - 2 = 4(x - 3)$	(point-slope form)
$y = 4(x - 3) + 2$	(add 2 to each side)
$y = 4x - 12 + 2$	(distributive law)
$y = 4x - 10$	(add -12 and 2)

Therefore $m = 4$ and $b = -10$. The graph is sketched in Figure 9.

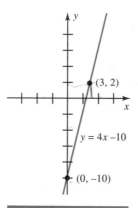

FIGURE 9 Line through (3, 2) with $m = 4$.

Example 7 shows how to determine the equation of a line when two points on the line are known.

EXAMPLE 7

Problem

Find an equation of the line that passes through the points $(1, 2)$ and $(-1, 5)$.

Solution The slope of the line is

$$m = \frac{5 - 2}{-1 - 1} = \frac{3}{-2} = -\frac{3}{2}$$

Using the point-slope form with the point $(1, 2)$ [we could have chosen the point $(-1, 5)$ instead] the equation is

$$y - 2 = -\frac{3}{2}(x - 1)$$

The equation in slope-intercept form is found by solving for y.

$$y = -\frac{3}{2}x + \frac{7}{2}$$

Linear Depreciation

The tax law requires businesses to determine the current value of every asset. Most assets lose value as they get older. It is said that they *depreciate* in value. There are several ways to compute the depreciation of an asset. One method, called **linear depreciation,** assumes that the current value y of the asset is related to the age x of the asset by a linear equation. This equation usually takes the form $y = b - mx$, where b is the purchase price of the asset, or the value when $x = 0$, and m is the rate of depreciation.

EXAMPLE 8

A firm purchases a machine for $50,000 and the machine has a useful life (determined by tax laws) of ten years. If the machine depreciates $4,000 per year, the value of the machine after x years is

$$y = 50,000 - 4,000x$$

The salvage value of an asset is its value when the asset reaches its useful life.

Problem

Determine the salvage value of the machine.

Solution Find the value of y when $x = 10$.

$$y = 50,000 - 4,000(10) = 50,000 - 40,000$$

$$= 10,000$$

The salvage value of the machine is $10,000.

(handwritten marginalia): 19. (Below) $\frac{-6-1}{3-2} = -7$ $m = -7$ $\frac{y_2 - y_1}{x_2 - x_1}$ $-7(2) + 1 = 15$ $y = mx + b$ $y = mx + 15$ $y = -7x + 15$

Here is a summary of the main items in this section.

1. The slope is $m = \dfrac{y_2 - y_1}{x_2 - x_1}$ (unless $x_1 = x_2$).

2. The slope measures steepness. As x increases from left to right, y increases if the slope is positive and y decreases if the slope is negative.

3. Horizontal lines have form $y = c$; their slope is 0. Vertical lines have form $x = k$; their slope is undefined.

4. Three principal forms for the equation of a line follow.

Name of Form	Equation	Variables
general	$Ax + By = C$	A, B, C real numbers A, B not both 0
slope-intercept	$y = mx + b$	$m = $ slope $b = y$-intercept
point-slope	$y - y_1 = m(x - x_1)$	$m = $ slope $(x_1, y_1) = $ point

EXERCISE SET 1–2

In Problems 1 to 4 sketch the graph of the line.

1. $5x - 2y = 10$ 2. $5x + 3y = 15$

3. $3x + 4y = 12$ 4. $3x - y = 6$

In Problems 5 to 10 find the slope of the line passing through the two points.

5. $(1, 2)$, $(2, 4)$ 6. $(3, 1)$, $(4, 3)$

7. $(-1, 2)$, $(3, -4)$ 8. $(-3, 8)$, $(3, -8)$

9. $(1.6, 4.1)$, $(2.5, 5.9)$ 10. $(\frac{3}{2}, 5)$, $(-1, -\frac{3}{2})$

In Problems 11 to 14 sketch the graph of the line.

11. $y = 4x - 5$ 12. $y = 2x + 1$

13. $y = -2x + 3$ 14. $y = -3x - 2$

In Problems 15 and 16 determine if the lines are parallel.

15. Line 1 passes through $(0, 1)$ and $(1, 2)$; line 2 passes through $(-1, 6)$ and $(6, -1)$.

16. Line 1 passes through $(1, 3)$ and $(5, -1)$; line 2 passes through $(1, 1)$ and $(2, 0)$.

In Problems 17 and 18 determine the slope of the line.

17. (a) $x = 2$ (b) $y = -7$

18. (a) $x = 15$ (b) $y = 2$

In Problems 19 and 20 find the slope-intercept form of the line passing through the given points.

19. $(2, 1)$, $(3, -6)$ 20. $(2, -1)$, $(-3, 6)$

In Problems 21 to 34 find an equation of the line from the information given. Sketch the graph of the line.

21. $m = 2$, $b = 7$ 22. $m = -2$, $b = 7$

23. The horizontal line that passes through the point $(-5, -3)$.

24. The vertical line that passes through $(-5, -3)$.

25. The slope is undefined and the x-intercept is -2.

26. The slope is 0 and $b = -4$.

27. $m = -2$; the line passes through $(6, -1)$.

28. $m = 1$; the line passes through $(6, -1)$.

29. $b = -1$ and the line is parallel to $y = 2x + 1$.

30. The line passes through $(1, 1)$ and is parallel to the line $x - 3y = 4$.

31. The line passes through $(-2, 0)$ and $(0, 2)$.

32. The line passes through $(2, 0)$ and $(0, -2)$.

33. The line passes through $(-2, 0)$ and $(-2, -2)$.

34. The line passes through $(0, 4)$ and $(2, 4)$.

In Problems 35 and 36 determine if the slope of the line is positive, negative, zero, or undefined.

35.

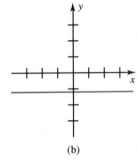

(a)   (b)

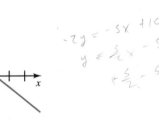

$-2y = -5x + 10$
$y = \frac{5}{2}x - 5$
$+\frac{5}{2} - 5$

(c)

36.

(a)   (b)

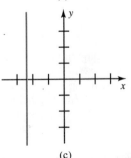

(c)

M IS THE SLOPE
B IS THE (Y INTERCEPT)

37. (a) Put the equation $3x + 2y = 12$ in slope-intercept form. (b) What is the slope? (c) What is the y-intercept?

38. (a) Put the equation $5x + 3y = 15$ in slope-intercept form. (b) What is the slope? (c) What is the y-intercept?

In Problems 39 and 40 determine the equation of the line passing through the two intercepts.

39.

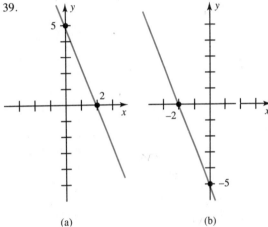

(a)   (b)

40.

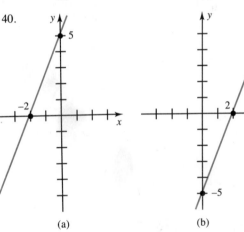

(a)   (b)

In Problems 41 and 42 write the equation (in point-slope form) of the line that has the given table.

$(0,1) (1,-3)$
$\frac{3-1}{1-0} = \frac{-4}{1}$
$= -4$

41.

x	−1	0	1
y	5	1	−3

$y-1$

42.

x	−2	0	2
y	5	1	−3

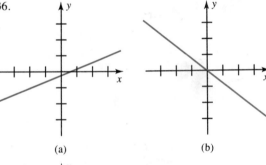

43. For tax purposes some firms assume that the current value y of an asset is related to the asset's age x by a linear equation. Suppose a firm buys a work station with ten microcomputers and assumes that after x years the work station's worth is given by

$$y = 60,000 - 15,000x$$

(a) Draw the graph of the equation.
(b) What is the work station's worth after three years?
(c) What economic interpretation can be given to the y-intercept?

Problems 44 to 47 are based on figures supplied by the U.S. Census Bureau.

44. The population of Texas was 14,229,191 in 1980. The average increase for each year during the 1980s was 283,061.4. (a) Write an equation to describe the population of Texas during that decade. (b) Use this equation to determine the population of Texas in 1990.

45. The population of New York was 17,558,072 in 1980. The average increase for each year during the 1980s was 48,643.3. (a) Write an equation to describe the population of New York during that decade. (b) Use this equation to determine the population of New York in 1990.

46. In Section 1–1 the equation for the population of California in the 1980s was given as $C = 617,134.8x + 23,667,902$. Explain the meaning of the slope of this equation.

47. The equation for the population of Iowa in the 1980s is $I = -12,638.4x + 2,913,808$. Explain the meaning of the negative slope.

In Problems 48 to 51 match the graphs in Figures a to d to the given linear equations.

48. $2x + 5y = 20$

49. $3x + 4y = 24$

50. $y = 5x + 4$

51. $y = 2x + 6$

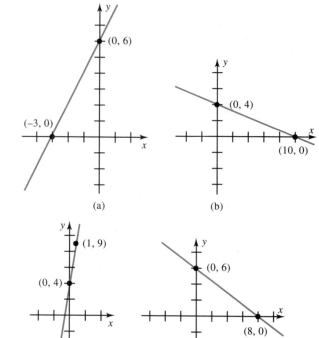

(a)   (b)

(c)   (d)

52. Prove that the slope of the line $y = mx + b$ is m. *Hint:* Let (x_1, y_1) and (x_2, y_2) be points on the line with $x_1 \neq x_2$. Show that

$$\frac{(y_2 - y_1)}{(x_2 - x_1)} = m$$

53. Is it true that every equation in the variables x and y in which the highest power of x is 1 and the highest power of y is 1 must be linear?

Referenced Exercise Set 1–2

1. The petroleum industry uses "decline curve analysis" to estimate reserves of oil and gas wells, even though more sophisticated techniques are available. The reason for the continued emphasis on this method is that the required data are almost always available. Recently two petroleum engineers developed a simple graphing technique for evaluating production data from a linear equation.* Let y represent the loss ratio (a technical term

*D. A. Rowland and C. Lin, "New Linear Method Gives Constants of Hyperbolic Decline," *Oil and Gas Journal*, January 14, 1985, pp. 86–90.

measured here in months) and x represent the time (also in months). The line which the engineers discovered for a particular set of data has slope 0.5 and y-intercept 7.5.

(a) What is the equation of the line? (b) Sketch the line on the interval of 0 months to 72 months. (c) What is the loss ratio after 6 years?

2. This problem deals with four "rules of thumb" relating the height and weight of adults to those of children.† Write each relationship as an equation. (This requires the introduction of appropriate variables.) (a) The height of a male adult is double the height at age 2. (b) The height of a female adult is double the height at age $1\frac{1}{2}$. (c) The weight of a male adult is five times the weight at age 2. (d) The weight of a female adult is five times the weight at age $1\frac{1}{2}$.

3. Figure 10 shows the power necessary to achieve loud peak volume levels (106 decibels) for a pair of Acoustic Research loudspeakers.‡ The three different types of rooms ("dead," "average," and "live") require amplifiers of different sizes to produce loud peak volume levels. Let y be the power (in watts) and x be the volume of the room (in cubic feet). For each type of room, express y in terms of x in slope-intercept form.

4. A carpenter asked University of Delaware Professor Richard J. Crouse the following question:§ "I am building roofs that are 30 feet (ft.) long at the base. The pitch of each roof rises 4 inches (in.) for every horizontal foot. I have to install vertical supports every 16 in. (See Figure 11.) This forces me to climb the ladder, measure 16 in. horizontally, measure the vertical height at that point, climb down the ladder, saw the support, and climb back up to put it in place. Is there a formula that I can use to determine the lengths of the supports in advance?"

(a) What are the coordinates of points A and B in Figure 11?

(b) Write the equation of the line through points A and B, where x is the horizontal distance and y is the height.

(c) Use the equation in part (b) to find the length of the support at 16 in.

(d) Use the equation in (b) to find the length of the support at 8 ft.

Problems 5 and 6 deal with the percentage of energy that is supplied to the body by fats and carbohydrates during treadmill running.‖ Figure 12 shows that each energy source is a linear function of time.

5. (a) Interpret the meaning of the fact that during the first few minutes of running the slope of the carbohydrate function is positive.

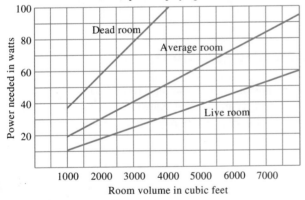

Power per channel required for a pair of acoustic, research loudspeakers playing at 106dB S.P.L.

FIGURE 10 Problem 3.

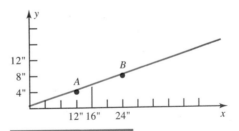

FIGURE 11 Problem 4.

†Wilton M. Krogman, *Child Growth*, Ann Arbor, Mich., University of Michigan Press, 1972, p. 42.

‡Source: Teledyne Acoustic Research of Canton, Massachusetts.

§R. J. Crouse, "Linear Equation Saves Carpenter's Time," *Mathematics Teacher*, May 1990, pp. 400–401.

‖Trevor Smith, "Chemistry, Exercise, and Weight Control," *Today's Chemist*, February 1990, pp. 10–11, 21.

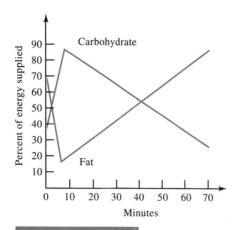

FIGURE 12 Problems 5 and 6.

(b) Interpret the meaning of the fact that after a few minutes of running the slope of the carbohydrate function is negative.

6. (a) Interpret the meaning of the fact that during the first few minutes of running the slope of the fat function is negative.
(b) Interpret the meaning of the fact that after a few minutes of running the slope of the fat function is positive.

Cumulative Exercise Set 1–2

In Problems 1 to 3 use the plotting method to sketch the graph of the equation.

1. $y = 2x - 3$

2. $x + 4y = 4$

3. $y = -x^2 + 1$

4. Sketch the graphs of the total cost function C and the total revenue function R.

$$C = 50 + 2x \qquad R = 12x$$

5. Find the slope-intercept form of the line passing through the points $(-1, 4)$ and $(1, -4)$.

6. Find the equations of the vertical line and the horizontal line passing through the point $(2, 5)$.

In Problems 7 and 8 find an equation of the line from the given information.

7. The line has y-intercept 5 and is parallel to the line $2x - 4y = 1$.

8. The line has x-intercept 7 and the slope is undefined.

9. A company has total cost $C = 0.4x + 15,000$, where x is the number of sales and the fixed cost is \$15,000. If the company sells \$30,000 worth of goods, what is the total cost? Sketch the graph of the equation.

1–3 Linear and Quadratic Functions

In Section 1–1 we sketched the graphs of linear equations. Here we introduce the fundamental concept of a function and regard a linear equation as a function. We also introduce quadratic functions and give a criterion for distinguishing between equations and functions.

Linear Functions

The equation $y = 1 - 4x$ has the property that for each value of x there corresponds precisely one value of y. In such a case y is called a *function* of x, written $y = f(x)$ or $f(x) = 1 - 4x$. [$f(x)$ is read "f of x."]

DEFINITION

A **function** f is a rule that assigns to each element x in a set exactly one element $f(x)$ in a (possibly different) set. The element $f(x)$ is called the **image** of x, and x is called the **preimage** of $f(x)$.

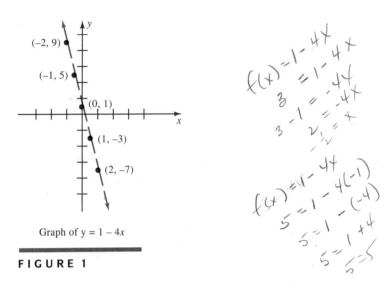

Graph of $y = 1 - 4x$

FIGURE 1

Consider the function $f(x) = 1 - 4x$. The **plotting method** for drawing its graph was described in Section 1–1. First, make a table of points satisfying the equation.

x	-1	$-\frac{1}{2}$	0	$\frac{1}{2}$	1
$f(x)$	5	3	1	-1	-3
point	$(-1, 5)$	$(-\frac{1}{2}, 3)$	$(0, 1)$	$(\frac{1}{2}, -1)$	$(1, -3)$

Then plot the points on a Cartesian coordinate system and connect them with the line drawn in Figure 1.

A **linear function** is a function that can be written in the form $f(x) = mx + b$, where m and b are given numbers. We saw in Section 1–2 that the graph of a linear function is a line with slope m and y-intercept b.

The table for the linear function $f(x) = 1 - 4x$ lists the images of five values of x. (Actually, only two images are necessary for drawing the graph of a line.) Since $y = 5$ when $x = -1$, the image of -1 is 5. We write $f(-1) = 5$. This image can be obtained directly by substituting the value of -1 for x wherever x occurs in the equation $f(x) = 1 - 4x$.

$$f(-1) = 1 - 4(-1) = 1 + 4 = 5$$

Example 1 computes some images of another function. Parts (a) and (b) show that it is possible for $f(a) = f(b)$ when $a \neq b$.

EXAMPLE 1

Let f be the function defined by the equation

$$f(x) = 2x^2 - 4x + 5$$

Problem

Find (a) $f(2)$, (b) $f(0)$, (c) $f(-\frac{1}{2})$.

Solution (a) Since $f(2)$ is the image of 2, it is calculated by substituting 2 for x wherever x occurs in the formula for $f(x)$.

$$f(2) = 2(2)^2 - 4(2) + 5 = 5$$

Thus

$$f(2) = 5$$

(b) $f(0) = 2(0)^2 - 4(0) + 5 = 5$, so

$$f(0) = 5$$

(c) $f(-\frac{1}{2}) = 2(-\frac{1}{2})^2 - 4(-\frac{1}{2}) + 5 = 2(\frac{1}{4}) + 2 + 5 = \frac{1}{2} + 7 = \frac{15}{2}$, so

$$f(-\frac{1}{2}) = \frac{15}{2}$$

Example 2 illustrates the procedure for finding the image of an algebraic expression.

EXAMPLE 2

Let

$$f(x) = 2x^2 - 4x + 5$$

Problem

Find and simplify each of these expressions.

(a) $f(a)$, (b) $f(x + 3)$, (c) $f(x + h)$, (d) $\dfrac{f(x + h) - f(x)}{h}$, for $h \neq 0$.

Solution (a) The value of $f(a)$ is found by substituting a for x wherever x occurs in the formula for $f(x)$.

$$f(x) = 2x^2 - 4x + 5$$
$$f(a) = 2a^2 - 4a + 5$$

(b) To obtain $f(x + 3)$, substitute $x + 3$ for x wherever x occurs in the formula for $f(x)$.

$$\begin{aligned}
f(x + 3) &= 2(x + 3)^2 - 4(x + 3) + 5 \\
&= 2x^2 + 12x + 18 - 4x - 12 + 5 \\
&= 2x^2 + 8x + 11
\end{aligned}$$

(c) To obtain $f(x + h)$, substitute $x + h$ for x wherever x occurs in the formula for $f(x)$.

$$\begin{aligned}
f(x + h) &= 2(x + h)^2 - 4(x + h) + 5 \\
&= 2x^2 + 4xh + 2h^2 - 4x - 4h + 5
\end{aligned}$$

(d) Using part (c) and the definition of $f(x)$ gives $f(x + h) - f(x)$

$$\begin{aligned}
&= 2x^2 + 4xh + 2h^2 - 4x - 4h + 5 - (2x^2 - 4x + 5) \\
&= 4xh + 2h^2 - 4h = h(4x + 2h - 4)
\end{aligned}$$

Then

$$\frac{f(x + h) - f(x)}{h} = \frac{h(4x + 2h - 4)}{h}$$

Since $h \neq 0$ the h-factors can be canceled, so

$$\frac{f(x + h) - f(x)}{h} = 4x + 2h - 4$$

In many applications it is advantageous to use letters other than f to represent the function. *Fortune* magazine, which is known for its many rankings of businesses, provides an example. Figure 2 displays the world's 10 biggest industrial corporations in 1986, with the 1985 ranking on the right side of the left-hand column. The list describes a function F (for *Fortune*) defined by the rule that $F(n)$ is the sales of the nth biggest corporation in 1986 (in \$thousands). It follows from Figure 2 that $F(1) = 102,813,700$ and $F(8) = 35,211,000$.

Domain and Range

A **function** is a correspondence between two sets in which each element x in one set, called the **domain,** has associated with it precisely one element in another set, called the **range.** Therefore a function consists of all ordered pairs (x,y), where x belongs to some set S (the domain) and y belongs to some set T (the range). The domain of the *Fortune* function F is the set $\{1, 2, \ldots, 10\}$. Most of the functions in this book will be defined by an equation, with the domain and range being sets of real numbers. Unless otherwise specified, the domain of a function that is defined by an equation consists of all numbers for which the equation is defined. For example, the domain of a linear function $f(x) = mx + b$ is the set of all real

Rank		Company	Headquarters	Industry	Sales ($ thousands)	Net Income ($ thousands)
1	1	General Motors	Detroit	Motor vehicles and parts	102,813,700	2,944,700
2	2	Exxon	New York	Petroleum refining	69,888,000	5,360,000
3	3	Royal Dutch/Shell Group	The Hague/London	Petroleum refining	64,843,217	3,725,779
4	6	Ford Motor	Dearborn, Mich.	Motor vehicles and parts	62,715,800	3,285,100
5	7	International Business Machines	Armonk, N.Y.	Computers	51,250,000	4,789,000
6	4	Mobil	New York	Petroleum refining	44,866,000	1,407,000
7	5	British Petroleum	London	Petroleum refining	39,855,564	731,954
8	12	General Electric	Fairfield, Conn.	Electronics	35,211,000	2,492,000
9	10	American Tel. & Tel.	New York	Electronics	34,087,000	139,000
10	8	Texaco	White Plains, N.Y.	Petroleum refining	31,613,000	725,000

FIGURE 2 Adapted from "The World's 50 Biggest Industrial Corporations," *Fortune*, August 3, 1987, pp. 23–25.

numbers. However, the domain of the function $f(x) = \dfrac{3}{2 - x}$ consists of all numbers different from 2 because division by 0 is undefined. A **function** f is **defined at a number** a if $f(a)$ exists. Notice that the function $f(x) = \dfrac{3}{2 - x}$ is defined at 0 because $f(0) = \dfrac{3}{2}$, but it is not defined at 2.

We are free to choose a value for x and substitute it into the equation to get the corresponding value $f(x)$, provided the equation is defined for that value of x. Thus, we refer to x as the **independent variable** and $f(x)$ as the **dependent variable.** These definitions allow us to give a more rigorous definition of the domain and range of a function.

DEFINITION

The **domain** of a function is the set of values of the independent variable for which the function is defined. The **range** of a function is the set of values of the dependent variable.

The range of a function is the set of images, while the domain of a function is the set of preimages. Example 3 computes the domain and range of another standard type of function.

EXAMPLE 3

Problem

Find the domain and range of the function $f(x)$ defined by the equation $f(x) = \sqrt{1 - x}.$

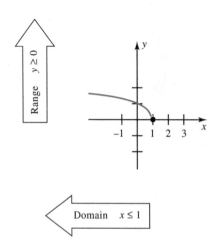

Range $y \geq 0$

Domain $x \leq 1$

Solution The square root is only defined when the radicand $1 - x$ is nonnegative, that is, when $1 - x \geq 0$. Therefore the domain of $f(x)$ is the set of all numbers x such that $x \leq 1$. As an interval, the domain is $(-\infty, 1]$.

To find the range, observe that $\sqrt{1 - x}$ is never negative, and as x takes on all values in the domain, $f(x)$ takes on all nonnegative numbers. Thus the range is the set of all numbers greater than or equal to 0. As an interval, the range is $[0, \infty)$.

Quadratic Functions

A function defined by the equation

$$f(x) = ax^2 + bx + c$$

where a, b, and c are given real numbers with $a \neq 0$, is called a **quadratic function.** A quadratic function is not a linear function; its graph is a **parabola,** not a line. Chapter 4 will show how to sketch the graphs of quadratic functions by using the techniques of calculus, but for now we rely on the plotting method.

EXAMPLE 4

Problem

Sketch the graph of the quadratic function $f(x) = 2x^2 - 4x + 1$ by finding the images of $-1, 0, 1, 2,$ and 3.

Solution Set $x = -1$. Then $f(-1) = 2(-1)^2 - 4(-1) + 1 = 7$, so the point $(-1, 7)$ lies on the parabola. Similarly, $f(0) = 1$, $f(1) = -1$, $f(2) = 1$, and $f(3) = 7$. Therefore the points $(0, 1)$, $(1, -1)$, $(2, 1)$, and $(3, 7)$ also lie on the parabola. The graph is sketched in Figure 3.

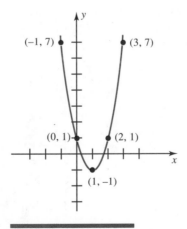

FIGURE 3 Graph of $f(x) = 2x^2 - 4x + 1$.

All parabolas are similar in shape to the one shown in Figure 3. Moreover, the graph of any quadratic function $f(x) = ax^2 + bx + c$ is a parabola that opens upward or downward, has a lowest or highest point (the **vertex**), and has a vertical line of symmetry passing through the vertex. Example 4 shows that the graph of $f(x) = 2x^2 - 4x + 1$ is a parabola that opens upward and has a vertical line of symmetry passing through the vertex (the lowest point on the graph) at $(1, -1)$.

Functions Versus Equations

There is a crucial distinction between functions and equations because not every equation represents a function. Consider the equations $y = x^2$ and $y^2 = x$. Their graphs, which are both parabolas, are drawn in Figure 4. The points $(4, 2)$ and $(4, -2)$ lie on the graph of the equation $y^2 = x$, so if a function $g(x)$ represented this equation, then we would have $g(4) = 2$ and $g(4) = -2$. By definition, a function of x cannot have a value x in the domain with two images. Therefore $y^2 = x$ is an equation that does not represent a function of x. However, for the equation $y = x^2$ each value of x has precisely one image, so it represents the function $f(x) = x^2$.

The **vertical line test** provides a visual means of determining whether a curve represents the graph of a function.

VERTICAL LINE TEST

A curve represents the graph of a function if and only if every vertical line intersects the curve at exactly one point or at no points.

Apply the vertical line test to the curves in Figure 4. In part (a) no vertical lines cross the curve in two or more points, so the curve represents the graph of the function $f(x) = x^2$. Many vertical lines intersect the curve in part (b) in two points,

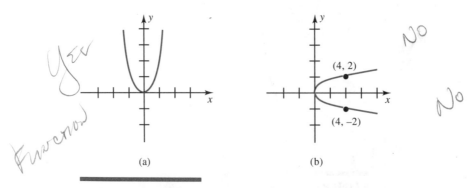

(a)   (b)

FIGURE 4 (a) The equation $y = x^2$ represents a function.
(b) The equation $y^2 = x$ does not represent a function.

so the curve does not represent a function. Example 5 applies the vertical line test to other curves.

EXAMPLE 5

Problem

Which of these curves represent graphs of functions?

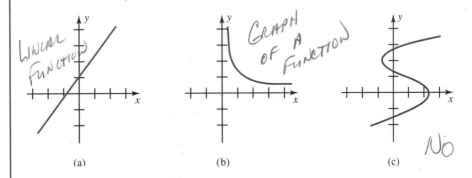

(a)   (b)   (c)

Solution The curve in (a) is a straight line. Each vertical line intersects the curve at precisely one point, so the curve represents the graph of a linear function.

The curve in (b) is drawn in the first quadrant only. The vertical lines through positive values of x meet the curve at one point. The vertical lines through the other values of x do not meet the curve. Thus the curve represents the graph of a function.

Look at the y-axis in (c). It is a vertical line that intersects the curve at more than one point, in fact, at three points. Therefore the curve does not represent the graph of a function.

Cost Analysis

Linear equations can be applied to some manufacturing problems in the following way. Let the cost of producing an item be denoted by C. If C depends on the number x of items produced, then C is called the **cost function,** denoted by $C(x)$. The cost function is composed of two parts, the **fixed cost,** which does not depend on x (such as the manufacturer's insurance), and the **variable cost,** which changes as x changes.

$$C(x) = \text{variable cost} + \text{fixed cost}$$
$$C(x) = (\text{variable cost per item})x + \text{fixed cost}$$

The **marginal cost** is the change in the cost when one more item is produced. If the variable cost per item is constant, then $C(x)$ is a linear function whose slope is the marginal cost.

As an illustration of these terms consider the cost of receiving cable reception where there is a one-time installation fee of $20 and a charge of $0.30 per day (including fractions of a day). To write the cost function, let x represent the time in days. Since the variable cost is $0.30x$ and the fixed cost is the flat rate of $20, the cost function is $C(x) = 0.3x + 20$, where $x \geq 0$. The cost for one regular year is $C(365) = 0.3(365) + 20 = 129.5$, or $129.50.

EXAMPLE 6

A printer quotes a price of $5 for producing a 200-page book and $8 for producing a 600-page book.

Problem

Assuming a linear relationship that depends on the number of pages, (a) what is the cost function? and (b) how much will it cost to produce a 300-page book?

Solution (a) Since the cost depends on the number of pages, let x denote the number of pages. There is a linear relationship between $C(x)$ and x, so $C(x) = mx + b$, where m is the variable cost per book and b is the fixed cost. We have $C(200) = 5$ and $C(600) = 8$, so two points on the line are (200, 5) and (600, 8). Therefore the variable cost is $m = (8 - 5)/(600 - 200) = 3/400$. By the point-slope form, using the point (200, 5), we get

$$C(x) - 5 = \left(\frac{3}{400}\right)(x - 200)$$

$$C(x) = \left(\frac{3}{400}\right)x + \left(\frac{7}{2}\right)$$

From this form we see that the fixed cost is $3.50 since $C(0) = 7/2$. (b) Once the cost function is known, the task of finding the cost of producing a book of any number of pages is straightforward. If $x = 300$, then the cost is $5.75 since $C(300) = (3/400)300 + (7/2) = 23/4 = 5.75$. (See Figure 5.)

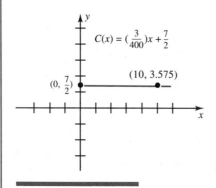

FIGURE 5

EXERCISE SET 1-3

In Problems 1 to 4 compute $f(2)$, $f(0)$, and $f(-\frac{1}{2})$ for the given function.

1. $f(x) = 2x - 3$.

2. $f(x) = -3x + 4$

3. $f(x) = -x^2 + x + 3$

4. $f(x) = x^2 + 4x - 5$

In Problems 5 to 8 find and simplify each of these expressions for the given function. (a) $f(a)$, (b) $f(x + 3)$, (c) $f(x + h)$, (d) $\dfrac{f(x + h) - f(x)}{h}$, for $h \neq 0$.

5. $f(x) = 2x - 3$

6. $f(x) = -3x + 4$

7. $f(x) = -x^2 + x + 3$

8. $f(x) = x^2 + 4x - 5$

Problems 9 and 10 refer to Figure 2 in the text, which lists the 10 biggest industrial corporations in the world in 1986. Let F be the function defined by the rule that $F(n)$ was the sales of the nth biggest industrial corporation that year.

9. Compute $F(2)$

10. Compute $F(3)$

In Problems 11 to 14 find the domain of the stated function.

11. $f(x) = \sqrt{x + 1}$

12. $f(x) = \sqrt{2x + 1}$

13. $f(x) = \sqrt{1 - 2x}$

14. $f(x) = \sqrt{2 - x}$

In Problems 15 to 18 sketch the graph of the quadratic function by finding the images of -2, -1, 0, 1, and 2.

15. $f(x) = x^2 + 2x - 5$

16. $f(x) = x^2 - 2x + 3$

17. $f(x) = 3x^2 - 5x - 4$

18. $f(x) = 2x^2 - 3x - 5$

In Problems 19 and 20 determine which curves represent the graph of a function.

19.

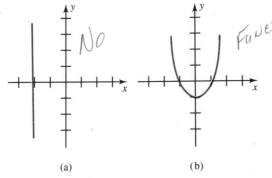

(a)   (b)

20.

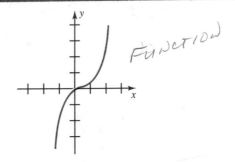

(c)

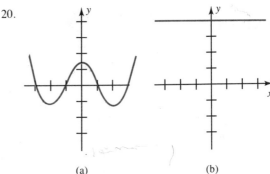

(a)   (b)

(c)

In Problems 21 to 24 evaluate $f(x + h) - f(x)$ for the given function.

21. $f(x) = 2x - 3$

22. $f(x) = -3x + 4$

23. $f(x) = -x^2 + x + 3$

24. $f(x) = x^2 + 4x - 5$

In Problems 25 to 28 evaluate $\dfrac{f(x + h) - f(x)}{h}$ for the given function, with $h \neq 0$.

25. $f(x) = 4 - x$

26. $f(x) = -1 - 2x$

27. $f(x) = -2x^2 - 4x$

28. $f(x) = 6x^2 + 3x$

In Problems 29 to 38 find the domain of the stated function.

29. $f(x) = \dfrac{3}{2 - x}$

30. $f(x) = \sqrt{x^2 + 1}$

31. $f(x) = \sqrt{1 - x^2}$

32. $f(x) = \dfrac{1}{x}$

33. $f(x) = \sqrt{x^2 - 1}$

34. $f(x) = \sqrt{x^3 - 1}$

35. $f(x) = \dfrac{1}{x(x - 2)}$

36. $f(x) = \dfrac{x}{x^2 - 9}$

37. $f(x) = \dfrac{\sqrt{x - 3}}{x - 4}$

38. $f(x) = \dfrac{x + 3}{\sqrt{x + 4}}$

In Problems 39 and 40 find all preimages of the given value for the function $f(x) = x^2 + 1$.

39. (a) 5   (b) 0   (c) 1.25

40. (a) 2   (b) -1 (c) 3.25

In Problems 41 and 42 sketch the graph of the equation and determine if it represents a function.

41. $y = \sqrt{x}$

42. $y^2 = 2x$

Referenced Exercise Set 1–3

1. Meteorologists use the equation $d^3 = 216t^2$ as a model to describe the size and intensity of four types of violent storms: tornadoes, thunderstorms, hurricanes, and cyclones.* Here d is the diameter of the storm (in miles) and t is the time (in hours) the storm travels before dissipating.
   (a) Is d a function of t? If so, solve for $d(t)$.
   (b) The world's worst recorded monsoon lasted for 24 hours on November 13–14, 1970, in the Ganges Delta Islands in Bangladesh. More than one million people died. What was the storm's diameter?
   (c) In the United States about 150 tornadoes occur each year, mostly in the central plains states, with the highest frequency in Iowa, Kansas, Arkansas, Oklahoma, and Mississippi. If a tornado's diameter is 2 miles, how long would it be expected to last?

2. Figure 6 lists the top 10 corporations in 1986 located outside the United States.† Let I be the function defined by the rule that $I(n)$ is the sales of the nth biggest corporation that year (in $thousands).
   (a) What is $I(1)$?
   (b) What is the image of 4?
   (c) What is the preimage of 22,668,085?

3. Would the Fortune 500 list of the biggest corporations in 1986 define a function if there is a tie for twelfth place?

Rank 1986	1985	Company	Country	Sales ($ thousands)
1	1	Royal Dutch/Shell Group	Neth./Britain	64,843,217
2	2	British Petroleum	Britain	39,855,564
3	5	IRI	Italy	31,561,709
4	3	Toyota Motor	Japan	31,553,827
5	16	Daimler-Benz	W. Germany	30,168,550
6	7	Matsushita Electric Industrial	Japan	26,459,539
7	6	Unilever	Neth./Britain	25,141,672
8	15	Volkswagen	W. Germany	24,317,154
9	8	Hitachi	Japan	22,668,085
10	4	ENI	Italy	22,549,921

FIGURE 6 The Top 10 Corporations Located Outside the United States, 1986.

*E. E. David and J. G. Truxal, *The Man-Made World*, New York, McGraw-Hill Book Company, 1971, pp. 176–177. The authors thank Glenn Allinger of Montana State University for supplying this excellent source of applications.

†Terence Paré and Wilton Woods, "The International 500: The Fortune Directory of the Biggest Industrial Corporations Outside the U.S.," *Fortune*, August 3, 1987, pp. 214–233.

4. This problem refers to the number of chirps that a cricket makes. Biologists have shown that although different species of crickets produce different numbers of chirps, the number of chirps produced by each species is a linear function of the temperature. For instance, the number of chirps per minute that the cricket *Gryllus pennsylvanicus* makes is given by $C(t) = 4t - 160$, where t is in degrees Fahrenheit.‡
   (a) What is the domain of this function? (Make sure to take into account physical considerations.)
   (b) How many chirps per minute does this cricket make at 68°?
   (c) If this cricket makes 160 chirps per minute, what is the temperature?

5. In 1820 the median age in the United States was 16.7 years. This means that half of the population was younger than that age, and half was older. Table 1 lists the median age in the United States for each year from 1980 to 1987.§
   (a) Define a function $y = f(x)$ in which x is a year and y is the corresponding median age.
   (b) Evaluate $f(1990)$.
   (c) For what value of x is $y = 29.1$?

Table 1

Year	Median Age
1987	32.1
1986	31.8
1985	31.5
1984	31.2
1983	30.9
1982	30.6
1981	30.3
1980	30.0

Cumulative Exercise Set 1–3

1. (a) Sketch the graph of the equation $3x - y = -4$.
   (b) Sketch the graph of the equation $y = x^2$.
   (c) Find where the graphs of the equations in parts (a) and (b) intersect.

2. An equation that predicts the winning time in the women's 100-meter dash in the Olympics is

   $$y = -.0189x + 48.4781$$

   where y is the time in seconds and x is the year.
   (a) What is the slope of this line?
   (b) What is the y-intercept of this line?
   (c) What was the winning time for this event at the 1992 Olympics in Barcelona based on this equation? (Round the answer to two decimal places.)

3. Find the slope-intercept form of the line whose x-intercept is -3 and y-intercept is -6.

4. Find the slope-intercept form of the line that has slope 2 and intersects the graph of $f(x) = x^2$ at $x = 1$.

5. Sketch the graphs of the functions $f(x) = x^2$ and $g(x) = x^2 + 1$ on the same coordinate axes.

6. (a) Find $f(-x)$ for $f(x) = x^2 - x - 4$.
   (b) Draw the graphs of $f(x)$ and $f(-x)$ on the same coordinate system.

7. Evaluate $\dfrac{f(x + h) - f(x)}{h}$ for $f(x) = 3 - x$ and $h \neq 0$.

8. Find the domain of the function $f(x) = \dfrac{x}{x^2 - 2x}$.

9. Figure 2 in the text lists the 10 largest industrial corporations in the world in 1986. Compute $N(5)$, where $N(x)$ is the net income of the xth largest industrial corporation.

‡Philip S. Callahan, *Insect Behavior*, New York, Four Winds Press, 1977.
§Randolf E. Schmid, "Median Age in U.S. Reaches 32.1 Years, Highest on Record," Associated Press, April 6, 1988.

1–4 Polynomial and Other Functions

In this section we define polynomial functions and sketch their graphs by the plotting method. Then we introduce rational functions and sketch their graphs. The concept of a vertical asymptote plays an important role in graphing rational functions. We also introduce functions that are defined by different formulas over different intervals and functions that are defined by a curve instead of a formula.

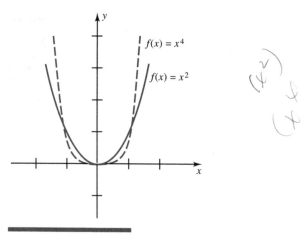

FIGURE 1 Contrast of $f(x) = x^2$ vs. $f(x) = x^4$.

Polynomial Functions

The graph of the quadratic function $f(x) = x^2$ is called a parabola. The graph of the function $f(x) = x^4$ has a very similar shape, although it is slightly different. The two curves are contrasted in Figure 1. The graphs of the functions $f(x) = x^n$ for $n = 6, 8, 10, \ldots$ are similar.

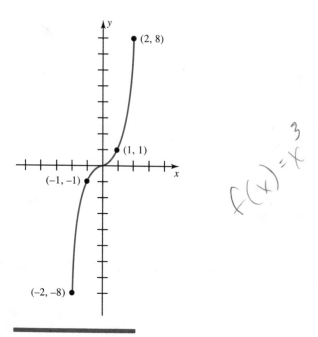

FIGURE 2 Graph of $f(x) = x^3$.

However, the graph of $f(x) = x^n$ is quite different when n is an odd number. For $n = 1$ the graph of $f(x) = x$ is a line. The graph of $f(x) = x^3$ is sketched in Figure 2. The graphs of the functions $f(x) = x^n$ for $n = 5, 7, 9, \ldots$ have similar shapes.

Collectively, functions of the form $f(x) = x^n$, where n can be any real number, are called **power functions.** Since n does not have to be a natural number, some additional examples of power functions are

$$n = 0 \qquad\qquad n = -1 \qquad\qquad n = \frac{1}{2}$$

$$f(x) = x^0 = 1 \quad\quad f(x) = x^{-1} = \frac{1}{x} \quad\quad f(x) = x^{1/2} = \sqrt{x}$$

A **polynomial function** of the variable x is defined by the formula

$$f(x) = a_n x^n + a_{n-1} x^{n-1} + \cdots + a_2 x^2 + a_1 x + a_0$$

where n is a nonnegative integer, the coefficients a_i are real numbers, and $a_n \neq 0$. The integer n is called the **degree of the polynomial.** The domain of a polynomial function consists of all real numbers. A polynomial function of degree 2 is a quadratic function, a polynomial function of degree 1 is a linear function, and a polynomial function of degree 0 is a constant function. The function

$$f(x) = x^4 - 8x^3 + 22x^2 + 9$$

is a polynomial function of degree 4 with $a_4 = 1$, $a_3 = -8$, $a_2 = 22$, $a_1 = 0$, and $a_0 = 9$. Some other examples of polynomial functions are

$$g(x) = 12 \qquad F(x) = 4x - 3.5 \qquad G(x) = 3x^2 - 1.5x + 2$$

Some functions which are not polynomial functions are

$$h(x) = \frac{2}{x} \qquad P(x) = \sqrt{x}$$

Example 1 shows how to sketch the graph of a polynomial function by the plotting method.

$a_4 = 1 \quad a_3 = -8 \quad a_2 = 22 \quad a_1 = 0 \quad a_0 = 9$

EXAMPLE 1

Problem

Sketch the graph of the polynomial function
$16 + 64 + 88 + 48 + 9$
$$f(x) = x^4 - 8x^3 + 22x^2 - 24x + 9$$

Solution Evaluate $f(x)$ at the values $x = -2, -1, 0, 1,$ and 2. The results are displayed in the table.

x	-2	-1	0	1	2
$f(x)$	225	64	9	0	1

These values suggest that the graph is the curve shown in Figure 3a. However,

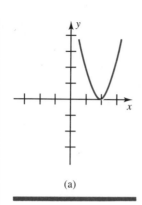 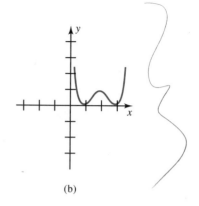

(a)   (b)

FIGURE 3   Two views of $f(x) = x^4 - 8x^3 + 22x^2 - 24x + 9$.

plotting additional points from an expanded table (rounded to one decimal place) leads to the correct graph of the polynomial function shown in Figure 3b.

x	0.5	1.5	2.5	3	3.5	4	5
$f(x)$	1.6	0.6	0.6	0	1.6	9	64

Example 1 shows that sometimes it is necessary to plot many points to obtain an accurate graph of a function. Yet the question persists: How do we know that the curve in Figure 3b is the correct graph of the given polynomial? The techniques of calculus will answer this question.

Rational Functions

Consider the power function $f(x)$ defined by the formula

$$f(x) = \frac{1}{x}$$

Its graph, which is called a **hyperbola,** consists of two disconnected branches. (See Figure 4.) The function is not defined at $x = 0$. Notice that as the values of x get closer and closer to 0, the values of $f(x)$ get larger and larger in absolute value. When a number v has the property that the values of a function $f(x)$ get larger and larger in absolute value as the values of x approach v, the line $x = v$ is called a **vertical asymptote** of the graph of $f(x)$. Figure 4 shows that the line $x = 0$ is a vertical asymptote of the function $f(x) = 1/x$.

A **rational function** is the quotient of two polynomial functions. Therefore $f(x)$ is a rational function if it can be written in the form

$$f(x) = \frac{g(x)}{h(x)}$$

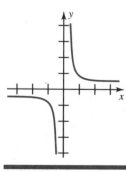

FIGURE 4   Graph of $f(x) = \dfrac{1}{x}$.

for some polynomials $g(x)$ and $h(x)$, where $h(x) \neq 0$. The function $f(x) = 1/x$ is a rational function with $g(x) = 1$ and $h(x) = x$, for $x \neq 0$.

The domain of a rational function $f(x) = g(x)/h(x)$ consists of all real numbers for which $h(x) \neq 0$. Each number v for which $h(v) = 0$ will produce a vertical asymptote of $f(x)$ at $x = v$ if $g(v) \neq 0$. Example 2 shows how to find vertical asymptotes.

EXAMPLE 2

Problem

Find the vertical asymptotes of the function

$$f(x) = \frac{x - 2}{x^2 - 1} \qquad f(x) = \frac{g(x)}{h(x)}$$

Solution Set $g(x) = x - 2$ and $h(x) = x^2 - 1$. If the rational function $f(x)$ has a vertical asymptote, it will occur where $h(x) = 0$. Set $h(x) = x^2 - 1 = 0$. Then $x = 1$ or $x = -1$. Also, $g(1) \neq 0$ and $g(-1) \neq 0$. Therefore the lines $x = 1$ and $x = -1$ are the vertical asymptotes of the graph of $f(x)$.

Consider the rational function $f(x) = \dfrac{x^2 - 4}{x + 2}$. Set $g(x) = x^2 - 4$ and $h(x) = x + 2$. The only possible vertical asymptote will be the line $x = -2$ since -2 is the only number with $h(-2) = 0$. Although $g(-2) = 0$ we cannot necessarily conclude that $x = -2$ is a vertical asymptote. In fact, in this case the line $x = -2$ is not a vertical asymptote of $f(x)$.

A knowledge of the vertical asymptotes of a rational function eases the task of sketching its graph. Example 3 illustrates the procedure.

EXAMPLE 3

Problem

Sketch the graph of the rational function

$$f(x) = \frac{x}{x - 1}$$

Solution Set $g(x) = x$ and $h(x) = x - 1$. The only vertical asymptote of $f(x)$ is the line $x = 1$ since $h(1) = 0$ but $g(1) \neq 0$. Draw it as a dashed line. Construct a table of ordered pairs for $x < 1$.

x	-2	-1	0	0.5	0.9	0.99
$f(x)$	$\frac{2}{3}$	$\frac{1}{2}$	0	-1	-9	-99

Notice that as the values of x approach 1, the images of $f(x)$ get larger and larger in absolute value. This shows the behavior of $f(x)$ to the left of the vertical asymptote. A table for $x > 1$ shows the behavior to the right of the vertical asymptote.

x	4	3	2	1.5	1.1	1.01
$f(x)$	$\frac{4}{3}$	$\frac{3}{2}$	2	3	11	101

Taken together, the graph consists of the two branches that result from plotting the points in these tables. (See Figure 5.)

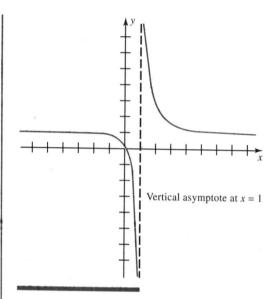

FIGURE 5 Graph of $f(x) = \dfrac{x}{x-1}$.

Other Functions

Next we introduce two additional kinds of functions. The first concerns functions that are defined by different formulas over different intervals.

EXAMPLE 4

Problem

Sketch the graph of the function defined by

$$f(x) = \begin{cases} 2x + 1 & \text{if } x < 2 \\ x^2 - 1 & \text{if } x \geq 2 \end{cases}$$

Solution The notation means that $f(x) = 2x + 1$ for all values of x such that $x < 2$. This is the linear function with slope 2 and y-intercept 1. The graph of $f(x)$ in the interval $(-\infty, 2)$ is part of this line. An open circle at the right end point of the line indicates that this point is not part of the function, but rather, serves as a boundary. The graph of $f(x)$ in the interval $[2, \infty)$ is part of the quadratic function $f(x) = x^2 - 1$. The entire graph consists of the two branches shown in Figure 6.

A function can be defined by a graph as well as by a formula. This sometimes occurs when the equation is either unknown or too complicated to express easily. For each value of x for which there is one point (x, y) on the curve, the image $f(x)$ is defined to be equal to y. This defines the function f implicitly. Example 5 examines one such function.

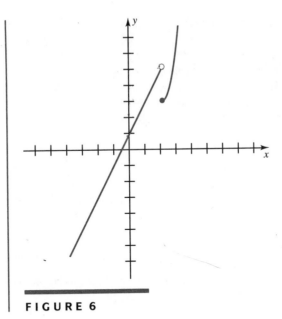

FIGURE 6

EXAMPLE 5

A firm's financial leverage is defined as its debt/assets ratio, or D/A ratio. The D/A ratio is usually given as a percent. There is a relation between the D/A ratio and the firm's expected earnings-per-share (EPS). Figure 7 gives the graph for Santa Clara Industries on January 1, 1985.

Problem

(a) Approximate $f(30)$. (b) What D/A ratio has the maximal expected EPS?

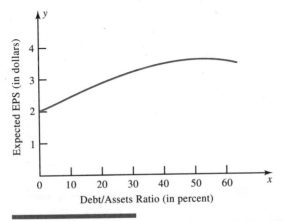

FIGURE 7 Financial Leverage vs. D/A Ratio for Santa Clara Industries. (Adapted from "Financial Leverage," in B. Campsey and E. Brigham, *Introduction to Financial Management*, Hinsdale, IL, The Dryden Press, 1985, pp. 534–541.)

Solution (a) The notation $f(30)$ stands for the image of $x = 30\%$. It corresponds to the value of y from the point $(30, y)$ on the graph. From Figure 7 it appears that $y = 3$, so $f(30) = 3$ and the expected EPS is $3. (According to Santa Clara Industries' tables, $f(30) = \$2.97$. This degree of accuracy cannot be inferred from the graph, however.)

(b) The maximal expected EPS of $3.36 occurs at the point $(50, 3.36)$. Thus $f(50) = 3.36$. The preimage of 3.36 is 50, so the D/A ratio is 50%.

Application

In the previous section we examined linear cost functions. Here we discuss a kind of nonlinear cost function that arises in manufacturing.

EXAMPLE 6

A manufacturer has determined that the cost of producing x items is $C(x) = \dfrac{x^2 + 25}{x}$, where fractional parts of an item can be produced.

Problem

(a) What is the domain of $C(x)$? (b) Sketch the graph of $C(x)$ to determine how many items should be produced in order to minimize cost.

Solution (a) The domain of $C(x)$ does not include $x = 0$ because division by 0 is undefined. In addition, since it makes no sense to talk about a negative num-

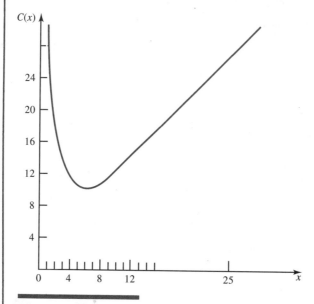

FIGURE 8 The cost function $C(x) = \dfrac{x^2 + 25}{x}$.

ber of items, it is meaningless to let x assume negative values. Therefore the domain of $C(x)$ consists of all positive real numbers, or the interval $(0, \infty)$.
(b) Set $g(x) = x^2 + 25$ and $h(x) = x$. Then $h(x) = 0$ and $g(x) \neq 0$, so the line $x = 0$ is a vertical asymptote of $f(x)$. The relevant graph of $C(x)$ is obtained by the plotting method from the table.

x	1	2	3	4	5	6	7	25
$C(x)$	26	$\frac{29}{2}$	$\frac{34}{3}$	$\frac{41}{4}$	10	$\frac{61}{6}$	$\frac{74}{7}$	26

The graph, which is sketched in Figure 8, indicates that the minimum cost occurs at $x = 5$ items.

In Chapter 4 we will describe how to use calculus to find the minimum cost in Example 6.

EXERCISE SET 1–4

In Problems 1 to 4 sketch the graph of the stated function by the plotting method.

1. $f(x) = 2x^3 - 9x^2 + 12x$

2. $f(x) = x^3 - 9x^2 + 24x - 20$

3. $f(x) = x^4 - 4x^3 + 4x^2 + 1$

4. $f(x) = x^4 - 6x^2$

5. What is the degree of the polynomial in Problem 1?

6. What is the degree of the polynomial in Problem 4?

In Problems 7 to 10 find the vertical asymptotes of each function.

7. $f(x) = \dfrac{x}{x - 5}$

8. $f(x) = \dfrac{3x}{2x - 5}$

9. $f(x) = \dfrac{x - 2}{x^2 - 9}$

10. $f(x) = \dfrac{3 - x}{x^2 - 4}$

In Problems 11 to 16 sketch the graph of the stated function.

11. The function defined in Problem 7.

12. The function defined in Problem 8.

13. $f(x) = \begin{cases} 3x + 2 & \text{if } x \leq 0 \\ x^2 + 2 & \text{if } x > 0 \end{cases}$

14. $f(x) = \begin{cases} 2x + 1 & \text{if } x \geq 3 \\ x^2 - 2 & \text{if } x < 3 \end{cases}$

15. $f(x) = \begin{cases} x^2 - 1 & \text{if } x < 0 \\ 2x - 1 & \text{if } x \geq 0 \end{cases}$

16. $f(x) = \begin{cases} 1 - x^2 & \text{if } x > 2 \\ 1 - 2x & \text{if } x \leq 2 \end{cases}$

Problems 17 to 20 refer to Figure 7, which shows the relation between the D/A ratio and the expected EPS for Santa Clara Industries.

17. Approximate $f(0)$.

18. Approximate $f(60)$.

19. What D/A ratio produces an expected EPS of approximately $3?

20. What D/A ratio produces an expected EPS of approximately $2.75?

In Problems 21 and 22 the cost of producing x items is given. Fractional parts of an item can be produced.
(a) What is the domain of $C(x)$?
(b) Sketch $C(x)$ to determine how many items should be produced in order to minimize cost.

21. $C(x) = \dfrac{x^2 + 16}{x}$

22. $C(x) = \dfrac{x^2 + 81}{x}$

In Problems 23 and 24 sketch the graphs of the two stated functions on the same coordinate axis.

23. $f(x) = x^2$ and $f(x) = x^6$

24. $f(x) = x^3$ and $f(x) = x^5$

In Problems 25 and 26 state the degree of the polynomial.

25. $f(x) = 1 - 2x - 5x^2$

26. $f(x) = 3 - 5x$

In Problems 27 to 34 find the vertical asymptotes, if any, and sketch the graph of the stated function.

27. $f(x) = \dfrac{x - 2}{x^2 - 9}$

28. $f(x) = \dfrac{3 - x}{x^2 - 4}$

29. $f(x) = \dfrac{x}{x^2}$

30. $f(x) = \dfrac{1}{x^2 + 1}$

31. $f(x) = \dfrac{1}{x(x^2 - 1)}$

32. $f(x) = \dfrac{x - 1}{x(x^2 - 4)}$

33. $f(x) = \dfrac{2x}{(3 + 2x)(x - 1)}$

34. $f(x) = \dfrac{x}{(1 + 2x)(x - 2)}$

In Problems 35 to 38 sketch the graph of the stated function.

35. $f(x) = \begin{cases} x & \text{if } x \geq 1 \\ x^2 + 1 & \text{if } x \leq 0 \end{cases}$

36. $f(x) = \begin{cases} 2x - 1 & \text{if } x \geq 3 \\ x^3 & \text{if } x \leq 1 \end{cases}$

37. $f(x) = \begin{cases} 1 - 2x - 5x^2 & \text{if } x \leq 1 \\ x & \text{if } x > 1 \end{cases}$

38. $f(x) = \begin{cases} -x^2 & \text{if } x \geq 0 \\ x^2 & \text{if } x < 0 \end{cases}$

39. Suppose the profit per item is given by the formula $P(x) = 20(3 - x)(x - 25)$, where x is the number of items that are produced. Draw a graph of $P(x)$ to determine the number of items that should be produced to yield the largest profit per item.

40. A store offers this special sale: If 10 compact disks (CDs) are purchased at the full price of $12, then additional CDs can be purchased at half price. There is a limit of 25 CDs per person. Express the cost of the CDs as a function of the number purchased and draw the graph.

In Problems 41 to 44 match the equation with the given graph.

41. $f(x) = \dfrac{x}{x + 1}$

42. $f(x) = \dfrac{x}{1 - x}$

43. $f(x) = \begin{cases} x^2 & x \geq 1 \\ -x^2 & x < 1 \end{cases}$

44. $f(x) = \begin{cases} x^2 & x \geq 0 \\ -x^2 & x < 0 \end{cases}$

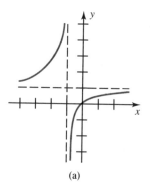

(a)

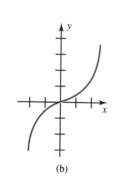

(b)

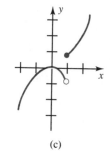

(c)

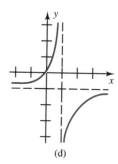

(d)

In the programming language BASIC there is a function INT that produces the integral part of a number. For instance, $INT(3.14) = 3$. In Problems 45 and 46 sketch the graph of the given function.

45. $f(x) = INT(x)$ for $x \geq 0$

46. $f(x) = INT(x^2)$ for $x \geq 0$

Referenced Exercise Set 1–4

1. The curve in Figure 9 shows the relation between the implied stock price and the D/A ratio for Santa Clara Industries. What percentage yields the maximal D/A ratio?

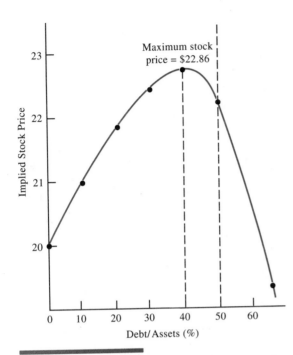

FIGURE 9 Relation between implied stock price and the D/A ratio for Santa Clara Industries. (Adapted from B. Campsey and E. Brigham, *Introduction to Financial Management*, Hinsdale, IL, The Dryden Press, 1985, p. 542.)

Stock Market Indexes

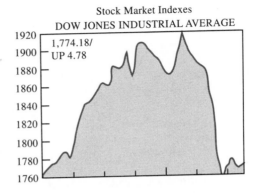

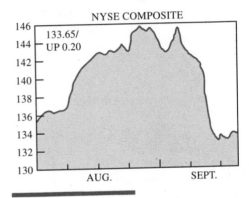

FIGURE 10 Stock market activity during two dramatic months in 1986. (Dow–Jones Index.)

2. Figure 10 shows two curves portraying stock market activity during two dramatic months in 1986.
   (a) What was the maximum value of the Dow Jones Industrial average during this period?
   (b) During which week did the Dow Jones Industrial average tumble precipitously?

3. An article in the *Wall Street Journal* analyzed the federal income tax that was signed into law on October 22, 1986.* Consider this statement as an example of the tax law: "If the income lies in the range from $0 through $29,300 (inclusive) then the tax is 15 percent of the income; if the income exceeds $29,300 then the tax equals $4350 plus 27 percent of the portion of the income above $29,300."

   (a) Write this statement as a function $y = f(x)$, where x is the income and y is the corresponding tax.
   (b) Sketch the graph of $f(x)$.
   (c) Evaluate $f(x)$ for $x = 29,300$.
   (d) Evaluate $f(x)$ for $x = 29,301$.
   (e) What do you conclude from parts (c) and (d)?

4. Public health officials must often determine the threshold level at which effects on health begin. The threshold level for certain environmental issues occurs when the introduction of one more unit of pollution becomes a detriment to public health. One study† made use of the

*Alan Murray, "Winners? Losers? Estimates Show How Impact of Tax Proposal Varies," *Wall Street Journal*, May 9, 1986, p. 25.

†Louis A. Cox, Jr., "A New Measure of Attributable Risk for Public Health Applications," *Management Science*, July 1985, pp. 800–813.

health function $H(x)$ defined by

$$H(x) = \begin{cases} 0 & \text{if } x < 5 \\ (x - 5)^2 & \text{if } x \geq 5 \end{cases}$$

where x denotes the number of units of pollution.
(a) Sketch the graph of this health function.
(b) At what value of x does the threshold level occur?

5. Human beings have raced over a large span of times and distances. Take women's swimming, for instance. Races have been contested from a scant 50 meters to the 80-kilometer (km) swim across the English Channel. It turns out that when world records for many athletic events are recorded for various distances at any given time, it is possible to derive a power function that relates the time to the distance. The power function for women's swimming up to 1980, with the time t in minutes and the distance x in kilometers (km), is:‡

$$t = 10.578x^{1.03256}$$

Use this equation to calculate the time of the world record for (a) 1500 meters and (b) 80 km.

6. Police departments use a manual when investigating accidents to establish the velocity v (in miles per hour) that the vehicle involved in the accident was traveling. Most manuals use the function

$$v = \sqrt{30\, f(x)}$$

where f is the coefficient of friction (a number that measures the friction between the vehicle's tires and the road's surface, usually between 0 and 1.5) and x is the length of the skid marks in feet.§ Set the coefficient of friction equal to 1. (a) If the skid marks are 200 feet, what was the vehicle's velocity? (b) What is the domain of $f(x)$? (c) Sketch the graph of $f(x)$.

Cumulative Exercise Set 1–4

In Problems 1 and 2 sketch the graph of the equation.

1. $y = x^2 + 1$

2. $y = 2 - x^3$

3. Find the slope-intercept form of the line passing through the points $(0,1)$ and $(-1,0)$.

4. Find the equation of the line that passes through the point $(1,0)$ and is (a) vertical, (b) horizontal, (c) parallel to $y = -3x + 1$.

5. Find the domain of the function $f(x) = (x - 4)^{1/2}$.

6. Sketch the graph of the function $f(x) = x^2 - 2x + 1$.

7. Evaluate $\dfrac{f(x + h) - f(x)}{h}$ for $f(x) = x^2 - 3$ and $h \neq 0$.

8. Find the domain of the function $f(x) = \dfrac{1}{9 - x^2}$.

9. Sketch the graph of the function $f(x) = x^3 - x^2 + 1$.

10. Sketch the graph of the function

$$f(x) = \begin{cases} 2x \text{ if } x \geq 0 \\ x \text{ if } x < 0 \end{cases}$$

In Problems 11 and 12 find the vertical asymptotes and sketch the graph of the function.

11. $f(x) = \dfrac{x}{x - 4}$.

12. $f(x) = \dfrac{x}{x^2 - 4}$.

‡Peter S. Riegel, "Athletic Records and Human Endurance," *American Scientist*, May-June 1981, pp. 285–290.

§David S. Daniels, "Fast brakes!" *Mathematics Teacher*, February 1989, pp. 104–107 and 111.

1–5 The Algebra of Functions

The four familiar operations on real numbers are addition, subtraction, multiplication, and division. In this section we define these operations on functions and introduce a fifth operation, composition. We then discuss additional types of functions that are defined in terms of these five operations.

Operations on Functions

Consider the polynomial functions f and g defined by

$$f(x) = 2x^2 - 3x - 7 \qquad \text{and} \qquad g(x) = 2x + 3$$

The **sum** of these functions, denoted by $f + g$, is defined by the formula

$$(f + g)(x) = f(x) + g(x)$$
$$= (2x^2 - 3x - 7) + (2x + 3)$$
$$= 2x^2 - x - 4$$

The **difference** $f - g$ is defined similarly.

$$(f - g)(x) = (2x^2 - 3x - 7) - (2x + 3)$$
$$= 2x^2 - 5x - 10$$

The **product** $f \cdot g$ is defined by carrying out the multiplication of the two polynomials.

$$(f \cdot g)(x) = f(x) \cdot g(x)$$
$$= (2x^2 - 3x - 7)(2x + 3)$$
$$= (2x^2 - 3x - 7)(2x) + (2x^2 - 3x - 7)(3)$$
$$= (4x^3 - 6x^2 - 14x) + (6x^2 - 9x - 21)$$
$$= 4x^3 - 23x - 21$$

The **quotient** f/g is defined by

$$\left(\frac{f}{g}\right)(x) = \frac{f(x)}{g(x)} = \frac{2x^2 - 3x - 7}{2x + 3}$$

Sometimes the quotient can be left in this form. At other times the division of polynomials must be carried out.

$$
\begin{array}{r}
x - 3 \\
2x + 3 \overline{)2x^2 - 3x - 7} \\
\underline{2x^2 + 3x} \\
-6x - 7 \\
\underline{-6x - 9} \\
2
\end{array}
$$

Then

$$\left(\frac{f}{g}\right)(x) = x - 3 + \frac{2}{2x + 3}$$

The operations of $+$, $-$, and $\cdot$ are defined whenever both $f(x)$ and $g(x)$ are defined, so the domain of each one is the intersection of the domain of f and the domain of g. The quotient f/g is somewhat different. It is defined whenever both

$f(x)$ and $g(x)$ are defined except where $g(x) = 0$. In the example above, the domain of f/g consists of all real numbers except $-\frac{3}{2}$ because $g(-\frac{3}{2}) = 0$. The display gives the general definitions.

DEFINITION

If f and g are any functions for which $f(x)$ and $g(x)$ are defined, then

$$(f + g)(x) = f(x) + g(x) \qquad \text{(sum)}$$

$$(f - g)(x) = f(x) - g(x) \qquad \text{(difference)}$$

$$(f \cdot g)(x) = f(x) \cdot g(x) \qquad \text{(product)}$$

$$\left(\frac{f}{g}\right)(x) = \frac{f(x)}{g(x)} \qquad \text{if } g(x) \neq 0 \quad \text{(quotient)}$$

The four operations $f + g$, $f - g$, $f \cdot g$, and f/g are referred to collectively as the **rational operations.**

EXAMPLE 1

Let f and g be functions defined by

$$f(x) = \frac{x}{x - 1} \quad \text{and} \quad g(x) = \frac{x - 3}{x + 1}$$

Problem

Find $f + g$, $f \cdot g$, and f/g. State their domains.

Solution By definition $f + g$ is defined by

$$(f + g)(x) = \frac{x}{x - 1} + \frac{x - 3}{x + 1}$$

The proper form of the sum is obtained by carrying out the addition, which involves finding the least common denominator.

$$(f + g)(x) = \frac{x(x + 1) + (x - 1)(x - 3)}{(x - 1)(x + 1)}$$

$$= \frac{x^2 + x + x^2 - 4x + 3}{x^2 - 1}$$

$$= \frac{2x^2 - 3x + 3}{x^2 - 1}$$

The product $f \cdot g$ is obtained by multiplying the numerators and denominators.

$$(f \cdot g)(x) = \frac{x(x - 3)}{(x - 1)(x + 1)}$$

Often it is convenient to write this expression in a more compact form.

$$(f \cdot g)(x) = \frac{x^2 - 3x}{x^2 - 1}$$

The quotient is obtained from the "invert and multiply" rule

$$\left(\frac{f}{g}\right)(x) = \frac{x}{x-1} \div \frac{x-3}{x+1} = \frac{x}{x-1} \cdot \frac{x+1}{x-3}$$

$$= \frac{x^2 + x}{x^2 - 4x + 3}$$

The domain of f consists of all real numbers except 1 since $f(1)$ is not defined. Similarly, $g(-1)$ is not defined. Therefore the domain of $f + g$ and $f \cdot g$ is the set of all real numbers except 1 and -1. Since $g(3) = 0$, the domain of f/g is the set of all real numbers except 1, -1, and 3.

If a number a lies in the domain of $f + g$, we write $(f + g)(a) = f(a) + g(a)$. The same notation applies to the other three rational operations on two functions. Example 2 shows how to find such images.

EXAMPLE 2

Let f and g be functions defined by

$$f(x) = \frac{x}{x-1} \qquad \text{and} \qquad g(x) = \frac{3x-2}{x+1}$$

Problem

Compute $(f + g)(\frac{2}{3})$, $(f \cdot g)(\frac{2}{3})$, and $(f/g)(\frac{2}{3})$, if possible.

Solution First $f(\frac{2}{3}) = -2$ and $g(\frac{2}{3}) = 0$. By definition, $(f + g)(\frac{2}{3}) = f(\frac{2}{3}) + g(\frac{2}{3})$ so $(f + g)(\frac{2}{3}) = -2 + 0 = -2$. Similarly, $(f \cdot g)(\frac{2}{3}) = f(\frac{2}{3}) \cdot g(\frac{2}{3}) = (-2) \cdot 0 = 0$. However, $(f/g)(\frac{2}{3})$ is not defined because $g(\frac{2}{3}) = 0$.

Composition of Functions

Consider once again the polynomial functions

$$f(x) = 2x^2 - 3x - 7 \quad \text{and} \quad g(x) = 2x + 3$$

Regard each function as an "input-output machine" such that each input value comes from the domain and the corresponding output lies in the range. Now put the machines in a line with g in front of f, so the output of g becomes the input of f. For example, take $x = 1$. The output of g is $g(1) = 5$. This value becomes the input of f, and since $f(5) = 28$, its output is 28. Altogether the sequence of "machines" has produced an image of 28 from the original input of $x = 1$. This

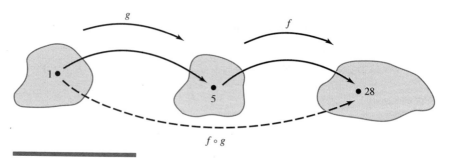

FIGURE 1 The composition of functions.

is shown in Figure 1, where the dashed curve represents the result of the two functions.

The operation of performing one function f on the image of another function g is called **composition** and is denoted by $f \circ g$. It is defined by

$$(f \circ g)(x) = f(g(x))$$

For the polynomials f and g above,

$$
\begin{aligned}
(f \circ g)(x) = f(g(x)) &= f(2x + 3) = 2[g(x)]^2 - 3g(x) - 7 \\
&= 2(2x + 3)^2 - 3(2x + 3) - 7 \\
&= 2(4x^2 + 12x + 9) - 6x - 9 - 7 \\
&= 8x^2 + 18x + 2
\end{aligned}
$$

The display provides the formal definition.

DEFINITION

If f and g are any functions, then the **composition** of f and g is the function $f \circ g$ defined by

$$(f \circ g)(x) = f(g(x))$$

The *domain* of $f \circ g$ is the set of all real numbers x in the domain of g whose image $g(x)$ lies in the domain of f.

If a is a number in the domain of $f \circ g$, then $(f \circ g)(a) = f(g(a))$. For $f(x) = 2x^2 - 3x - 7$, $g(x) = 2x + 3$, and $a = 1$, we get $(f \circ g)(1) = f(g(1)) = f(5) = 28$. However, $(f \cdot g)(1) = f(1) \cdot g(1) = (-8) \cdot 5 = -40$, so $(f \circ g)(x) \neq (f \cdot g)(x)$.

 Warning: Do not confuse $f \circ g$ with $f \cdot g$. The composition of two functions is entirely different from the product of two functions.

EXAMPLE 3

Let f and g be the functions defined by

$$f(x) = \frac{1}{x + 1} \quad \text{and} \quad g(x) = \sqrt{x}$$

Problem

Find $f \circ g$ and $g \circ f$.

Solution

$$(f \circ g)(x) = f(g(x))$$

$$= f(\sqrt{x})$$

$$= \frac{1}{\sqrt{x} + 1}$$

Similarly

$$(g \circ f)(x) = g(f(x)) = g\left(\frac{1}{x + 1}\right) = \sqrt{\frac{1}{x + 1}} = \frac{1}{\sqrt{x + 1}}$$

When forming the composition of two functions the order of the functions is important. Notice, for instance, that for the functions f and g in Example 3, $(f \circ g)(1) = 1/2$ but $(g \circ f)(1) = 1/\sqrt{2}$, so $(g \circ f)(1) \neq (f \circ g)(1)$. We single this out as a warning.

  Warning: In general $f \circ g \neq g \circ f$.

One way to determine whether $f \circ g \neq g \circ f$ for functions $f(x)$ and $g(x)$ is to see if their domains are unequal. For the functions defined in Example 3, the domain of $f \circ g$ consists of all numbers x with $x \geq 0$, whereas the domain of $g \circ f$ consists of all numbers x with $x > -1$. Therefore $f \circ g \neq g \circ f$.

When applying the techniques of calculus, often we find it advantageous to regard a given function as the composition of two functions. For example, consider the function $h(x) = \sqrt{x^2 - 3x + 5}$. Since the polynomial $x^2 - 3x + 5$ is evaluated first, set $g(x) = x^2 - 3x + 5$. Then $h(x) = \sqrt{g(x)}$. Now put $f(x) = \sqrt{x}$. Then $(f \circ g)(x) = f(g(x)) = \sqrt{g(x)} = h(x)$, so $h = f \circ g$. In this way $h(x)$ is written as the composition of the two simpler functions f and g.

EXAMPLE 4

Problem

Write the function

$$k(x) = \frac{(x^2 - 3x + 5)^5 - 1}{(x^2 - 3x + 5)^4 + 1}$$

as a composition of two simpler functions.

Solution The polynomial $x^2 - 3x + 5$ is evaluated first, so set $g(x) = x^2 - 3x + 5$. Then $k(x)$ can be written as

$$k(x) = \frac{[g(x)]^5 - 1}{[g(x)]^4 + 1}$$

Define $f(x)$ by substituting x for $g(x)$ in this expression.

$$f(x) = \frac{x^5 - 1}{x^4 + 1}$$

Then $k = f \circ g$ since

$$(f \circ g)(x) = f(g(x)) = \frac{[g(x)]^5 - 1}{[g(x)]^4 + 1} = k(x)$$

Quality Control

For many companies quality is the most important product. Quality control often depends on the production level; a low production level can cause problems with a worker's concentration and morale, while a high production level can lead to fatigue and nervousness. The usual assembly line offers a prime example.

It is possible to measure quality Q in terms of the production level P, where Q ranges from 0 to 100 (100 being a "perfect" product) and P represents the number of units produced per day where P ranges from 0 units (meaning that the assembly line is shut down) to 18 units (meaning it is operating at peak speed). Production, in turn, often depends on the time of day t, measured in hours. This means that ultimately the quality Q can be regarded as a function of the time of day t. The underlying concept is the composition of functions. Example 5 addresses this issue.

EXAMPLE 5

Let the quality of a product be defined by

$$Q(P) = 50 + 8P - 0.4P^2$$

where P is the number of units produced per day, with $0 \leq P \leq 18$. Let the daily production be defined by

$$P(t) = 6t - 0.5t^2$$

where t is the time of day in hours, with $0 \leq t \leq 8$.

Problem

Express the quality Q as a function t.

Solution Write Q as a function of $P(t)$.

$$Q(P(t)) = 50 + 8P(t) - 0.4P(t)^2$$

Substitute the expression for $P(t)$.

$$Q(P(t)) = 50 + 8(6t - 0.5t^2) - 0.4(6t - 0.5t^2)^2$$

Collect similar terms.

$$Q(P(t)) = 50 + 48t - 4t^2 - 0.4(36t^2 - 6t^3 + 0.25t^4)$$
$$= 50 + 48t - 18.4t^2 + 2.4t^3 - 0.1t^4$$

In this way quality Q is defined as a function of time t. (See Figures 2 and 3.)

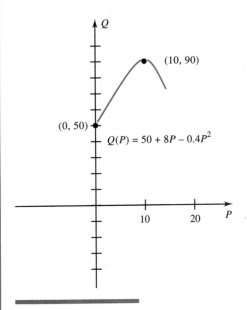

(10, 90)

(0, 50)

$Q(P) = 50 + 8P - 0.4P^2$

10   20 P

FIGURE 2

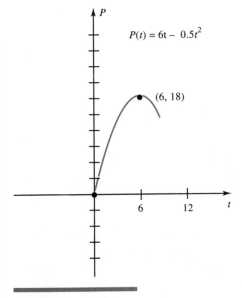

$P(t) = 6t - 0.5t^2$

(6, 18)

6   12 t

FIGURE 3

An important type of relationship between functions is defined by the inverse of functions.

DEFINITION

Two functions $f(x)$ and $g(x)$ are **inverses** of each other if

$$f(g(x)) = x = g(f(x))$$

for each value of the domain of $f(x)$ and $g(x)$.

We denote the inverse of $f(x)$ as $f^{-1}(x)$, read "$f(x)$ inverse." Do not confuse the notation for f inverse with the exponent "-1," meaning $1/f(x)$. By $f^{-1}(x)$ we will always mean $f(x)$ inverse and never the reciprocal of $f(x)$.

EXAMPLE 6

Problem

Show that the two functions $f(x) = \sqrt[3]{2x - 5} = (2x - 5)^{1/3}$ and $g(x) = (x^3 + 5)/2$ are inverses of each other.

Solution Find $f(g(x))$ and $g(f(x))$:

$$f(g(x)) = f[(x^3 + 5)/2]$$
$$= [2(x^3 + 5)/2 - 5]^{1/3}$$
$$= [x^3 + 5 - 5]^{1/3} = (x^3)^{1/3}$$
$$= x$$

$$g(f(x)) = g[(2x - 5)^{1/3}]$$
$$= \{[(2x - 5)^{1/3}]^3 + 5\}/2$$
$$= [(2x - 5) + 5]/2 = 2x/2$$
$$= x$$

Intuitively, the inverse function undoes what the function does. For example, the function $f(x) = x^3$ cubes a number and its inverse function $f^{-1}(x) = x^{1/3}$ takes the cube root of the number. So if you compute the composition of the two functions, you first cube x and then find the cube root of that number, so you get back to x, meaning that $f(g(x)) = x$.

To compute the inverse function from a given function, interchange the roles of x and y and solve the equation for x. This reverses each of the operations that constitute the formula for $f(x)$. Then again reverse the roles of x and y. Example 7 illustrates this procedure.

E X A M P L E 7

Problem

Find the inverse of the function $f(x) = \sqrt{3x - 5}$.

Solution Substitute y for $f(x)$ and get $y = (3x - 5)^{1/2}$. To find the inverse function, we solve this equation for x:

$$(3x - 5)^{1/2} = y$$

$$3x - 5 = y^2$$

$$3x = y^2 + 5$$

$$x = (y^2 + 5)/3$$

This means that the inverse function starts with a value y, squares it, adds five to this result, and then divides this number by three. If we write the independent variable as x, we get

$$f^{-1}(x) = (x^2 + 5)/3$$

EXERCISE SET 1–5

In Problems 1 to 4 find $f + g$, $f \cdot g$, and f/g.

1. $f(x) = \dfrac{x + 2}{x - 1}$ and $g(x) = \dfrac{x}{x + 2}$

2. $f(x) = \dfrac{x - 2}{x + 3}$ and $g(x) = \dfrac{x}{x - 3}$

3. $f(x) = \dfrac{x}{x + 1}$ and $g(x) = \dfrac{1}{x}$

4. $f(x) = \dfrac{x}{x - 1}$ and $g(x) = \dfrac{x}{x + 2}$

5. Find the domain of $f + g$ for the functions f and g defined in Problems 1 and 3.

6. Find the domain of $f - g$ for the functions f and g defined in Problems 2 and 4.

7. Find the domain of f/g for the functions f and g defined in Problems 1 and 3.

8. Find the domain of f/g for the functions f and g defined in Problems 2 and 4.

In Problems 9 to 12 compute (a) $(f + g)(a)$, (b) $(f \cdot g)(a)$, and (c) $(f/g)(a)$ for the functions f and g in the stated problem and the given value of a.

9. f and g in Problem 1, $a = 2$

10. f and g in Problem 2, $a = 4$

11. f and g in Problem 3, $a = -2$

12. f and g in Problem 4, $a = 0$

In Problems 13 to 16 find $f \circ g$ and $g \circ f$.

13. $f(x) = \sqrt{x}$ $g(x) = \dfrac{x}{1 - x}$

14. $f(x) = \sqrt{2x}$ $g(x) = \dfrac{1}{1 - x}$

15. $f(x) = 2x$ $g(x) = \dfrac{x - 2}{x + 2}$

16. $f(x) = \dfrac{1}{x}$ $g(x) = \sqrt{x}$

In Problems 17 to 20 write the function as the composition of two functions.

17. $f(x) = \sqrt{x^3 + 2x - 5}$

18. $f(x) = \sqrt{2x + 3}$

19. $f(x) = \dfrac{1}{(x - 2)^2} + (x - 2)^3$

20. $f(x) = \dfrac{1 + (x + 1)^2}{3 + (x + 1)^3}$

In Problems 21 and 22 the quality of a product is defined by $Q(P)$, where P is the number of units produced per day and the daily production is defined by $P(t)$, with t the time of day in hours. Express Q as a function of t.

21. $Q(P) = 20 + 4P - 0.02P^2$
    $P(t) = 8t - 0.5t^2$

22. $Q(P) = 25 + 2P - 0.05P^2$
    $P(t) = 6t - 0.8t^2$

In Problems 23 and 24 let $f(x)$ and $g(x)$ be the functions defined in Problem 1 and $h(x) = \dfrac{1}{x - 2}$.

23. Compute (a) $f + g + h$,   (b) $f \cdot g \cdot h$.

24. Compute (a) $(f/g)/h$,   (b) $f/(g/h)$.
    (c) Is $(f/g)/h = f/(g/h)$?

In Problems 25 and 26 let $f(x)$ and $g(x)$ be the functions defined in Problem 2 and $h(x) = \dfrac{1}{x + 3}$.

25. Compute (a) $f + g + h$,   (b) $f \cdot g \cdot h$.

26. Compute (a) $(f/g)/h$,   (b) $f/(g/h)$.
    (c) Is $(f/g)/h = f/(g/h)$?

27. Evaluate $(f \cdot g \cdot h)(0)$ for the functions in Problem 23.

28. Evaluate $(f \cdot g \cdot h)(0)$, where $f(0)$ and $g(0)$ are defined and $h(0) = 0$.

In Problems 29 to 32 compute $f \circ g$.

29. $f(x) = \dfrac{1}{\sqrt{x}}$ $g(x) = \dfrac{1}{x^2}$

30. $f(x) = 1 - x$ $g(x) = \dfrac{1}{x^3}$

31. $f(x) = \dfrac{x}{x + 1}$ $g(x) = \dfrac{x + 1}{x}$

32. $f(x) = \sqrt{1 - x}$ $g(x) = 1 - x^2$

33. Let $f(x)$ and $g(x)$ be the functions defined in Problem 31. Is it true that $(f \circ g)(1) = (g \circ f)(1)$?

34. Let $f(x)$ and $g(x)$ be the functions defined in Problem 30. Is it true that $(f \circ g)(-1) = (g \circ f)(-1)$?

In Problems 35 and 36 write the given function as the composition of two functions.

35. $f(x) = \sqrt[3]{(x - 1)^2}$ 36. $f(x) = \dfrac{5}{(2 - x)^3}$

In Problems 37 and 38 let f and g be the functions defined in Example 1 in the text.

37. Compute $(f + g)(a)$, where a is a given value of x.

38. Compute $(f + g)(x + h)$.

In Problems 39 and 40 let f and g be the functions defined in Example 3 in the text.

39. Compute $(f + g)(a)$, where a is a given value of x.

40. Compute $(f + g)(x + h)$.

41. For the functions f and g in Problem 1, is it true that $f \cdot g = f \circ g$?

42. For the functions f and g in Problem 3, is it true that $f \cdot g = f \circ g$?

43. Suppose the number of hours required to stock x percent of a shipment is $S(x) = g(x)/h(x)$, where $g(x) = 40x$ and $h(x) = 200 - x$.
    (a) What is the domain of the function $S(x)$?
    (b) For what values of x does $S(x)$ have a practical interpretation in this context?
    (c) How many hours were required to stock 40% of the shipment?
    (d) How many hours were required to stock the entire shipment?
    (e) What percentage of the shipment was stocked after 20 hours?

44. Suppose the cost of conducting a political survey of x percent of the eligible voters in a precinct is $C(x) = g(x)/h(x)$ dollars, where $g(x) = 105x$ and $h(x) = 110 - x$.
    (a) What is the domain of the function $C(x)$?
    (b) For what values of x does $C(x)$ have a practical interpretation in this context?
    (c) How much did it cost to poll 40% of the precinct?
    (d) How much did it cost to poll the entire precinct?
    (e) What percentage of the precinct could be polled with a fund of $500?

45. This problem refers to the profit function $P(x)$ on the sales of x units of an item: $P(x) = R(x) - C(x)$, where $R(x) = 480x - 12x^2 - 2100$ is the revenue function and $C(x) = 96x + 204$ is the cost function.
    (a) Draw the graphs of $R(x)$ and $C(x)$ on the same coordinate axis.
    (b) Use the graphs in part (a) to determine the numbers of units of the item that should be sold so the profit is a positive amount.

In Problems 46 and 47 let f, g, and h be the functions

$$f(x) = \frac{1}{x} \qquad g(x) = \sqrt{x} \qquad h(x) = x + 1$$

46. Compute $\dfrac{f \circ g}{h}$.

Table 1

letter	A	B	C	D	...	L	...	X	Y	Z
value	65	66	67	68	...	76	...	88	89	90

47. (a) Compute $(f \circ g) \circ h$.
    (b) Compute $f \circ (g \circ h)$.
    (c) Compute $f \circ g \circ h$.

Referenced Exercise Set 1–5

1. The coding of messages using a computer involves the composition of three functions.* The first function, $h(x)$, assigns a number, called the ASCII value, to each letter in the alphabet. (ASCII stands for the American Standard Code for Information Interchange.) It is partially defined in Table 1.

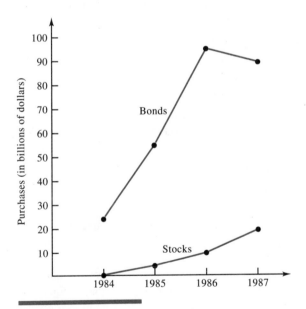

FIGURE 4

The second function, $g(x)$, is called the encoding function. Let $g(x)$ be defined by

$$g(x) = \begin{cases} x + 10 & \text{if } 65 \leq x \leq 80 \\ x - 16 & \text{if } 81 \leq x \leq 90 \end{cases}$$

The third function, $f(x)$, uses Table 1 to assign a letter to each ASCII value x, where $65 \leq x \leq 90$. The code function $c(x)$, which replaces each letter in the alphabet by another letter, is defined by $c = f \circ g \circ h$. For instance, $c(L) = (f \circ g \circ h)(L) = f(g(h(L))) = f(g(76)) = f(86) = V$, so the letter L is always replaced in the code by the letter V. Find the code for each letter in the word FUN.

2. Figure 4 traces Japanese net purchases of foreign stocks and bonds from 1984 through 1987.† Let $B(x)$ denote the purchases of bonds and $S(x)$ denote the purchases of stocks in year x. Write an equation for $B(x) + S(x)$ during the years 1984–1987.

3. When the weather is hot and humid, what is the best time of the day to perform physical activity? Some people think the best time is when the temperature is at its lowest point. Others think it is when the humidity is lowest. Actually the best time to exercise under such conditions is when the sum of the two measures is minimal.‡ The accompanying chart lists hourly measurements of temperature and humidity in Philadelphia on August 27, 1990. The chart shows that the best time to exercise might not occur when the temperature or the humidity is minimal. Most newspapers carry such a chart, so you can verify the same conclusion using records from your own locale.
    (a) If $T(x)$ stands for the temperature at x o'clock, at what value of x is $T(x)$ the lowest?
    (b) If $H(x)$ stands for the humidity at x o'clock, at what value of x is $H(x)$ the lowest?
    (c) At what value of x is $T(x) + H(x)$ the lowest?

*James Reagan, "Get the Message? Cryptographs, Mathematics, and Computers," *The Mathematics Teacher,* October 1986, pp. 547–553.

†Mark Beauchamp and John Hines, "We'll Send You VCRs—You Send Us Stocks," *Forbes Magazine,* August 10, 1987, pp. 60–62.

‡You can consult any book on exercise to confirm the results of this problem, for instance, Jim Fixx's *The Complete Book of Running.*

Yesterday in Philadelphia §

A.M.	Temp./Hum.	P.M.	Temp./Hum.
1:00	69/96	1:00	87/62
2:00	70/93	2:00	89/54
3:00	69/100	3:00	90/55
4:00	68/100	4:00	89/53
5:00	68/100	5:00	89/53
6:00	68/100	6:00	87/54
7:00	67/100	7:00	84/62
8:00	71/100	8:00	82/69
9:00	76/90	9:00	79/78
10:00	79/84		
11:00	85/75		
Noon	86/67		

§Source: *The Philadelphia Inquirer,* August 28, 1990.

Cumulative Exercise Set 1–5

In Problems 1 and 2 use the plotting method to sketch the graph of each equation.

1. $3y - 5x = 0$

2. $y = (x - 3)^2$

3. (a) What is the equation of the line that passes through the points $(2, 3)$ and $(2, 4)$?
   (b) What is the y-intercept of the line that passes through the points $(2, 3)$ and $(2, 4)$?

4. Find an equation of the line that passes through the origin and the point $(-4, -2)$.

5. Compute each of the following for the function $f(x) = -2x^2$.
   (a) $f(-2)$
   (b) $f(x + h)$
   (c) $f(x + h) - f(x)$
   (d) $\dfrac{f(x + h) - f(x)}{h}$
   for $h \neq 0$

6. What is the domain of the function $f(x) = 1 - x^2$?

7. (a) Sketch the graph of $f(x) = 8 - x^3$ by the plotting method.
   (b) What is the degree of $f(x)$?

8. Find the vertical asymptotes of the function
   $$f(x) = \frac{2}{4 - x^2}$$

9. Sketch the graph of the function
   $$f(x) = \begin{cases} \sqrt{x} & x \geq 0 \\ -x & x < 0 \end{cases}$$

10. Write $f + g$ as a function with denominator $1 - x^2$, where
    $$f(x) = \frac{x}{1 - x} \qquad g(x) = \frac{x}{1 + x}$$

11. Find $(f \circ g)(2)$ for the functions f and g in Problem 10.

12. For what function $h(x)$ can $f(x) = (x^2 - 1)^3$ be written in the form $f = g \circ h$, where $g(x) = x^3$?

1–6   Solving Polynomial Equations

Many applications of mathematics require a knowledge of points where the graph of a function $f(x)$ crosses the x-axis. Algebraically this is equivalent to solving the equation $f(x) = 0$.

  We begin this section by reviewing quadratic equations and the quadratic formula. Then we apply the method of factoring to solving polynomial equations. We end by using these procedures to find the points of intersection of two graphs.

The Quadratic Formula

The graph of a function $f(x)$ crosses the x-axis at $x = a$ if and only if $f(a) = 0$. Geometrically the point $(a, 0)$ is an x-intercept of the graph. Algebraically the value $x = a$ is called a **solution** (or a **root**) **of the equation** $f(x) = 0$. Therefore solutions of the equation $f(x) = 0$ yield x-intercepts of the graph of f. A **polynomial equation** is an equation $f(x) = 0$, where $f(x)$ is a polynomial function. To "solve a polynomial equation" means to find all of its solutions. Recall that the **degree of a polynomial equation** is the highest power of the polynomial.

A **quadratic equation** is a polynomial equation of degree 2. The general form of a quadratic equation is

$$ax^2 + bx + c = 0$$

where a, b, and c are real numbers with $a \neq 0$. We have seen that the graph of a quadratic function $f(x) = ax^2 + bx + c$ is a parabola, so it has either zero, one, or two x-intercepts. Equivalently, a quadratic equation has either zero, one, or two solutions. The solution of a quadratic equation can be computed from the **quadratic formula.** The quadratic formula also shows when a solution does not exist.

Quadratic Formula
If $ax^2 + bx + c = 0$ and $a \neq 0$, then

$$x = \frac{-b + \sqrt{b^2 - 4ac}}{2a}$$

or

$$x = \frac{-b - \sqrt{b^2 - 4ac}}{2a}$$

The number $D = b^2 - 4ac$, which is called the **discriminant,** determines whether there are zero, one, or two solutions. If $D > 0$, then the quadratic equation has two distinct solutions,

$$x = \frac{-b + \sqrt{D}}{2a} \qquad \text{or} \qquad x = \frac{-b - \sqrt{D}}{2a}$$

If $D = 0$, there is one solution $x = -b/2a$. If $D < 0$, there is no solution because the square root of a negative number is not a real number. Example 1 shows how the three possibilities arise.

Problem

For each function $f(x)$, solve the quadratic equation $f(x) = 0$ and sketch the graph of $f(x)$.
(a) $f(x) = 2x^2 - x - 10$
(b) $f(x) = x^2 + 4x + 4$
(c) $f(x) = x^2 + 4x + 5$

Solution (a) To use the quadratic formula, set $a = 2$, $b = -1$, and $c = -10$. Then $D = b^2 - 4ac = 81$. Since $D > 0$, the two solutions of the equation $f(x) = 0$ are

$$x = \frac{-(-1) + \sqrt{81}}{2(2)} = \frac{5}{2} \qquad x = \frac{-(-1) - \sqrt{81}}{2(2)} = -2$$

Therefore the graph of $f(x)$ has x-intercepts at the points $(-2, 0)$ and $(\frac{5}{2}, 0)$. Since the graph of a quadratic equation is a parabola that opens upward or downward, it suffices to plot a point (x, y) where $-2 < x < (\frac{5}{2})$. Since $f(0) = -10$, the point $(0, -10)$ lies on the graph so that the parabola opens upward. It is sketched in Figure 1a.

(b) To use the quadratic formula, set $a = 1$, $b = 4$, and $c = 4$. Then $D = 4^2 - 4(1)(4) = 0$, so there is one solution of the equation $f(x) = 0$: $x = -b/2a = -\frac{4}{2} = -2$. This means that the graph of $f(x)$ has one x-intercept, the point $(-2, 0)$, which must be the vertex of the parabola. Since $f(0) = 4$, the point $(0, 4)$ lies on the graph, so the parabola opens upward. It is sketched in Figure 1b.

(c) Set $a = 1$, $b = 4$, and $c = 5$. Then $D = 4^2 - 4(1)(5) = -4$, so there is no solution of the equation $f(x) = 0$. This means that the graph of $f(x)$ has no x-intercepts. The only means at our disposal for sketching the graph is the plot-

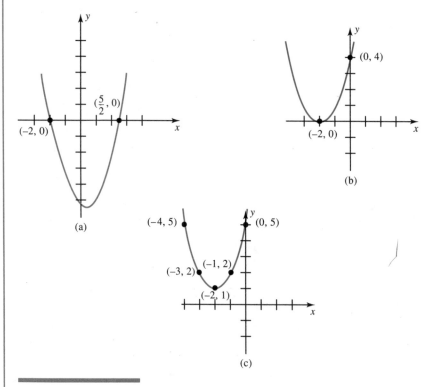

(a)

(b)

(c)

FIGURE 1

ting method using a table like the one here. The resulting graph is sketched in Figure 1c.

x	-4	-3	-2	-1	0
y	5	2	1	2	5

Example 2 illustrates how to use the quadratic formula to determine the domain of a rational function whose denominator is a quadratic function.

EXAMPLE 2

Problem

Determine the domain of the function

$$f(x) = \frac{x}{12x^2 - 11x + 2}$$

Solution The domain consists of all real numbers except where the denominator is 0. Set $12x^2 - 11x + 2 = 0$. Since $D = (-11)^2 - 4(12)(2) = 25$, the values of x where $12x^2 - 11x + 2 = 0$ are

$$x = \frac{-b + \sqrt{D}}{2a} = \frac{11 + 5}{24} = \frac{2}{3}$$

or

$$x = \frac{-b - \sqrt{D}}{2a} = \frac{11 - 5}{24} = \frac{1}{4}.$$

Therefore the domain of $f(x)$ is all real numbers except $\frac{2}{3}$ and $\frac{1}{4}$.

Factoring

A number r is a root of a polynomial equation $f(x) = 0$ if and only if $f(x) = (x - r)g(x)$ for some polynomial $g(x)$. Each of the functions $(x - r)$ and $g(x)$ is called a **factor** of $f(x)$. The method of decomposing a polynomial into a product of polynomials of lesser degree is called **factoring.**

Consider the general quadratic equation $ax^2 + bx + c = 0$. Since $a \neq 0$, we can divide both sides of the equation by a and get $x^2 + Bx + C = 0$, where $B = b/a$ and $C = c/a$. If r and s are roots of $x^2 + Bx + C = 0$, then

$$x^2 + Bx + C = (x - r)(x - s)$$

Expand the right side of the equation.

$$(x - r)(x - s) = x^2 - (r + s)x + rs$$

Therefore

$$x^2 + Bx + C = x^2 - (r + s)x + rs$$

Set the coefficients equal: $B = -(r + s)$, $C = rs$. The sum of the roots equals $-B$ (since $r + s = -B$) and the product of the roots equals the constant term C. For instance, if $f(x) = x^2 - x - 2$, then $B = -1$ and $C = -2$, so $r + s = 1$ and $rs = -2$. We are searching for two numbers whose sum is 1 and product is -2. The solution by trial and error is $r = 2$ and $s = -1$. Therefore the linear factors of $x^2 - x - 2$ are $(x - 2)$ and $(x + 1)$, so

$$x^2 - x - 2 = (x - 2)(x + 1)$$

Example 3 uses factoring to determine the domain and x-intercepts of a rational function.

EXAMPLE 3

Consider the function

$$f(x) = \frac{x^2 - 4}{x^2 - 2x - 8}$$

Problem

(a) Find the domain of $f(x)$. (b) Find all x-intercepts of the graph of $f(x)$.

Solution (a) The domain of $f(x)$ consists of all real numbers except those where the denominator is 0. Factor the denominator.

$$x^2 - 2x - 8 = (x - 4)(x + 2)$$

The roots are $x = 4$ or $x = -2$. Therefore the domain of $f(x)$ consists of all real numbers except 4 and -2.

(b) The x-intercepts of the graph of $f(x)$ occur where the numerator is 0 and the denominator is not 0. Set $x^2 - 4 = 0$. Since $x^2 - 4 = (x + 2)(x - 2)$, the solutions are $x = 2$ or $x = -2$. However, -2 does not lie in the domain of $f(x)$. Thus the only possible x-intercept occurs at $x = 2$. Since $x^2 - 2x - 8 \neq 0$ when $x = 2$, $f(x)$ has one x-intercept at $(2, 0)$.

The next two examples show other instances where the solution of a polynomial equation is required. Example 4 makes use of the fact that if a and b are numbers with $ab = 0$, then either $a = 0$ or $b = 0$. This fact extends to functions, in which case if $f(x)$ and $g(x)$ are functions with $f(x)g(x) = 0$, then either $f(x) = 0$ or $g(x) = 0$. Therefore if a polynomial has been factored, its roots are found by setting each of the factors equal to 0.

EXAMPLE 4

Problem

Solve $f(x) = 0$, where

$$f(x) = (2x + 3)(x^2 + 3)(x^2 - 3)$$

Solution The solutions to $f(x) = 0$ are found by setting each of the factors equal to 0. The solution to the linear equation $2x + 3 = 0$ is $x = -\frac{3}{2}$. The equation $x^2 = -3$ has no real solutions, so the factor $(x^2 + 3)$ adds no solutions to $f(x) = 0$. However, $x^2 - 3 = (x + \sqrt{3})(x - \sqrt{3})$, so $\sqrt{3}$ and $-\sqrt{3}$ are solutions to $x^2 - 3 = 0$, and hence are solutions to $f(x) = 0$. Altogether the solutions to $f(x) = 0$ are $-\frac{3}{2}$, $\sqrt{3}$, and $-\sqrt{3}$.

Example 5 makes repeated use of one of the basic rules of algebra.

$$a^2 - b^2 = (a + b)(a - b)$$

In particular, $x^2 - 1 = (x + 1)(x - 1)$.

EXAMPLE 5

Problem

Find the x-intercepts of the function

$$f(x) = x^4 - 1$$

Solution The x-intercepts are found by solving $f(x) = 0$. First factor $f(x)$ by making the substitution $y = x^2$.

$$f(x) = x^4 - 1 = y^2 - 1 = (y + 1)(y - 1)$$

Next substitute for y.

$$f(x) = (x^2 + 1)(x^2 - 1)$$

Then factor $x^2 - 1$.

$$f(x) = (x^2 + 1)(x + 1)(x - 1)$$

The roots of $f(x) = 0$ are found by setting each factor equal to 0: This yields $x = 1$, $x = -1$. The factor $x^2 + 1$ is never equal to 0 so it does not yield a root. These are the x-intercepts of $f(x)$.

Points of Intersection

A point at which two graphs meet is called a **point of intersection** of the graphs. We find the intersection of the graphs of two polynomial functions by a method quite different from the one described in Section 1–3 for two linear functions.

Let f and g be any two functions. The intersection of their graphs is obtained by solving the equation $f(x) = g(x)$, or, equivalently, by solving $f(x) - g(x) = 0$. If there are no solutions to this equation, then the graphs have no points in common. Otherwise each solution x, along with the corresponding value of $y = f(x)$ or $y = g(x)$, leads to a point of intersection (x, y).

For example, let $f(x) = 3x$ and $g(x) = 4 - x^2$. Then

$$f(x) - g(x) = 0$$
$$3x - (4 - x^2) = 0$$
$$x^2 + 3x - 4 = 0$$
$$(x + 4)(x - 1) = 0$$

The roots are $x = -4$ and $x = 1$. The corresponding values are $f(-4) = g(-4) = -12$ and $f(1) = g(1) = 3$. Each root leads to a point of intersection, so the points of intersection of $f(x)$ and $g(x)$ are $(-4, -12)$ and $(1, 3)$. The graphs are sketched in Figure 2.

An application of this concept comes from a field called **break-even analysis.** We have already introduced the cost function $C(x)$, where x denotes the number of items being produced. The revenue R is also a function of x, called the **revenue function** and denoted by $R(x)$. The **profit function** $P(x)$ is then defined by

$$P(x) = R(x) - C(x)$$

The **break-even points** occur when cost equals revenue: $R(x) = C(x)$. Much of the information can be conveyed visually by drawing the graphs of $R(x)$ and $C(x)$ on the same coordinate system and finding the points of intersection.

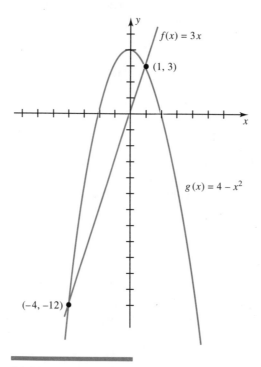

FIGURE 2

EXAMPLE 6

A manufacturer produces compact disks. The variable cost per disk is $2 and the fixed cost is $12. If x disks are manufactured, the revenue function is given by $R(x) = x(10 - x)$.

Problem

What are the break-even points?

Solution The cost function is

$$C(x) = (\text{variable cost per disk})x + (\text{fixed cost})$$

$$C(x) = 2x + 12$$

The revenue function is

$$R(x) = x(10 - x) = 10x - x^2$$

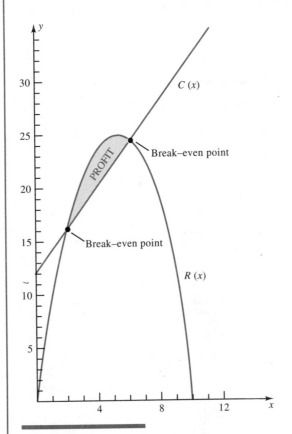

FIGURE 3

The break-even points occur when cost equals revenue.

$$C(x) = R(x)$$

$$2x + 12 = 10x - x^2$$

$$x^2 - 8x + 12 = 0$$

The possible solutions to this quadratic equation are $x = 2$ or $x = 6$, so the break-even points are $(2, 16)$ and $(6, 24)$. (See Figure 3.)

It is informative to view Figure 3 for a deeper understanding of break-even analysis. The functions $C(x)$ and $R(x)$ are sketched on the same coordinate axis. If either two disks or six disks are produced, the manufacturer will break even. A profit is realized if the number of disks produced lies between these numbers. The profit function is

$$P(x) = R(x) - C(x) = x^2 + 8x - 12$$

What production level do you think will maximize profit?

EXERCISE SET 1-6

In Problems 1 to 6 solve the quadratic equation $f(x) = 0$ and sketch the graph of $f(x)$.

1. $f(x) = x^2 + 2x - 8$

2. $f(x) = x^2 - 4x + 4$

3. $f(x) = x^2 + 6x + 9$

4. $f(x) = x^2 - 2x + 5$

5. $f(x) = x^2 + 4$

6. $f(x) = x^2 - 2x - 8$

In Problems 7 to 10 determine the domain and x-intercepts of each function.

7. $f(x) = \dfrac{x - 2}{2x^2 - x - 3}$

8. $f(x) = \dfrac{x - 3}{x^2 + x - 2}$

9. $f(x) = \dfrac{x + 2}{4x^2 + 5x - 6}$

10. $f(x) = \dfrac{x - 3}{5x^2 - 13x - 6}$

In Problems 11 and 12 solve $f(x) = 0$ for the given $f(x)$.

11. $f(x) = (2x - 3)(x^2 - 4)(x^2 + 3)$

12. $f(x) = x(x^2 - 1)(x^2 + 1)$

In Problems 13 to 16 find the x-intercepts of each function.

13. $f(x) = x^4 - 16$

14. $f(x) = x^4 - 81$

15. $f(x) = (x^2 + 3)(x - 4)(x + 3)\cdot$

16. $f(x) = (x^2 + 4)(x + 7)(x + 3)$

In Problems 17 to 20 a cost function and a revenue function are given. Find the break-even points.

17. $R(x) = x(10 - x)$ $C(x) = x + 8$

18. $R(x) = x(10 - x)$ $C(x) = 18 - x$

19. $R(x) = x(8 - x)$ $C(x) = 14 - x$

20. $R(x) = x(8 - x)$ $C(x) = x + 6$

In Problems 21 to 24 solve the equation $f(x) = 0$ and sketch the graph of $f(x)$.

21. $f(x) = \dfrac{x - 2}{4x^2 - 12x + 9}$

22. $f(x) = \dfrac{x - 3}{9x^2 + 6x - 1}$

23. $f(x) = \dfrac{x + 1}{x^2 + 1}$

24. $f(x) = \dfrac{x - 1}{5x^2 + 6}$

In Problems 25 to 28 determine the domain and x-intercepts of each function.

25. $f(x) = \dfrac{x - 2}{x^3 - 4x^2 + 4x}$

26. $f(x) = \dfrac{x - 3}{x^3 + 2x^2 + x}$

27. $f(x) = \dfrac{x^2 - x + 2}{x^3 - 4x^2 + 4x}$

28. $f(x) = \dfrac{x^2 - x - 6}{x^3 - 6x^2 + 9x}$

In Problems 29 to 34 find where the graphs of the functions intersect.

29. $f(x) = x + 2, \quad g(x) = x^2$

30. $f(x) = 6x - 5, \quad g(x) = x^2$

31. $f(x) = x^3, \quad g(x) = x^2$

32. $f(x) = x^2, \quad g(x) = \sqrt{x}$

33. $f(x) = x^2 + 4, \quad g(x) = 22 - x^2$

34. $f(x) = x^3, \quad g(x) = x$

In Problems 35 to 38 an object is thrown vertically upward with height $h(t) = 28t - 4t^2$ (in feet) at time t (in seconds). Determine when the object will reach the stated height.

35. 40 feet

36. 48 feet

37. 49 feet

38. 0 feet

39. Let the revenue function on the sales of x units of an item be $R(x) = 480x - 12x^2 - 2100$ and the cost function be $C(x) = 96x + 204$. Determine the break-even points.

In Problems 40 to 43 determine the break-even points and the fixed cost given the graphs of the cost function $C(x)$ and revenue function $R(x)$.

41.

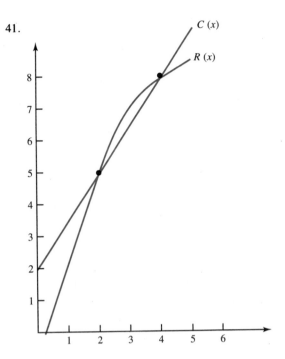

42.

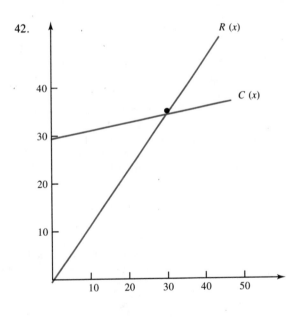

40.

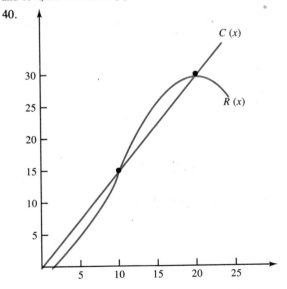

43.

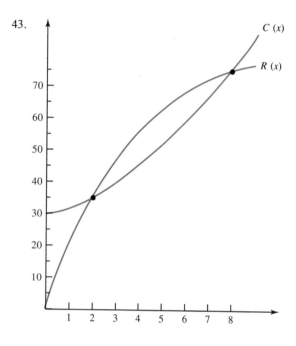

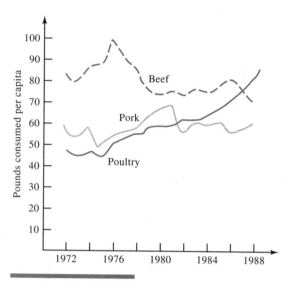

FIGURE 4

Referenced Exercise Set 1–6

Problems 1 and 2 are taken from a lecture that was given by Robert L. Devaney to a sectional meeting of the Mathematical Association of America.* It was accompanied by an exhilarating collection of computer graphics that illustrated the area of mathematics called dynamical systems.

1. Enter any positive number into a calculator. Tap the square-root key. Tap it again. Continue tapping the key about 15 or 20 more times.
   (a) What do you observe?
   (b) Repeat this procedure for several more positive numbers, including numbers between 0 and 1. What general conclusion can be drawn?

2. (a) Find all points of intersection of the graphs of the functions $f(x) = \sqrt{x}$ and $g(x) = x$ for $x > 0$.
   (b) What does Problem 1 have to do with part (a)?

3. Figure 4 illustrates the dramatic shift in the American public's eating habits from red meats to poultry. It has

been drawn from data supplied by the U. S. Department of Agriculture.†
   (a) Interpret the point of intersection of the poultry curve and the pork curve.
   (b) Interpret the point of intersection of the poultry curve and the beef curve.

Cumulative Exercise Set 1–6

1. Find the equation of the line passing through $(-2,0)$ and $(-1,-1)$.

2. Evaluate $f(x + h) - f(x)$ for $f(x) = x^2 + 2x - 1$.

3. Sketch the graph of the function $f(x) = x^2 + 2x - 1$.

4. Find the vertical asymptotes and sketch the graph of the function $f(x) = 1/(1 - x^2)$.

5. Find $f + g$ and $f \cdot g$ for $f(x) = x^2 + x$ and $g(x) = x^3 - 1$.

6. Find $f \circ g$ and $g \circ f$ for $f(x) = 1/x$ and $g(x) = x^3 - x$.

7. Write the function $f(x) = (x^3 + 2)^4$ as the composition of two functions.

*Robert L. Devaney, "Computer Graphics Experiments in Complex Dynamical Systems," Eastern Pennsylvania and Delaware Section of the Mathematical Association of America, Franklin & Marshall College, November 21, 1987.

†Drawn from data extracted from Neill Boroski, "Poultry's Prospects Far from Paltry," *The Philadelphia Inquirer,* June 15, 1987, pp. C1 and C6.

8. Express $Q(P) = 10 + 2P - 3P^2$ as a function of t where $P(t) = 5t - t^2$.

9. Find the domain and the x-intercepts of the function
$$f(x) = x/(x^2 - 4x + 4)$$

10. Solve $f(x) = 0$ for $f(x) = x(x - 3)(x^2 - 9)$.

11. Find the break-even points for the revenue function
$$R(x) = x(8 - x) \text{ and the cost function } C(x) = x + 6.$$

12. Find where the graphs of the functions intersect:
$$f(x) = x^3 - x \qquad g(x) = x^2 - x$$

CASE STUDY   Powerlifting*

Powerlifting is one of the world's youngest sports, having attained international status in 1968 when the strong local programs from the United States and Great Britain joined together with the developing programs in Canada, Japan, and the Scandinavian countries. It consists of three types of lifts: the squat (the bar is set on the back of the neck, and the lifter does a deep knee bend, returning to a standing position); the bench press (the lifter lies on a bench and pushes the weight away from the chest until the arms are locked); and the deadlift (the weight is lifted waist high from the floor, with the arms kept straight). A lifter's total is the sum of the highest weight lifted in each of the three categories.

Surprisingly, a sixth-degree polynomial plays a crucial role in determining the winner of a competition.

The winner of a powerlifting meet is not necessarily the competitor who has lifted the most weight. In fact, there are many winners. Like other sports in which a competitor's weight is one of the crucial determining factors, powerlifting competitions are divided into several weight classes, and there is a champion in each weight class. However, most meets crown a "Champion of Champions." What criteria should be used to decide this distinction?

Let us consider two lifters, Amos Jones and Bob Brown. Amos weighs 167 pounds (lb) so he competes in the 181-lb class. Bob is a behemoth, who, at 272 lb, competes in the 275-lb class. (There is a super heavyweight class beyond this one.) In a particular meet Amos squatted 710 lb, bench pressed 450 lb, and deadlifted 640 lb, while Bob's best three lifts were 850, 570, and 790 lb, respectively. Therefore Amos's total was 1800 (since $710 + 450 + 640 = 1800$) and Bob's total was 2210. Each lifter was a weight class champion in the meet. Which one should win the Champion of Champions award?

Mathematics to the rescue!

The Schwartz Formula   There are several ways to approach the question of who is stronger, Amos or Bob. If only the totals are taken into account, then Bob wins by 410 lb. Yet Amos's total is about 11 times his bodyweight, while Bob's total is about 8 times his bodyweight.

*This Case Study was inspired by Joseph A. Gallian's note "Handicapping Bodyweights," which appeared in the August–September 1986 issue of the *American Mathematical Monthly*, pp. 583–584. We are indebted to Lyle Schwartz and Linc Gotshalk for supplying details and to Walt Evans for supplying the photograph that introduces this chapter.

The International Powerlifting Federation (IPF) sought a handicapping scheme that would compensate for the widely differing body weights of the lifters. In 1976 the IPF chose a scheme that was developed by Lyle Schwartz to fit the empirical data that were available at the time. It has had the effect of taking into account not only body weights and muscular cross-sectional areas, but also body size versus fixed barbell plate, bar diameter, and muscle-ligament attachment leverage.

Dr. Schwartz had developed the scheme in 1967–1968 while he was a Professor of Materials Science at Northwestern University. (Today he is the Director of the Institute for Materials Science and Engineering at the National Bureau of Standards.) His method was published first in the magazine *Muscular Development* in 1970 and adopted by the U.S. Powerlifting Federation in 1975. The method uses a function of the lifter's bodyweight to produce a number that is then multiplied by the total number of pounds lifted to give the lifter's score. The competitor with the highest score wins the Champion of Champions trophy.

The *Schwartz formula* is the following intimidating sixth-degree polynomial function.

$$S(x) = 6.31926 - 1.189995 \cdot 10^{-1} x + 1.052494 \cdot 10^{-3} \cdot x^2$$
$$- 4.850446 \cdot 10^{-6} \cdot x^3 + 1.132899 \cdot 10^{-8} \cdot x^4$$
$$- 1.037119 \cdot 10^{-11} \cdot x^5 - 6.348189 \cdot 10^{-16} \cdot x^6$$

Here x is the bodyweight in pounds. The domain of $S(x)$ is $114 \leq x \leq 275$, so the Schwartz formula only applies to weights in this interval.

Let us evaluate the Schwartz formula for Amos, who weighs 167 lb. The details for computing $S(167)$ are shown in Table 1. They reinforce the fact that all seven terms in $S(x)$ must be included. Since the IPF rounds all numbers to four decimal places, we see that $S(167) = 0.6593$.

The **Schwartz formula total** (denoted SFT) is equal to $S(x)$ when $114 \leq x \leq 275$. Other methods are used to determine the SFT for lifters who weigh less than 114 lb or more than 275 lb. (See Exercises 9 and 10.) The SFT for Amos is 0.6593. Similar computations show that the SFT for Bob is $S(272) = 0.5232$.

Each lifter's score is the product of the total weight lifted and the SFT. For Amos, the score is $1800 \cdot 0.6593 = 1186.74$. For Bob, the score is $2210 \cdot 0.5232 =$

Table 1

Term	Value with $x = 167$	Cumulative Total
6.319260	6.319260	
$-1.189995 \cdot 10^{-1} x$	-19.872917	-13.553657
$1.052494 \cdot 10^{-3} x^2$	29.353005	15.799348
$-4.850446 \cdot 10^{-6} x^3$	-22.590773	-6.791425
$1.132899 \cdot 10^{-8} x^4$	8.811647	2.020222
$-1.037119 \cdot 10^{-11} x^5$	-1.347134	0.673088
$-6.348189 \cdot 10^{-16} x^6$	-0.013770	0.659318

Table 2

	Amos	Bob
Body weight x	167	272
$S(x)$	0.6593	0.5232
Total	1800	2210
Score	1186.7	1156.3

1156.272. Therefore Amos is declared the Champion of Champions. Their statistics are summarized in Table 2. If you have a programmable calculator it is possible to write a short program that would duplicate this table.

EXAMPLE 1

Jack London won the Pennsylvania state teenage powerlifting championship in 1987. He squatted 701 lb, bench pressed 451 lb, and deadlifted 585 lb. He weighed a mere 275 lb at the time.

Problem

What is his score?

Solution London's total is $701 + 451 + 585 = 1737$. The Schwartz formula total is $S(275) = 0.5214$. Therefore London's score is $1737 \cdot 0.5214 = 905.67$.

It is not necessary for a powerlifting meet director to employ someone with a calculator to evaluate $S(x)$ for each lifter. Instead, the director uses a table like the one in Figure 1, which also contains the Malone formula total (MFT) for female lifters and the Schwartz master's formula (SMF) for lifters of age 40 or more. However, neither the MFT nor the SMF is defined by a polynomial function, so they will not be described here.

Case Study Exercises

1. Use the Schwartz formula $S(x)$ to determine the Schwartz formula total for a 200-pound (lb) male powerlifter.
2. Use the Schwartz formula $S(x)$ to determine the Schwartz formula total for a a 125-lb male powerlifter.
3. Use Figure 1 (page 74) to verify your answer to Problem 1.
4. Use Figure 1 to verify your answer to Problem 2.
5. Suppose Bob had lost 4 lb and lifted 10 lb less. Would he have won the Champion of Champions award?

Problems 6 to 8 require the use of Figure 1.

6. What is the score of a female powerlifter who weighs 140 lb and lifts 950 lb?
7. Suppose that a 120-lb female powerlifter lifts 640 lb. If she gains 6 lb, how much more weight will she have to lift to maintain the same score?
8. What is the score of a 63-year-old person who lifts 580 lb and weighs 180 lb?

Problems 9 and 10 define the Schwartz formula in terms of body weight x (in kilograms) for certain male behemoths. Sketch the graph of each formula.

9. $S(x) = \begin{cases} 0.5208 - 0.0012(x - 125) & \text{if } 125 < x \leq 135 \\ 0.5088 - 0.0011(x - 135) & \text{if } 135 < x \leq 145 \end{cases}$

10. $S(x) = \begin{cases} 0.4978 - 0.0010(x - 145) & \text{if } 145 < x \leq 155 \\ 0.4878 - 0.0009(x - 155) & \text{if } 155 < x \leq 165 \end{cases}$

Schwartz/Malone Formula (in pounds)*

BWT	Schwartz	Malone	BWT	Schwartz	Malone	BWT	Schwartz	Malone	BWT	Schwartz	Malone	BWT	Schwartz	Malone
90	1.2803	1.1756	123	.8783	.9110	155	.7004	.7565	187	.6077	.6595	219	.5556	.6008
91	1.2627	1.1645	124	.8706	.9086	156	.6967	.7520	188	.6056	.6566	220	.5545	.5993
92	1.2455	1.1557	125	.8630	.9019	157	.6930	.7490	189	.6036	.6543	221	.5535	.5981
93	1.2287	1.1450	126	.8556	.8980	158	.6893	.7453	190	.6014	.6521	222	.5524	.5965
94	1.2124	1.1365	127	.8483	.8902	159	.6857	.7431	191	.5994	.6492	223	.5514	.5953
95	1.1965	1.1261	128	.8412	.8851	160	.6822	.7387	192	.5978	.6464	224	.5504	.5938
96	1.1809	1.1180	129	.8343	.8788	161	.6787	.7358	193	.5954	.6442	225	.5494	.5926
97	1.1657	1.1079	130	.8276	.8738	162	.6753	.7322	194	.5935	.6415	226	.5485	.5911
98	1.1509	1.0980	131	.8210	.8676	163	.6720	.7293	195	.5916	.6387	227	.5476	.5896
99	1.1365	1.0903	132	.8146	.8628	164	.6688	.7258	196	.5897	.6366	228	.5467	.5884
100	1.1223	1.0807	133	.8083	.8568	165	.6656	.7230	197	.5879	.6339	229	.5458	.5869
101	1.1086	1.0732	134	.8022	.8508	166	.6624	.7196	198	.5861	.6317	230	.5449	.5858
102	1.0952	1.0657	135	.7961	.8462	167	.6593	.7168	199	.5843	.6300	231	.5441	.5843
103	1.0821	1.0566	136	.7903	.8401	168	.6563	.7134	200	.5826	.6286	232	.5433	.5831
104	1.0693	1.0494	137	.7846	.8358	169	.6533	.7107	201	.5809	.6269	233	.5426	.5817
105	1.0569	1.0405	138	.7790	.8302	170	.6504	.7074	202	.5792	.6256	234	.5418	.5805
106	1.0448	1.0336	139	.7735	.8257	171	.6475	.7040	203	.5776	.6239	235	.5411	.5791
107	1.0329	1.0250	140	.7682	.8202	172	.6447	.7014	204	.5760	.6226	236	.5405	.5779
108	1.0214	1.0165	141	.7630	.8159	173	.6420	.6981	205	.5744	.6209	237	.5398	.5765
109	1.0101	1.0098	142	.7579	.8105	174	.6392	.6956	206	.5729	.6196	238	.5391	.5754
110	.9991	1.0016	143	.7528	.8052	175	.6365	.6923	207	.5714	.6180	239	.5385	.5740
111	.9884	.9952	144	.7479	.8010	176	.6339	.6898	208	.5700	.6167	240	.5379	.5725
112	.9779	.9872	145	.7432	.7959	177	.6313	.6866	209	.5685	.6151	241	.5373	.5714
113	.9677	.9809	146	.7385	.7918	178	.6288	.6837	210	.5670	.6134	242	.5367	.5700
114	.9578	.9731	147	.7339	.7867	179	.6262	.6810	211	.5657	.6122	243	.5362	.5693
115	.9481	.9670	148	.7294	.7827	180	.6238	.6786	212	.5643	.6109	244	.5357	.5686
116	.9385	.9595	149	.7250	.7769	181	.6214	.6755	213	.5630	.6093	245	.5352	.5681
117	.9293	.9536	150	.7207	.7737	182	.6190	.6731	214	.5617	.6077	246	.5347	.5671
118	.9203	.9462	151	.7165	.7697	183	.6167	.6701	215	.5604	.6064	247	.5342	.5669
119	.9115	.9390	152	.7124	.7666	184	.6144	.6671	216	.5592	.6049	248	.5337	.5662
120	.9029	.9333	153	.7083	.7627	185	.6121	.6639	217	.5580	.6036	249	.5333	.5656
121	.8946	.9263	154	.7044	.7596	186	.6099	.6618	218	.5568	.6021	250	.5328	.5649
122	.8863	.9208												

*Updated (4/84) Schwartz formula for men, Malone formula for women with bodyweights in pounds. To determine the "Best Lifter," multiply each lifter's coefficient (to the right of each body weight listed) by his or her total. The resulting factor is his Schwartz or her Malone formula total (SFT/MFT). The lifter with the highest SFT/MFT is considered the "Best Lifter." The weight class winner with the highest (SFT/MFT) is the Champion of Champions.

F I G U R E 1 Schwartz/Malone formula. (Courtesy of the United States Powerlifting Federation.)

Schwartz Formula (*Continued*)

BWT	Schwartz	BWT	Schwartz	BWT	Schwartz	BWT	Schwartz
251	.5325	279	.5192	307	.5043	335	.4909
252	.5320	280	.5186	308	.5037	336	.4905
253	.5316	281	.5180	309	.5032	337	.4901
254	.5312	282	.5175	310	.5027	338	.4896
255	.5308	283	.5169	311	.5022	339	.4891
256	.5304	284	.5164	312	.5017	340	.4887
257	.5300	285	.5158	313	.5013	341	.4883
258	.5296	286	.5154	314	.5007	342	.4878
259	.5292	287	.5147	315	.5002	343	.4874
260	.5289	288	.5142	316	.4998	344	.4870
261	.5284	289	.5137	317	.4992	345	.4866
262	.5281	290	.5132	318	.4988	346	.4862
263	.5276	291	.5126	319	.4982	347	.4858
264	.5273	292	.5121	320	.4978	348	.4854
265	.5268	293	.5115	321	.4973	349	.4850
266	.5263	294	.5109	322	.4968	350	.4845
267	.5259	295	.5104	323	.4964	351	.4841
268	.5254	296	.5098	324	.4959	352	.4837
269	.5248	297	.5094	325	.4955	353	.4833
270	.5243	298	.5088	326	.4950	354	.4829
271	.5239	299	.5083	327	.4946	355	.4825
272	.5232	300	.5077	328	.4941	356	.4821
273	.5227	301	.5072	329	.4937	357	.4817
274	.5220	302	.5067	330	.4932	358	.4813
275	.5214	303	.5062	331	.4928	359	.4809
276	.5208	304	.5057	332	.4924	360	.4805
277	.5203	305	.5053	333	.4919	361	.4801
278	.5197	306	.5047	334	.4914	362	.4796

Problems 11 to 13 refer to performances at the 1988 National Collegiate Power-lifting Championships.

11. There were four session champions chosen at this meet. For the two lifters listed here, determine their score based on the Schwartz formula. (a) Sheridan Suttles of Middle Tennessee State College, who weighed 129 lb, squatted 440 lb, bench pressed 303 lb, and deadlifted 518 lb. (b) Ty Stapleton Jr. of the University of Oklahoma, who weighed 164 lb, squatted 573 lb, bench pressed 413 lb, and deadlifted 562 lb.

12. Jack London of Temple University weighed 289 lb at this competition, so he entered the super heavyweight class. He squatted 771 lb, bench pressed 458 lb, and deadlifted 661 lb. (a) Use Figure 1 to determine his score. (b) Is this score better than his score in the competition described in Example 1?

13. The Champion of Champions was based on the results of Sheridan Suttles, Ty Stapleton Jr., and Jack London. Who won this prestigious award?

Schwartz Master's Formula†

Age	SMF	Age	SMF
40	1.000	61	1.700
41	1.003	62	1.755
42	1.009	63	1.810
43	1.018	64	1.865
44	1.031	65	1.920
45	1.048	66	1.970
46	1.069	67	2.010
47	1.092	68	2.030
48	1.117	69	2.048
49	1.144	70	2.062
50	1.173	71	2.070
51	1.204	72	2.076
52	1.239	73	2.080
53	1.281	74	2.082
54	1.330	75	2.083
55	1.380	76	2.084
56	1.430	77	2.085
57	1.480	78	2.086
58	1.535	79	2.087
59	1.590	80	2.088
60	1.645		

†Use after the regular formula has been used.

CHAPTER REVIEW

Key Terms

1–1   The Cartesian Coordinate System
Real Line

Origin

Unit Distance

x-Axis; y-Axis

x-Coordinate; y-Coordinate

Coordinates

Cartesian Coordinate System

Quadrants

Graph

The Plotting Method

Linear Equation

1–2   Lines
General Form

x-Intercept

y-Intercept

Slope

Slope-Intercept Form

Parallel Lines

Horizontal Lines

Vertical Lines

Point-Slope Form

1–3   Linear and Quadratic Functions
Function

Image

Preimage

Plotting Method

Linear Function
Domain
Range
Function Defined at a Number
Quadratic Function
Vertex

Vertical Line Test
Cost Function
Fixed Cost
Variable Cost
Marginal Cost

1–4 Polynomial and Other Functions

Parabola
Power Function
Polynomial Function
Degree of a Polynomial

Hyperbola
Vertical Asymptote
Rational Function

1–5 The Algebra of Functions

Sum of Functions: $f + g$
Difference of Functions: $f - g$
Product of Functions: $f \cdot g$

Quotient of Functions: f/g
Rational Operations
Composition of Functions: $f \circ g$

1–6 Solving Polynomial Equations

Solutions of an Equation
Root of an Equation
Polynomial Equation
Degree of a Polynomial Equation
Quadratic Equation
Quadratic Formula
Discriminant

Factor
Factoring
Point of Intersection
Break-Even Analysis
Revenue Function
Profit Function
Break-Even Point

Summary of Important Concepts

Quadratic Formula

If $ax^2 + bx + c = 0$ and $a \neq 0$, then

$$x = \frac{-b + \sqrt{b^2 - 4ac}}{2a}$$

or

$$x = \frac{-b - \sqrt{b^2 - 4ac}}{2a}$$

REVIEW PROBLEMS

Problems 1 to 3 refer to the function $f(x) = x^2 - 4x - 6$.

1. Compute $f(3)$, $f(2)$, $f(1)$, $f(0)$, and $f(-2)$.

2. Draw the graph of f.

3. Evaluate $f(x + h) - f(x)$.

In Problems 4 and 5 determine the domain of each function.

4. $f(x) = \sqrt{2x - 8}$

5. $f(x) = \dfrac{1}{1 - x}$

In Problems 6 and 7 determine whether each curve represents the graph of a function.

6.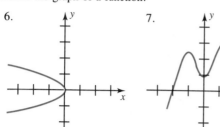
7.

8. Use the plotting method to draw the graph of the function $f(x) = 2x^3 - 3x^2 - 12x + 1$.

9. Find the vertical asymptotes of the graph of the function $f(x) = \dfrac{x}{(2 - x)(x^2 - 9)}$.

10. Draw the graph of the function
$$f(x) = \begin{cases} x^2 + 4 & \text{if } x \geq 1 \\ 3x + 2 & \text{if } x < 1 \end{cases}$$

In Problems 11 to 13 let
$$f(x) = \frac{x + 2}{x - 1} \quad \text{and} \quad g(x) = \frac{x}{x + 2}$$

11. Find (a) $f + g$, (b) $f \cdot g$, (c) f/g.

12. State the domains of (a) $f + g$, (b) $f \cdot g$, (c) f/g.

13. Evaluate (a) $(f + g)(2)$, (b) $(f \cdot g)(2)$, (c) $(f/g)(2)$.

14. Find $f \circ g$ and $g \circ f$ for $f(x) = \sqrt{\dfrac{3}{x}}$ and $g(x) = x^2$.

15. Write $f(x)$ as the composition of two functions:
$$f(x) = x + 2 - \sqrt{x + 2}$$

16. Express Q as a function of t, where
$$Q(P) = 40 - 0.1P^2 \quad \text{and} \quad P(t) = 5t - 0.2t^2$$

In Problems 17 to 19 find the x-intercepts of each polynomial function.

17. $f(x) = x^2 + 4x + 5$

18. $f(x) = 6x^2 - 13x + 6$

19. $f(x) = 3(x - 1)(x^2 + 4)(x^2 - 4)$

20. State the domain of the function
$$f(x) = \frac{x - 3}{4x^2 - 12x + 9}$$

21. Determine the domain and x-intercepts of the function
$$f(x) = \frac{x^2 - 1}{x^2 + x - 2}$$

22. Find all points where the graphs of $f(x) = x^4$ and $g(x) = x^2$ intersect.

▦ GRAPHICS CALCULATOR EXPLORATIONS

Exploration 1

Graphing functions is an indispensible tool in calculus. You can use your graphics calculator to get started on this important skill. The first step is to distinguish among the various types of functions mentioned in this chapter. The more familiar you become with these simple functions, the easier it will be to understand more complex functions in later chapters. Example 1 gives several functions whose graphs you should master to understand later sections. This example will help you become proficient in determining their graphs.

Example 1

Sketch the graph of each function on a piece of paper and then use your calculator to see how close your graph is to the graph on the calculator screen.

1. $y = x^2$
2. $y = -x^2$
3. $y = x^3$

4. $y = x^3$
5. $y = 2x^2 + 5$
6. $y = -3x^2 - 2$
7. $y = 0.5x^2 - 3$
8. $y = 0.5x^3 + 1$

You should familiarize yourself with each of these functions and be able to picture the general shape of each graph from its equation without plotting any points.

Exploration 2

Another way to picture graphs of equations is to build up the graph of a more complicated function by starting with an easier function. For example, if you forget what the graph of the equation $y = x^2 + 3$ looks like, start with the graph of the easier function $y = x^2$ and then recognize that the algebraic constant "$+ 3$" has the geometric effect of raising the graph three units. Hence the graph of $y = x^2 + 3$ is identical to the graph of $y = x^2$ raised three units.

Example 2

Sketch the graph of the three functions in the same screen:

$$f(x) = x^2, \qquad g(x) = 0.5x^2, \qquad h(x) = 0.5x^2 + 1$$

CASIO

1. Press GRAPH ALPHA x x² EXE .

2. Press GRAPH .5 ALPHA x x² EXE .

3. Press GRAPH .5 ALPHA x x² + 1 EXE .

TI-81

1. Press Y = .

2. Press X|T x² ENTER .

   Press . 5 X|T x² ENTER .

   Press . 5 X|T x² + 1 ENTER .

3. Press GRAPH .

Explain how to build (**a**) the graph of $g(x)$ from $f(x)$ and then (**b**) the graph of $h(x)$ from $g(x)$.

Solution (**a**) The graph of $f(x)$ is "narrower" than the graph of $g(x)$, which means that if you picture the graph of $f(x)$ as a flexible piece of wire, the graph of $g(x)$ can be obtained from it by bending the wire outward. (**b**) Then raise the graph of $g(x)$ one unit to obtain the graph of $h(x)$.

Example 3 uses the ideas in Example 2 to graph a more complex function given the graph of an easier function. Before reading Example 3, use your calculator to sketch the graph of $f(x) = x^3 - 3x$. Then try to graph on a piece of paper the function obtained from $f(x)$ by multiplying $f(x)$ by 2. Its graph will be obtained by "stretching" the graph of $f(x)$ by a factor of 2. Then try to graph on a piece of paper the graph of $h(x) = 2x^3 - 6x + 1$.

Example 3

Sketch the graph of the three functions in the same screen:

$$f(x) = x^3 - 3x, \qquad g(x) = 2x^3 - 6x, \qquad h(x) = 2x^3 - 6x + 1$$

Explain how to build (**a**) the graph of $g(x)$ from $f(x)$ and then (**b**) the graph of $h(x)$ from $g(x)$.

Solution (**a**) The graph of $g(x)$ is obtained from the graph of $f(x)$ by "stretching" it by a factor of two, that is, each functional value of $g(x)$ is twice the height of the corresponding value of $f(x)$. (**b**) Then raise the graph of $g(x)$, or, equivalently, raise the wire one unit to obtain the graph of $h(x)$.

Further Explorations

1. Write a program that graphs the functions $f(x) = ax^2$ for the following fixed values of a: $a = 1$, $a = 2$, $a = 3$, $a = \frac{1}{2}$, $a = \frac{1}{3}$, $a = .1$.
2. Write a program that graphs the functions $f(x) = -ax^2$ for the following fixed values of a: $a = .01$, $a = .2$, $a = .3$, $a = \frac{1}{2}$, $a = \frac{4}{3}$, $a = 1.1$.
3. Write a program that graphs the functions $f(x) = x^2 + b$ for the following fixed values of b: $b = 1$, $b = 1.1$, $b = 1.5$, $b = -1$ and $b = 1$, $b = -1$, $b = -\frac{1}{3}$, $b = -.1$.
4. Write a program that graphs the functions $f(x) = ax^2 + b$ for the following fixed values of a and b: $a = 1$ and $b = 1$, $a = 1$ and $b = -1$, $a = -1$ and $b = 1$, $a = -1$ and $b = -1$.
5. Graph the following functions on the same screen and note the similarities and the differences: $y = x$, $y = x^2$, $y = x^3$, $y = x^4$, $y = x^{1/2}$.
6. Use the zoom feature to approximate the roots of the following equations: (a) $x^2 + x - 1 = 0$, (b) $x^2 - x - 1 = 0$, (c) $x^3 - 3x + 1$.

The Derivative

Economists use differentiation to chart and describe the complexities of worldwide grain production. (H. Armstrong Roberts/M. Roessler)

Chapter Overview

In this chapter we introduce the part of calculus called differentiation. The basic notion is the limit of a function. In Section 2–1 we motivate limits by considering the velocity of a moving object, the rate of change of a function, and lines which are tangent to the graph of a given function. In Section 2–2 we develop limits intuitively in terms of "closer and closer" and "approaching." We also consider limits at infinity.

The derivative is introduced in Section 2–3, which relies on the properties of functions developed in Chapter 1 and the notion of a limit developed in Sections 2–1 and 2–2. The derivatives of several functions are computed from the definition, then the definition is applied to tangent lines and rates of change. Section 2–4 defines and contrasts continuous functions and differentiable functions.

2–1 Rates of Change and Tangents

There are two broad categories of applications of differential calculus: the rate of change and optimization. In this section the rate of change of a function is introduced by means of the velocity of an object that moves in a straight line. Then the tangent line to the graph of a function is presented by analogy with the rate of change. The tangent forms the basis for many optimization problems. Also discussed is the relationship between average rates of change and secant lines, as well as the relationship between the instantaneous rate of change and the tangent line.

Average Rate of Change

The **velocity** of an object that moves in a straight line is the rate of change of its distance with respect to time. The direction in which the object is traveling must also be taken into account, with a plus sign indicating one direction and a minus sign the opposite direction. We begin by computing the velocity in a specific setting.

Let $d(t)$ be the distance (in miles) that a car travels due east in time t (in hours after noon). If the car starts at 1 P.M., then $d = 0$ at $t = 1$, so that $d(1) = 0$. If the car travels 100 miles (mi) by 3 P.M., then $d(3) = 100$. The average velocity over the interval of time from 1 to 3 P.M. is 50 miles per hour (mph) since the car traveled 100 mi in 2 hours. This is shown in Figure 1.

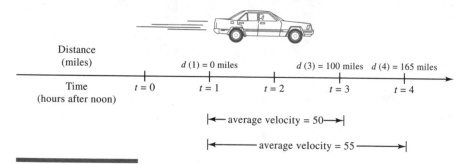

FIGURE 1

Suppose that the car traveled 165 mi by 4 P.M., so that $d(4) = 165$. Then the average velocity over the interval from $t = 1$ to $t = 4$ is 55 mph since

$$\frac{165 - 0}{4 - 1} = 55$$

In general, if a and b denote points in time with $a < b$, then the average velocity over the interval from $t = a$ to $t = b$ is given by the formula

$$\frac{d(b) - d(a)}{b - a}$$

This definition generalizes to the average rate of change of an arbitrary function.

DEFINITION

The **average rate of change** of the function $y = f(x)$ over the interval from $x = a$ to $x = b$, with $a < b$, is

$$\frac{f(b) - f(a)}{b - a}$$

EXAMPLE 1

$f(x) = mx + b =$

Consider the function $f(x) = 3x - 1$.

Problem

(a) What is the average rate of change of $f(x)$ over the interval from $x = -1$ to $x = 4$? (b) What is the average rate of change of $f(x)$ from $x = 1$ to $x = 2$?

Solution (a) First $f(4) = 11$ and $f(-1) = -4$. By definition the average rate of change over the interval from $x = -1$ to $x = 4$ is

$$\frac{f(4) - f(-1)}{4 - (-1)} = \frac{11 - (-4)}{5} = \frac{15}{5} = 3$$

(b) The average rate of change over the interval from $x = 1$ to $x = 2$ is

$$\frac{f(2) - f(1)}{2 - 1} = \frac{5 - 2}{1} = 3$$

LINEAR

In Example 1 the average rate of change of $f(x) = 3x - 1$ over each interval is equal to 3. The exercises will show that the average rate of change of a linear function $f(x) = mx + b$ over any interval is always equal to the slope m. The exercises will also show that linear functions are the only functions whose average rates of change are constant over all intervals. Example 2 indicates that nonlinear functions can have different rates of change over different intervals.

EXAMPLE 2

Problem

Compute the average rate of change of $f(x) = x^3$ over the intervals (a) from 1 to 3 and (b) from 1 to 1.5.

Solution (a) Since $f(3) = 27$ and $f(1) = 1$, the average rate of change over the interval from $x = 1$ to $x = 3$ is

$$\frac{f(3) - f(1)}{3 - 1} = \frac{27 - 1}{2} = 13$$

(b) The average rate of change over the interval from $x = 1$ to $x = 1.5$ is

$$\frac{f(1.5) - f(1)}{1.5 - 1} = \frac{3.375 - 1}{0.5} = 4.75$$

Instantaneous Rate of Change

Table 1 lists the average rates of change of the function $f(x) = x^3$ over the intervals from $x = 1$ to $x = 1 + h$ for various values of h. Example 2 covered $h = 2$ and $h = 0.5$.

Table 2 shows what happens to the average rate of change of $f(x) = x^3$ as the values of h approach 0 for negative values of h.

We can define the instantaneous rate of change of the function at the value $x = 1$ by viewing the numbers in the column for the average rates of change in both Tables 1 and 2. As h approaches 0 from either direction (positive or negative values), the average rates of change approach 3. The number 3 is called the *instantaneous rate of change* of $f(x)$ at $x = 1$, or just the *rate of change* of $f(x)$ at $x = 1$. The general definition is given here.

Table 1

h	Interval	Average Rate of Change
2	from 1 to 3	13
1	from 1 to 2	7
0.5	from 1 to 1.5	4.75
0.1	from 1 to 1.1	3.31
0.01	from 1 to 1.01	3.0301
0.001	from 1 to 1.001	3.003001

Table 2

h	Interval	Average Rate of Change
−0.1	from 0.9 to 1	2.71
−0.01	from 0.99 to 1	2.9701
−0.001	from 0.999 to 1	2.997001

$3(2)^3$

DEFINITION

The **instantaneous rate of change** of the function $y = f(x)$ at a value of x is the number (if it exists) approached by the average rates of change from x to $x + h$ as h approaches 0 and $h \neq 0$.

EXAMPLE 3

Problem

$x - h$

Find the instantaneous rate of change of the function $f(x) = x^2 - 1$ at $x = 2$.

Solution Table 3 shows the average rates of change of $f(x) = x^2 - 1$ over various intervals for $h > 0$ and $h < 0$. Consider, for instance, $h = 0.1$. The interval is from 2 to 2.1, so the average rate of change is

$$\frac{f(2.1) - f(2)}{2.1 - 2} = \frac{(4.41 - 1) - (4 - 1)}{0.1} = \frac{0.41}{0.1} = 4.1$$

Now let h get closer and closer to 0. For $h = 0.01$, the average rate of change is 4.01. For $h = 0.001$, the average rate of change is 4.001. It seems apparent that as h gets closer to 0, the average rate of change approaches 4. Therefore the instantaneous rate of change at $x = 2$ is equal to 4.

Table 3

h	Interval	Average Rate of Change
0.1	from 2 to 2.1	4.1
0.01	from 2 to 2.01	4.01
0.001	from 2 to 2.001	4.001
−0.1	from 1.9 to 2	3.9
−0.01	from 1.99 to 2	3.99
−0.001	from 1.999 to 2	3.999

Limit is 4

Instantaneous rate of change at the point x = 2 is 4 also means slope of the curve at that point

Secant and Tangent Lines

Thus far we have used algebra to define the instantaneous rate of change in terms of the average rates of change. There is also a geometric analog to this idea.

In general a line that passes through two points on a curve is called a **secant line**. (Recall that the secant line of a circle passes through two points on the circle.) Consider the function $f(x) = x^2$, whose graph is a parabola. The line through the points $(2, 4)$ and $(3, 9)$ is a secant line. (See Figure 2.) By definition the slope of this secant line is equal to

$$m = \frac{9 - 4}{3 - 2} = 5$$

rise / run

Then slope of the curve will be the derivative evaluated at that point.

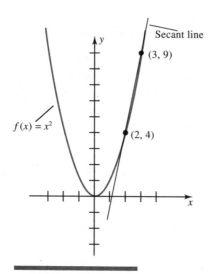

FIGURE 2

Notice that the average rate of change of $f(x) = x^2$ over the interval from $x = 2$ to $x = 3$ is defined similarly.

$$\frac{f(3) - f(2)}{3 - 2} = \frac{9 - 4}{1} = 5$$

These equalities illustrate the fact that *the slope of the secant line is equal to the average rate of change.*

What geometric entity corresponds to the instantaneous rate of change? To answer this question, look at Figure 3, showing some secant lines of the graph of

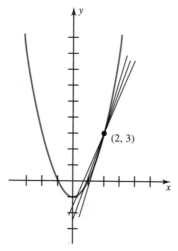

Secant lines of $f(x) = x^2 - 1$

FIGURE 3

$f(x) = x^2 - 1$ from $x = 2$ to $x = 2 + h$ for values of h that approach 0 and $h \neq 0$. Observe that the secant lines get closer and closer to one particular line. This line is called a **tangent line.** (Recall that the tangent to a circle is a line that touches the circle at precisely one point.) Because of this, we define the slope of the line tangent to a given point on a curve to be the number (if it exists) that the slopes of the secant lines approach. From our earlier discussion it follows that *the slope of the line tangent to a point corresponds to the instantaneous rate of change at that point.*

The equation of the tangent line through a given point on the graph of a function can be derived once the slope is known. Example 4 makes use of Example 3 to illustrate the method that is used to derive the equation of the tangent line.

EXAMPLE 4

Problem

Find the slope-intercept form of the line which is tangent to the graph of $f(x) = x^2 - 1$ at $x = 2$.

Solution Example 3 showed that the instantaneous rate of change of $f(x)$ at $x = 2$ is equal to 4. Therefore the slope of the tangent line is 4. Since $f(2) = 3$, the tangent line passes through the point $(2, 3)$. Therefore the equation of the tangent line in point-slope form is $y - 3 = 4(x - 2)$. Solving this equation for y yields the slope-intercept form $y = 4x - 5$. See Figure 4.

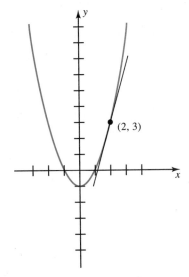

Tangent line of $f(x) = x^2 - 1$ at $x = 2$

FIGURE 4

The **slope of a function** $f(x)$ at a point on the graph of $f(x)$ is defined to be the slope of the tangent line through that point. The procedure for computing the slope of a function $f(x)$ at any value of x involves three steps:

1. Form the **difference quotient:**

$$\frac{f(x + h) - f(x)}{h}$$

2. Simplify the difference quotient, using the fact that $h \neq 0$.
3. The slope of $f(x)$ is the expression that results when h approaches 0 and $h \neq 0$.

EXAMPLE 5

Problem

Find the slope of $f(x) = x^3$ at an arbitrary value of x.

Solution It is instructive to refer to Figure 5 while carrying out the three steps. By step 1 the difference quotient is

$$\frac{f(x + h) - f(x)}{h} = \frac{(x + h)^3 - x^3}{h}$$

By step 2

$$\frac{f(x + h) - f(x)}{h} = \frac{(x^3 + 3x^2h + 3xh^2 + h^3) - x^3}{h}$$

$$= \frac{h(3x^2 + 3xh + h^2)}{h}$$

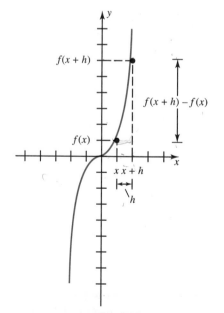

FIGURE 5

$$(x+h)^3 = (x+h)(x+h)(x+h)$$
$$= (x^2 + 2xh + h^2)(x+h)$$
$$= (x^2)(x+h) + (x+h)(2xh) + (x+h)(h^2)$$

The factor h can be canceled since $h \neq 0$, so

$$\frac{f(x + h) - f(x)}{h} = 3x^2 + 3xh + h^2$$

By step 3 let h approach 0. The terms $3xh$ and h^2 also approach 0, so $3x^2 + 3xh + h^2$ approaches $3x^2$. Thus the slope of $f(x) = x^3$ at any value of x is $3x^2$.

The tangent to a curve measures the steepness of the curve. An intuitive way of visualizing this is to put a roller coaster car, with a long board attached to the hubs of the wheels, on the curve. The car is the point, the board is the tangent. (See Figure 6.) As the car moves along the curve from left to right, the board reflects the steepness of the climb and the descent. As the car ascends, the board points upward so the tangent is a positive number. When the car is on the top of the hill, the board is horizontal so the tangent is 0. As the car descends, the board points downward so the tangent is a negative number. We will develop these ideas in greater detail in Chapter 4.

Supply and Demand

The ideas in this section can be applied to the part of economics that deals with supply and demand curves. The **demand** for a product is the number $D(x)$ of items that consumers are willing and able to buy at a given price per unit x. For example, suppose the demand for a new toy called *Cubic* is $D(x) = 5 - x^2$, where x is in dollars per Cubic and $D(x)$ is in thousands of Cubics per week. Notice that the demand slackens as the price rises. For instance, if the price of each Cubic is raised from \$1 to \$2 the average rate of change of the demand is

$$\frac{D(2) - D(1)}{2 - 1} = \frac{1 - 4}{1} = -3$$

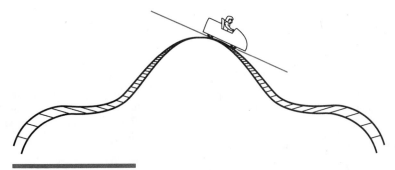

FIGURE 6   The board attached to the roller coaster car approximates the tangent line.

This means that the demand will decrease by 3 ($= 3000$ Cubics per week) if the price of the Cubics is raised from \$1 to \$2. The instantaneous rate of change of the demand at any price x is called the **marginal demand.** Example 6 computes the marginal demand of this function.

EXAMPLE 6

Problem

If the demand function is $D(x) = 5 - x^2$, what is the marginal demand?

Solution The marginal demand can be obtained from the three-step procedure. By step 1 the difference quotient is

$$\frac{D(x + h) - D(x)}{h} = \frac{[5 - (x + h)^2] - (5 - x^2)}{h}$$

By step 2

$$\frac{D(x + h) - D(x)}{h} = \frac{5 - x^2 - 2xh - h^2 - (5 - x^2)}{h}$$

$$= \frac{h(-2x - h)}{h}$$

The factor h can be canceled since $h \neq 0$, so

$$\frac{D(x + h) - D(x)}{h} = -2x - h$$

By step 3, as h gets close to 0, the term $-2x - h$ gets close to $-2x$. Thus the marginal demand at an arbitrary price x is equal to $-2x$. This means that the demand for Cubics decreases at a rate that is double the price of each Cubic.

EXERCISE SET 2–1

In Problems 1 to 8 compute the average rate of change of the function $f(x)$ over the stated intervals.

1. $f(x) = 2x - 3$
   (a) from 1 to 2   (b) from -1 to 4

2. $f(x) = 4x - 3$
   (a) from 1 to 2   (b) from -1 to 4

3. $f(x) = 3 - 4x$
   (a) from 0 to 5   (b) from -3 to -2

4. $f(x) = 3 - 2x$
   (a) from 0 to 5   (b) from -3 to -2

5. $f(x) = x^3$
   (a) from 0 to 1   (b) from 0 to 0.1

6. $f(x) = x^3$
   (a) from -1 to 0   (b) from -0.1 to 0

7. $f(x) = x^4$
   (a) from 0 to 1   (b) from 0 to 0.1

8. $f(x) = x^4$
   (a) from -1 to 0   (b) from -0.1 to 0

In Problems 9 to 12 use the method shown in Example 3 to compute the instantaneous rate of change of the function $f(x)$ at the stated value of x.

9. $f(x) = x^2 - 5$ at $x = 1$

10. $f(x) = x^2 - 5$ at $x = -2$

11. $f(x) = 10 - 2x^2$ at $x = -2$

12. $f(x) = 10 - 2x^2$ at $x = 1$

In Problems 13 to 16 find the slope-intercept form of the line tangent to the graph of the function in the stated problem above.

13. Problem 9

14. Problem 10

15. Problem 11

16. Problem 12

In Problems 17 to 20 find the marginal demand at an arbitrary price x, where the demand function is given.

17. $D(x) = 10 - 2x^2$

18. $D(x) = x^2$

19. $D(x) = x^2 + 2$

20. $D(x) = 1 - 2x^2$

In Problems 21 to 26 compute the average rate of change of the function $f(x)$ over the stated intervals.

21. $f(x) = x^2 - 5$ (a) from 2 to 2.1, (b) from 2 to 2.01, (c) from 2 to 2.001

22. $f(x) = 10 - x^2$ (a) from 2 to 2.1, (b) from 2 to 2.01, (c) from 2 to 2.001

23. $f(x) = 2x^2 - 4x - 3$ (a) from 3 to 3.1, (b) from 3 to 3.01, (c) from 3 to 3.001

24. $f(x) = 3x^2 + 2x - 29$ (a) from 3 to 3.1, (b) from 3 to 3.01, (c) from 3 to 3.001

25. $f(x) = \sqrt{x}$ (a) from 4 to 4.1, (b) from 4 to 4.01, (c) from 4 to 4.001

26. $f(x) = 1/x$ (a) from 0.5 to 0.51, (b) from 0.5 to 0.501, (c) from 0.5 to 0.5001

In Problems 27 and 28 compute the instantaneous rate of change of the function $f(x)$ at the stated value of x.

27. $f(x) = x^3 + x + 1$ at $x = 1$

28. $f(x) = x^3 + x^2 + 1$ at $x = 1$

In Problems 29 and 30 find the slope-intercept form of the line tangent to the graph of the function in the preceding stated problem.

29. Problem 27

30. Problem 28

In Problems 31 to 34 find the slope of the line tangent to the graph of the function $f(x)$ at an arbitrary value of x.

31. $f(x) = 2x^2 - 4x - 3$

32. $f(x) = 3x^2 + 2x - 29$

33. $f(x) = x^3 + x + 1$

34. $f(x) = x^3 + x^2 + 1$

In Problems 35 to 38 the height of a ball thrown into the air is given by $d(t)$. Find the slope of the line tangent to the graph of the function $d(t)$ at an arbitrary value of t.

35. $d(t) = -16t^2 + 192t$

36. $d(t) = 70t - 16t^2$

37. $d(t) = -16t^2 + 128t + 64$

38. $d(t) = -16t^2 + 32t + 64$

In Problems 39 to 42 choose one of the numbers, -1, 0, 1, 4, that most closely approximates the slope of the tangent line at the point $(a, f(a))$ for the given graph of $y = f(x)$.

39.

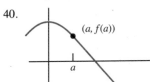

40.

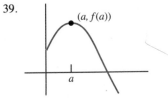

41.

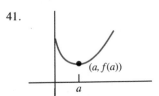

42.

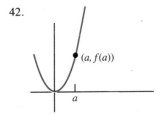

43. Let a, b, and c be arbitrary real numbers. Compute the slope of the line tangent to the graph of $f(x) = ax^2 + bx + c$ at an arbitrary value of x.

44. Demonstrate that if $f(x)$ is a function having the property that the average rate of change is the same number over all intervals, then $f(x) = mx + b$ for some m and b.

45. Demonstrate that if $f(x)$ is a function having the property that the slope of the tangent line at each point on the graph of $f(x)$ is the same number, then $f(x) = mx + b$ for some m and b.

46. Let m, a, b, and c be real numbers and $f(x) = mx + b$.
    (a) Compute the average rate of change of $f(x)$ over the interval from a to c.
    (b) Compute the instantaneous rate of change of $f(x)$ at $x = a$.

Referenced Exercise Set 2–1

Problems 1 and 2 refer to Figure 7, which shows the personal consumption expenditures (in billions of dollars) in the United States on durable goods over the first part of the 1980s.*

1. Estimate the average rate of change of the expenditures
    (a) from 1980 to 1984.
    (b) from 1982 to 1984.

2. Estimate the instantaneous rate of change of the expenditures in 1981.

Problems 3 and 4 refer to Figure 8, which shows the U.S. Federal Budget Deficit as a percentage of the Gross National Product (GNP) for the first half of the 1980s.†

3. Estimate the average rate of change of the percentage of the GNP
    (a) from 1981 to 1983.
    (b) from 1984 to 1985.

4. Estimate the instantaneous rate of change of the percentage of the GNP in 1985.

5. An article by Alan Murray in the *Wall Street Journal* used the fictitious Business Corporation to illustrate the tax law of 1986.‡ The tax law permitted Business Corporation to depreciate its new $6.3 million building over a period of 31.5 years by the "straightline method." What is the rate of depreciation? (The rate of depreciation is defined as the slope of the straight line where the x-axis is time in years and the y-axis is the cost in millions of dollars. The rate is expressed as a percent.)

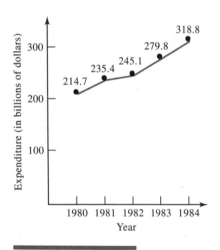

FIGURE 7

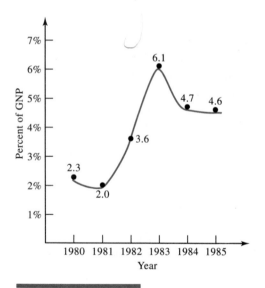

FIGURE 8

Source: Bureau of Economic Analysis, U.S. Commerce Department.

†*The World Almanac and Book of Facts,* New York, Newspaper Enterprise Association Inc., 1986, p. 102.

‡Alan Murray, "Individuals' Top Rate Would Plunge to 28 Percent; Tax-Break Curbs Offset Benefits to Wealthy," *Wall Street Journal,* August 18, 1986, p. 6.

Problems 6 and 7 refer to an editorial by Paul Craig Roberts, who holds the William E. Simon chair of political economy at the Center for Strategic and International Studies in Washington.§ The editorial compared the economy under presidents Carter and Reagan during two 58-month periods using these facts: Under President Carter the unemployment rate fell 27% and the CPI (consumer price index) rose 48%, while under President Reagan the unemployment rate fell 45% and the CPI rose 17%.

6. (a) If graphs for the unemployment rate under each administration are sketched with years on the *x*-axis and unemployment rates on the *y*-axis, are the slopes of the functions (which are defined by the graphs) positive or negative?

    (b) Which administration had a better record regarding unemployment?

7. (a) If graphs for the CPI under each administration are sketched with years on the *x*-axis and the CPI on the *y*-axis, are the slopes positive or negative?

    (b) Which administration had a better record regarding the CPI?

§Paul Craig Roberts, "While Critics Carp, the Supply-Side Revolution Sweeps On," *Business Week,* April 25, 1988, p. 15.

2–2 Limits

In the previous section the problem of finding the instantaneous rate of change entailed evaluating the difference quotient as *h* approached 0. In this section we give a more formal definition of this process. It is called "finding the *limit* of $f(x)$ as *x* approaches *a*," which intuitively means that we find the value that $f(x)$ should be when *x* approaches *a*. Of course, part of the problem is to discuss what is meant by "should be" and "approaches."

As an example, consider $f(x) = x^2$. As *x* gets close to 2, $f(x)$ gets close to $2^2 = 4$, so we say that $f(x)$ approaches 4 as *x* approaches 2. Likewise, as *x* approaches 3, the function $g(x) = x^3$ approaches $3^3 = 27$. Expressed in terms of limits, we say the limit of $f(x) = x^2$ as *x* approaches 2 is 4 and the limit of $g(x) = x^3$ as *x* approaches 3 is 27.

It is easy to find these particular limits because in each case the limit is equal to the functional value, $f(a)$; that is, to find the limit we simply substitute *a* into the formula of the function. The difficulty arises when the limit is not equal to the functional value. This was the case in the previous section when we had to find the limit of the difference quotient as *h* approached 0. It could not be done by simply substituting 0 into the expression because the expression was not defined at 0. Another method had to be used. It is this method that we will discuss in a more rigorous way in this section.

The problem is approached from three perspectives. First we look at a *geometric* approach by describing the problem in terms of the graph of $f(x)$. The limit of $f(x)$ as *x* approaches *a*, if it exists, is the second coordinate of a point that the function "should" go through. For example, the graph of $f(x) = x^2$ "should" go through (2, 4). That it actually does pass through this point is immaterial in this discussion because we are more concerned with evaluating limits when the function does not go through this point.

Then we will consider an *arithmetic* approach to the problem, where we define "approaches $x = a$" by actually computing functional values of $f(x)$ for numbers within smaller and smaller distances of a. For example, if we want to compute the limit as x approaches 1, we compute functional values at $x = 1.1, 1.01$, and 1.001 and at $x = 0.9, 0.99$, and 0.999, and then see if these functional values are getting close to a specific number. Neither of these two approaches is very rigorous, so, to be more precise, we use a third method, the *algebraic* approach to the problem.

Geometric Approach

The graphs of the two functions $f(x) = x + 2$ and $g(x) = \dfrac{x^2 - 4}{x - 2}$ are identical except for one point: The graph of $f(x)$ includes the point $(2, 4)$ while the graph of $g(x)$ does not include it. This is because $f(2) = 2 + 2 = 4$ while $g(2)$ is $\dfrac{0}{0}$, which is not defined. It is clear that the limit of $f(x)$ as x approaches 2 is 4, because the graph passes through $(2, 4)$ in an unbroken manner. The situation for $g(x)$ is similar, but with a little catch. There is a "hole" in the graph at $(2, 4)$, meaning that the curve would be unbroken if we included $(2, 4)$ in the graph. In an intuitive sense, this is what is meant by saying that $g(x)$ "should" pass through $(2, 4)$, that its graph would be unbroken if that point were included in the graph. For this reason we say the limit of $g(x)$ as x approaches 2 is 4, since 4 is the second coordinate of the point that $g(x)$ should go through to be unbroken.

    As another example, consider the function

$$f(x) = \frac{x^2 - 8x + 15}{3 - x} = \frac{(x - 3)(x - 5)}{3 - x}$$

Its graph is the line with the "hole" in it drawn in Figure 1. The open circle indicates that $f(3)$ is not defined. However, even though the function is not defined

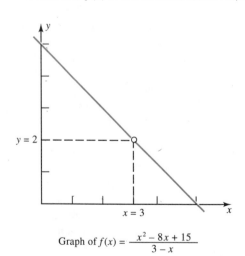

Graph of $f(x) = \dfrac{x^2 - 8x + 15}{3 - x}$

FIGURE 1

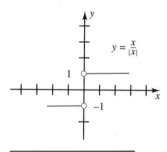

FIGURE 2

at $x = 3$, the limit is defined there because the functional values of $f(x)$ get close to 2 as x gets close to 3. The geometric interpretation of the limit is the height of the graph when the values of x get close to 3.

To see when a function does not have a limit at a particular value of x, consider the function

$$f(x) = \frac{x}{|x|} = \begin{cases} 1 \text{ if } x > 0 \\ -1 \text{ if } x < 0 \end{cases}$$

The function is not defined at $x = 0$. The graph of the $f(x)$ is given in Figure 2. There is a "gap" in the graph at $x = 0$, and so there is no point that we could put into the graph at $x = 0$ so that the function would be unbroken. This means that for $f(x) = \dfrac{x}{|x|}$ **the limit does not exist** at $x = 0$.

Another example of a more familiar function that does not have a limit at particular values of x is the *quality points function*, which many universities use to define grade point average (GPA). If a student's test average is 90 or above, the student gets an A and four quality points are awarded. The chart shows the other values.

Test average	≥ 90	80s	70s	60s	<60
Quality points	4	3	2	1	0

EXAMPLE 1

Let $Q(x)$ denote the quality points function whose graph is drawn in Figure 3. An open circle means that the function is not defined at that end point.

Problem

Explain why the limit of $Q(x)$ as x approaches 80 does not exist.

Solution Note that $Q(80) = 3$ since an average of 80 yields three quality points. In fact, for any number x that is greater than 80 but close to 80, $Q(x) = 3$. However, for values of x that are close to 80 but less than 80, $Q(x) = 2$.

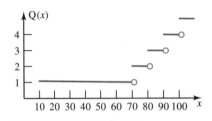

FIGURE 3 Graph of the quality points function $Q(x)$.

As x gets closer to 80, we cannot choose whether $Q(x)$ gets closer to 3 or 2. Since the limit must be a unique number, we conclude that the limit does not exist. For example, $Q(79.9) = 2$ and $Q(80) = 3$.

Arithmetic Approach

In Section 2–1 we derived the instantaneous rate of change of the function $f(x) = x^3$ at $x = 1$ from a table whose essential ingredients are reproduced here (rounded to three places).

h	-0.1	-0.01	-0.001	0.001	0.01	0.1
Average rate	2.71	2.970	2.997	3.003	3.030	3.31

The average rates of change approach 3 as h approaches 0. This leads to an intuitive definition of the limit of a function.

DEFINITION

If there is a unique number L with the property that $f(x)$ approaches L as x approaches the number a and $x \neq a$, then L is called the **limit** of $f(x)$ as x approaches a. In symbols

$$\lim_{x \to a} f(x) = L$$

The notation $\lim_{x \to a} f(x)$ is read "the limit of f of x as x approaches a." If the limit L exists, then it is the only number with the defining property.

To illustrate the definition, consider the function

$$f(x) = x^3 - 3x^2 + 5$$

for values of x that are *near* 1, but are *not equal to* 1. Construct Table 1 with values of x that approach 1 in the top row and the corresponding values of $f(x)$ below them (rounded to three decimal places). It is important to observe that the values of x approach 1 from two directions: from the left (numbers that are less than 1) and from the right (numbers that are greater than 1).

Table 1

	Approach 1 from the left				Approach 1 from the right		
x	0.9	0.99	0.999	1	1.001	1.01	1.1
$f(x)$	3.299	3.030	3.003		2.997	2.970	2.701

It follows from Table 1 that $f(x)$ approaches 3 as x approaches 1 from both directions, so

$$\lim_{x \to 1} f(x) = 3$$

It is possible to obtain this limit directly by evaluating $f(1)$ since $f(1) = 3$. However, such a substitution is not always possible. For example, consider again

$$f(x) = \frac{x^2 - 8x + 15}{3 - x} = \frac{(x - 3)(x - 5)}{3 - x}$$

What is the behavior of $f(x)$ when x assumes values that are *near* 3 but *not equal to* 3? To answer the question, construct Table 2. Notice that $f(x)$ is not defined at $x = 3$ because both the numerator and denominator are 0.

Table 2

	Approach 3 from the left				Approach 3 from the right		
x	2.9	2.99	2.999	3	3.001	3.01	3.1
$f(x)$	2.1	2.01	2.001		1.999	1.99	1.9

It follows from Table 2 that $f(x)$ approaches 2 as x approaches 3 from both directions, so

$$\lim_{x \to 3} f(x) = 2$$

EXAMPLE 2

Problem

Compute $\lim_{x \to 1} f(x)$ for $f(x) = \dfrac{4x^2 - 4x}{x - 1} = \dfrac{4x(x - 1)}{x - 1}$.

Solution Construct a table with values of x approaching 1 from both sides in the top row. Compute the corresponding values of $f(x)$ and place them in the bottom row.

	Approach 1 from the left				Approach 1 from the right		
x	0.9	0.99	0.999	1	1.001	1.01	1.1
$f(x)$	3.6	3.96	3.996		4.004	4.04	4.4

The table shows that $f(x)$ approaches 4 as x approaches 1 from both sides, so

$$\lim_{x \to 1} f(x) = 4$$

In Example 2 the function is defined for values of x that are near 1 but are not equal to 1, yet the limit exists at $x = 1$ (and is equal to 4). Example 3 examines an entirely different situation.

EXAMPLE 3

Problem

Compute $\lim\limits_{x \to 2} f(x)$ where $f(x) = \dfrac{x + 1}{x - 2}$.

Solution The appropriate table is shown here.

	Approach 2 from the left				Approach 2 from the right		
x	1.9	1.99	1.999	2	2.001	2.01	2.1
$f(x)$	-29	-299	-2999		3001	301	31

The table shows that as x approaches 2 from the left, the values of $f(x)$ get smaller and smaller, and as x approaches 2 from the right, the values of $f(x)$ get larger and larger. Therefore $f(x)$ does not have a limit when x approaches 2, meaning that there is no fixed number that the values of $f(x)$ approach as x approaches 2.

Example 3 illustrates another way in which the limit of a function may fail to exist. Unlike the graph in Example 1, where there was a gap in the graph, the graph of $f(x)$ in Example 3 does not approach any number as x approaches 2 because $f(x)$ **decreases without bound** to the left of $x = 2$ and **increases without bound**

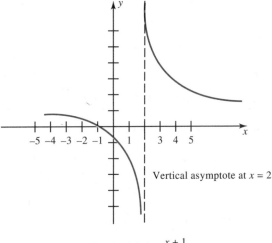

Graph of $f(x) = \dfrac{x+1}{x-2}$

FIGURE 4

to the right of $x = 2$. When a function either increases or decreases without bound near a value $x = a$, we write

$$\lim_{x \to a} f(x) \quad \text{does not exist}$$

Figure 4 shows that the graph of the function $f(x) = \dfrac{x+1}{x-2}$ has a vertical asymptote at $x = 2$. As x approaches 2 from the left, $f(x)$ decreases without bound, and as x approaches 2 from the right, $f(x)$ increases without bound. Therefore

$$\lim_{x \to 2} f(x) \quad \text{does not exist}$$

In general, whenever a function $f(x)$ has a vertical asymptote at the line $x = a$ then $\lim\limits_{x \to a} f(x)$ does not exist.

Algebraic Approach

We have evaluated limits by arithmetic and geometric methods. The most rigorous method, however, is to use algebraic simplifications based on the following properties of limits. These properties can be derived from the definition of the limit of a function.

Properties of Limits

Let $f(x)$ and $g(x)$ be functions for which $\lim_{x \to a} f(x)$ and $\lim_{x \to a} g(x)$ exist.

Then

1. $\lim_{x \to a} f(x) = \lim_{x \to a} g(x)$ if $f(x) = g(x)$ for all $x \neq a$

2. $\lim_{x \to a} cf(x) = c \lim_{x \to a} f(x)$ for any real number c

   (Limit of a constant times a function)

3. $\lim_{x \to a} [f(x) \pm g(x)] = \lim_{x \to a} f(x) \pm \lim_{x \to a} g(x)$

   (Limit of the sum or difference)

4. $\lim_{x \to a} [f(x)g(x)] = \lim_{x \to a} f(x) \lim_{x \to a} g(x)$ (Limit of a product)

5. $\lim_{x \to a} \dfrac{f(x)}{g(x)} = \dfrac{\lim_{x \to a} f(x)}{\lim_{x \to a} g(x)}$, provided that $\lim_{x \to a} g(x) \neq 0$

   (Limit of a quotient)

6. $\lim_{x \to a} [f(x)]^r = \left[\lim_{x \to a} f(x) \right]^r$ for any real number r

   (Limit of a function raised to a power)

(where x must be restricted to avoid even roots of negative numbers)

The next example makes use of two properties that can be proved from the definition of a limit.

$$\lim_{x \to a} x = a \quad \text{and} \quad \lim_{x \to a} c = c$$

EXAMPLE 4

Problem

Evaluate $\lim_{x \to 0.5} (2x^2 - x - 4)$

Solution It follows from the stated properties that

$$\lim_{x \to 0.5} x = 0.5 \quad \text{and} \quad \lim_{x \to 0.5} 4 = 4$$

By properties 2 and 4

$$\lim_{x \to 0.5} 2x^2 = 2 \left(\lim_{x \to 0.5} x^2 \right) = 2 \left(\lim_{x \to 0.5} x \right) \left(\lim_{x \to 0.5} x \right)$$

$$= 2(0.5)(0.5) = 0.5$$

Using property 3

$$\lim_{x \to 0.5} (2x^2 - x - 4) = \lim_{x \to 0.5} 2x^2 - \lim_{x \to 0.5} x - \lim_{x \to 0.5} 4$$

$$= 0.5 - 0.5 - 4 = -4$$

Property 1 is a powerful tool for evaluating many of the limits that appear in calculus. For instance, let

$$f(x) = \frac{x^2 - 8x + 15}{3 - x}$$

We have seen that $\lim_{x \to 3} f(x) = 2$. The algebraic approach is to reduce $f(x)$ to a function $g(x)$ such that $g(3)$ is defined and $g(x) = f(x)$ for all x except $x = 3$. For this function

$$\frac{x^2 - 8x + 15}{3 - x} = \frac{(x - 3)(x - 5)}{-(x - 3)} = 5 - x$$

The common factor $(x - 3)$ can be canceled because $x \neq 3$. Set $g(x) = 5 - x$. Then $g(x) = f(x)$ for all x except $x = 3$. Since $g(3) = 2$, it follows that $\lim_{x \to 3} f(x) = g(3)$, so

$$\lim_{x \to 3} \frac{x^2 - 8x + 15}{3 - x} = \lim_{x \to 3} (5 - x) = 2$$

Examples 5, 6, and 7 exploit property 1.

EXAMPLE 5

Problem

Evaluate $\lim_{x \to 2} \dfrac{x^2 - 2x}{x - 2}$.

Solution $\lim_{x \to 2} \dfrac{x^2 - 2x}{x - 2} = \lim_{x \to 2} \dfrac{x(x - 2)}{x - 2} = \lim_{x \to 2} x = 2$

EXAMPLE 6

Problem

Evaluate $\lim_{x \to 3} \dfrac{3 - x}{x^2 - 6x + 9}$.

Solution $\lim_{x \to 3} \dfrac{3 - x}{x^2 - 6x + 9} = \lim_{x \to 3} \dfrac{-(x - 3)}{(x - 3)^2} = \lim_{x \to 3} \dfrac{-1}{x - 3}$

This limit does not exist because the function $f(x) = \dfrac{-1}{x - 3}$ has a vertical asymptote at $x = 3$.

Example 7 shows that it is sometimes necessary to rationalize the numerator in order to divide out the factor that results in an expression of the form $\dfrac{0}{0}$.

EXAMPLE 7

Problem

Evaluate $\lim\limits_{x \to 9} \dfrac{\sqrt{x} - 3}{x - 9}$

Solution $\lim\limits_{x \to 9} \dfrac{\sqrt{x} - 3}{x - 9} = \lim\limits_{x \to 9} \dfrac{(\sqrt{x} - 3)(\sqrt{x} + 3)}{(x - 9)(\sqrt{x} + 3)}$

$$= \lim\limits_{x \to 9} \dfrac{x - 9}{(x - 9)(\sqrt{x} + 3)}$$

$$= \lim\limits_{x \to 9} \dfrac{1}{\sqrt{x} + 3} = \dfrac{1}{6}$$

Limits at Infinity

Consider the function $f(x) = \dfrac{1 + 2x}{x}$. Table 3 shows that as x assumes larger and larger values in the positive direction, the corresponding values of $f(x)$ approach the number 2. This is called taking a **limit at infinity.** The notation is

$$\lim\limits_{x \to \infty} f(x) = 2$$

It is read "the limit of $f(x)$ as x approaches infinity is 2," where the symbol for infinity is ∞.

Table 3

x	10	100	1000	10,000
$f(x)$	2.1	2.01	2.001	2.0001

Table 4 shows that for values of x that get smaller and smaller in the negative direction, the corresponding values of $f(x)$ also approach the number 2, so

$$\lim\limits_{x \to -\infty} f(x) = 2$$

This is read "the limit of $f(x)$ as x approaches minus infinity is 2."

Table 4

x	-10	-100	-1000	$-10,000$
$f(x)$	1.9	1.99	1.999	1.9999

The display contains the general definitions.

DEFINITION

$\lim\limits_{x\to\infty} f(x) = L$ if there is a unique number L such that the values of $f(x)$ approach L as x assumes larger and larger values. $\lim\limits_{x\to-\infty} f(x) = L$ if there is a unique number L such that the values of $f(x)$ approach L as x assumes smaller and smaller values. If either $\lim\limits_{x\to\infty} f(x) = L$ or $\lim\limits_{x\to-\infty} f(x) = L$, then the horizontal line $y = L$ is called a **horizontal asymptote** of $f(x)$.

Horizontal asymptotes of rational functions can be found by dividing the numerator and the denominator by x^n, where n is the larger of the degrees of the polynomials in the numerator and the denominator. Example 8 illustrates the method.

EXAMPLE 8

Let

$$f(x) = \frac{5x^2 - 1}{2x^2 - x + 5}$$

Problem

(a) Evaluate $\lim\limits_{x\to\infty} f(x)$.   (b) Find all horizontal asymptotes of $f(x)$.

Solution (a) Both the numerator and denominator are polynomials of degree 2. Divide each one by x^2.

$$f(x) = \frac{5x^2 - 1}{2x^2 - x + 5} = \frac{\dfrac{5x^2 - 1}{x^2}}{\dfrac{2x^2 - x + 5}{x^2}} = \frac{\dfrac{5x^2}{x^2} - \dfrac{1}{x^2}}{\dfrac{2x^2}{x^2} - \dfrac{x}{x^2} + \dfrac{5}{x^2}}$$

$$= \frac{5 - \dfrac{1}{x^2}}{2 - \dfrac{1}{x} + \dfrac{5}{x^2}}$$

As the values of x get larger and larger, $\dfrac{1}{x^2}$ approaches 0. Therefore the numerator $5 - \dfrac{1}{x^2}$ approaches 5. Similarly the denominator $2 - \dfrac{1}{x} + \dfrac{5}{x^2}$ approaches 2. Thus the desired limit is $\frac{5}{2}$.

(b) Part (a) shows that $\lim\limits_{x\to\infty} f(x) = \frac{5}{2}$. Similarly $\lim\limits_{x\to-\infty} f(x) = \frac{5}{2}$. Therefore the line $y = \frac{5}{2}$ is the only horizontal asymptote of $f(x)$. See Figure 5.

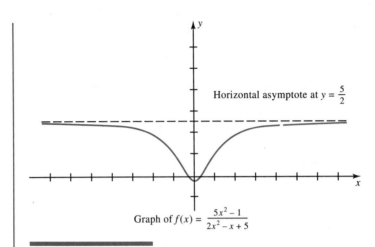

Horizontal asymptote at $y = \dfrac{5}{2}$

Graph of $f(x) = \dfrac{5x^2 - 1}{2x^2 - x + 5}$

FIGURE 5

EXERCISE SET 2–2

In Problems 1 to 8 compute the given limits by constructing tables.

1. $\lim\limits_{x \to 1} f(x)$ for $f(x) = \dfrac{3x^2 - 3x}{x - 1}$

2. $\lim\limits_{x \to 1} f(x)$ for $f(x) = \dfrac{2x^2 - 2x}{x - 1}$

3. $\lim\limits_{x \to 3} f(x)$ for $f(x) = \dfrac{2x^2 - 6x}{x - 3}$

4. $\lim\limits_{x \to 3} f(x)$ for $f(x) = \dfrac{3x^2 - 9x}{x - 3}$

5. $\lim\limits_{x \to 2} f(x)$ for $f(x) = \dfrac{x + 1000}{x - 2}$ *NO LIMIT*

6. $\lim\limits_{x \to 2} f(x)$ for $f(x) = \dfrac{x + 2}{x - 2}$

7. $\lim\limits_{x \to 3} f(x)$ for $f(x) = \dfrac{3 + x}{3 - x}$ *NO LIMIT*

8. $\lim\limits_{x \to 3} f(x)$ for $f(x) = \dfrac{x + 1000}{3 - x}$

In Problems 9 to 12 evaluate the limit, if it exists, of the quality points function $Q(x)$.

9. $\lim\limits_{x \to 70} Q(x)$

10. $\lim\limits_{x \to 90} Q(x)$

11. $\lim\limits_{x \to 81} Q(x)$

12. $\lim\limits_{x \to 79} Q(x)$

In Problems 13 to 26 evaluate each limit, if it exists, by algebraic methods.

13. $\lim\limits_{x \to 2} x^2 - 2x$

14. $\lim\limits_{x \to 4} x^2 - 4x$

15. $\lim\limits_{x \to 1} 1 - x$

16. $\lim\limits_{x \to 2} 2 - x$

17. $\lim\limits_{x \to 2} \dfrac{x^2 - 2x}{x - 2}$

18. $\lim\limits_{x \to 4} \dfrac{x^2 - 4x}{x - 4}$

19. $\lim\limits_{x \to 1} \dfrac{1 - x}{x^2 - 2x + 1}$

20. $\lim\limits_{x \to 2} \dfrac{2 - x}{x^2 - 4x + 4}$

21. $\lim\limits_{x \to -2} \dfrac{x + 2}{x^2 + 4x + 4}$

22. $\lim\limits_{x \to -1} \dfrac{x + 1}{x^2 + 2x + 1}$

23. $\lim\limits_{x \to 1} \dfrac{\sqrt{x} - 1}{x - 1}$

24. $\lim\limits_{x \to 4} \dfrac{\sqrt{x} - 2}{x - 2}$

25. $\lim\limits_{x \to 9} \dfrac{3 - \sqrt{x}}{9 - x}$

26. $\lim\limits_{x \to 1} \dfrac{1 - \sqrt{x}}{1 - x}$

In Problems 27 to 32 evaluate $\lim\limits_{x \to \infty} f(x)$ for each function.

27. $f(x) = \dfrac{5}{x^2}$

28. $f(x) = \dfrac{2}{5x^3}$

29. $f(x) = \dfrac{1 - 4x}{2 + x}$

30. $f(x) = \dfrac{9 + 2x}{1 - 3x}$

31. $f(x) = \dfrac{2x^2 + 3x - 5}{6x^2 - 5}$

32. $f(x) = \dfrac{2x^2 + 12}{5x^3 - 6}$

In Problems 33 to 36 compute each limit, if it exists, by constructing tables.

33. $\lim\limits_{x \to 1} f(x)$ for $f(x) = \dfrac{3x^3 - 3x^2}{x^2 - x}$

34. $\lim\limits_{x \to 1} f(x)$ for $f(x) = \dfrac{2x^3 - 2x^2}{x^2 - x}$

35. $\lim\limits_{x \to 2} f(x)$ for $f(x) = \dfrac{x - 2}{x + 2}$

36. $\lim\limits_{x \to 3} f(x)$ for $f(x) = \dfrac{3 - x}{3 + x}$

In Problems 37 to 40 compute each limit, if it exists, by the geometric approach.

37. $\lim\limits_{x \to 0} |x|$

38. $\lim\limits_{x \to 0} \dfrac{1}{x^2}$

39. $\lim\limits_{x \to 0} f(x)$, where $f(x) = \begin{cases} -x & \text{if } x < 0 \\ 3x & \text{if } x \ge 0 \end{cases}$

40. $\lim\limits_{x \to 1} f(x)$, where $f(x) = \begin{cases} x + 1 & \text{if } x < 1 \\ x & \text{if } x \ge 1 \end{cases}$

In Problems 41 to 52 compute each limit, if it exists, by any method.

41. $\lim\limits_{x \to 1} \dfrac{x^2 - 9}{x - 3}$

42. $\lim\limits_{x \to 1} \dfrac{x^2 - 4}{x - 2}$

43. $\lim\limits_{x \to -2} \dfrac{x^2 + 3x + 2}{x^2 - 4}$

44. $\lim\limits_{x \to 1} \dfrac{x^2 + 3x + 2}{x^2 - 1}$

45. $\lim\limits_{x \to 3} \dfrac{x^3 - 9x}{x - 3}$

46. $\lim\limits_{x \to -3} \dfrac{x^3 - 9x}{x + 3}$

47. $\lim\limits_{x \to 2} \dfrac{x^3 - 4x}{x - 2}$

48. $\lim\limits_{x \to 0} \dfrac{x^2 - 3x}{x}$

49. $\lim\limits_{x \to -2} \dfrac{x^2 + 4x + 4}{x + 2}$

50. $\lim\limits_{x \to -1} \dfrac{x^2 + 2x + 1}{x + 1}$

51. $\lim\limits_{x \to 1} \dfrac{x - 1}{\sqrt{x} - 1}$

52. $\lim\limits_{x \to 4} \dfrac{x - 2}{\sqrt{x} - 2}$

In Problems 53 to 56 evaluate $\lim\limits_{x \to -\infty} f(x)$ for each function.

53. $f(x) = \dfrac{4x + 3}{2x - 6}$

54. $f(x) = \dfrac{x - 3}{5x + 1}$

55. $f(x) = \dfrac{3x^2 - 4x - 5}{x^2 + 5}$

56. $f(x) = \dfrac{7x^2 + 12}{3x^2 - 6}$

In Problems 57 to 60 evaluate $\lim\limits_{x \to \infty} f(x)$ for each function.

57. $f(x) = \dfrac{2 - 3x}{x^3}$

58. $f(x) = \dfrac{x^3}{4 + 5x}$

59. $f(x) = \dfrac{x^2 + 6}{x - 1}$

60. $f(x) = \dfrac{9 + 2x}{x^2 - 3}$

Problems 61 to 64 depict situations where the limit of a function is determined by examining values close to the limiting value.

61. A small midwestern newspaper services 20 small towns within a 20-mile radius of the city where the newspaper is published. A reporter who is charged with recording the daily temperature table notices that 19 of the towns report a noon temperature of 90°F, but the remaining town is missing. What number do you suppose the reporter filled in?

62. The table below lists the population of a bacterium culture, where t is measured in hours and $P(t)$ in millions of bacteria. The measurement is missing at $t = 5$. What value should be inserted?

t	4.6	4.7	4.8	4.9	5	5.1	5.2
$P(t)$	12	14	18	26		74	138

63. A rocket is designed to explode precisely 5 seconds after it has been fired into the air. An instrument measures the distance the rocket travels up to the point of the explosion, but it cannot measure the distance at the point of explosion. The table gives the distance $s(t)$ traveled in miles after t seconds. How far did the rocket travel before the explosion?

t	4.9	4.99	4.999
$s(t)$	199	199.9	199.99

64. A laboratory tests a new adhesive to see how many pounds it will hold for 1 minute. The exact weight cannot be measured because if the weight falls, it was too heavy, and if the weight does not fall, it was too light. The laboratory compiled the data in the table. What is the maximum weight that the adhesive can hold?

	Held			Did not hold		
Weight (lb)	198	199	199.9	200.1	201	202
Time (min)	60	60	60	59.8	59.5	59

Referenced Exercise Set 2–2

A study by K. Rohrbach proposed a new accounting strategy for auditing called *monetary unit acceptance sampling*.* Figure 6 describes the so-called power functions of the four different statistical distributions which he used. (The nominal power function is based on the binomial distribution.) Problems 1 and 2 refer to Figure 6.

1. Complete the following table for the gamma – 100 power function $f(x)$ and use it to evaluate $\lim\limits_{x \to .08} f(x)$.

x	.05	.06	.07	.08
$f(x)$				

2. Repeat Problem 1 for the uniform power function.

3. In an article concerning various demands and price structures of finished goods inventories, F. Arcelus and G. Srinivasan† introduced Figure 7 to show how

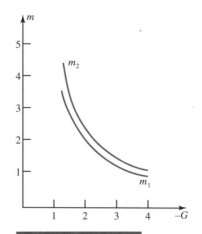

FIGURE 7

increases in the price elasticity of demand G affect the markup rate m. Use Figure 6 to evaluate $\lim\limits_{-G \to \infty} f(x)$ for each of these functions.

(a) $f(x) = m_1(x)$

(b) $f(x) = m_2(x)$

4. This problem illustrates limits at infinity using concepts from probability. It was suggested by Jerry Johnson of Oklahoma State University.‡

(a) Suppose you have a huge sock drawer which holds 100 blue socks and 100 brown socks. One morning you are half asleep while dressing for school, and you blindly reach into the drawer to pick out a pair of socks. What is the probability that they match? (*Hint:* Remove one sock. What fraction of the remaining socks match in color?)

(b) Repeat part (a) for a drawer with n socks of each color.

(c) Repeat part (a) for a drawer with infinitely many socks of each color.

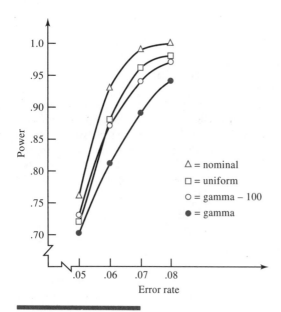

FIGURE 6

Cumulative Exercise Set 2–2

1. Compute the average rate of change for $f(x) = x^2$ from $x = -2$ to $x = -1.99$.

*Kermit J. Rohrbach, "Monetary Unit Acceptance Sampling," *Journal of Accounting Research,* Spring 1986, 127–150.

†F. J. Arcelus and G. Srinivasan, "Inventory Policies Under Various Optimizing Criteria and Variable Markup Rates," *Management Science,* June 1987, pp. 756–762.

‡Jerry A. Johnson, "An Illustration of Limits," *Mathematics Teacher,* December 1986, pp. 722–723.

2. Use the method shown in Example 3 of Section 2–1 to compute the instantaneous rate of change of $f(x) = x^2$ at $x = -2$.

3. Use the result of Problem 2 to find the slope-intercept form of the line tangent to the graph of $f(x) = x^2$ at $x = -2$.

4. The height of a ball thrown into the air is given by $d(t) = -16t^2 + 64t + 96$, where t is the time in seconds after the ball is released. What is the instantaneous rate of change of the ball one second after it is released?

5. Compute the limit by constructing a table.

$$\lim_{x \to -1} f(x) \text{ for } f(x) = \frac{2x^2 + 4x + 2}{x + 1}$$

In Problems 6 and 7 evaluate each limit, if it exists, by algebraic methods.

6. $\displaystyle\lim_{x \to -1} \frac{x + 1}{2x^2 + 4x + 2}$

7. $\displaystyle\lim_{x \to -3} \frac{(x + 3)^{99}}{98(x + 3)^{98}}$

8. The table below lists the population of a city (in thousands) in the 20th century. If this trend continues, what will the population be in the 21st century?

Year	1900	1910	1920	1930	1940	1950	1960	1970	1980	1990
Pop.	800	480	320	240	200	180	170	165	162.5	161.25

2–3 Definition of the Derivative

Section 2–1 provided the motivation and Section 2–2 laid the foundation for the definition of the derivative. Here we define the derivative, show how to compute it, and apply it to graphing techniques and rates of change. Then we discuss continuous and differentiable functions.

Procedure

The methods of computing the instantaneous rate of change of a function $f(x)$ at a value $x = a$ and the line tangent to the graph of $f(x)$ at the point $(a, f(a))$ involve the same procedure: Form a difference quotient and evaluate a limit as the denominator approaches 0. This motivates the following fundamental definition.

DEFINITION

The **derivative of a function** $f(x)$ is a function, denoted by $f'(x)$, defined by

$$f'(x) = \lim_{h \to 0} \frac{f(x + h) - f(x)}{h}$$

The domain of $f'(x)$ consists of all x in the domain of $f(x)$ for which the limit exists.

The notation $f'(x)$ is read "f prime of x." The method of computing $f'(x)$ is the same method that was introduced in Section 2–1 to compute the slope of the line tangent to the graph of $f(x)$. The crucial step in evaluating the limit is to perform algebraic simplifications that enable the denominator h to be canceled.

EXAMPLE 1

Problem

Compute $f'(x)$ for $f(x) = x^3 - 2x$.

Solution The difference quotient is

$$\frac{f(x + h) - f(x)}{h} = \frac{[(x + h)^3 - 2(x + h)] - (x^3 - 2x)}{h}$$

$$= \frac{x^3 + 3x^2h + 3xh^2 + h^3 - 2x - 2h - x^3 + 2x}{h}$$

$$= \frac{h(3x^2 + 3xh + h^2 - 2)}{h}$$

Cancel the common factor h.

$$\frac{f(x + h) - f(x)}{h} = 3x^2 + 3xh + h^2 - 2$$

Then let the values of h approach 0:

$$f'(x) = \lim_{h \to 0} (3x^2 + 3xh + h^2 - 2) = 3x^2 - 2$$

The display contains the important fact that the derivative of a function can be interpreted as a rate of change.

> The instantaneous rate of change of a function $f(x)$ at a value $x = a$ is the derivative $f'(a)$, provided that $f'(a)$ exists.

EXAMPLE 2

Problem

Find the instantaneous rate of change of

$$f(x) = \sqrt{x} \qquad \text{at } x = 9$$

Solution First find $f'(x)$. The difference-quotient is

$$\frac{f(x + h) - f(x)}{h} = \frac{\sqrt{x + h} - \sqrt{x}}{h}$$

The algebraic steps that enable h to be canceled are

$$\frac{\sqrt{x + h} - \sqrt{x}}{h} = \frac{(\sqrt{x + h} - \sqrt{x})}{h} \cdot \frac{(\sqrt{x + h} + \sqrt{x})}{(\sqrt{x + h} + \sqrt{x})}$$

$$= \frac{(x + h) - x}{h(\sqrt{x + h} + \sqrt{x})} = \frac{1}{\sqrt{x + h} + \sqrt{x}}$$

Then

$$f'(x) = \lim_{h \to 0} \frac{1}{\sqrt{x + h} + \sqrt{x}} = \frac{1}{\sqrt{x} + \sqrt{x}} = \frac{1}{2\sqrt{x}}$$

Therefore the instantaneous rate of change at $x = 9$ is

$$f'(9) = \frac{1}{2\sqrt{9}} = \frac{1}{6}$$

Tangents

In Section 2–1 the slope of the line tangent to the graph of a function $f(x)$ was shown to correspond to the instantaneous rate of change of $f(x)$. This means that if the derivative $f'(x)$ can be evaluated at $x = a$, then $f'(a)$ is the slope of the line tangent to the curve at that point. The equation of the tangent line can then be written because one point $(a, f(a))$ and the slope $f'(a)$ are known.

The equation of the line tangent to the graph of a function $f(x)$ at the point $(a, f(a))$ is

$$y - f(a) = f'(a)(x - a)$$

provided that $f'(a)$ exists.

EXAMPLE 3

Problem

Find the slope-intercept form of the line that is tangent to the graph of $f(x) = 2/x$ at $x = -3$.

Solution The approach is to find $f'(x)$ and then to evaluate it at $x = -3$. First find the difference quotient:

$$\frac{f(x + h) - f(x)}{h} = \frac{\dfrac{2}{x + h} - \dfrac{2}{x}}{h}$$

$$= \frac{\dfrac{2x - 2(x + h)}{(x + h)x}}{h}$$

$$= \frac{2x - 2(x + h)}{(x + h)x} \cdot \frac{1}{h}$$

$$= \frac{-2h}{(x + h)xh} = \frac{-2}{(x + h)x}$$

Then find the limit:

$$f'(x) = \lim_{h \to 0} \frac{-2}{(x + h)x} = \frac{-2}{(x)x} = \frac{-2}{x^2}$$

Therefore $f'(-3) = \dfrac{-2}{(-3)^2} = \dfrac{-2}{9}$. The equation of the tangent line is

$$y - f(-3) = f'(-3)(x + 3)$$

$$y - \left(\frac{2}{-3}\right) = \left(\frac{-2}{9}\right)(x + 3)$$

$$y = \frac{-2}{9} x - \frac{4}{3}$$

Velocity

In Section 2–1 the instantaneous velocity of an object moving in a straight line was shown to correspond to the instantaneous rate of change. But the instantaneous rate of change is the derivative. Therefore the derivative of the distance function can also be interpreted as the velocity of a moving object at a given time.

> If $d(t)$ is the distance function of a moving object, then the velocity $v(t)$ is the derivative, provided that $d'(t)$ exists. In symbols, $v(t) = d'(t)$.

If a ball is thrown into the air, it will reach a high point before returning to earth. The ball's velocity is a positive number on the way up, and a negative number on the way down. At the instant when the ball reaches its zenith, the velocity is 0. This observation is crucial for solving problems like Example 4.

EXAMPLE 4

The height of a ball thrown from ground level into the air with an initial velocity of 128 feet per second is given by the equation

$$d(t) = -16t^2 + 128t$$

where d is measured in feet and t in seconds.

Problem

(a) When will the ball reach its maximum height? (b) What is the maximum height?

Solution (a) The velocity at any time t is given by

$$d'(t) = \lim_{h \to 0} \frac{d(t + h) - d(t)}{h}$$

$$= \lim_{h \to 0} \frac{[-16(t + h)^2 + 128(t + h)] - (-16t^2 + 128t)}{h}$$

$$= \lim_{h \to 0} \frac{-16t^2 - 32th - 16h^2 + 128t + 128h + 16t^2 - 128t}{h}$$

$$= \lim_{h \to 0} \frac{h(-32t - 16h + 128)}{h}$$

$$= \lim_{h \to 0} (-32t - 16h + 128)$$

$$= -32t + 128$$

The ball will reach its maximum height when its velocity is 0, so $d'(t) = 0$. Then $-32t + 128 = 0$, so $t = 4$. Therefore the ball will reach its maximum height after 4 seconds. (b) By part (a) the maximum height occurs at $t = 4$. Since $d(4) = -16 \cdot 4^2 + 128 \cdot 4 = 256$, the maximum height is 256 feet. Note that we substitute $t = 4$ into the formula for $d(t)$ to find $d(4)$. Don't be misled into substituting $t = 4$ into the formula for $d'(t)$ to find $d(4)$.

Marginal Cost

Marginal cost functions were introduced in Chapter 1. The **marginal cost** of producing an item is the rate of change in cost when that item is produced. It is an approximation to the change in cost for the next item. For instance, let the cost (in dollars) of producing x liters of distilled water be

$$C(x) = x^2 - 2x + 6 \quad 1 \le x \le 6$$

The additional cost of producing the third liter of distilled water is \$3, the difference between the cost of producing 3 liters and the cost of producing 2 liters, since $C(3) - C(2)/(3 - 2) = 9 - 6 = 3$. This can be seen in Figure 1 as the slope of the secant line from (2, 6) to (3, 9).

However, a liter is an arbitrary unit. If deciliters are used instead, the interval from 2 to 3 liters becomes 20 to 30 deciliters. Then the marginal cost of producing the third liter becomes $C(3) - C(2.9)/(3 - 2.9)$ because the interval runs from 29 to 30 deciliters. If centiliters are used, the marginal cost is $C(3) - C(2.99)/(3 - 2.99)$, while if milliliters are used, the marginal cost is $C(3) - C(2.999)/(3 - 2.999)$. This limit motivates the following definition.

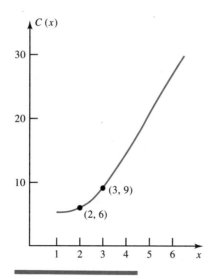

FIGURE 1

DEFINITION

If $C(x)$ is the cost of producing x units of a product, the **marginal cost** is $C'(x)$. The marginal cost of producing the nth unit is $C'(n)$.

EXAMPLE 5

Problem

What is the marginal cost of producing the third unit if the cost function is $C(x) = x^2 - 2x + 6$?

Solution The aim is to compute $C'(x)$ and then to substitute $x = 3$.

$$C'(x) = \lim_{h \to 0} \frac{[(x + h)^2 - 2(x + h) + 6] - (x^2 - 2x + 6)}{h}$$

$$= \lim_{h \to 0} \frac{x^2 + 2xh + h^2 - 2x - 2h + 6 - x^2 + 2x - 6}{h}$$

$$= \lim_{h \to 0} \frac{2xh + h^2 - 2h}{h} = \lim_{h \to 0} \frac{h(2x + h - 2)}{h}$$

$$= \lim_{h \to 0} (2x + h - 2)$$

$$= 2x - 2$$

When $x = 3$, $C'(x) = 4$ so the marginal cost of producing the third item is \$4.

EXERCISE SET 2–3

In Problems 1 to 10 compute the derivative of the function.

1. $f(x) = x^2 + 3$
2. $f(x) = x^2 - 5$
3. $f(x) = \sqrt{2x}$
4. $f(x) = -\sqrt{x}$
5. $f(x) = 1 - \sqrt{x}$
6. $f(x) = 1 + \sqrt{x}$
7. $f(x) = 1/x$
8. $f(x) = 3/x$
9. $f(x) = 1 + 1/x$
10. $f(x) = 1 - 1/x$

In Problems 11 to 14 find the equation of the tangent line to the function at the stated value of x.

11. $f(x) = x^2 - 2x + 6$ at $x = 0$
12. $f(x) = x^2 - 2x + 6$ at $x = 1$
13. $f(x) = x^2 + x - 5$ at $x = -1$
14. $f(x) = x^2 + x - 5$ at $x = 1$

In Problems 15 to 18 the height of a ball thrown from ground level into the air is given by $d(t)$, where d is measured in feet and t in seconds. Find (a) the amount of time it takes for the ball to reach its maximum height and (b) the maximum height.

15. $d(t) = -16t^2 + 192t$
16. $d(t) = -16t^2 + 48t$
17. $d(t) = 70t - 16t^2$
18. $d(t) = 96t - 16t^2$

In Problems 19 to 22 determine the marginal cost of producing the third item for the given cost function $C(x)$.

19. $C(x) = x^2 + x + 2$
20. $C(x) = x^2 - x + 5$
21. $C(x) = x^2 - 1$
22. $C(x) = x^2 + 4$

In Problems 23 to 26 compute the derivative of the function.

23. $f(x) = 1/\sqrt{x}$
24. $f(x) = x^3$
25. $f(x) = x^{-2}$
26. $f(x) = x^{-3}$

In Problems 27 to 32 find the equation of the tangent line to the function at the stated value of x.

27. $f(x) = 1/x$ at $x = 2$
28. $f(x) = 6/x$ at $x = -2$
29. $f(x) = 1/x^2$ at $x = -2$
30. $f(x) = \sqrt{x}$ at $x = 4$

31. $f(x) = 3x^2 - 5$ at $x = 0$
32. $f(x) = 4 - 2x^2$ at $x = 1$

In Problems 33 and 34 the height of a ball thrown from the top of a 64-foot building into the air is given by $d(t)$, where d is measured in feet and t in seconds. Find (a) the amount of time it takes for the ball to reach its maximum height and (b) the maximum height.

33. $d(t) = -16t^2 + 128t + 64$
34. $d(t) = -16t^2 + 32t + 64$

35. For how many seconds will the ball in Problem 15 be in the air?

36. For how many seconds will the ball in Problem 18 be in the air?

In Problems 37 and 38 let a, b, and c be real numbers. Find the derivatives of the stated function.

37. $f(x) = ax^2 + bx + c$
38. $f(x) = ax + b$

39. What is the marginal cost of producing the third unit if the cost function is $C(x) = x^2 - 3x + 20$?

40. What is the marginal cost of producing the fourth unit if the cost function is $C(x) = x^2 - 4x + 30$?

Referenced Exercise Set 2–3

In the context of income taxes, the *marginal tax rate* is the derivative of the tax function as a percent. Also, *tax brackets* are intervals of income where the derivative is constant. Problems 1 and 2 refer to the following tax function $T(x)$, where x represents a family's income in dollars.*

$T(x) =$

$$\begin{cases} .15x & \text{if } 0 \le x \le 29{,}750 \\ .28(x - 29{,}750) + 4463 & \text{if } 29{,}750 < x \le 71{,}900 \\ .33(x - 71{,}900) + 16{,}265 & \text{if } 71{,}900 < x \le 227{,}250 \\ .28(x - 227{,}250) + 67{,}531 & \text{if } x > 227{,}250 \end{cases}$$

1. What is the marginal tax rate for a family whose income is
   (a) $25,000
   (b) $50,000
   (c) $100,000
   (d) $500,000

*Alan Murray, "Individuals' Top Rate Would Plunge to 28 Percent; Tax-Break Curbs Offset Benefits to Wealthy," *Wall Street Journal*, August 18, 1986, p. 6.

2. What interval (or intervals) represents each of these tax brackets?
   (a) 15 percent   (b) 28 percent
   (c) 33 percent

The authoritative *Automobile Magazine* declared the Porsche 959 sports car to be the car of the future.† Problems 3 and 4 refer to Figure 2, which gives some indication of the Porsche's astounding power and technology by listing the units for power on the left and the units for torque on the right.

3. (a) At what rpm is the derivative of the power function equal to 0?
   (b) What is the power at that point?

4. (a) At what rpm is the derivative of the torque function equal to 0?
   (b) What is the torque at that point?

5. Figure 3 shows the rate of change of certain economic indicators for 1987. Let $f(x)$ denote the function defined by the graph.
   (a) Interpret the meaning of the expression "percent change" in mathematical terms.
   (b) What month had the highest value of the derivative $f'(x)$?
   (c) What month had the lowest value of the derivative $f'(x)$?

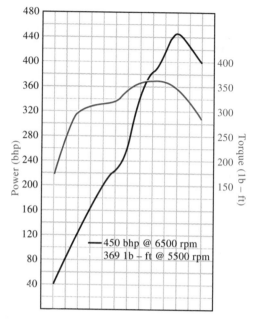

RPM x 1000

FIGURE 2

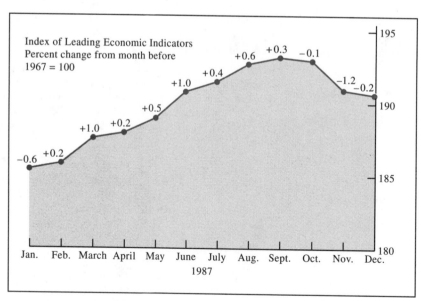

FIGURE 3 *Source:* Commerce Department.

†Mel Nichols, "Review: Porsche 959," *Automobile Magazine*, November 1987, pp. 60–71.

Cumulative Exercise Set 2–3

1. Compute the instantaneous rate of change of $f(x) = x^2 + 1$ at $x = 2$.

2. Compute the slope-intercept form of the tangent line to the graph of $f(x) = 2x^2 - 1$ at $x = 1$.

3. Compute the average rate of change of $f(x) = x^2 + 1$ from 2 to 2.1 and from 2 to 2.01.

In Problems 4 to 7 find $\lim_{x \to a} f(x)$ for the given function $f(x)$ and $x = a$.

4. $f(x) = \dfrac{x^2 - 16}{x - 4}$, $a = 4$

5. $f(x) = \dfrac{x^2 - 16}{x + 4}$, $a = 4$

6. $f(x) = \dfrac{x^3 - 4x}{2 - x}$, $a = 2$

7. $f(x) = \dfrac{x - 4}{x - 1}$, $a = 1$

8. Compute the slope-intercept form of the tangent line to the graph of $f(x) = x^2 + x - 1$ at $x = 1$.

9. The height of a ball thrown from ground level into the air is given by $d(t) = -16t^2 + 64t$, where d is measured in feet and t in seconds. Find the amount of time it takes for the ball to reach its maximum height.

In Problems 10 to 12 compute the derivative of the function.

10. $f(x) = x^3$

11. $f(x) = x^{-3}$

12. $f(x) = \dfrac{4}{x}$

2–4 Continuity and Differentiability

This section introduces continuous functions by geometric and algebraic means. It also considers differentiable functions and relates them to continuous functions.

Geometric Approach

From a geometrical standpoint, a *function* is *continuous* if a person's pencil need not leave the paper when the graph of the function is being sketched. Consider the graphs of the six functions shown in Figure 1. The graph in part (a) can be sketched without the pencil leaving the paper, so the function $f(x) = 5 - x^2$ is continuous. Similarly, $f(x) = x^3$ is a continuous function as shown in part (b).

However, the other four graphs require the pencil to be lifted from the paper, so they are discontinuous at each value of x where the pencil is lifted. For instance, the function

$$f(x) = \frac{(x - 1)(x + 1)}{x - 1}$$

in part (c) is discontinuous at $x = 1$ because the pencil must be lifted to jump over the hole in the graph at $x = 1$. The hole is indicated by an open circle. Points of discontinuity can also occur at gaps like those in parts (d) and (e) and at jumps like those in part (f).

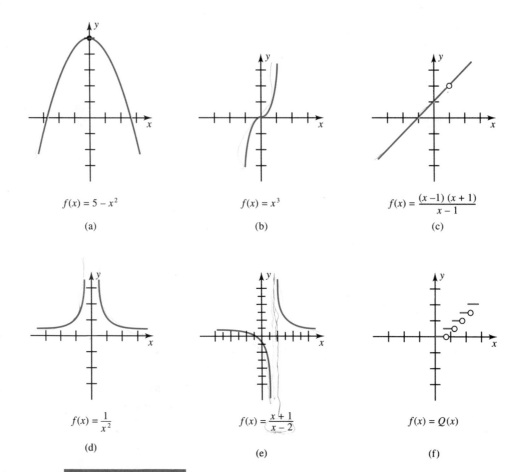

$f(x) = 5 - x^2$

(a)

$f(x) = x^3$

(b)

$f(x) = \dfrac{(x-1)(x+1)}{x-1}$

(c)

$f(x) = \dfrac{1}{x^2}$

(d)

$f(x) = \dfrac{x+1}{x-2}$

(e)

$f(x) = Q(x)$

(f)

FIGURE 1

Example 1 examines the continuity of functions defined by different expressions on different intervals.

EXAMPLE 1

Problem

Determine whether $f(x)$ is a continuous function, where

$$f(x) = \begin{cases} 5 - x^2 & \text{if } x < 2 \\ x - 1 & \text{if } x \geq 2 \end{cases}$$

Solution The graph of $f(x)$ is sketched in Figure 2. It consists of two sides; the left side is part of a parabola and the right side is part of a line. Because the two sides meet at the point (2, 1) the graph can be sketched without the pencil leaving the paper. Therefore $f(x)$ is a continuous function.

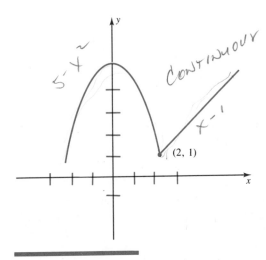

Handwritten annotations on figure: ~5·x, CONTINUOUS, x−1, (2, 1)

FIGURE 2

Algebraic Approach

The geometric approach to continuous functions leads to the following algebraic definition.

DEFINITION

A function $f(x)$ is **continuous at a value** $x = a$ if

Handwritten: WHEN →

1. $f(x)$ is defined at $x = a$,   *FIND $f(a)$ x i.y VALUE*
2. $\lim\limits_{x \to a} f(x)$ exists, and
3. $\lim\limits_{x \to a} f(x) = f(a)$.   *limit whatever the y value*

Handwritten: limit is y value

If any one of these conditions is not satisfied then $f(x)$ is a **discontinuous function** at $x = a$. $f(x)$ is **continuous on an interval** if $f(x)$ is continuous at every point in that interval. $f(x)$ is a *continuous function* if $f(x)$ is continuous at every real number.

Although the points of discontinuity of a function can easily be found from its graph, general statements can be made about some special kinds of functions without appealing to their graphs.

1. Every polynomial function is continuous.
2. Every rational function $\dfrac{f(x)}{g(x)}$ is continuous at all values of x for which $g(x) \neq 0$.

Parts (a) and (b) in Figure 1 illustrate the first statement, while parts (c), (d), and (e) illustrate the second statement. Example 2 shows how to determine points of continuity of a function without sketching its graph.

EXAMPLE 2

Problem

Determine all points where the function $f(x)$ is discontinuous:

$$f(x) = \frac{x + 2}{x^2 + 2x}$$

Solution $f(x)$ is a rational function so its continuity can be determined without sketching the graph. According to the second statement in the display, the points of discontinuity occur where $x^2 + 2x = 0$. Since $x^2 + 2x = x(x + 2)$, $f(x)$ is discontinuous at $x = 0$ and at $x = -2$.

Notice that the discontinuity of the rational function $f(x)$ in Example 2 was determined without sketching its graph. The next example shows how both the geometric and algebraic approaches can be utilized in certain cases.

EXAMPLE 3

Problem

Determine where the function $f(x) = \dfrac{x}{|x|}$ is continuous.

Solution The graph of $f(x)$, sketched in Figure 3, shows that $f(x)$ is discontinuous at $x = 0$ and is continuous at every other value of x.

Another way to proceed is to write $f(x)$ as a function defined in different ways on different intervals. Since the absolute value function $|x|$ is equal to x for

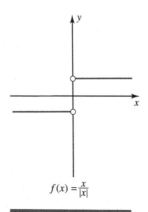

$f(x) = \frac{x}{|x|}$

FIGURE 3

$x > 0$, the function $f(x)$ is equal to x/x for $x > 0$. Similarly $|x|$ is equal to $-x$ for $x < 0$, so $f(x)$ is equal to $-x/x$ for $x < 0$. Therefore

$$f(x) = \begin{cases} 1 \text{ if } x > 0 \\ -1 \text{ if } x < 0 \end{cases}$$

Clearly $f(x)$ is continuous for all $x < 0$ and for all $x > 0$. However, when x is negative, $f(x) = -1$, but when x is positive, $f(x) = 1$. So there is a jump from one side of $x = 0$ to the other. This means that $f(x)$ is discontinuous at $x = 0$; $f(x)$ is continuous at all other values of x.

The discontinuity of $f(x) = \dfrac{x}{|x|}$ at $x = 0$ is sometimes called a "gap discontinuity" because it cannot be removed by defining $f(x)$ at $x = 0$. A contrasting type of discontinuity is a "hole discontinuity," at which the point of discontinuity can be removed. For instance, consider the function

$$g(x) = \frac{x^2 - 9}{x - 3}$$

By factoring the numerator and canceling common terms, $g(x)$ is equal to the function $x + 3$ for all values of x except $x = 3$. The graph of $g(x)$ is shown in Figure 4, where the open circle depicts the hole in the curve at $x = 3$. This discontinuity can be removed by defining a new function $h(x)$ such that $h(x) = g(x)$ for all $x \neq 3$ and $h(3) = 6$.

Differentiable Functions

A **function** $f(x)$ is **differentiable** at a value $x = a$ if the derivative exists at $x = a$, that is, if $f'(a)$ exists. The function is **differentiable on an interval** if $f'(x)$

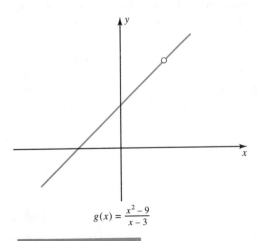

$$g(x) = \frac{x^2 - 9}{x - 3}$$

FIGURE 4

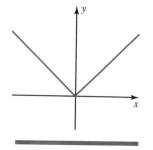

FIGURE 5

exists at all numbers in that interval, and $f(x)$ is differentiable if $f'(x)$ exists at all real numbers.

Upon first glance the definition of a differentiable function might not seem to carry any geometric connotation. However, the graph of a differentiable function can be viewed as a smooth curve. Historically there was no need to distinguish between continuous functions and differentiable functions until the difference became vital in applications during the first part of the 19th century, when the following fact was proved.

If a function is differentiable then it is also continuous.

The connection between differentiable and continuous functions helps us list some ways in which a function $f(x)$ can fail to be differentiable, because if $f(x)$ is not continuous at a point then it is not differentiable there either. Figure 1 thus shows four ways in which functions can fail to be differentiable.

Figure 5 shows another example of a way in which a function can fail to be differentiable. This function also reveals that the converse of the statement above does not hold for all functions, a result worthy of being singled out.

A continuous function need not be differentiable.

The classic example illustrating this fact is the absolute value function $f(x) = |x|$, which is continuous at all values of x but which is not differentiable at $x = 0$. Figure 5 shows that the graph can be sketched without the pencil leaving the paper, so it is continuous, but it is not a smooth curve since it has a sharp corner at $x = 0$, so it is not differentiable there. In other words, if you consider all values of x to be less than zero, the tangent line to this curve is $y = -x$, so the slope of the tangent line is -1. But the tangent line is $y = x$ for all values of x greater than 0, so the slope of the tangent line is 1. What is the slope of the tangent line at $x = 0$? It looks like it should be -1 and it should be 1. Since we don't get a unique value we say the slope of the tangent line, and hence the derivative, at $x = 0$ is undefined.

Example 4 uses the criterion of smoothness to determine where a function is differentiable.

EXAMPLE 4

Problem

Determine whether the function $f(x) = \sqrt[3]{x^2}$ is differentiable.

Solution Use the plotting method to sketch the graph of $f(x)$ by first squaring the value of x and then taking the cube root of that value. Thus $f(-8)$ is obtained by squaring -8 to get 64, then taking the cube root of 64 to get 4.

x	-8	-1	0	1	8
$f(x)$	4	1	0	1	4

The graph of $f(x)$ is sketched in Figure 6. It is not smooth at $x = 0$ so $f(x)$ is not differentiable there. However, the graph is smooth at all other values of x, so $f(x)$ is differentiable at all points except $x = 0$.

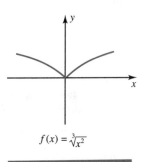

$$f(x) = \sqrt[3]{x^2}$$

FIGURE 6

Example 5 used a geometric way to test for differentiability at the value $x = 0$. There is an algebraic test for this too, but it involves one-sided derivatives, which are not considered in this book.

EXERCISE SET 2–4

The figures in Problems 1 to 10 show the graphs of ten functions. Determine whether each function is continuous.

1.

2.

3.

4.

5.

6.

7.

8.

9.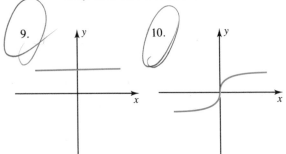

10.

In Problems 11 to 14 determine whether the function $f(x)$ is continuous by sketching its graph.

11. $f(x) = \begin{cases} x^2 & \text{if } x \geq 0 \\ -x & \text{if } x < 0 \end{cases}$

12. $f(x) = \begin{cases} x^3 & \text{if } x \geq 0 \\ -x^2 & \text{if } x < 0 \end{cases}$

13. $f(x) = \begin{cases} x^2 - 5 & \text{if } x \geq 2 \\ x - 1 & \text{if } x < 2 \end{cases}$

14. $f(x) = \begin{cases} x + 1 & \text{if } x \geq 1 \\ -x & \text{if } x < 1 \end{cases}$

In Problems 15 to 24 determine all points where the function $f(x)$ is discontinuous.

15. $f(x) = \dfrac{x - 1}{x^2 - x}$

16. $f(x) = \dfrac{2 - x}{x^2 - 2x}$

17. $f(x) = \dfrac{1}{x^2 - 16}$

18. $f(x) = \dfrac{1}{(x - 2)(x - 3)}$

19. $f(x) = \dfrac{x + 4}{(x + 4)(x - 2)}$

20. $f(x) = \dfrac{x + 2}{x^2 - x - 6}$

21. $f(x) = \dfrac{5}{|x|}$

22. $f(x) = \dfrac{|x|}{x}$

23. $f(x) = \dfrac{x - 3}{x^2 - 3x}$

24. $f(x) = \dfrac{x^2 + 3x}{x + 3}$

In Problems 25 to 34 determine whether the function whose graph is sketched in the stated problem is differentiable.

25. Problem 1

26. Problem 2

27. Problem 3

28. Problem 4

29. Problem 5

30. Problem 6

31. Problem 7

32. Problem 8

33. Problem 9

34. Problem 10

In Problems 35 to 40 determine whether the function $f(x)$ is differentiable by sketching its graph.

35. $f(x) = \sqrt[3]{(1 - x)^2}$

36. $f(x) = \sqrt[3]{(-x)^2}$

37. $f(x) = \sqrt[3]{x^2 - 2x + 1}$

38. $f(x) = \sqrt[3]{4 - 4x + x^2}$

39. $f(x) = 1 - |x|$

40. $f(x) = |1 - x|$

In Problems 41 to 44 determine if the given function is continuous at $x = 2$.

41. $f(x) = x - 2$

42. $f(x) = x^2 - 2$

43. $f(x) = \dfrac{1}{x - 2}$

44. $f(x) = \dfrac{x - 2}{x^2 - 4}$

In Problems 45 to 48 draw the graph of the function and determine whether it is continuous or differentiable.

45. $f(x) = \sqrt{|x|}$

46. $f(x) = \begin{cases} x + 6 & \text{if } x \geq 3 \\ x^2 & \text{if } x < 3 \end{cases}$

47. $f(x) = \dfrac{x^2 - 1}{x - 1}$

48. $f(x) = \begin{cases} x^2 & \text{if } x \leq 0 \\ -x^2 & \text{if } x > 0 \end{cases}$

In Problems 49 to 54 determine whether $f(x)$ is (a) continuous at $x = 2$ and (b) differentiable at $x = 2$, from the given figure.

49.

50.

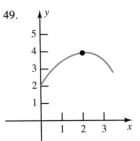

51.

52.

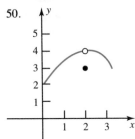

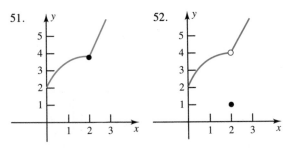

53.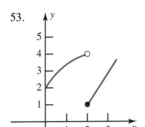

54.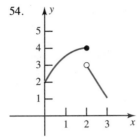

Cumulative Exercise Set 2–4

1. What is the instantaneous rate of change of $f(x) = 1 - x$ at $x = -1$?

2. Find the slope-intercept form of the line tangent to the graph of $f(x) = 1 - x^2$ at $x = -1$.

In Problems 3 and 4 evaluate each limit, if it exists.

3. $\displaystyle\lim_{x \to 1} \frac{x^2 - 3x + 2}{x^2 + x - 2}$

4. $\displaystyle\lim_{x \to -1} \left(1 + \frac{1}{x}\right)$

In Problems 5 and 6 compute the derivative of the function.

5. $f(x) = 3x^2$

6. $f(x) = x^3$

In Problems 7 and 8 find the equation of the line tangent to the graph of $f(x)$ at the stated value of x.

7. $f(x) = 3x^2$ at $x = -1$

8. $f(x) = x^3$ at $x = -1$

9. Determine whether the function whose graph is shown is continuous or differentiable.

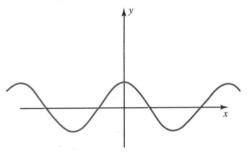

10. Determine whether the function $f(x)$ is continuous by sketching its graph.

$$f(x) = \begin{cases} 2x & \text{if } x \geq 0 \\ x & \text{if } x < 0 \end{cases}$$

11. Determine all points where the function $f(x) = 1 - \dfrac{1}{x}$ is discontinuous.

12. Determine whether the function $f(x) = 1 + \sqrt[3]{x^2}$ is differentiable.

CHAPTER REVIEW

Key Terms

2–1   Rate of Change and Tangents
Velocity
Rate of Change
Average Rate of Change
Instantaneous Rate of Change
Secant Line

Tangent Line
Slope of a Function
Difference Quotient
Demand
Marginal Demand

2–2   Limits
Limit
Limit Does Not Exist
Decreases Without Bound

Increases Without Bound
Limit at Infinity
Horizontal Asymptote

2–3   Definition of the Derivative
Derivative of a Function
Marginal Cost

2–4 Continuity

Continuous Function

Continuous at a Value

Discontinuous Function

Differentiable Function

Differentiable at a Value

Continuous on an Interval

Summary of Important Concepts

Formula

$$f'(x) = \lim_{h \to 0} \frac{f(x + h) - f(x)}{h}$$

Interpretations of the derivative

1. Instantaneous rate of change.
2. Tangent line.
3. Velocity of a moving object.

Steps in computing the derivative of $f(x)$

1. Form the difference quotient $\dfrac{f(x + h) - f(x)}{h}$.

2. Perform algebraic manipulations to factor h from the numerator of the difference quotient.
3. Cancel the h factors.
4. Take the limit as h approaches 0.

REVIEW PROBLEMS

1. Compute the average rate of change of the function $f(x) = x^2 + 1$ over the stated intervals.
   (a) from 1 to 2   (b) from 1 to 1.1

2. Compute the instantaneous rate of change of the function $f(x) = 2x^2 + 1$ at $x = -1$.

3. Find the slope of the tangent line to the graph of the function $f(x) = x^2 - x + 4$ at an arbitrary value of x.

4. Compute the limit by constructing a table.

   $$\lim_{x \to 3} f(x) \quad \text{for } f(x) = \frac{x^2 - x - 6}{3 - x}$$

In Problems 5 and 6 evaluate the limit, if it exists, by algebraic methods.

5. $\displaystyle\lim_{x \to 1} \frac{2x^2 - 2x}{x - 1}$

6. $\displaystyle\lim_{x \to 4} \frac{2 - \sqrt{x}}{4 - x}$

In Problems 7 to 10 compute each limit, if it exists, by any method.

7. $\displaystyle\lim_{x \to 1} \frac{x^2 + 2x - 3}{x^2 + x - 2}$

8. $\displaystyle\lim_{x \to 0} \frac{1}{x}$

9. $\displaystyle\lim_{x \to 0} x^2 + x$

10. $\displaystyle\lim_{x \to 1} \frac{x - 1}{(x - 1)^2}$

In Problems 11 and 12 compute the derivative of the function from the definition of the derivative.

11. $f(x) = x^2 - 3x - 7$

12. $f(x) = \dfrac{1}{x - 2}$

13. Find the equation of the tangent line to the function $f(x) = x^2 - 3x - 7$ at the value $x = 4$.

14. The height of a ball thrown from the top of a 64-foot building into the air is given by $d(t) = -16t^2 + 16t + 64$, where d is measured in feet and t in seconds. Find
    (a) the amount of time it takes for the ball to reach its maximum height.
    (b) the maximum height.

GRAPHICS CALCULATOR EXPLORATIONS

Your graphics calculator can be of great help to visualize whether a function has a limit or whether it fails to have a limit. If the graph of a function is a continuous curve through a point (a,b), then the limit of the function at $x = a$ is b. The limit of the function at $x = a$ is also b if (a,b) is missing from the graph, but the graph would be a continuous curve through (a,b) if it were part of the graph. If the latter is the case, we will intuitively describe this by stating that the graph has a "hole" at $x = a$. An example of such a function is

$$f(x) = \frac{x^2 - 1}{x - 1}$$

For all values of x except $x = 1$ this function is identical to $y = x + 1$ but $f(x)$ is not defined at $x = 1$, so its graph has a hole at $(1,2)$. The limit at $x = 1$ is 2.

However, if a function has a "gap" at $x = a$, then the limit does not exist. Your graphics calculator can help explain the difference between a hole and a gap in the graph of a function. As an example define the piecewise function

$$f(x) = \begin{cases} 2x & \text{if} & x > 0 \\ x + 1 & \text{if} & x < 0 \end{cases}$$

Notice that the functional value at $x = 0$ is not defined. Graph this function on your calculator.

CASIO

Note: The simplest way to sketch the graph of a step function on the Casio calculator is to graph both functions and only consider the particular graph over its stated domain.

1. Press **GRAPH** 2 **ALPHA** **x** **EXE** .

2. Press **GRAPH** **ALPHA** **x** **+** 1 **EXE** .

TI-81

1. Press **Y =** .

2. Enter the expression $(X<0)$ $(X+1)$ $+$ $(X>0)$ $(2X)$.

   Press **ENTER** .

Note: The expression $(X<0)$ is entered by the following sequence of keystrokes:

(**X|T** **2nd** **TEST** 5   0)

3. Press **GRAPH** .

From the graph of this function you can see there is a gap at $x = 0$, meaning that the functional value at $x = 0$ looks like it *should* be two different values depending on whether you look at the part of the graph to the left or to the right of the line $x = 0$. That is, the functional values approach 0 when you look at that part of the function to the right of the y-axis, whereas they approach 1 when you look at that part of the function to the left of the y-axis.

Now change the function so the gap becomes smaller. Define

$$f(x) = \begin{cases} 2x & \text{if} & x > 0 \\ x + 0.1 & \text{if} & x < 0 \end{cases}$$

Sketch the graph of this function on your calculator. The two lines appear to almost meet at the origin. Now use the zoom feature to see that there is still a significant gap in the graph of the function at $x = 0$.

CASIO

1. Press $\boxed{\textbf{SHIFT}}$ $\boxed{\textbf{FACTOR}}$ $\boxed{5}$ $\boxed{:}$

   $\boxed{\textbf{GRAPH}}$ $\boxed{\textbf{ALPHA}}$ $\boxed{\textbf{x}}$ $\boxed{+}$ $\boxed{.1}$ $\boxed{\textbf{EXE}}$.

2. Press $\boxed{\textbf{GRAPH}}$ $\boxed{2}$ $\boxed{\textbf{ALPHA}}$ $\boxed{\textbf{x}}$ $\boxed{\textbf{EXE}}$.

TI-81

1. Press $\boxed{\textbf{Y =}}$.

2. Place the cursor on the 1 in X + 1.

   Press $\boxed{\textbf{INS}}$ $\boxed{.}$ $\boxed{\textbf{ENTER}}$.

3. Press $\boxed{\textbf{TRACE}}$.

4. Press $\boxed{\textbf{ZOOM}}$ $\boxed{\textbf{IN}}$ $\boxed{\textbf{ENTER}}$.

Once again, the limit does not exist. Now define a function with an even smaller gap.

$$f(x) = \begin{cases} 2x & \text{if} & x > 0 \\ x + 0.01 & \text{if} & x < 0 \end{cases}$$

To see that there is still a gap you will have to use the zoom feature several times. Is there a positive value of a that is small enough so that the graph of the following function does not have a gap?

$$f(x) = \begin{cases} 2x & \text{if} & x > 0 \\ x + a & \text{if} & x < 0 \end{cases}$$

No matter how small we choose a to be, if it is positive the zoom feature can be used to see that the graph has a gap at $x = 0$. In fact any value of a that is negative will also produce a gap in the graph at $x = 0$.

What happens when the value of a is chosen to be 0? The function then becomes

$$f(x) = \begin{cases} 2x & \text{if} & x > 0 \\ x & \text{if} & x < 0 \end{cases}$$

No matter how many times you use the zoom feature you will not see a gap in this function. Yet we know that the graph is not defined at $x = 0$, so it is not continuous at $x = 0$. We describe this situation graphically by putting an open circle at the point (0,0). By this we mean that if the point (0,0) were inserted into the graph it would be a continuous curve. This means that the limit of this function at $x = 0$ is 0.

Further Explorations

In Problems 1 to 4 write a program to evaluate the given function at values $h = 1, .1, .01,$ $.001,$ and $.0001.$ Use these values to find $\lim\limits_{h \to 0} f(h)$. Then graph $f(h)$ and approximate $f(0)$.

1. $f(h) = \dfrac{2h^2 - 5h + 1}{3h^2 + h - 1}$

2. $f(h) = \dfrac{3h^2 - 7h - 2}{5h^2 + 2h - 1}$

3. $f(h) = \dfrac{4h^3 - 2h - 3}{5h^3 + 3h - 4}$

4. $f(h) = \dfrac{h^4 - h^2 - 3}{h^4 - h^2 + 3h - 4}$

Differentiation Rules

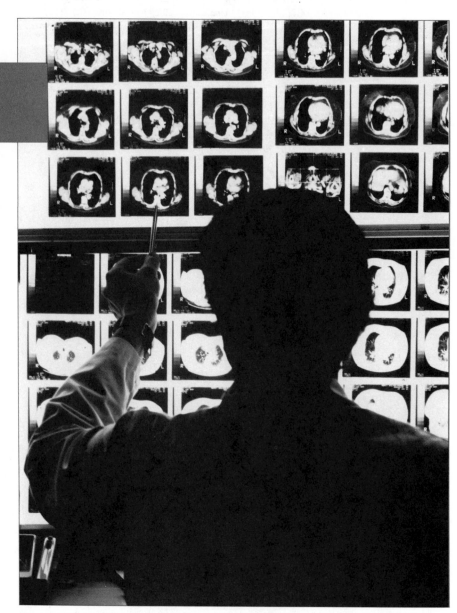

The case study discusses mathematical models and cancer therapy. (J. Nettis/H. Armstrong Roberts)

Chapter Overview

Calculus can be applied to a wide variety of real-world problems. Many examples will be given throughout the rest of the book. In each case the application will be described by a function. The problem is solved by analyzing the derivative of the function because the derivative gives information about the original function. Thus we will need to calculate the derivative of many types of functions. This chapter considers various rules for calculating derivatives.

The first section deals with elementary rules that govern functions that can be expressed as the sum of simpler functions or the product of a constant times a function. In the next section more complicated functions are handled, those that can be expressed as the product or quotient of other functions. In the third section the chain rule is presented. The chain rule governs composite functions. The fourth section deals with higher derivatives.

CASE STUDY PREVIEW

The primary treatments for cancer are surgery and radiotherapy. More than half of all tumors can be treated by radiation therapy. The treatment consists of several beams directed at various parts of the tumor. Each beam has a specific intensity depending on the mass of the tumor and the amount of healthy cells the beam will affect. There are many variables that affect the design of an individual patient's treatment. Medical researchers are using mathematical optimization techniques to maximize the amount of cancer cells killed while also minimizing the number of healthy cells affected. The case study shows how the derivative plays an essential role in balancing these two objectives.

3–1 A Few Elementary Rules

We have calculated the derivative of several functions using the definition of the derivative. The definition is cumbersome, so in this section some elementary differentiation rules are developed to simplify the task.

Marginal Product

Figure 1 shows a typical production function from economics. It measures the output of a firm at a given level of input. The input is usually labor or land. For example, the production function given by Samuelson* can be represented by the equation

$$f(x) = -0.16x^2 + 1.6x$$

Here $f(x)$ is the total product, or amount of output, of the firm with a labor force of size x, where x is measured in hundreds of workers and $f(x)$ is measured in units of $10,000. An important economic concept derived from the production function is **marginal product.** It is defined as the rate of increase of production per unit of

*Paul A. Samuelson, *Economics*, New York, McGraw-Hill Book Co., 1987.

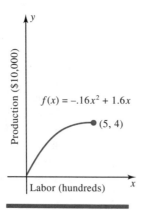

$f(x) = -.16x^2 + 1.6x$

(5, 4)

Production ($10,000)

Labor (hundreds)

FIGURE 1   *Source:* Paul A. Samuelson, *Economics,* New York, McGraw-Hill Book Co., 1987.

input. Thus marginal product is the derivative of the production function. In this section we study how to find the derivative of functions like $f(x)$ above. At the end of the section we show how marginal product is a key ingredient that economists use to determine a fair wage.

Other Notation for the Derivative

There are several notations for expressing the derivative of $y = f(x)$. The three most common notations are

$$f'(x) \quad D_x f(x) \quad \frac{df}{dx}$$

The last notation can also be written as

$$\frac{dy}{dx} \quad \text{or} \quad \frac{d}{dx} f(x)$$

Each notation has an advantage in a particular situation so that no one notation is preferred over the others in every context. An advantage to the $f'(x)$ notation is that it is more convenient to express the derivative when evaluated at a particular value. For example, it is easier to write $f'(3)$ with this notation than the others. Using the $\frac{dy}{dx}$ and $D_x f(x)$ notations, we express $f'(3)$ as

$$D_x f(x)\big|_{x=3} \qquad \frac{d}{dx} f(x)\bigg|_{x=3}$$

The D_x notation is more convenient to use in formulas and when expressing the derivative of more complicated functions in terms of the derivative of simpler functions.

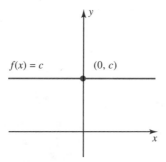

The constant function $f(x) = c$

FIGURE 2

Constant Functions

Consider the constant function $f(x) = c$. Its graph is a horizontal straight line. (See Figure 2.) The slope of such a line is 0. Since the derivative measures the slope of the tangent line and the tangent to a line is the line itself, **the derivative of a constant** function is 0. This can be remembered as "the derivative of a constant is 0."

Constant Rule
If $f(x) = c$ for c a real number, then $f'(x) = 0$.
This is also expressed as

$$D_x(c) = 0 \qquad \text{for any constant } c$$

This rule is actually a theorem whose proof is straightforward.

Proof of Constant Rule

If $f(x) = c$ for some real number c, then

$$f'(x) = \lim_{h \to 0} \frac{f(x + h) - f(x)}{h} = \lim_{h \to 0} \frac{c - c}{h}$$

$$= \lim_{h \to 0} \frac{0}{h} = 0$$

Hence $f'(x) = 0$ for every real number x.

Power Rule

The following derivatives were calculated in the previous chapter:

$$D_x(x) = 1 \quad D_x(x^2) = 2x \quad D_x(x^3) = 3x^2 \quad D_x(x^{1/2}) = \tfrac{1}{2}x^{-1/2}$$

Each of these functions has the form $y = x^r$, where r is 1, 2, 3, and $\tfrac{1}{2}$, respectively. There is a pattern to the derivatives. It suggests an important rule called the *power rule*.

Power Rule

$$D_x(x^r) = rx^{r-1} \qquad \text{for any real number } r$$

The exercises indicate how to prove this theorem for r a positive integer. The proof for other real numbers is beyond the scope of this text. However, we emphasize that the power rule works for all real numbers as illustrated by Example 1.

E X A M P L E   1

Problem

Find $f'(x)$ for (a) $f(x) = x^{-5}$, (b) $f(x) = \dfrac{1}{\sqrt{x}}$.

Solution (a) Use the power rule with $r = -5$.

$$f'(x) = D_x[x^{-5}] = -5x^{-5-1} = -5x^{-6}$$

(b) Express the function as $f(x) = x^{-1/2}$. Then use the power rule with $r = -\tfrac{1}{2}$.

$$f'(x) = D_x[x^{-1/2}] = (-\tfrac{1}{2})x^{(-1/2)-1} = (-\tfrac{1}{2})x^{-3/2}$$

Constant Times a Function

From Section 2–3 we compute $D_x[4x^2] = 8x = 4 \cdot 2x = 4 \cdot D_x[x^2]$. The general rule that governs this derivative is "the **derivative of a constant times a function is the constant times the derivative of the function**."

Constant Times a Function Rule

$$D_x[cf(x)] = cD_x[f(x)] \qquad \text{for any real number } c$$

Proof of the Constant Times a Function Rule

If $g(x) = cf(x)$ for some constant c, then

$$g'(x) = \lim_{h \to 0} \frac{g(x + h) - g(x)}{h} = \lim_{h \to 0} \frac{cf(x + h) - cf(x)}{h}$$

$$= \lim_{h \to 0} \frac{c[f(x + h) - f(x)]}{h} = \lim_{h \to 0} c \frac{f(x + h) - f(x)}{h}$$

$$= c \lim_{h \to 0} \frac{f(x + h) - f(x)}{h}$$

$$= cf'(x)$$

EXAMPLE 2

Problem

Find $f'(x)$ for (a) $f(x) = -3x$, (b) $f(x) = 2x^5$.

Solution (a) Use the constant times a function rule with $c = -3$ and $f(x) = x$.

$$f'(x) = D_x[-3x] = -3D_x[x] = -3 \cdot 1 = -3$$

(b)

$$f'(x) = D_x[2x^5] = 2D_x[x^5] = 2 \cdot 5x^4 = 10x^4$$

Derivative of a Sum

The next formula states that "the **derivative of a sum** is the sum of the derivatives." The rule states that this is true for the sum of two functions, but it easily extends to the sum of any finite number of functions.

Sum Rule

DO DERIVATIVE OF EACH TERM &

ADD TOGETHER

$$D_x[f(x) + g(x)] = D_x[f(x)] + D_x[g(x)]$$

Since the difference of two functions, $f(x) - g(x)$, can be regarded as the sum, $f(x) + [-g(x)]$, the derivative of the difference of two functions is the difference of their derivatives.

The exercises will indicate how to prove the sum rule.

EXAMPLE 3

Problem

Find $f'(x)$ for (a) $f(x) = 4x - 3/x$, (b) $f(x) = 8 + 2x^{10} + 3x^{-5} - 7x^{5/3}$.

Solution (a) Recall that $3/x = 3x^{-1}$.

$$f'(x) = D_x(4x - 3x^{-1}) = D_x(4x) + D_x(-3x^{-1})$$

$$= 4D_x(x) - 3D_x(x^{-1}) = 4\cdot 1 - 3(-1)x^{-2}$$

$$= 4 + \frac{3}{x^2}$$

(b)

$$f'(x) = D_x(8 + 2x^{10} + 3x^{-5} - 7x^{5/3})$$

$$= D_x(8) + D_x(2x^{10}) + D_x(3x^{-5}) + D_x(-7x^{5/3})$$

$$= 0 + 2\cdot 10x^9 + 3(-5)x^{-6} - 7\left(\frac{5}{3}\right)x^{(5/3 - 1)}$$

$$= 20x^9 - 15x^{-6} - \left(\frac{35}{3}\right)x^{2/3}$$

The next example computes the equation of the tangent line to a curve.

EXAMPLE 4

Problem

Find an equation of the tangent line to the graph of the function $f(x) = x^2 - 5x + 3$ at $x = 1$, that is, at the point $(1, -1)$.

Solution The slope of the tangent line is measured by $f'(1)$. Since $f'(x) = 2x - 5$, $f'(1) = -3$. Using the point-slope form of the equation of the tangent line yields

$$y - (-1) = -3(x - 1)$$

Simplifying gives the equation $y = -3x + 2$.

Applications

The derivative measures instantaneous rate of change. If an object is dropped so that gravity is the only force acting on it, then the relationship between the distance traveled, d, and time, t, is $d(t) = -16t^2$, where d is measured in feet and t in seconds. A negative sign is in the expression because the object is falling and hence is moving in the negative direction. The average velocity is the average rate of change of $d(t)$ with respect to t. The instantaneous velocity measures how fast the object is falling at that instant. It is measured by $d'(t)$. Thus the instantaneous velocity is $v(t) = d'(t) = -32t$. Similarly the instantaneous acceleration of the object is the instantaneous rate of change of its velocity with respect to t. Thus the object's instantaneous acceleration is $a(t) = v'(t) = -32$.

EXAMPLE 5

An object is thrown upward and its position is given by $d(t) = -16t^2 + 64t$.

Problem

Find (a) the velocity and acceleration of the object, (b) the time at which it reaches its maximum height, (c) how many feet it travels upward.

Solution (a) The velocity is $v(t) = d'(t) = -32t + 64$ and the acceleration is $a(t) = v'(t) = -32$.
(b) The object will reach its maximum height when $v(t) = 0$ because it stops at that instant. Letting $v(t) = 0$ yields $-32t - 64 = 0$, which implies $t = 2$ seconds.
(c) The object travels upward until it stops, at which time it starts downward. Thus it travels upward from $t = 0$ to $t = 2$ seconds. Hence the object travels $d(2)$ feet upward.

$$d(2) = -16(2)^2 + 64(2) = 64$$

The object travels 64 feet upward.

Marginal Product Revisited

The production function from Samuelson given at the beginning of the section is

$$f(x) = -0.16x^2 + 1.6x$$

The marginal product of the firm is the derivative $f'(x)$.

$$f'(x) = -0.32x + 1.6$$

The marginal product measures the rate of change of output as input changes. It makes no sense to let x take on values less than 0, and it is assumed that the largest number of labor units available to the firm is $x = 5$. This function is valid only for values of x in the interval [0, 5]. It is a quadratic function whose graph is given in Figure 1. It is typical of production functions because as x increases $f'(x)$ decreases. For example, $f'(0) = 1.6$ and $f'(1) = 1.28$. Another way to see this is to note that the marginal product is a linear function with a negative slope. Thus as x increases, $f'(x)$ decreases. Economists refer to this as the "law of diminishing returns." It means that as more labor is used, more output is generated, but the rate of increase of output, the marginal product, decreases.

Economists use marginal product to determine fair wages. The fundamental principle is that *wages = marginal product*. This assumption is derived from the fact that if an extra worker were hired, the worker would increase the firm's productivity only by an amount equal to that worker's marginal product. For instance, consider a particular segment of the work force, so that all of these workers are paid the same wage. If there are 100 workers, thus $x = 1$, then the last one hired, the 100th worker hired, would increase the firm's productivity by the marginal product $f'(1)$. Therefore, in theory, all workers would receive the same fair wage, $f'(1)$, the marginal product of the last worker hired.

EXAMPLE 6

Consider the firm with the production function

$$p(x) = -0.16x^2 + 1.6x$$

Suppose the firm has 100 employees with the same wage.

Problem

Find the theoretical fair wage of the employees.

Solution The theoretical fair wage is the marginal product at $x = 1$ (hundred employees). It is $p'(1)$ where $p'(x) = -0.32x + 1.6$. Thus $p'(1) = -0.32(1) + 1.6 = 1.28$. Each worker is paid $12,800.

Are workers always paid their marginal product? Economist Robert H. Frank answers this question in an article appearing in the *American Economic Review*. His conclusion is that some workers, such as salespersons, do command wages equal to their marginal product, but others, such as college professors, are paid less. Referenced Exercise 3 deals with some of the data from this study.

EXERCISE SET 3–1

In Problems 1 to 26 find $f'(x)$.

1. $f(x) = x^3 + 5$

2. $f(x) = x^6 - 2$

3. $f(x) = 2x^4 - 7$

4. $f(x) = 3x^5 + 1$

5. $f(x) = 5x^{-3} + 5x$

6. $f(x) = 4x^{-5} - x^3 + 1$

7. $f(x) = 2x^{1/3} + 3x^{1/2} - 4x$

8. $f(x) = 7x^{1/5} - x^{2/5} + 2x$

9. $f(x) = 2x^{-1/3} + 7x^{-1/4} - 5$

10. $f(x) = 6x^{-1/6} - 3x^{-3/7} - 6x$

11. $f(x) = x^{0.3} + 3x^{0.4}$

12. $f(x) = 1.2x^{-1.5} - 3.1x^{-2.2}$

13. $f(x) = 1/x^3 + 3x^4$

14. $f(x) = 4x^{0.5/5} - 3x^{2.2/7}$

15. $f(x) = 1/x^{1.5} + 4/x^{1/4} + 4x^{2.4}$

16. $f(x) = 4/x^{2.5} - 2/x^{2/5} - 2x^{2/5}$

17. $f(x) = 2/x^{-3.5} + 4/x^{1.4} + 4x^{-1.3}$

18. $f(x) = 1/x^{-2.5} - 2x^{-3/5} - x^{-2.7/3}$

19. $f(x) = 0.25/x^{-0.5} + 0.48/x^{0.4}$

20. $f(x) = 1.5x^{-0.5} - 2.75x^{-0.05}$

21. $f(x) = 2\sqrt{x} - 2x$

22. $f(x) = \sqrt{x}/9 + 9/x$

23. $f(x) = 5\sqrt{x} - 2/\sqrt{x}$

24. $f(x) = \sqrt{9/x} + 2/\sqrt{7x}$

25. $f(x) = 18/\sqrt{3x} + 2/\sqrt[3]{4x}$

26. $f(x) = 8/\sqrt[4]{3x} + \frac{2}{3}\sqrt[4]{x}$

In Problems 27 to 30 find the equation of the tangent line to $y = f(x)$ at the given value of x.

27. $f(x) = x^2 + 2x$ at $x = 1$

28. $f(x) = x^3 - 4x$ at $x = 2$

29. $f(x) = x^{-2} + 2x^2$ at $x = 1$

30. $f(x) = x^{-3/2} - x$ at $x = 1$

31. An object is thrown upward, and its position is given by $d(t) = -16t^2 + 32t$ with $d(t)$ measured in feet and t in seconds. Find (a) the velocity and acceleration of the object, (b) the time at which it reaches its maximum height, and (c) how many feet it travels upward.

32. An object is thrown upward, and its position is given by $d(t) = -16t^2 + 50t + 6$ with $d(t)$ measured in feet and t in seconds. Find (a) the velocity and accel-

eration of the object, (b) the time at which it reaches its maximum height, and (c) how many feet it travels upward.

33. A firm determines that the relationship of the quantity x that the public is willing to purchase at price p is given by

$$p(x) = 5000 - 300x^{-1}$$

Find the rate of change of $p(x)$ when $x = 5$.

34. The profit P in thousands of dollars from the sales volume of x records is found to be

$$P(x) = 0.2x^2 + 10x - 100$$

Find the marginal profit when $x = 10$.

35. The cost C in thousands of dollars to produce x million batteries is determined to be

$$C(x) = 10 + 20x - 0.1x^2$$

Find the marginal cost when $x = 50$.

36. Business analysts call a product a "fad" if its total sales increase rapidly and then level off after a short period of time. Let S represent total sales in thousands of units and let x represent time in weeks. A function that represents the behavior of a fad is

$$S(x) = 100 - 50x^{-1}$$

Show that the rate of change of $S(x)$ with respect to x is decreasing as x increases by computing $S'(x)$ for (a) $x = 1$, (b) $x = 5$, and (c) $x = 10$.

37. The binomial theorem states that for any real numbers p and q and any positive integer n

$$(p + q)^n = p^n + np^{n-1}q + n(n - 1)/2p^{n-2}q^2$$

$$+ \cdots + npq^{n-1} + q^n$$

Use the binomial theorem to prove the power rule for positive integers.

38. Use the definition of the derivative to prove the sum rule.

Referenced Exercise Set 3–1

1. In an article illustrating the benefits of quality control to executives, A. W. Whitton, Jr.,* Senior Corporate Quality Consultant for Abbott Laboratories, uses Figure 3 to demonstrate how to reduce total quality cost. Locate point a on the graph of $y = TQC(x)$. It represents the lowest point on the TQC curve because its y-coordinate is the smallest of all points on the graph. The slope of the tangent line at a is 0. From the graph show that the following equations are true:

$$TQC(x) = FC(x) + AP(x) + PC(x)$$

$$TQC'(a) = FC'(a) + AP'(a) + PC'(a)$$

2. In the study of optics the subjective brightness a viewer senses is compared with the actual brightness from the source.† If x measures the actual brightness, then the subjective brightness, S, is a function of x. The functional relationship is usually described by a power function $S(x) = kx^n$, where k and n are constants depending on the light intensity and the type of light the viewer is coming from before the experiment. The *subjective sensitivity* is defined as the rate of change of the subjective brightness. Thus subjective sensitivity is

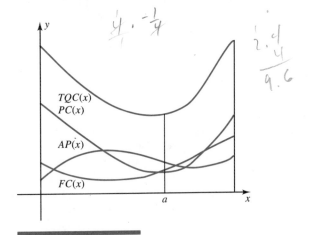

FIGURE 3 *Source:* A. W. Whitton, Jr., "Methods for Selling Total Quality Cost Systems," *American Society of Quality Control – Technical Conference Transactions,* 1972, pp. 365–371.

*A. W. Whitton, Jr., "Methods for Selling Total Quality Cost Systems," *American Society of Quality Control—Technical Conference Transactions,* 1972, pp. 365–371.

†Robert Sekuler et al., "Structural Modeling of Spatial Vision," *Vision Research,* Vol. 24, 1984, pp. 689–700, and Heywood M. Petry et al., "Spatial Contrast Sensitivity of the Tree Shrew," *Vision Research,* Vol. 24, 1984, pp. 1037–1042.

$S'(x)$. Suppose for those in relative darkness before an experiment that the subjective brightness is given by $S_1(x) = 0.001x^{1/4}$ and for those in light the subjective brightness $S_2(x) = 0.002x^{1/2}$. At what value of x are the subjective sensitivities equal?

3. In an article studying whether certain segments of the work force are paid their theoretical fair wage, defined by economists as the marginal product of the last worker hired, Robert H. Frank computes the following production function for real estate sales agents, where $f(x)$ represents the earnings for a sales force of size x.‡

$$f(x) = -0.024x^2 + 0.575x$$

Compute the marginal product function.

‡Robert H. Frank, "Are Workers Paid Their Marginal Products?" *American Economic Review*, Vol. 74, No. 4, 1984, pp. 549–571.

3–2 Product and Quotient Rules

The formulas developed in the previous section are used to compute the derivatives of polynomial functions and functions that are sums of terms of the form ax^r. Some functions do not fit one of these forms. Examples of functions whose derivatives cannot be found directly by these rules are

$$f(x) = x^2(5x + 11)$$

$$h(x) = \frac{x^3}{(4x + 1)}$$

We develop two formulas in this section. The first handles functions that are the product of simpler functions, like $f(x)$, and the second is applied to quotients, like $h(x)$.

The Product Rule

At first it might seem that the derivative of a product is the product of the derivatives, similar to the derivative of a sum. A quick inspection shows that this is not the case. Let $f(x) = x$ and $g(x) = x^2$. Then $f(x) \cdot g(x) = x^3$, so the derivative of the product is $3x^2$. But the product of the derivatives is $f'(x) \cdot g'(x) = 1 \cdot 2x = 2x$. Thus the derivative of the product is *not* the product of the derivatives. The formula for the derivative of a product is more complicated.

> **The Product Rule**
> If $f(x) = g(x) \cdot k(x)$, then
>
> $$f'(x) = g(x)k'(x) + k(x)g'(x)$$

The 1st times deriva of 2nd + the 2nd times deriv of

The **product rule** states that "the derivative of a product is the first function times the derivative of the second plus the second times the derivative of the first." We prove this theorem at the end of the section.

E X A M P L E   1

Problem

Use the product rule to find $D_x x^2(5x + 11)$.

Solution Let $f(x) = x^2(5x + 11)$ with $g(x) = x^2$ and $k(x) = 5x + 11$. Then $f(x) = g(x) \cdot k(x)$. Since $g'(x) = 2x$ and $k'(x) = 5$, we have

$$f'(x) = g(x)k'(x) + k(x)g'(x) = x^2(5) + (5x + 11)(2x)$$
$$= 5x^2 + 10x^2 + 22x = 15x^2 + 22x$$

The function in Example 1 could be expressed as $f(x) = 5x^3 + 11x^2$ by multiplying through by x^2. Using the rules of the previous section, we have $f'(x) = 15x^2 + 22x$, which is the same derivative computed by the product rule. In all of the examples of this section the functions that are products can be expanded in this way, but in the next section some functions require the product rule to compute their derivatives.

For example, care is needed when computing the derivative of functions that are raised to a power. Consider $f(x) = (x^2 + 1)^2$. It might seem from the power rule that $f'(x) = 2(x^2 + 1)$. But by using the product rule with $g(x) = k(x) = x^2 + 1$, we get

$$f'(x) = (x^2 + 1)2x + (x^2 + 1)2x = 4x(x^2 + 1)$$

In the next section we will learn how to handle power functions that cannot be expanded like this function.

Let us look at one more example of the product rule.

E X A M P L E   2

Problem

Find $D_x(x^2 - 1)(x^3 + x)$.

Solution To apply the product rule let $f(x) = (x^2 - 1)(x^3 + x)$ with $g(x) = x^2 - 1$ and $k(x) = x^3 + x$. Since $g'(x) = 2x$ and $k'(x) = 3x^2 + 1$, we have

$$f'(x) = g(x)k'(x) + k(x)g'(x)$$
$$= (x^2 - 1)(3x^2 + 1) + (x^3 + x)(2x)$$
$$= 3x^4 - 3x^2 + x^2 - 1 + 2x^4 + 2x^2 = 5x^4 - 1$$

The Quotient Rule

Sometimes functions can be expressed as the quotient of simpler functions, such as $h(x) = x^3/(4x + 1)$. To find the derivative of such functions use the following rule.

The Quotient Rule

If $f(x) = \dfrac{g(x)}{k(x)}$, then

$$f'(x) = \frac{k(x)g'(x) - g(x)k'(x)}{[k(x)]^2}$$

bottom times deriv. of top minus top times deriv of bottom

bottom squared

The numerator in the quotient rule is similar to the expression in the product rule, but there is a significant difference. The two terms cannot be interchanged as they can in the product rule because the first term has a positive sign and the second has a negative sign. Examples 3 and 4 illustrate how to apply the quotient rule.

EXAMPLE 3

Problem

Find $D_x \dfrac{x^3}{4x + 1}$

Solution Let $f(x) = x^3/(4x + 1)$ with $g(x) = x^3$ and $k(x) = 4x + 1$. Then $f(x) = g(x)/k(x)$. Since $g'(x) = 3x^2$ and $k'(x) = 4$, we have

$$f'(x) = \frac{k(x)g'(x) - g(x)k'(x)}{[k(x)]^2}$$

$$= \frac{(4x + 1)(3x^2) - x^3(4)}{(4x + 1)^2} = \frac{8x^3 + 3x^2}{(4x + 1)^2}$$

EXAMPLE 4

Problem

Find $D_x \dfrac{x^3}{x^2 + 5}$

Solution

$$D_x \frac{x^3}{x^2 + 5} = \frac{(x^2 + 5)D_x x^3 - x^3 D_x(x^2 + 5)}{(x^2 + 5)^2}$$

$$= \frac{(x^2 + 5)3x^2 - x^3(2x)}{(x^2 + 5)^2}$$

$$= \frac{3x^4 + 15x^2 - 2x^4}{(x^2 + 5)^2} = \frac{x^4 + 15x^2}{(x^2 + 5)^2}$$

An Application: The Lerner Index

There are many methods to measure a particular firm's power and influence in the marketplace. One measure, called the **Lerner index,** named after economist A. P. Lerner of England and Michigan State University, depends on the difference between price P and marginal cost, expressed as a proportion of P.* It is defined as

$$\text{Lerner index} = \frac{P - MC}{P}$$

(handwritten: PRICE − MARGINAL COST over P)

The Lerner index is sometimes referred to as a measure of a firm's monopoly power. If an industry has perfect competition, then $P = MC$, which means that the price each firm establishes for its product is equal to the rate of change of cost, marginal cost. Under perfect competition the Lerner index is 0, meaning the firm has no monopoly power. If a firm has a large amount of monopoly power, it can set a price that is much larger than marginal cost, so the Lerner index will increase. The closer the index is to 1 (its limiting value), the more monopoly power the firm has. Once a firm calculates its price and cost functions it is interested in the rate of change of the Lerner index L, measured by L'.

EXAMPLE 5

(handwritten: SEU → PRICE   PRODUCE → COST)

A firm determines that its price $P(x)$ and cost $C(x)$ per level of production x are

$$P(x) = x^2 + 4$$
$$C(x) = 3x + 10$$

(handwritten: Price, MC = C'x = 3, L(x) = (P − mc)/P =, L(x) = ((x²+4) − 3)/(x²+4) = (x²+1)/(x²+4))

Problem

Compute the firm's Lerner index $L(x)$ and find the rate of change of $L(x)$.

Solution First calculate the marginal cost function, $MC(x) = C'(x)$.

$$MC(x) = C'(x) = 3$$

Substituting this equation along with the equation of $P(x)$ into the definition of $L(x)$ yields

$$L(x) = \frac{P - MC}{P} = \frac{x^2 + 4 - 3}{x^2 + 4} = \frac{x^2 + 1}{x^2 + 4}$$

Applying the quotient rule gives

(handwritten: NEXT →)

$$L'(x) = \frac{2x(x^2 + 4) - (x^2 + 1)(2x)}{(x^2 + 4)^2}$$

(handwritten: DERIVATIVE OF LERNER INDEX)

$$= \frac{2x^3 + 8x - (2x^3 + 2x)}{(x^2 + 4)^2}$$

$$= \frac{6x}{(x^2 + 4)^2}$$

(handwritten: SLOPE OF THE LINE AT A GIVEN POINT)

*Milton H. Spenser, *Contemporary Macroeconomics*, 6th ed., New York, Worth Publishers, 1986.

(handwritten: RATE OF CHANGE)

Proof of the Product Rule

The product rule is not as readily apparent as the derivative of a sum. In fact, intuition leads to an incorrect formula; the derivative of a product looks like it should be the product of the derivatives. Why intuition fails is answered in the proof of the theorem. The proof uses an algebraic ploy—adding and subtracting the same quantity to and from an expression. This results in adding 0 to the expression, so that it remains the same. This is done so the expression can be factored in a certain way. This is why the formula contains two terms, each containing one derivative.

Proof of the Product Rule

If $f(x) = g(x) \cdot k(x)$, then $f(x + h) = g(x + h) \cdot k(x + h)$. From the definition of the derivative of $f(x)$

$$f'(x) = \lim_{h \to 0} \frac{f(x + h) - f(x)}{h}$$

$$= \lim_{h \to 0} \frac{g(x + h)k(x + h) - g(x)k(x)}{h}$$

Here we add and subtract $g(x + h)k(x)$ in the numerator.

$$f'(x) = \lim_{h \to 0} \frac{g(x + h)k(x + h) - g(x + h)k(x) + g(x + h)k(x) - g(x)k(x)}{h}$$

$$= \lim_{h \to 0} \frac{g(x + h)[k(x + h) - k(x)] + k(x)[g(x + h) - g(x)]}{h}$$

$$= \lim_{h \to 0} g(x + h) \left[\frac{k(x + h) - k(x)}{h} \right] + \lim_{h \to 0} k(x) \left[\frac{g(x + h) - g(x)}{h} \right]$$

$$= \lim_{h \to 0} g(x + h) \lim_{h \to 0} \frac{k(x + h) - k(x)}{h} + \lim_{h \to 0} k(x) \lim_{h \to 0} \frac{g(x + h) - g(x)}{h}$$

$$= g(x)k'(x) + k(x)g'(x)$$

provided that $g'(x)$ and $k'(x)$ exist.

EXERCISE SET 3-2

In Problems 1 to 18 find $f'(x)$.

1. $f(x) = x^4(x + 5)$ 2. $f(x) = x^6(x - 2)$

3. $f(x) = 2x^3(3x - 7)$ 4. $f(x) = 3x^5(2x + 1)$

5. $f(x) = (x^{-3} + 5)(3x^2 - 17)$

6. $f(x) = (x^{-5} - 9)(12x^3 + 1)$

7. $f(x) = (x^{1/3} + x)(3x^{1/2} - 4x)$

8. $f(x) = (x^{1/5} - x)(x^{3/5} + 2x)$

9. $f(x) = (x^{-1/3} + 7x^{-1/4})(4x^2 - 5)$

10. $f(x) = (2x^{-1/6} - x^{-3/7} + 2)(x^{1/2} - 6)$

11. $f(x) = (2x + 1)^2$ 12. $f(x) = (4x - 3)^2$

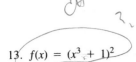

13. $f(x) = (x^3 + 1)^2$

14. $f(x) = (x^4 - 2)^2$

15. $f(x) = x/(2x + 1)$

16. $f(x) = 2x/(5x - 4)$

17. $f(x) = (x^3 + 5)/3x^2$

18. $f(x) = (x^5 - 11)/x^3$

In Problems 19 to 22 find the equation of the tangent line to $y = f(x)$ at the given point

19. $f(x) = (x^{1/2} - 4)^2$, $(1, 9)$

20. $f(x) = (x^{4/5} + 1)^3$, $(1, 8)$

21. $f(x) = (x^3 - 2)/(x^2 - 1)$, $(0, 2)$

22. $f(x) = (x^{1/2} - 1)/(x^{1/2} + 8)$, $(1, 0)$

In Problems 23 and 24 a firm determines that its price $P(x)$ and cost $C(x)$ per level of production x are the given functions. Compute the firm's Lerner index $L(x)$ and find the rate of change of $L(x)$.

23. $P(x) = 0.5x^2 + 5$, $C(x) = 2x + 7$

24. $P(x) = 0.22x^2 + 4.1$, $C(x) = 4x + 3$

In Problems 25 to 32 find $f'(x)$.

25. $f(x) = x^2(x^3 + 15)(x^2 - 17)$

26. $f(x) = x^{1/2}(x^{-5} - 9)(12x^3 + 1)$

27. $f(x) = x^{-1}(x^{1/3} + 2)(x^2 - 3)$

28. $f(x) = x^{-1/2}(x^{1.5} - 1)(2x^3 - 3)$

29. $f(x) = x^2(x + 5)/(x^2 - 7)$

30. $f(x) = x^{1/2}(x^2 - 1)/(x^3 + 1)$

31. $f(x) = \dfrac{x^2 + 3}{x(x^2 - 1)}$

32. $f(x) = \dfrac{x^3 - 2}{x^2(x^{-1} + 1)}$

33. The total cost of producing x amount of books is given by $C(x) = 100 + 23x(1 - 2x^2)$. Find the marginal cost.

The *average cost per good* is defined to be the total cost to produce x amount of goods divided by x. If $C(x)$ measures the total cost to produce x amount of goods and $AC(x)$ is the average cost per good, then $AC(x)$ is given by

$$AC(x) = \frac{C(x)}{x}$$

In Problems 34 and 35 find the marginal average cost, $AC'(x)$.

34. $C(x) = 10 + 25x - 2x^2$

35. $C(x) = 100 + 3.5x - 0.4x^2$

36. Use the definition of the derivative and an argument similar to the proof of the product rule to prove the quotient rule.

37. Show why the Lerner index $L = \dfrac{P - MC}{P}$, for price P and marginal cost MC, is never more than 1.

Referenced Exercise Set 3–2

1. With the breakup of AT&T, the transition from monopoly to competition necessitated the use of different tools for business decisions. In an article* describing how GTE Telephone Operations decided on prices for various products, one model centered on the Styleline phone. Of key interest is the total quantity sold q after x years, where q is measured in millions of units. The data can be approximated by

$$q(x) = 0.13x(x - 10)^2$$

GTE was concerned with finding when $q(x)$ would attain its maximum rate of increase. In the next chapter we will show how this is done by using $q'(x)$. Compute $q'(x)$.

2. The Atomic Energy Commission has demanded extensive quality control in nuclear power plants, especially since the Three Mile Island and Chernobyl incidents. In an article describing the significance of AEC regulations, even for "minor" equipment, John B. Silverwood,† Reliability and Assurance Manager for United Engineers and Constructors, outlines various procedures that are necessary in performance testing. In one example he describes the efficiency rating of a centrifugal pump by using a graph (Figure 1) whose equation is approximated by

$$f(x) = \left(\frac{x}{4}\right)\left(5 - \frac{x}{32}\right)$$

$f(x)$ measures the efficiency, in percent, at which the pump is operating when x gallons are being pumped

*Arvind G. Jadhav, "Market Simulation for Telecommunications Services," *Forecasting Public Utilities*, New York, North-Holland Publishing Co., pp. 165–186.

†John B. Silverwood, "Objective Quality Evidence of Product Integrity," *American Society of Quality Control—Technical Conference Transactions*, 1972, pp. 276–283.

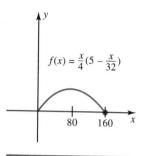

$f(x) = \frac{x}{4}(5 - \frac{x}{32})$

FIGURE 1 *Source:* John B. Silverwood, "Objective Quality Evidence of Product Integrity," *American Society of Quality Control – Technical Conference Transactions*, 1972, pp. 276–283.

per minute. The function is defined in [0, 100]. From the graph it can be seen that as x increases, the efficiency increases for a while and then it starts to decrease. At that point the tangent line has slope 0. Find the point where $f(x)$ changes from increasing to decreasing by finding the value of x that solves $f'(x) = 0$.

3. In a paper‡ studying the effects of private pond fishing on preexisting common property fishing, researchers centered their attention on the crawfish market. The cost function for wild crawfish, those caught in com-

mon property waters, can be approximated by

$$C(x) = \frac{7.12}{13.7x - x^2}$$

Here x is the quantity of crawfish supplied to the market, measured in millions of pounds, and C is measured in millions of dollars. Find the marginal cost $C'(x)$.

Cumulative Exercise Set 3–2

In Problems 1 and 2 find $f'(x)$.

1. $f(x) = 3x^{-2} - x^4 + 5$

2. $f(x) = x^{-1/2} - 2x^{-3}$

3. Find the equation of the tangent line to $f(x) = x^3 - 4x$ at $x = 1$.

4. The profit P in thousands of dollars from the sales volume of x thousands of records is found to be $P(x) = 0.3x^2 + 20x - 200$. Find the marginal profit when $x = 3$.

In Problems 5 to 7 find $f'(x)$.

5. $f(x) = (x^{-5} - 9)(12x^3 + 1)$

6. $f(x) = (x^5 - 1)^2$

7. $f(x) = x^2(x - 5)(x^2 + 7)$

8. The total cost of producing radios is given by $C(x) = 200 + 15x(1 - 3x^2)$. Find the marginal cost.

‡Frederick W. Bell, "Competition from Fish Farming in Influencing Rent Dissipation: The Crawfish Fishery," *American Journal of Agricultural Economics*, Vol. 68, pp. 95–101.

3–3 The Chain Rule

Sometimes more complicated functions can be differentiated by separating the function into two parts, an "inner" and "outer" part, and then computing the derivative of each. The rule that allows us to find a derivative this way is called the chain rule. It has several forms, each governing a different type of function. In this section we study the form of the chain rule called the generalized power rule, so-called because it deals with the derivative of functions raised to a power. In general the chain rule is used to find the derivative of a composite function of the form $f(g(x))$. With this notation, $g(x)$ is the "inner" function and $f(x)$ is the "outer" function. At the end of the section the chain rule is studied in this more general setting so that it can be applied later to other types of functions.

The Generalized Power Rule

The differentiation rules developed so far do not allow us to find the derivative of more complicated functions, such as

$$f(x) = (x^3 + x)^5 \quad s(x) \doteq (x^2 + 1)^{-1} \quad h(x) = (3x^2 + x)^{1/2}$$

Each of these functions is an expression in x raised to a power. The general form of such a function is

$$f(x) = g(x)^n$$

This type of function is called a **general power function.** In each case $g(x)$ is referred to as the "inside part" of the function or the inside function. For example, by letting $g(x) = x^3 + x$ and $n = 5$, we can form the previous function $f(x) = (x^3 + x)^5$. Similarly $s(x)$ has $g(x) = x^2 + 1$ and $n = -1$, while $h(x)$ has $g(x) = 3x^2 + x$ and $n = \frac{1}{2}$.

Let us find the derivative of a specific power function.

EXAMPLE 1

Problem

Use the product rule to find the derivative of (a) $f(x) = (x^3 + x)^2$, (b) $g(x) = (x^3 + x)^3$.

Solution (a) Consider $f(x)$ as the product of two functions, each being $x^3 + x$. That is, $f(x) = (x^3 + x)(x^3 + x)$. By the product rule

$$\begin{aligned} f'(x) &= (x^3 + x)D_x(x^3 + x) + (x^3 + x)D_x(x^3 + x) \\ &= 2(x^3 + x)D_x(x^3 + x) \\ &= 2(x^3 + x)(3x^2 + 1) \end{aligned}$$

(b) Consider $g(x)$ as the product of the two functions $f(x) = (x^3 + x)^2$ from (a) and $x^3 + x$. Thus

$$g(x) = (x^3 + x)^2(x^3 + x)$$

Use the product rule, with $f'(x) = 2(x^3 + x)(3x^2 + 1)$ from (a).

$$\begin{aligned} g'(x) &= (x^3 + x)^2 D_x(x^3 + x) + (x^3 + x)D_x(x^3 + x)^2 \\ &= (x^3 + x)^2(3x^2 + 1) + (x^3 + x)[(2)(x^3 + x)(3x^2 + 1)] \end{aligned}$$

Factor out $3x^2 + 1$ and get

$$\begin{aligned} g'(x) &= [(x^3 + x)^2 + 2(x^3 + x)^2](3x^2 + 1) \\ &= 3(x^3 + x)^2(3x^2 + 1) \end{aligned}$$

Recall the power rule:

$$D_x x^n = nx^{n-1}$$

It governs only functions of the form $f(x) = x^n$. In Example 1, when finding $f'(x)$, it is tempting to use just the power rule, in which case we would ·get $2(x^3 + x)$.

While this expression is part of the derivative, it is not precisely the derivative. It varies by a factor of $3x^2 + 1$, which is the derivative of the inside function, $x^3 + x$. The generalized power rule requires that the derivative of a general power function contain the derivative of the inside part of the function as a factor.

Example 1 motivates the chain rule for general power functions, called the **generalized power rule.**

The Generalized Power Rule

$$D_x[g(x)]^n = ng(x)^{n-1}D_x[g(x)]$$

Notice that the power rule is a special case of the generalized power rule when $g(x) = x$. Let us apply the rule to the function in Example 1(a). The function is $f(x) = (x^3 + x)^2$ where $g(x) = x^3 + x$ and $n = 2$. Then

$$f'(x) = 2(x^3 + x)D_x(x^3 + x)$$
$$= 2(x^3 + x)(3x^2 + 1)$$

The next three examples illustrate the generalized power rule for functions that cannot be differentiated by the product rule.

Example 2 illustrates the generalized power rule for a power function whose exponent is a positive fraction.

EXAMPLE 2

Problem

Find $D_x f(x)$ where $f(x) = (x^4 + 2x)^{1/2}$.

Solution In the generalized power rule we have $g(x) = x^4 + 2x$ and $n = \frac{1}{2}$. Therefore

$$D_x(x^4 + 2x)^{1/2} = (\tfrac{1}{2})(x^4 + 2x)^{(1/2)-1}D_x(x^4 + 2x)$$
$$= (\tfrac{1}{2})(x^4 + 2x)^{-1/2}(4x^3 + 2)$$
$$= (2x^3 + 1)(x^4 + 2x)^{-1/2}$$

Notice when simplifying the expression for the derivative, the term corresponding to the derivative of the "inside" part of the function appears on the left of the expression corresponding to the power function.

Example 3 finds the derivative of a power function whose exponent is a negative integer.

EXAMPLE 3

Problem

Find $f'(x)$ where $f(x) = (x^3 + 6x^2)^{-4}$.

Solution In the generalized power rule we have $g(x) = x^3 + 6x^2$ and $n = -4$. Therefore

$$
\begin{aligned}
f'(x) &= -4(x^3 + 6x^2)^{-4-1}D_x(x^3 + 6x^2) \\
&= -4(x^3 + 6x^2)^{-5}(3x^2 + 12x) \\
&= -12x(x + 4)(x^3 + 6x^2)^{-5}
\end{aligned}
$$

We look at one more illustration of the generalized power rule in Example 4, this time with the exponent of the power function a negative fraction.

EXAMPLE 4

Problem

Find $f'(x)$ where $f(x) = (x^4 + 12x + 4)^{-1/4}$.

Solution Let $g(x) = x^4 + 12x + 4$ and $n = -\frac{1}{4}$. Then

$$
\begin{aligned}
f'(x) &= (-\tfrac{1}{4})(x^4 + 12x + 4)^{-5/4}D_x(x^4 + 12x + 4) \\
&= (-\tfrac{1}{4})(x^4 + 12x + 4)^{-5/4}(4x^3 + 12) \\
&= -(x^3 + 3)(x^4 + 12x + 4)^{-5/4}
\end{aligned}
$$

Example 5 demonstrates how to use the generalized power rule in conjunction with the product rule. First regard the function as a product, separating the first function from the second. Then express the derivative in terms of the product rule, the first function times the derivative of the second plus the second function times the derivative of the first. Then apply the generalized power rule where appropriate.

EXAMPLE 5

Problem

Compute $f'(x)$ where $f(x) = x^3(x^2 + 5)^{-1}$.

Solution Use the product rule, with the first function being x^3 and the second function being $(x^2 + 5)^{-1}$.

$$
\begin{aligned}
f'(x) &= x^3 D_x(x^2 + 5)^{-1} + (x^2 + 5)^{-1}D_x x^3 \\
&= x^3(-1)(x^2 + 5)^{-2}(2x) + (x^2 + 5)^{-1}(3x^2) \\
&= -2x^4(x^2 + 5)^{-2} + 3x^2(x^2 + 5)^{-1} \\
&= (x^2 + 5)^{-2}[-2x^4 + 3x^2(x^2 + 5)] \\
&= (x^2 + 5)^{-2}(-2x^4 + 3x^4 + 15x^2) \\
&= (x^4 + 15x^2)(x^2 + 5)^{-2} \\
&= x^2(x^2 + 15)(x^2 + 5)^{-2}
\end{aligned}
$$

Notice that the quotient rule could have been used in Example 5 by expressing $f(x)$ as a quotient, $x^3/(x^2 + 5)$.

The General Form of the Chain Rule

The generalized power rule is a special case of the **chain rule.** The general form of the chain rule is stated in terms of the composition of functions.

> **The Chain Rule**
> If $f(x) = h(g(x))$, then $f'(x) = h'(g(x))g'(x)$.

To illustrate, suppose $f(x) = (x^2 + 1)^3$. Let $h(x) = x^3$ and $g(x) = x^2 + 1$ so that $f(x) = h(g(x))$. To find $f'(x)$ by the general form of the chain rule, we use $h'(x) = 3x^2$ and $g'(x) = 2x$. Then $h'(g(x)) = 3[g(x)]^2 = 3(x^2 + 1)^2$ so that

$$f'(x) = h'(g(x))g'(x) = 3(x^2 + 1)^2(2x) = 6x(x^2 + 1)^2$$

The chain rule can be expressed in another form, which is easy to memorize. This form uses the $\dfrac{dy}{dx}$ notation of the derivative. Suppose $y = f(g(x))$. Make the substitution of variable $u = g(x)$ so that $y = f(u)$. Then y can be expressed as a function of u or of x. With this notation, we write the derivatives as $g'(x) = \dfrac{du}{dx}$, $f'(u) = \dfrac{dy}{du}$, and $f'(x) = \dfrac{dy}{dx}$. Then the chain rule states

$$\frac{dy}{dx} = \frac{dy}{du}\frac{du}{dx}$$

In the illustration above, $y = f(x) = (x^2 + 1)^3$ and $u = g(x) = x^2 + 1$ so that $y = f(u) = u^3$. Then $\dfrac{dy}{du} = 3u^2$ and $\dfrac{du}{dx} = 2x$. Therefore

$$\frac{dy}{dx} = \frac{dy}{du}\frac{du}{dx} = 3u^2(2x) = 3(x^2 + 1)^2(2x) = 6x(x^2 + 1)^2$$

This agrees with the preceding expression for $f'(x)$.

The generalized power rule is a special case of the chain rule. We now derive the special case, the generalized power rule, from the more general theorem, the chain rule.

Derivation of the Generalized Power Rule from the Chain Rule

Consider the power function $f(x) = g(x)^n$. Let $h(x) = x^n$. Then $f(x) = h(g(x))$ and $h'(x) = nx^{n-1}$, so $h'(g(x)) = ng(x)^{n-1}$. From the general form of the chain rule we have

$$f'(x) = h'(g(x))g'(x) = ng(x)^{n-1}g'(x)$$

This is the statement of the generalized power rule.

An Application

There are many situations in business where one variable depends upon another, and the second variable depends on a third. An example is the important business tool, quality control. Quality $Q(x)$ is measured in terms of level of production x, and x is measured in terms of time t. Q is measured on a scale from 0 to 100 (with 100 being a product with no flaws), x is measured in units of 100, and t is measured in hours. The next example shows how the rate of change of Q can be computed in two ways, by expressing Q in terms of t directly and by using the chain rule.

EXAMPLE 6

A company determines that

$$Q(x) = 90 - 0.4(x - 10)^2 = 50 + 8x - 0.4x^2$$

where $Q(x)$ is defined for $0 < x < 18$. The graph of $Q(x)$ is given in Figure 1. Production level, measured in hours, depends on the time of day, t, and is given by

$$x(t) = 18 - 0.5(t - 6)^2 = 6t - 0.5t^2$$

where $x(t)$ is defined for $0 \le t \le 8$. This means that at time t the factory is producing goods at a rate such that if that rate were held constant all day $100x(t)$ units would be produced. The graph of $x(t)$ is given in Figure 2.

Problem

Find $D_t Q$ by expressing Q as a function of t and differentiating, and also by using the chain rule. Show that the two expressions are equal.

Solution Since Q is a function of x and x is a function of t, Q is a function of t; namely

$$\begin{aligned} Q(t) &= Q(x(t)) = Q(6t - 0.5t^2) \\ &= 50 + 8(6t - 0.5t^2) - 0.4(6t - 0.5t^2)^2 \end{aligned}$$

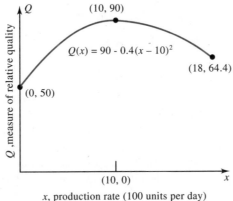

FIGURE 1

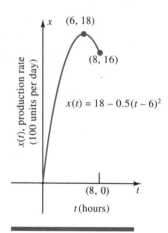

x(t), production rate (100 units per day)

(6, 18)

(8, 16)

$x(t) = 18 - 0.5(t - 6)^2$

(8, 0)   t

t (hours)

FIGURE 2

To find $Q'(t)$ directly from this formula, differentiate with respect to t.

$$Q'(t) = 8(6 - t) - 0.8(6t - 0.5t^2)(6 - t)$$
$$= 48 - 36.8t + 7.2t^2 - 0.4t^3$$

Using the chain rule to find $Q'(t)$ yields

$$Q'(t) = Q'(x(t))x'(t)$$
$$= D_x[Q(x)]D_t[x(t)]$$
$$= D_x[50 + 8x - 0.4x^2]D_t[6t - 0.5t^2]$$
$$= [8 - 0.8x][6 - t]$$

Now substitute the expression $x(t) = 6t - 0.5t^2$ for x.

$$Q'(t) = [8 - 0.8(6t - 0.5t^2)](6 - t)$$
$$= (8 - 4.8t + 0.4t^2)(6 - t)$$
$$= 48 - 36.8t + 7.2t^2 - 0.4t^3$$

Thus the two expressions for $Q'(t)$ are equal.

The graph of $Q(t)$ is given in Figure 3. In Section 4–2 we will see how to obtain this graph. The quality is lowest at the start of the shift when it is 50; that is, $Q(0) = 50$. The graph of Q then gradually increases until it reaches its largest value, 90, when $t = 2$, meaning that two hours into the shift quality is at its peak. At this point $Q'(t)$ is 0 since the rate of change of Q goes from positive when Q is increasing to negative when it is decreasing. Then Q decreases to the value 64.4 at $t = 6$ hours, at which time it then starts to increase again. Thus when $t = 6$, $Q'(t) = 0$. We will discuss these concepts further in the next chapter.

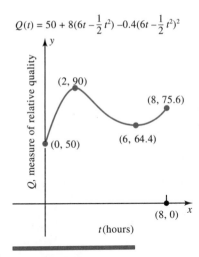

$$Q(t) = 50 + 8(6t - \frac{1}{2}t^2) - 0.4(6t - \frac{1}{2}t^2)^2$$

FIGURE 3

An Application: The Lerner Index

The measure of a firm's monopoly power, called the Lerner index L, was presented in the previous section. It is defined by

$$L = \frac{P - MC}{P}$$

P is the price of the firm's product and MC is the marginal cost. If MC is fixed, then L is a function of P. If P is a function of the units produced x, then L is also a function of x. In the previous section we expressed L as a function of x and then computed $L'(x)$. The next example shows that the chain rule can also be used to find $L'(x)$. The price and cost functions are from Example 5 in the previous section.

EXAMPLE 7

A firm determines that its price $P(x)$ and cost $C(x)$ per level of production x are

$$P(x) = x^2 + 4$$
$$C(x) = 3x + 10$$
$$MC(x) = C'(x) = 3$$

Problem

Use the chain rule to compute $L'(x)$.

Solution Since MC is the constant 3, we can express L as a function of P.

$$L(P) = \frac{P - MC}{P} = \frac{P - 3}{P} = 1 - 3P^{-1}$$

Then we have

$$\frac{dL}{dP} = 3P^{-2}$$

Also,

$$\frac{dP}{dx} = 2x$$

Applying the chain rule yields

$$\frac{dL}{dx} = \frac{dL}{dP}\frac{dP}{dx} = (3P^{-2})(2x) = 3(x^2 + 4)^{-2}(2x)$$

$$= \frac{6x}{(x^2 + 4)^2}$$

This is the same expression for the derivative that was computed in the last section.

EXERCISE SET 3–3

In Problems 1 to 10 find $f'(x)$ by using the generalized power rule.

1. $f(x) = (5x + 2)^3$

2. $f(x) = (15x + 2)^4$

3. $f(x) = (x^3 + 4)^3$

4. $f(x) = (5x^2 + 2)^6$

5. $f(x) = (x^3 + 3x^2)^3$

6. $f(x) = (2x^4 + 3x^3)^6$

7. $f(x) = (x^{3/2} + 4)^{1/3}$

8. $f(x) = (x^{3/5} - 2)^{2/3}$

9. $f(x) = (x^{1/2} + 2x^{-1})^{-6}$

10. $f(x) = (5x^{1/3} + x^{-2})^{-1/2}$

In Problems 11 to 16 find $f'(x)$ using the generalized power rule in conjunction with the product rule or the quotient rule.

11. $f(x) = x(5x^2 + 2)^2$

12. $f(x) = 3x(3x^2 + 5)^4$

13. $f(x) = x^2(x^3 + 2x)^2$

14. $f(x) = 3x^2(x^3 + 3x)^3$

15. $f(x) = \dfrac{x^2}{(x^3 - x)^3}$

16. $f(x) = 4x^4(x^2 - 2x)^{1/2}$

In Problems 17 to 20 find the equation of the tangent line at the given value of x.

17. $f(x) = (x^2 + 2)^3$ at $x = 0$

18. $f(x) = (x^3 - 1)^2$ at $x = 0$

19. $f(x) = (x^3 + 4x)^{-1}$ at $x = 1$

20. $f(x) = (3x^2 - x - 1)^{-2}$ at $x = 0$

A company determines that its quality control $Q(x)$, measured in terms of level of production x, measured in terms of time t, is given by the functions in Problems 21 and 22. Find $D_t Q$ by expressing Q as a function of t and differentiating and also by using the chain rule. Show that the two expressions are equal.

21. $Q(x) = 95 - 0.2(x - 5)^2$,
    $x(t) = 10 - 0.2(t - 6)^2$

22. $Q(x) = 100 - 0.3(x - 8)^2$,
    $x(t) = 10 - 0.1(t - 6)^2$

A company determines that its price $P(x)$ and cost $C(x)$ per level of production x are given by the functions in Problems 23 and 24. Find the Lerner index $L(x)$ and $L'(x)$.

23. $P(x) = x^2 + 10$, $C(x) = 5x + 20$

24. $P(x) = 100 + 15x - 2x^2$,
    $C(x) = 4x + 25$

In Problems 25 to 34 find $f'(x)$ by using the chain rule together with the rules in the previous section.

25. $f(x) = x^2(x^3 + 2x^2)^4$

26. $f(x) = 3x^4(2x^4 + 3x^3)^6$

27. $f(x) = 2x^{1/2}(x^{3/2} + 4)^{-3}$

28. $f(x) = 4x^{-2}(x^{5/2} + 2x^{-1})^{1/6}$

29. $f(x) = (x^2 + 1)^2(x^3 + 1)^3$

30. $f(x) = (x^2 - 3)^{-1}(x^3 - 1)^2$

31. $f(x) = (x^2 + 1)^{1/2}(x^3 + 3)^{1/3}$

32. $f(x) = (x^{1/2} + 3)^{1/3}(x^2 + 3x)^{-1}$

33. $f(x) = \dfrac{2x^{1/2}}{(x^2 + 2x^{-2})^3}$

34. $f(x) = \dfrac{(x^4 + 1)^2}{(3x^4 + 3)^{1/3}}$

In Problems 35 to 38 find $f'(x)$ by using the chain rule and also by expressing the function as a quotient and using the quotient rule.

35. $f(x) = x(x^2 + 1)^{-1}$ 36. $f(x) = x^2(x^3 - 1)^{-1}$

37. $f(x) = (x - 2)(x^2 + 5)^{-1}$

38. $f(x) = (x + 3)^2(4x - 5)^{-1}$

39. The total profit $p(x)$ for x amount of goods sold is

$$p(x) = 500(200 - x)^2$$

Find the marginal profit.

40. The strength $S(x)$ of an individual's reaction to the quantity x of a drug administered into the bloodstream is given by

$$S(x) = 5x(20 - x/2)^{1/2}$$

Find the rate of change of $S(x)$ with respect to x when $x = 8$.

41. The chain rule can be used in conjunction with the product rule to derive the quotient rule. Suppose $f(x)$ is a quotient function $f(x) = \dfrac{g(x)}{h(x)}$. Express $f(x)$ as the product function $f(x) = g(x)h(x)^{-1}$. First apply the product rule and then the chain rule for the derivative of $h(x)^{-1}$. Factor $h(x)^{-2}$ from each term to get the quotient rule.

Referenced Exercise Set 3–3

1. Many-variables affect quality control of production. In describing different quality control procedures for various products, Bartlett and Provost* used as an example the net weight of a #303 can of whole kernel corn.

This type of can should weigh 16 ounces. They show that the acceptance of a lot by a retailer from the producer depends on the average net weight of all cans in the lot, and the average net weight depends on the size of the sample of cans selected from the lot to be inspected. Let P represent the likelihood that a lot will be accepted by the retailer, W the average net weight of all the cans in the lot, and x the sample size. The relationships between the variables can be expressed as follows:

$$P(W) = 0.007W^2 - 0.33W + 100$$
$$W(x) = 16 - 0.41x^{-1/2}$$

Express P as a function of x and find $P'(x)$.

2. Franchised businesses account for over 38% of all retail sales in the United States and originate more than 12% of the gross national product. In an article analyzing methods to optimize franchise contractual agreements, the researchers† derive the following function that relates franchisee's profits P in terms of the quantity Q of goods sold by the equation

$$P = aQ - bQ^2 - F$$

where F is the initial lump-sum franchise fee paid by the franchisee, and a and b are constants whose values depend on the type of industry to which the model is being applied. It is also shown that Q is a function of R where R is the royalty rate charged by the parent company to the franchisee

$$Q(R) = c - dR$$

The constants c and d also depend on the particular industry. Q is measured in thousands of units and P in thousands of dollars. Typical values for the constants in the fast food industry are $a = 0.35$, $b = 0.11$, $c = 0.1$, and Q is measured in 1000 units. Find the marginal profit with respect to R via the chain rule and then express P as a function of R and find the marginal profit with respect to R directly. Show that the two expressions are equal.

3. (a) Find $f'(x)$ for the function
$$f(x) = \sqrt{2x - x^2 - 1}$$

   (b) The answer to part (a) is "Does not exist." Use a graphing calculator or a graphing program for a

*Richard P. Bartlett and Lloyd P. Provost, "Tolerances in Standards and Specifications," *Quality Progress*, 1973, pp. 14–18.

†Roger D. Blair and David L. Kaserman, "Optimal Franchising," *Southern Economics Journal*, Vol. 49, 1983, pp. 495–505.

microcomputer to sketch the graph of $f(x)$ to see why the derivative does not exist.‡ [In **Graph 20/30,** the software program that accompanies this text, use the module "Graphing" and define the function $y = \text{sqrt } (2x - x^2 - 1)$ between the limits of -5 and 5.]

(c) What is the domain of $f(x)$?

Cumulative Exercise Set 3–3

In Problems 1 to 5 find $f'(x)$.

1. $f(x) = 2x^3 + \sqrt{x} - 7 - \dfrac{1}{x}$

2. $f(x) = \dfrac{3}{\sqrt{2x}}$

3. $f(x) = x(x - 4)$ 4. $f(x) = \dfrac{x^2 - 3x + 5}{x^3 - 4x^2 - 9}$

5. $f(x) = (x^2 - 1)^3 - x(x^2 + 1)$

In Problems 6 to 9 find the equation of the line tangent to $y = f(x)$ at the given value of x.

6. $f(x) = x^2 - 4$, $x = -2$

7. $f(x) = \sqrt{9 - x^2}$, $x = 0$

8. $f(x) = \dfrac{x}{1 - x}$, $x = \dfrac{1}{2}$

9. $f(x) = (4 - x)^3$, $x = 2$

10. If the cost C, in hundreds of thousands of dollars, of producing x million batteries is

$$C(x) = 20 + 5.4x - 0.09x^2$$

what is the marginal cost of producing 2 million batteries?

11. Let $f(x) = (x^3 - x - 2)(x^3 - x - 2)$.
    (a) Find $f'(x)$ via the product rule.
    (b) Find $f'(x)$ via the generalized power rule.
    (c) Are the answers in parts (a) and (b) the same?

12. (a) Use the product rule and the generalized power rule to find $f'(x)$ for $f(x) = x(1 - x)^{-1}$.
    (b) Use the quotient rule to find $f'(x)$ for

    $$f(x) = \dfrac{x}{1 - x}.$$

    (c) Are the answers in parts (a) and (b) the same?

‡Y. L. Cheung, "Examples for Graphing Calculators," *Mathematics Teacher,* February, 1989, pp. 82–83.

3–4   Higher Derivatives and Other Notation

Just as the derivative $f'(x)$ of a function gives information about $f(x)$, so too the derivative of the derivative gives information about $f'(x)$. The derivative of the derivative is called the "second derivative." This section shows how to compute the second derivative and higher derivatives, as well as discusses a few additional points about alternate notation.

The Second Derivative

The derivative $f'(x)$ of a function is another function that measures the slope of the tangent line of $f(x)$. In the same way the derivative of $f'(x)$ is another function, called the **second derivative** of $f(x)$, and is denoted by $f''(x)$. Using alternate notation, $D_x(D_x f(x)) = f''(x)$, which says that the derivative of the derivative is the second derivative. The first example illustrates how to compute the second derivative.

E X A M P L E 1

Problem

Find $f''(x)$ for (a) $f(x) = 2x^3 + 5x$, (b) $f(x) = \sqrt{x} - 1/x$.

Solution (a) $f'(x) = 6x^2 + 5$ and so

$$f''(x) = 12x$$

(b) First write the function as $f(x) = x^{1/2} - x^{-1}$. Then

$$f'(x) = (\tfrac{1}{2})x^{-1/2} - (-x^{-2})$$
$$= (\tfrac{1}{2})x^{-1/2} + x^{-2}$$
$$f''(x) = -(\tfrac{1}{4})x^{-3/2} - 2x^{-3}$$

The second derivative provides important information about $f(x)$. In particular, it helps determine the shape of the curve of $f(x)$ in the vicinity of a given point. This topic will be amplified in the next chapter.

Higher Derivatives

In some applications it is necessary to compute higher derivatives. For example, if $f(x) = x^3$, then $f'(x) = 3x^2$, $f''(x) = 6x$, and the derivative of $f''(x)$, called the **third derivative,** denoted by $f'''(x)$, is the derivative of $6x$. Hence $f'''(x) = 6$. For even **higher derivatives** the notation changes. We write $f^{(n)}(x)$ for $n > 3$ to represent higher derivatives after the third one. In this example $f^{(4)}(x) = 0$ as do all higher derivatives.

EXAMPLE 2

Problem

Find $f^{(n)}(x)$ for $f(x) = x^{-1}$.

Solution Find the first, second, and third derivatives to determine a pattern.

$$f'(x) = -x^{-2}$$
$$f''(x) = -(-2)x^{-3} = 2x^{-3}$$
$$f'''(x) = 2(-3)x^{-4} = -6x^{-4}$$

Successive derivatives change sign so a factor $(-1)^n$ is necessary. The absolute value of the coefficient of each derivative has the form $1 \cdot 2 \cdot 3 \cdot \cdots \cdot n = n!$. The power of x in each derivative is negative and its absolute value is one more than the number of the derivative. For example, the third derivative has the power -4. Hence the nth derivative has the factor x to the power $-(n + 1)$. Therefore the nth derivative is

$$f^{(n)}(x) = (-1)^n n! x^{-(n+1)}$$

In particular, $f^{(4)}(x) = (-1)^4 4! x^{-5} = 24x^{-5}$ and $f^{(5)}(x) = -5! x^{-6} = -120x^{-6}$.

Other Notation

There are several notations for expressing higher derivatives of $y = f(x)$. If the function is $y = f(x)$, then the second derivative is expressed by each of these.

$$f''(x) \qquad \frac{d^2y}{dx^2} \qquad \frac{d^2f}{dx^2} \qquad \frac{d^2}{dx^2} f(x) \qquad D_x^2 f(x)$$

These notations might seem to contain powers, but they are merely contractions of more cumbersome notation. Referring to the second derivative as the "derivative of the derivative," the D_x notation is

$$f''(x) = D_x[D_x f(x)]$$

This is shortened to $D_x^2[f(x)]$. The contraction is more obviously preferred when trying to write the third and higher derivatives with this notation. The third derivative is written $D_x^3 f(x)$ instead of $D_x[D_x(D_x f(x))]$. In a similar way the notation $\frac{d^2y}{dx^2}$ is a shortened form of $\frac{d}{dx} \frac{dy}{dx}$. Similarly the notation for the third derivative is $\frac{d^3y}{dx^3}$.

The second derivative of $y = f(x)$ at $x = 3$ is expressed as $f''(3)$. Using the $\frac{dy}{dx}$ and $D_x f(x)$ notations, $f''(3)$ is expressed as

$$\left. \frac{d^2y}{dx^2} \right|_{x=3} \qquad \left. D_x^2[f(x)] \right|_{x=3}$$

The next example provides practice with these different notations.

EXAMPLE 3

Problem

If $f(x) = x^3 - 2x^2 + 5$, find (a) $\left. \dfrac{d^2y}{dx^2} \right|_{x=3}$, (b) $\left. D_x^2 f(x) \right|_{x=4}$.

Solution $f'(x) = 3x^2 - 4x$ and $f''(x) = 6x - 4$. Therefore

(a) $\left. \dfrac{d^2y}{dx^2} \right|_{x=3} = 6(3) - 4 = 14$

(b) $\left. D_x^2 f(x) \right|_{x=4} = 6(4) - 4 = 20$

Other Variables

In some applications it is convenient to use more appropriate variables than x and y, such as t and v when discussing time and velocity. When the variable is not x, the notation for the derivative is also changed. For instance, if the function is given as $v(t) = t^4$, then we write $v'(t) = D_t v = \dfrac{dv}{dt} = 4t^3$. The name of the variable does not affect the meaning of the derivative. Thus if $f(x) = 2x$ and $p(s) = 2s$, then $f'(x) = 2$ and $p'(s) = 2$ mean the same thing as far as the derivative is concerned—the slope of the tangent line at any value of the variable is 2. This illustrates that any letter can be used as a variable.

EXAMPLE 4

Problem

Compute the derivative of each function: (a) $G(u) = (3u + 5)^{1/2}$, (b) $z(y) = y^{-3} - y$.

Solution

(a) $G'(u) = (\frac{1}{2})(3u + 5)^{-1/2} \cdot 3$
$= (\frac{3}{2})(3u + 5)^{-1/2}$

(b) $D_y \, z(y) = -3y^{-4} - 1$

One of the most important applications of calculus is in the movement of objects. Many modern inventions are based on this application, from radar screens to television broadcasts. In Section 3–1 we considered an object moving in a straight line given by the function $y = f(t)$ at time t. Its velocity, $v(t)$, is given by $v(t) = f'(t)$. The rate of change of velocity is defined to be *acceleration, a(t)*. Thus if $f(t)$ is the position function, then $f'(t) = v(t)$ is the velocity and $f''(t) = a(t)$ is the acceleration. Examples 5 and 6 illustrate how to compute acceleration.

EXAMPLE 5

An object is moving in a straight line with its position at time t given by $f(t) = t^3 - 12t^2 + 36t + 1$.

Problem

Find the value of t where (a) its velocity is 0 and (b) its acceleration is 0.

Solution Find the first and second derivatives.

$f'(t) = 3t^2 - 24t + 36 = 3(t^2 - 8t + 12)$
$= 3(t - 6)(t - 2)$

$f''(t) = 6t - 24 = 6(t - 4)$

(a) Set $v(t) = f'(t) = 3(t - 6)(t - 2) = 0$. Then $v(t) = 0$ when $t = 6$ and when $t = 2$.

(b) Set $a(t) = 6(t - 4) = 0$. Then $a(t) = 0$ when $t = 4$.

An object comes to rest when $v(t) = 0$. So the object in Example 5 comes to rest at $t = 2$ and $t = 6$. In the next chapter we will learn that when the acceleration is 0, the object's speed changes from slowing down to speeding up.

When studying the velocity and acceleration of an object that is in motion in a vertical direction, usually the positive direction is upward and the negative direction is downward. This convention is used in the next example.

EXAMPLE 6

A ball is thrown straight upward. The ball's vertical distance $d(t)$ feet from the ground after t seconds is measured by the function

$d(t) = -16t^2 + 128t + 6$

Problem

(a) Find the velocity after $t = 1$ and $t = 6$ seconds.
(b) Find the acceleration when $t = 1$ second.
(c) At what value of t is the velocity equal to -32 ft/sec?
(d) When is the acceleration equal to -32 ft/sec^2?
(e) As far as the motion of the ball is concerned, what does it mean to say that the acceleration is a constant function?

Solution (a) First find the velocity function $v(t) = d'(t)$.

$$v(t) = d'(t) = -32t + 128$$

Evaluate $v(t)$ at $t = 1$ and $t = 6$.

$$v(1) = -32 + 128 = 96$$
$$v(6) = -192 + 128 = -64$$

When $t = 1$ second the velocity of the ball is 96 ft/sec. When $t = 6$ seconds the velocity is -64 ft/sec. The negative sign means that the ball is in a downward direction; that is, it is falling.

(b) Compute the acceleration function $a(t) = d''(t)$.

$$a(t) = d''(t) = -32$$

Evaluate $a(t)$ at $t = 1$.

$$a(1) = -32$$

When $t = 1$ second the acceleration is -32 ft/sec^2.

(c) Let $v(t)$ equal -32 and solve for t.

$$v(t) = -32$$
$$-32t + 128 = -32$$
$$-32t = -32 - 128 = -160$$
$$t = 5$$

When $t = 5$ seconds the velocity is -32 ft/sec.

(d) The acceleration function is the constant function $a(t) = -32$. This means that $a(t)$ is -32 ft/sec^2 when evaluated at every value of t.

(e) The second derivative measures the rate of change of the first derivative. Therefore the acceleration measures the rate of change of the velocity. When the ball is thrown in the air the only force acting on it is the force of gravity. It is a constant acceleration of -32 ft/sec^2. Gravity causes the velocity to decrease at this constant rate.

The second derivative is used in business to measure the rate of change of various quantities. For example, the derivative of cost is marginal cost. The rate of change of marginal cost is the derivative of marginal cost, which is the second derivative of cost.

EXAMPLE 7

A firm determines that its cost $C(x)$ of producing x units is given by the function

$$C(x) = 24 - 5x + 0.4x^2 - 0.01x^3$$

Problem

Find the derivative of the marginal cost function.

Solution The marginal cost function $MC(x)$ is

$$MC(x) = C'(x) = -5 + 0.8x - 0.03x^2$$

The derivative of $MC(x)$ is

$$MC'(x) = C''(x) = 0.8 - 0.06x$$

In the next chapter we will see that $MC'(x)$ gives information about $C(x)$.

In an example showing how firms such as fast food chains use price and demand to determine maximum profit, McCarty* uses the second derivative of a profit function. The next example shows how to compute the second derivative of this profit function.

EXAMPLE 8

The profit $P(Q)$ of a firm producing a quantity of Q goods is estimated to be

$$P(Q) = -50 + 390Q - 70.8Q^2$$

Problem

Find the second derivative of $P(Q)$.

Solution Compute $P'(Q)$ and then $P''(Q)$.

$$P'(Q) = 390 - 141.6Q$$
$$P''(Q) = -141.6$$

*Marilu Hurt McCarty, *Managerial Economics*, Glenview, IL, Scott, Foresman and Co., 1986, pp. 350–352.

EXERCISE SET 3–4

In Problems 1 to 14 find $f''(x)$.

1. $f(x) = x^2 + 1$

2. $f(x) = x^3 - 2$

3. $f(x) = 2x^4 - 7$

4. $f(x) = 3x^5 + 1$

5. $f(x) = 4x^{-1} + 5$

6. $f(x) = 2x^{-2} - 1$

7. $f(x) = x^{-3} + 2$

8. $f(x) = 3x^{-4} - 4$

9. $f(x) = x^4 - x^2$

10. $f(x) = x^5 + 3x^2$

11. $f(x) = 2x^{-1/3} + 7x^{-1/4} - 5$

12. $f(x) = 6x^{-1/6} - 3x^{-3/7} - 6x$

13. $f(x) = x^{0.3} + 3x^{0.4}$

14. $f(x) = 1.2x^{-1.5} - 3.1x^{-2.2}$

In Problems 15 to 18 compute $D_x^2 f(x)\Big|_{x=2}$.

15. $f(x) = x^2 + 2x^3$

16. $f(x) = 2x^2 + x^4 - 5$

17. $f(x) = x^{-2} - 2$

18. $f(x) = 2x^{-3} + x^{-4}$

In Problems 19 to 22 compute the second derivative.

19. $g(t) = t^2 + 2t^3$

20. $s(u) = 2u^2 + u^3 - 5$

21. $F(g) = g^{1/2} - 2g$ 22. $A(r) = 3r^{-1/3} + r^{-1}$

In Problems 23 to 26 an object is moving in a straight line with its position at time t given by $f(t)$. Find the value of t where (a) its velocity is 0 and (b) its acceleration is 0.

23. $f(t) = t^2 - 2t + 3$ 24. $f(t) = 2t^2 - 4t + 5$

25. $f(t) = t^3 - 2t^2 + t + 5$

26. $f(t) = 3t^3 - t^2 + 4t$

In Problems 27 and 28 a ball is thrown straight upward. The function $d(t)$ is the distance the ball is from the ground after t seconds. Find the acceleration function.

27. $d(t) = -16t^2 + 132t + 10$

28. $d(t) = -16t^2 + 164t + 5$

In Problems 29 and 30 the cost $C(x)$ of a firm producing x units is given. Find the derivative of the marginal cost function.

29. $C(x) = 120 - 22x + 0.9x^2 - 0.003x^3$

30. $C(x) = 233 - 32x + 0.5x^2 - 0.04x^3 + 0.001x^4$

In Problems 31 and 32 the profit $P(x)$ of a firm producing x units is given. Find the second derivative of the profit function.

31. $P(x) = -134 + x(122 - 0.4x)^2$

32. $P(x) = -533 + x(455 - 3.5x)^2$

In Problems 33 to 38 find $f''(x)$.

33. $f(x) = (x^2 + 1)^2$ 34. $f(x) = (x^3 + 4)^2$

35. $f(x) = (2x^2 + 1)^{-1}$ 36. $f(x) = (x^3 + 4)^{-2}$

37. $f(x) = (x^2 + 1)^{1/2}$ 38. $f(x) = (x^3 + 4)^{2/3}$

In Problems 39 to 42 find the values of x for which $f''(x) = 0$.

39. $f(x) = x^4 - 4x^3$ 40. $f(x) = x^4 + 2x^3$

41. $f(x) = x^4 - 6x^2$ 42. $f(x) = x^5 + 10x^3$

In Problems 43 to 46 find $f^{(3)}(x)$ and $f^{(4)}(x)$.

43. $f(x) = x^4 + 3x$

44. $f(x) = 2x^4 - x^3 + x$

45. $f(x) = 2x^{-1} + x^{-2}$

46. $f(x) = (1 + 2x)^{-1}$

In Problems 47 and 48 find $f^{(n)}(x)$.

47. $f(x) = x^{-2}$ 48. $f(x) = (1 - x)^{-1}$

Referenced Exercise Set 3–4

1. When a vendor supplies a business with a large amount of a product, the recipient often selects a small sample to test the quality. If the sample has an acceptable quality level, then the entire lot is accepted. There are two costs involved: the cost of using too many defectives when the lot was accepted but should have been rejected and the cost of rejecting the lot. The breakeven point is defined to be that fraction of defectives where the two costs are equal. If the fraction of defectives in the sample is greater than the breakeven point, then the lot is rejected. Edgar Dawes,* Director of Quality for the Dictaphone Corporation, illustrates how the breakeven point can be calculated by using the second derivative. In one particular sampling plan he defines the operating characteristic curve $y = f(x)$, where f measures the fraction of times the lot should be accepted when x defectives are found. Thus $f(0) = 1$ implies that when no defectives are found the lot should always be accepted. The function is approximated by

$$f(x) = 960x^3 - 180x^2 + 1$$

The function is defined in $[0, 0.16]$. In the next chapter we will see that the breakeven point occurs when $f''(x) = 0$. Determine the breakeven point for this sampling plan.

2. In an article studying local public goods, such as police and fire protection, Hayes† used empirical data from 90 Illinois municipalities. The cost $C(x)$ of these services can be estimated by the function

$$C(x) = 0.5x^{-0.7}$$

Find the derivative of the marginal cost function.

Cumulative Exercise Set 3–4

1. Find $f'(x)$ for $f(x) = 40 - 3x^{-2} - x^{0.4} + 5x$.

2. Find the equation of the tangent line to $f(x) = (x^3 - 4)^2$ at $x = 1$.

3. Find $f'(x)$ for $f(x) = x^3(2 - x)(1 - x^2)$.

*Edgar W. Dawes, "Optimizing Attribute Sampling Costs—A Case Study," *American Society of Quality Control—Technical Conference Transactions,* 1972, pp. 181–190.

†K. Hayes, "Local Public Goods Demands and Demographic Effects," *Applied Economics,* Vol. 18, 1986, pp. 1039–1045.

4. The total cost of producing stereos is given by $C(x) = 5000 + 25x(1 - 2x^2)$. Find the marginal cost.

In Problems 5 to 8 find $f'(x)$.

5. $f(x) = (3x^4 + x^5)^3$

6. $f(x) = (x^{1/3} + 2x^{-1} + 1)^{-2}$

7. $f(x) = (7x^3 + 2)(1 - 3x)^2$

8. $f(x) = x^3(1 - x^2)^{-1}$

In Problems 9 and 10 find $f''(x)$.

9. $f(x) = x^4 - 3x^{-2}$.

10. $f(x) = (x^4 + x)^2$

11. A ball is thrown straight upward. The function $d(t) = -16t^2 + 164t + 10$ is the distance the ball is from the ground after t seconds. Find the acceleration function.

12. Find $f^{(3)}(x)$ for $f(x) = x^5 - x^{-5}$.

CASE STUDY   **Radiotherapy Treatment Design***

This case study describes a new type of weapon used in the treatment of cancer. It discusses the mathematical models developed by David Sonderman and Philip Abrahamson to help cancer therapists devise a comprehensive treatment for their patients.

Cancer Therapy   Cancer is second only to heart disease as a cause of human mortality. One-fourth of all deaths of Americans are caused by cancer. After surgery the most important treatment of cancer is radiation therapy. More than half of all tumors can be treated by radiation therapy, which consists of directing ionizing radiation from external beams through a patient's body to the cancer cells. However, the radiation damages healthy cells along with cancerous tissue. Therefore it is necessary to develop a specific strategy of intensity and direction of the radiation for each individual patient in order to minimize the effect of the treatment on healthy cells. The process of selecting the intensity and direction of the beams is called *radiotherapy treatment design.*

It is common practice to develop a treatment design by using a trial-and-error process. The designer studies the patient's anatomy and proposes a treatment by using a series of several beams with various intensities and directions. Computer software is then used to calculate the effect of the radiation at specific points in the patient's body. If the designer is not satisfied with the predicted outcome of the plan, the design is modified or an entirely different treatment is proposed. The new treatment design is then evaluated, and the procedure continues until the predicted outcome is acceptable. This process is time-consuming and offers little guarantee that the design finally implemented will satisfy all the important constraints of the treatment, especially eliminating the cancer cells and minimizing the damage to healthy tissue.

Today many researchers are developing treatment design techniques that use mathematical optimization methods. Mathematical programming and calculus optimization techniques are replacing the trial-and-error process to produce better

*This case study is based on David Sonderman and Philip Abrahamson, "Radiotherapy Treatment Design Using Mathematical Programming Models," *Operations Research,* Vol. 33, No. 4, 1985, pp. 705–725.

treatment designs in much less time. These new rigorous techniques also ensure that all important constraints are satisfied.

Ionizing Radiation   Radiation from external beams is administered by sophisticated machines that accurately direct radiation that can vary in energy from several thousand electron volts to more than 40 million electron volts. Radiation dosage is measured in units called rads. While different patients require different dose levels, a common range is 3000 to 7000 rads. The human body cannot withstand such a large dose in one treatment so it is delivered in several treatments of smaller doses over a span of three to six weeks. However, the treatment design is based only on the total dose and its distribution throughout the body. Once the treatment design is formulated, the patient's requirements are determined to decide how often treatments will be given and what fractions of the total dose will be administered at various intervals.

Malignant tissue is slightly more sensitive to radiation than healthy tissue, so the radiation dose at each point of the treatment region must be large enough to kill the cancerous cells but small enough to spare the healthy cells. Tumor cells are usually microscopically interspersed among healthy cells. If the total dose given at a particular point of the tumor is too low, then cancerous tissue will survive and the disease will recur. Too large a dose can cause serious side effects. Therefore the treatment design must include a homogeneous dose distribution over the tumor region.

The Total Integral Dose Versus Homogeneity   The sum of the individual doses over a finite set of grid points throughout the treatment region is called the *total integral dose* and is denoted by A. The term *homogeneity* is defined to be the percentage of the tumor region that is covered by the radiation beams. If the only concern was to kill the cancerous tissue, then the homogeneity would be set at 100%. But as the homogeneity increases so does the number of healthy cells that are killed. This means that the total integral dose, A, is a function of the homogeneity, u. As u increases so does A.

For almost all patients there is a very gradual increase in A when u ranges between 80% and 85%. However, there is a dramatic increase in A when u approaches 95%. Therefore most treatment designers set u between 80% and 95%. The rule of thumb when designing a treatment is to increase u as much as possible until the increase in A is too great. This means that the rate of increase of A with respect to u, which is measured by the derivative, $A'(u)$, must not be greater than a specific value, usually set at $\frac{2}{3}$ or 0.67. That is, u can be increased as long as $A'(u)$ is less than 0.67.

A Specific Case   To illustrate their methods, Sonderman and Abrahamson present a specific case of a patient suffering from lung cancer. After studying the patient's anatomy, they used their linear programming techniques to devise a treatment based on maximizing u and minimizing A, subject to various constraints, including the direction and intensity of the beams, the location of the tumor, and the ability of the patient's anatomy to withstand a certain level of radiation. The final step is to determine the proper value of u. Their model

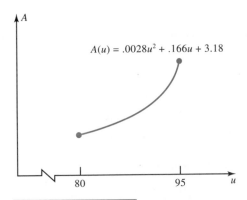

$A(u) = .0028u^2 + .166u + 3.18$

FIGURE 1 *Source:* David Sonderman and Philip Abrahamson, *Operations Research,* Vol. 33, No. 4, 1985, pp. 705–725.

produced the relationship between A and u, graphed in Figure 1, where the domain of A is [80, 100]; that is, u ranges from 80% to 100%. The graph can be approximated in the interval [80, 95] by the function

$$A(u) = 0.0028u^2 + 0.166u + 3.18$$

Figure 1 shows that there is a gradual increase in A when u increases from 80% to 85% but when u is larger than 95% A shows a dramatic rise. This case is typical. To find the value of u for the treatment, first find the derivative of A.

$$A'(u) = 0.0056u + 0.166$$

To demonstrate the behavior of the rate of change of A, we calculate some values of A': $A'(80) = 0.614$, $A'(85) = 0.642$, $A'(90) = 0.670$, and $A'(95) = 0.698$. The rule of thumb shows that the designer should set u at a value of about 90%. At this level of homogeneity the total integral dose of radiation to healthy cells is $A(90) = 40.8$.

The article closes with a comparison of the standard practice method of design treatment versus the mathematical optimization method. The standard practice design would either have yielded a smaller value of u than 90% to produce the same value of A, or, if the same value of u was used, the value of A would have been 49.6. Thus the mathematical model produced a savings of about 22% of A for the same homogeneity.

Case Study Exercises

Suppose a particular patient's profile yielded the stated function of A with respect to u. Find the value of u such that $A'(u) = \frac{2}{3}$.

1. $A(u) = 0.003u^2 + 0.18u + 3.5$
2. $A(u) = 0.003u^2 + 0.15u + 3.1$
3. $A(u) = 0.002u^2 + 0.33u + 2.5$
4. $A(u) = 0.002u^2 + 0.302u + 1.1$

CHAPTER REVIEW

Key Terms

3–1   A Few Elementary Rules
Marginal Product
Derivative of a Constant

Derivative of a Sum
Derivative of a Constant Times a Function

3–2   Product and Quotient Rules
Product Rule
Quotient Rule

Lerner Index

3–3   The Chain Rule
General Power Function
Generalized Power Rule

Chain Rule

3–4   Higher Derivatives
Second Derivative
Third Derivative

Higher Derivatives

Summary of Important Concepts

Formulas

$D_x(c) = 0$ for any constant c

$D_x(x^r) = rx^{r-1}$

$D_x[cf(x)] = cD_x[f(x)]$ for any constant c

$D_x[f(x) + g(x)] = D_x[f(x)] + D_x[g(x)]$

If $f(x) = g(x)k(x)$ then
$f'(x) = g(x)k'(x) + k(x)g'(x)$

If $f(x) = \dfrac{g(x)}{k(x)}$ then

$$f'(x) = \frac{k(x)g'(x) - g(x)k'(x)}{[k(x)]^2}$$

$D_x[g(x)]^n = ng(x)^{n-1} D_x[g(x)]$

REVIEW PROBLEMS

In Problems 1 to 4 find $f'(x)$.

1. $f(x) = 4x^2 + x^3 + 1$

2. $f(x) = 5x^4 - 2x^{-2}$

3. $f(x) = 4x^{-5} - x^{1/3} + 1$

4. $f(x) = \dfrac{5}{x^2} - \dfrac{2}{x^{1/2}}$

5. Find the equation of the tangent line to
$f(x) = x^{-3/2} - x$ at $x = 1$.

In Problems 6 to 16 find $f'(x)$.

6. $f(x) = (x^4 - 7)(11x^3 + 1)$

7. $f(x) = \dfrac{x^3}{x^2 + 2}$

8. $f(x) = \dfrac{x^3 - 2}{x^2 + 2x}$

9. $f(x) = \dfrac{x^{1/3}}{x^{4/5} + 2}$

10. $f(x) = (x^3 + 2)^3$

11. $f(x) = (2x^3 + x)^4$

12. $f(x) = (2 - 3x^{-1})^{-2}$

13. $f(x) = 3x^2 + 3(x^3 + 4x)^3$

14. $f(x) = x^2(x^3 + 2x)^3$

15. $f(x) = (x^2 + 1)^2(x^3 + 1)^3$

16. $f(x) = \dfrac{x^2}{(x^2 + 3x)^2}$

17. Find an equation of the tangent line to
$f(x) = (x^3 + 3)^2$ at $x = 1$.

In Problems 18 and 19 find $f''(x)$.

18. $f(x) = 6x^{1/2} - 3x^{-3} - 6x$

19. $f(x) = (x^3 + 4)^{-2}$

20. Find the values of x for which $f''(x) = 0$ where
$f(x) = x^3 - 4x^2$.

21. Find $f'''(x)$ for $f(x) = x^{-3}$.

GRAPHICS CALCULATOR EXPLORATIONS

Your graphics calculator can help you visualize the geometric connection between a function's graph and the graph of the tangent line to the graph at a point. This Exploration will help you understand what is meant by the slope of a curve by having you make a guess at the slope of the curve and then plot the tangent line to see how close your guess was.

Exploration 1

First we will investigate how to plot a function and one of its tangent lines in the same viewing screen. Consider the function $f(x) = x^3 - 3x^2$. Graph the function and then choose any convenient point on the graph. If we choose $x = 1$, the corresponding point is $(1, -2)$. Find the derivative of $f(x)$: $f'(x) = 3x^2 - 6x$. Then $f'(1) = -3$. The tangent line has slope -3 and passes through $(1, -2)$. Its equation is $y = -3x + 1$. Now look at the graph of $f(x)$ in the viewing screen and predict where the graph of the tangent line will appear. Guess where the tangent line will intersect the graph of $f(x)$ and its slope. Then graph $y = -3x + 1$ and see how close your guess was.

CASIO

1. Press GRAPH ALPHA x x^y 3 − 3 ALPHA
   x x² EXE .

2. Press GRAPH (−) 3 ALPHA x + 1 EXE .

TI-81

1. Press Y = .

2. Press X|T ^ 3 − 3 X|T ^ 2 ENTER .
   Press (−) 3 X|T + 1 ENTER .

3. Press GRAPH .

Do the same procedure with several more points and then do it again for another function until you can easily predict where the tangent line will appear.

Exploration 2

Now we will look at the same problem from a different perspective. Instead of choosing a point on a given graph, choose a particular numerical value for the slope of the curve. Then try to choose a point on the graph of the function so that the slope of the tangent line to the graph is equal to, or at least close to, the given value.

To demonstrate, start with the function $f(x) = x^3 - x^2$, sketch its graph and choose a particular value for the slope of the curve, 1, for example. Picture a tangent line with slope 1 and try to guess a point on the curve where the tangent line is 1. Suppose you choose $x = 1$. Now find the equation of the tangent line and graph it in the same viewing screen as the function. Since $f'(x) = 3x^2 - 2x$, we have $f'(1) = 1$ and so the tangent line at $x = 1$, that is, at the point (1,0), is 1 as desired. The equation of the tangent line is $y = x$. Graph the function and the tangent line in the same viewing screen and you will have a visual depiction of what it means to say that the slope of the curve is 1.

If you choose a different value for the slope, it may not be easy to find a point with a tangent line that has that slope. For instance, choose the slope to be 0.25. There are two points where the slope of the tangent line is 0.25—both are between $x = 0$ and $x = 1$. Use the trace feature to move along the curve until you find a point where the slope of the tangent line is close to 0.25, and then graph the tangent line to see how close you came. Of course, you can use the quadratic formula to find these points algebraically, but it is also beneficial to investigate how to solve this problem geometrically.

Exploration 3

It will become apparent in the next chapter that the most useful value to choose when looking for tangent lines is 0. It will be helpful if you can use your graphics calculator to build a firm geometric understanding of points on a curve that have slope of tangent lines equal to zero, that is, points where the tangent line is horizontal.

Consider the function $f(x) = x^3 - x^2$. From its graph try to find the two points that have horizontal tangent lines. Use the procedure in Exploration 2 above to see how close your guess is. You might be able to answer this question algebraically, which is an important skill covered in the next chapter. An example of a function with points that have horizontal tangent lines and are harder to determine algebraically is $f(x) = x^3 + 3x^2 - 6x$. For this function and many like it, the procedure outlined in these Explorations is a convenient way to approximate the horizontal tangents of the function. For this function, picture a tangent line with slope -1 and guess at a point on the curve whose tangent line is -1.

Further Explorations

In Problems 1 to 4 write a program to compute the derivative of the given type of function.

1. $f(x) = x^n$
2. $f(x) = kx^n + b$
3. $f(x) = (a + bx)^m$
4. $f(x) = x^n(a + bx)^m$

Applications of the Derivative

4

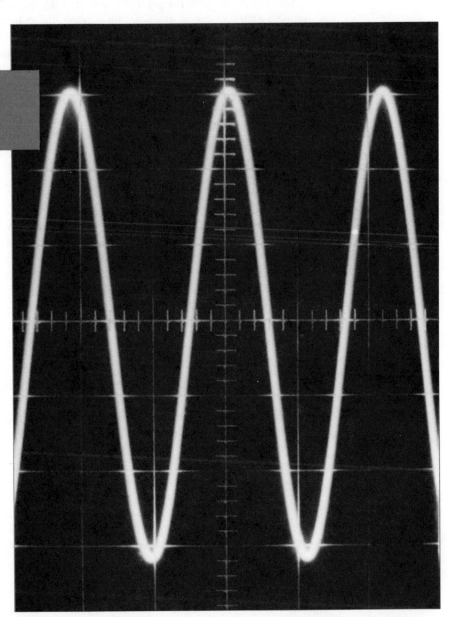

Graphs can refer to maximum profits at minimum cost, as described in this chapter.
(Zefa/H. Armstrong Roberts)

Chapter Overview

When a real-world application uses rate of change, the natural tool is the derivative. This chapter shows how the derivative measures various properties about functions that arise in applications. These include rate of inflation, maximum profit, the most efficient route, minimum cost, and minimum traffic noise.

The first section deals with the effect the derivative has in determining where a function is increasing and where it is decreasing. Section 4–2 covers relative extrema, that is, the point on a graph that is higher or lower than any other point on the graph in the vicinity. The derivative is used to determine where a function has a relative maximum and where it has a relative minimum. The second derivative also gives useful information about the function. It helps to determine where a function is concave up or concave down, which means, intuitively, where the function is bending up or bending down.

The next section applies the ideas on relative extrema to problems where an absolute extremum is sought. Section 4–5 deals with problems in which two variables are changing. These are called related rates problems and they occur often in applications.

CASE STUDY PREVIEW

One of the most important considerations of highway operation is safety. Obtaining optimum flow of traffic is also desired, but frequently these two goals work at odds with each other. The case study tells the story of how highway administrators in Houston used calculus to minimize acceleration noise, which is defined as the disturbance of a vehicle's speed from uniform speed. Acceleration noise is a good measure of the threat of hazardous conditions. So minimizing acceleration noise produces safer traffic conditions. Researchers found that the optimum speed for minimizing acceleration noise is very close to the speed that maximizes traffic volume.

4–1 Increasing and Decreasing Functions

One of the most important applications of calculus centers on when a function is increasing and when it is decreasing. A firm wants to know when profit is increasing and when cost is decreasing. A political party studies the latest polls to find out if its percentage of the vote is increasing or decreasing. In this section we study when graphs of functions are increasing and decreasing. The definition of increasing and decreasing functions has a firm geometrical foundation. However, it is also necessary to have an algebraic means for determining where a function is increasing and decreasing. The primary tool is the derivative of the function.

An Application: Laffer's Curve

Let us start by studying a concept from business that depends on when a function is increasing or decreasing.

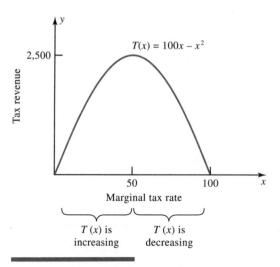

FIGURE 1 *Source:* Milton H. Spenser, *Contemporary Economics,* 6th ed., New York, Worth Publishers, 1986, p. 293.

Supply-side economics states that governments should stimulate production by giving incentives to business in order to increase economic growth. Other economists believe the best way to increase economic growth is to increase demand by giving incentives to consumers in the way of tax benefits. At the heart of the matter is the concept of *marginal tax rates,* the taxes paid on the last increment of income. Should the government increase or decrease marginal tax rates? Economist Arthur Laffer answers the question with a graph, called the *Laffer curve.* * It measures a tax revenue function, $T(x)$, where x is the marginal tax rate measured in percent. A typical Laffer curve is given in Figure 1, where $T(x) = 100x - x^2$. For this curve, tax revenue is highest when $x = 50\%$. From $x = 0\%$ to $x = 50\%$, $T(x)$ is increasing. This means that if the marginal tax rate were at a value from 0% to 50%, an increase in x that is not beyond 50% would generate larger tax revenue. But if x is greater than 50%, an increase in the marginal tax rate would generate less tax revenue. It is this concept of when a function is increasing and decreasing that we study in this section.

Increasing and Decreasing Functions

Consider the graph in Figure 2. Intuitively, the graph is increasing from point A to point B because it goes up as x increases from left to right. The curve is decreasing from B to C because it goes down from left to right. Thus the property of an increasing or decreasing function implies the convention that x always moves from left to right. From the point of view of the x-axis, the curve in Figure 2 is increasing from $x = a$ to $x = b$ and decreasing from $x = b$ to $x = c$.

*Milton H. Spenser, *Contemporary Economics,* 6th ed., New York, Worth Publishers, 1986, p. 293.

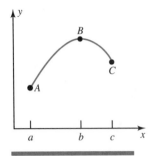

FIGURE 2

DEFINITION

A function $f(x)$ is said to be **increasing** from $x = a$ to $x = b$ if for any x_1 and x_2 in the interval (a, b)

when $x_1 < x_2$,   then $f(x_1) < f(x_2)$

A function $f(x)$ is **decreasing** from $x = a$ to $x = b$ if for any x_1 and x_2 in the interval (a, b)

when $x_1 < x_2$,   then $f(x_1) > f(x_2)$

A function $f(x)$ is said to be *increasing at $x = a$* if $f(x)$ is increasing in an interval containing a, and $f(x)$ is said to be *decreasing at $x = a$* if $f(x)$ is decreasing in an interval containing a.

EXAMPLE 1

Problem

Consider $f(x) = x^2$ whose graph is in Figure 3. Find the intervals where $f(x)$ is increasing and where $f(x)$ is decreasing.

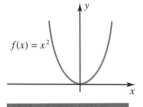

$f(x) = x^2$

FIGURE 3

Solution Because $f(x)$ is going down from left to right for all $x < 0$, $f(x)$ is decreasing in the interval $(-\infty, 0)$. For any x_1 and x_2 in $(-\infty, 0)$, if $x_1 < x_2$, then $f(x_1) > f(x_2)$ because $x_1^2 > x_2^2$. For example, $-2 < -1$ and $f(-2) = (-2)^2 = 4 > 1 = f(-1)$. Similarly $f(x)$ is increasing in $(0, \infty)$.

A Test for Increasing and Decreasing Functions

The primary algebraic tool for determining the intervals where a function is increasing and where it is decreasing is the derivative. For example, if $f(x) = x^2$, then $f'(x) = 2x$. This says that the slope of the tangent line of $f(x)$ is negative for all $x < 0$ and the slope of the tangent line is positive for all $x > 0$. Coupling this fact with Example 1 means that the function $f(x) = x^2$ is increasing when $f'(x) > 0$ and decreasing when $f'(x) < 0$. This is true in general. If the derivative is positive for every value in an interval, then the function is increasing in that interval. A negative derivative means the function is decreasing.

> **Test for Increasing and Decreasing Functions**
> If $f(x)$ is differentiable in an interval (a, b), then
>
> 1. $f(x)$ is increasing in (a, b) if $f'(x) > 0$ for all $x \in (a, b)$.
> 2. $f(x)$ is decreasing in (a, b) if $f'(x) < 0$ for all $x \in (a, b)$.

If $f'(x) = 0$ at $x = a$, then the slope of the tangent line of $y = f(x)$ at $x = a$ is 0. The tangent line is horizontal at such a point. These are the points that often divide the intervals where $f(x)$ changes from increasing to decreasing.

The function's derivative can also change sign at points where $f'(x)$ does not exist. For example, the derivative of $f(x) = x^{2/3}$ is $f'(x) = \frac{2}{3}x^{-1/3}$, and so $f'(0)$ does not exist. Since $f'(x) < 0$ for all x in $(-\infty, 0)$, $f(x)$ is decreasing to the left of $x = 0$; and since $f'(x) > 0$ in $(0, \infty)$, $f(x)$ is increasing to the right of $x = 0$. Therefore, at $x = 0$, the function's derivative changes sign and the graph changes from decreasing to increasing. The graph of $f(x) = x^{2/3}$ is given in Figure 4. The graph has a "cusp" at $x = 0$. Note that from Figure 4 the tangent line of $f(x) = x^{2/3}$ at $x = 0$ exists—it is the y-axis. But the slope of the tangent line is undefined because the tangent line is a vertical line which has no slope. Algebraically, we have $f'(0) = \dfrac{2}{3 \cdot 0} = \dfrac{2}{0}$, which is undefined.

This suggests how to determine the intervals where $f(x)$ is increasing and intervals where $f(x)$ is decreasing. First find the values $x = a$ where $f'(a) = 0$ or $f'(a)$ does not exist (but $f(a)$ does exist). These values of x are called the **critical**

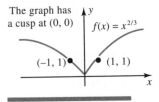

The graph has a cusp at (0, 0)

$f(x) = x^{2/3}$

$(-1, 1)$ $(1, 1)$

FIGURE 4

values of $f(x)$. The critical values divide the domain of the function into distinct intervals where $f(x)$ is either increasing or decreasing. To determine the sign of $f'(x)$ in an interval, it is enough to select one value in the interval, say, $x = c$, and then to find $f'(c)$. If $f'(c) > 0$, then $f'(x)$ is positive for all values in the interval. Thus $f(x)$ is increasing in that interval. We treat $f'(c) < 0$ in a similar manner.

Procedure for Determining Where $f(x)$ Is Increasing and Decreasing

1. Find the critical values of $f(x)$, that is, the values $x = a$ where $f'(a) = 0$ or $f'(a)$ does not exist (but $f(a)$ does exist).
2. Determine the distinct intervals where $f(x)$ is either increasing or decreasing using the values in step 1 as end points.
3. For each interval select a value c in the interval and find $f'(c)$.

   i. If $f'(c) > 0$, then $f'(x) > 0$ for all x in that interval.

   ii. If $f'(c) < 0$, then $f'(x) < 0$ for all x in that interval.

4. Apply the test for increasing and decreasing functions.

EXAMPLE 2

Problem

Find the intervals where $f(x) = x^2 - 4x + 3$ is increasing and the intervals where $f(x)$ is decreasing.

Solution Let us follow the steps in the procedure.

1. Compute $f'(x) = 2x - 4$. Solve $f'(x) = 2x - 4 = 0$. This implies $x = 2$. Thus $f'(2) = 0$. There are no values where $f'(x)$ does not exist.
2. This divides the real line into $(-\infty, 2)$ and $(2, \infty)$.
3. Select $0 \in (-\infty, 2)$ and $3 \in (2, \infty)$. (Remember, any suitable values can be chosen.)
   a. For $x = 0$, $f'(0) \doteq -4$, so $f'(x) < 0$ in $(-\infty, 2)$.
   b. For $x = 3$, $f'(3) = 2$, so $f'(x) > 0$ in $(2, \infty)$.
4. a. $f(x)$ is decreasing in $(-\infty, 2)$ because $f'(x) < 0$ in $(-\infty, 2)$.
   b. $f(x)$ is increasing in $(2, \infty)$ because $f'(x) > 0$ in $(2, \infty)$.

This information is recorded in the chart below.

Test point		
$f'(x) = 0$ at		
	$f'(0) = -4$	$f'(3) = 2$
Sign of $f'(x)$	$-----------$	$++++++++++$
$f(x)$ is	decreasing	increasing

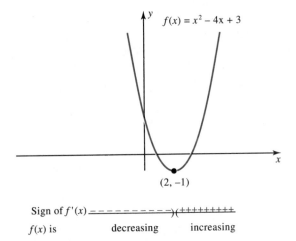

$f(x) = x^2 - 4x + 3$

$(2, -1)$

Sign of $f'(x)$ ‒ ‒ ‒ ‒ ‒ ‒ ‒ ‒ ‒ ‒)(+++++++++

$f(x)$ is   decreasing   increasing

FIGURE 5

The graph of $f(x) = x^2 - 4x + 3$ is given in Figure 5. Example 2 shows where $f(x)$ is increasing and where $f(x)$ is decreasing. Also note that there is a horizontal tangent line at $x = 2$ because $f'(2) = 0$. From these facts the graph can be obtained by plotting just a few points. This demonstrates that knowing where a function is increasing and where it is decreasing is a useful tool in graphing the function. In the next few sections we discuss other properties of the graphs of functions.

Example 3 looks at a more complicated function.

EXAMPLE 3

Problem

Find the intervals where $f(x) = x^3 - 6x^2 + 9x$ is increasing and where $f(x)$ is decreasing. Locate the points where the graph has a horizontal tangent line. Graph the function.

Solution

1. Solve $f'(x) = 0$ where

$$f'(x) = 3x^2 - 12x + 9 = 3(x^2 - 4x + 3) = 3(x - 1)(x - 3)$$

Then $3(x - 1)(x - 3) = 0$ implies $x = 1$ or $x = 3$. Therefore $f'(1) = f'(3) = 0$. There are no values where $f'(x)$ does not exist.
2. This divides the real line into the three intervals $(-\infty, 1)$, $(1, 3)$, and $(3, \infty)$.
3. Select $0 \in (-\infty, 1)$, $2 \in (1, 3)$, and $4 \in (3, \infty)$.
   a. For $x = 0$, $f'(0) = 9$, so $f'(x) > 0$ in $(-\infty, 1)$.
   b. For $x = 2$, $f'(2) = -3$, so $f'(x) < 0$ in $(1, 3)$.
   c. For $x = 4$, $f'(4) = 9$, so $f'(x) > 0$ in $(3, \infty)$.
4. a. $f(x)$ is increasing in $(-\infty, 1)$ because $f'(x) > 0$ in $(-\infty, 1)$.
   b. $f(x)$ is decreasing in $(1, 3)$ because $f'(x) < 0$ in $(1, 3)$.
   c. $f(x)$ is increasing in $(3, \infty)$ because $f'(x) > 0$ in $(3, \infty)$.

$f(x)$ has a horizontal tangent line when $f'(x) = 0$. Since $f'(1) = f'(3) = 0$, $f(x)$ has horizontal tangent lines at $x = 1$ and at $x = 3$, that is, at the points $(1, 4)$ and $(3, 0)$. This information is recorded in the following chart.

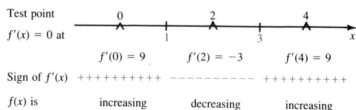

Test point

$f'(x) = 0$ at

$f'(0) = 9$ $f'(2) = -3$ $f'(4) = 9$

Sign of $f'(x)$ +++++++++++   -----------   +++++++++++

$f(x)$ is   increasing   decreasing   increasing

The graph is in Figure 6.

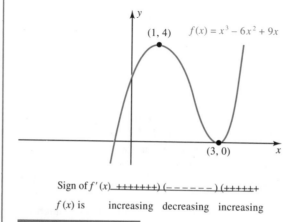

$(1, 4)$ $f(x) = x^3 - 6x^2 + 9x$

$(3, 0)$

Sign of $f'(x)$ +++++++) (------) (+++++++

$f(x)$ is   increasing   decreasing   increasing

FIGURE 6

EXAMPLE 4

Problem

Let $f(x) = x + \dfrac{4}{x}$. Find the intervals where $f(x)$ is increasing and where $f(x)$ is decreasing. Locate the points where the graph has a horizontal tangent line. Graph the function.

Solution

1. Set $f'(x) = 1 - 4/x^2$ equal to 0. Then $1 = 4/x^2$, so $x^2 = 4$. Hence $x = \pm 2$ are critical values. Also, $f'(x)$ does not exist at $x = 0$, but $x = 0$ is not a critical value because $f(0)$ does not exist.
2. This divides the real line into the four intervals $(-\infty, -2)$, $(-2, 0)$, $(0, 2)$, and $(2, \infty)$.
3. Select $-3 \in (-\infty, -2)$, $-1 \in (-2, 0)$, $1 \in (0, 2)$, and $3 \in (2, \infty)$.
   a. For $x = -3$, $f'(-3) = \frac{5}{9}$, so $f'(x) > 0$ in $(-\infty, -2)$.
   b. For $x = -1$, $f'(-1) = -3$, so $f'(x) < 0$ in $(-2, 0)$.
   c. For $x = 1$, $f'(1) = -3$, so $f'(x) < 0$ in $(0, 2)$.
   d. For $x = 3$, $f'(3) = \frac{5}{9}$, so $f'(x) > 0$ in $(2, \infty)$.

4. a. $f(x)$ is increasing in $(-\infty, -2)$ because $f'(x) > 0$ in $(-\infty, -2)$.
   b. $f(x)$ is decreasing in $(-2, 0)$ because $f'(x) < 0$ in $(-2, 0)$.
   c. $f(x)$ is decreasing in $(0, 2)$ because $f'(x) < 0$ in $(0, 2)$.
   d. $f(x)$ is increasing in $(2, \infty)$ because $f'(x) > 0$ in $(2, \infty)$.

$f(x)$ has a horizontal tangent line when $f'(x) = 0$. Since $f'(-2) = f'(2) = 0$, $f(x)$ has horizontal tangent lines at $x = -2$ and at $x = 2$; that is, at the points $(-2, -4)$ and $(2, 4)$. The graph approaches the y-axis asymptotically both to the left and right of 0. This information is recorded in the chart below.

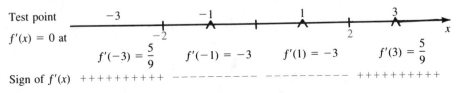

Test point	-3	-1	1	3
$f'(x) = 0$ at				
	$f'(-3) = \dfrac{5}{9}$	$f'(-1) = -3$	$f'(1) = -3$	$f'(3) = \dfrac{5}{9}$
Sign of $f'(x)$	$+++++++++$	$---------$	$--------$	$++++++++$
$f(x)$ is	increasing	decreasing	decreasing	increasing

The graph is in Figure 7. It shows that the line $x = 0$, which is the y-axis, is a vertical asymptote of $f(x)$.

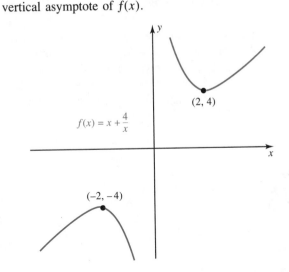

$f(x) = x + \dfrac{4}{x}$

$(2, 4)$

$(-2, -4)$

Sign of $f'(x)$ $+++++)(------)(------)(++++++$
$f(x)$ is   increasing decreasing   decreasing   increasing

FIGURE 7

An Application from Economics

The revenue of a firm depends on many factors, one of which is the quantity q of goods produced. A typical revenue function is dome-shaped as in Figure 8. When

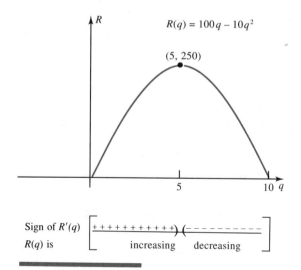

Sign of $R'(q)$ $[+ + + + + + + + + + + +)(- - - - - - - - - - -]$

$R(q)$ is   increasing   decreasing

FIGURE 8 *Source:* Based on Paul A. Samuelson, *Economics,* 10th ed., New York, McGraw-Hill Book Co., 1985, p. 479.

q is 0, that is, no goods are produced, the revenue $R(q)$ is 0. As q increases, so does $R(q)$ until it reaches a peak where maximum revenue occurs. In Figure 8 this occurs at $q = 5$. Then the values of $R(q)$ decrease for values of q greater than 5. There are several reasons that this decrease in $R(q)$ takes place. One reason is that the firm might not be able to sell such a large amount of goods. The next example computes the intervals where the revenue function in Figure 1 is increasing and where it is decreasing. This function is defined only in the interval [0, 10].

EXAMPLE 5

Problem

Determine the intervals where $R(q)$ is increasing and where $R(q)$ is decreasing.

$$R(q) = 100q - 10q^2, \qquad 0 \le q \le 10$$

Solution Compute $R'(q)$.

$$R'(q) = 100 - 20q$$

1. Solve $R'(q) = 0$. Then $100 - 20q = 0$ implies $q = 5$. There are no values where $R'(q)$ does not exist.
2. This divides the domain into [0, 5) and (5, 10).
3. Select $1 \in [0, 5)$ and $6 \in (5, 10]$.
   a. For $q = 1$, $R'(1) = 80$, so $R'(q) > 0$ in [0, 5).
   b. For $q = 6$, $R'(6) = -20$, so $R'(q) < 0$ in (5, 10].
4. a. $R(q)$ is increasing in [0, 5) because $R'(q) > 0$ in [0, 5).
   b. $R(q)$ is decreasing in (5, 10] because $R'(q) < 0$ in (5, 10].

EXERCISE SET 4–1

In Problems 1 to 28 find intervals where $f(x)$ is increasing and where $f(x)$ is decreasing. Find the point(s) where $f(x)$ has a horizontal tangent line. Graph the function.

1. $f(x) = x^2 + 5$

2. $f(x) = 3x^2 - 2$

3. $f(x) = 7 - 2x^2$

4. $f(x) = 1 - 3x^2$

5. $f(x) = x^2 - 9$

6. $f(x) = 4x^2 - 25$

7. $f(x) = x(x - 4)$

8. $f(x) = x(x - 8)$

9. $f(x) = x^2 + 4x - 1$

10. $f(x) = x^2 - 8x + 3$

11. $f(x) = x^3 + 3x^2 - 9x$

12. $f(x) = 2x^3 + 3x^2 - 12x + 1$

13. $f(x) = x^3 - 3x^2 - 9x$

14. $f(x) = 2x^3 + 3x^2 - 36x + 3$

15. $f(x) = -x^3 + 3x^2 + 9x$

16. $f(x) = -x^3 + 6x^2$

17. $f(x) = x + 9/x$

18. $f(x) = 4x + 25/x$

19. $f(x) = (x^2 + 12)/x$

20. $f(x) = (4x^2 + 9)/x$

21. $f(x) = x(x^2 - 4)$

22. $f(x) = x(9 - x^2)$

23. $f(x) = x^2(x - 2)$

24. $f(x) = x^2(3 - x)$

25. $f(x) = x/(x + 1)$

26. $f(x) = x/(x - 4)$

27. $f(x) = (x - 1)/(x + 1)$

28. $f(x) = (x + 3)/(x - 5)$

29. The average cost $AC(x)$ to produce the xth item of production, where $AC(x)$ is measured in dollars and x is measured in hundreds of items, is given by

$$AC(x) = 100 + 48x - 7x^2 + \frac{x^3}{3}$$

If $AC(x)$ is defined for $0 < x < 10$, find the intervals where $AC(x)$ is increasing and where it is decreasing.

30. The amount of glucose $G(t)$ in a patient's bloodstream, measured in tenths of a percent, is measured at time t in minutes during a stress test. The function is

$$G(t) = 12t - 2.1t^2 + 0.1t^3, \text{ where } 0 < t < 12$$

Find the intervals where $G(t)$ is increasing and where $G(t)$ is decreasing.

In Problems 31 to 34 use the information given to sketch the graph of $f(x)$, assuming that $f(x)$ is continuous for all real numbers x.

31. Values where $f'(x) = 0$

Sign of $f'(x)$

$f(0) = 0, f(1) = 2, f(2) = 0, f(3) = 2$

32. Values where $f'(x) = 0$

Sign of $f'(x)$

$f(-2) = 1, f(-1) = 0, f(3) = 1, f(4) = 0$

33. Values where $f'(x) = 0$

Sign of $f'(x)$

$f(0) = 1, f(1) = 0, f(3) = 2, f(5) = 1$

34. Values where $f'(x) = 0$

Sign of $f'(x)$

$f(0) = -1, f(3) = 2, f(6) = 4$

In Problems 35 to 42 find the intervals where $f(x)$ is increasing and where $f(x)$ is decreasing. Graph the functions.

35. $f(x) = x^2/(x - 2)$ 36. $f(x) = x^2/(1 - 2x)$

37. $f(x) = x^2(x^2 - 4)$ 38. $f(x) = x^2(9 - x^2)$

39. $f(x) = (x^2 - 1)(x^2 - 4)$

40. $f(x) = (x^2 - 3)(9 - x^2)$

41. $f(x) = (x - 5)^2(x + 1)^2$

42. $f(x) = (x + 6)^2(x - 4)^2$

In Problems 43 to 46 from the given figure find (a) the intervals where $f(x)$ is increasing, (b) the intervals where $f(x)$ is decreasing, and (c) the sign of the derivative in each interval.

43.

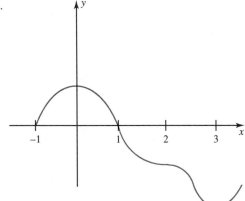

44.

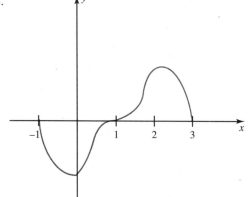

45.

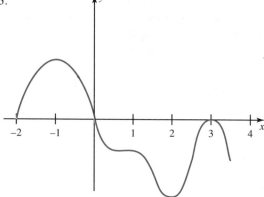

46.

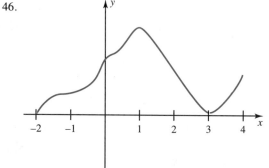

Referenced Exercise Set 4–1

1. In a paper studying the Environmental Protection Agency's motor vehicle emission testing procedures, researchers measured the risk, r, of an incorrect decision on a part versus the percentage, x, of actual deterioration of the part.* That is, laboratory standards were developed for the deterioration level of each part of the emission system. When a part reached the deterioration level where it should fail the EPA test, x was assigned the value 1. For $x < 1$, the part should pass, and it should fail if $x > 1$. (The units of the values assigned to r were arbitrary, but the higher the value the greater the risk, meaning the greater the error.) Field agents were then given the parts to test. There were two types of failure: The part should be passed and the agent failed it, and the part should fail and the agent passed it. The graph in Figure 9 was given measuring the risk r versus the deterioration x. Ex-

*George Miller et al., "Application of Sequential Testing to Motor Vehicle Emission Certification," *Decision Sciences*, Vol. 31, 1985, pp. 249–263.

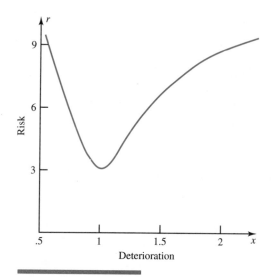

FIGURE 9 *Source:* George Miller et al., "Application of Sequential Testing to Motor Vehicle Emission Certification," *Decision Sciences,* Vol. 31, 1985, pp. 249–263.

plain why the curve is decreasing for $x < 1$ and increasing for $x > 1$.

2. In the theory of traffic science a "bottleneck" is defined as a section of roadway with a flow capacity less than the road ahead. When the volume reaches the capacity of the bottleneck, the velocity in the bottleneck is much less than that ahead of the bottleneck. A further increase in volume creates a queue in back of the bottleneck. In an article describing why congestion lasts longer than the interval in which demand exceeds capacity, a graph measuring the volume during rush-hour traffic

on a large city freeway is studied.[†] The data are approximated by the function $f(t) = -25t^3 + 75t + 150$, where $f(t)$ is the volume of automobiles passing through the bottleneck at time t, and where $t = 0$ represents 7:00 A.M. and $t = 1$ represents 7:30 A.M. The function is defined in the interval $[0, 2]$, that is, from 7:00 A.M. until 8:00 A.M.

(a) Find where $f(t)$ is increasing and where $f(t)$ is decreasing.

(b) The capacity of the bottleneck is $f(t) = 170$. From the graph estimate when demand exceeds capacity.

3. In an article[‡] appearing in the *Journal of Quality Progress* two administrators in the U.S. Department of Agriculture defined "tolerance" as any limit or allowance that is specified by a contract to measure quality. One example referred to the common practice of produce shippers, such as egg packers, of specifying that "no more than 20% of the eggs may be below A quality." The primary question addressed in the article is when should a shipment with this specification be rejected? The function used is $y = f(x)$, where f measures the percent of times the shipment was accepted with $x\%$ of the eggs below A quality. The function is approximated by

$$f(x) = \frac{x^3}{135} - \frac{x^2}{3} + 100$$

(a) Find where $f(x)$ is increasing and where $f(x)$ is decreasing.

(b) From the definition of $f(x)$, why is it reasonable to assume $f(x)$ is decreasing there?

(c) Find $f(20)$.

(d) This function models the data only in the interval $[0, 30]$. Give a reason for this.

[†]Karl Moskowitz and Leonard Newman, "Notes on Freeway Capacity," *Highway Resources Board Record,* Vol. 27, 1963.

[‡]Richard P. Bartlett and Lloyd P. Provost, "Tolerances in Standards and Specifications," *Journal of Quality Progress,* 1973, pp. 14–18.

4–2 Relative Extrema: The First Derivative Test

A common type of rate of change problem requires that a variable be maximized or minimized. A business wants to maximize profit; a physician needs to know the maximum amount of a drug that can be administered safely; a physiologist studies the minimum rate of stimulus needed to achieve a given response from an organism. The derivative plays a fundamental role in the solution of such problems.

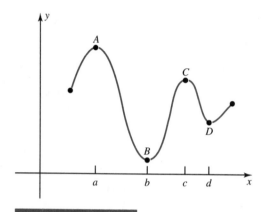

FIGURE 1

Relative Extrema

Consider the graph in Figure 1, which we use to describe the geometric meaning of extrema. Point A is the highest point on the curve because its second coordinate, $f(a)$, is greater than all others. We call $f(a)$ an *absolute maximum*. Similarly B is the lowest point of the graph so $f(b)$ is called an *absolute minimum*. Point C is not the absolute highest point, but it is higher than any of the points in its vicinity. For this reason $f(c)$ is called a *relative maximum*. Similarly $f(d)$ is a *relative minimum*. The term **extremum** refers to either a **maximum** or **minimum.** We also say that $f(x)$ has a relative maximum at $x = c$ and $f(x)$ has a relative minimum at $x = d$. This section deals with relative extrema and the next section considers absolute extrema.

DEFINITIONS

1. $f(a)$ is a **relative maximum** of $f(x)$ if there is an interval (c, d) containing a such that

   $f(x) \leq f(a)$ for all x in (c, d)

2. $f(a)$ is a **relative minimum** of $f(x)$ if there is an interval (c, d) containing a such that

   $f(x) \geq f(a)$ for all x in (c, d)

The first example ties together the definition and the geometric meaning of extrema. It considers a function that was graphed in Example 3 of the previous section.

EXAMPLE 1

Problem

Find the extrema of $f(x) = x^3 - 6x^2 + 9x$ by looking at the graph in Figure 2. Then show that the definition is satisfied for each extremum.

Solution The graph has a "peak" at $(1, 4)$ and a "valley" at $(3, 0)$. Thus the graph has a relative maximum of $f(1) = 4$ when $x = 1$ and a relative minimum of $f(3) = 0$ when $x = 3$. To see that $f(1)$ satisfies the definition of a relative maximum, we need an interval containing 1 such that the functional values of all other points in the interval are less than $f(1) = 4$. From Figure 2 it is clear that $f(x) < 4$ for all $x < 3$ (except $x = 1$), so the interval can be chosen to be $(-\infty, 3)$. Any smaller interval containing 1 will also work. (Can a larger interval be selected?) Similarly the largest interval for the relative minimum $f(3) = 0$ is $(0, \infty)$ because the $f(x) > 0$ for all $x > 0$ except $x = 3$.

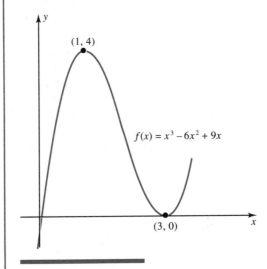

FIGURE 2

At a relative extremum a curve either has a horizontal tangent line or the derivative does not exist. For instance, in Example 1, $f(x) = x^3 - 6x^2 + 9x$ has horizontal tangent lines at $x = 1$ and at $x = 3$. The graph in Figure 2 has a horizontal tangent line at $(1, 4)$ and $(3, 0)$. In the previous section it was pointed out that the graph of a function has a horizontal tangent line when $f'(x) = 0$. In Example 2, $f'(x) = 3x^2 - 12x + 9 = 3(x - 1)(x - 3)$. By solving $f'(x) = 0$, we get the values $x = 1$ and $x = 3$ where the graph has a horizontal tangent line. This shows that a relative extremum can occur at a value of $f(x)$ where $f'(x) = 0$.

The other type of value $x = a$ that sometimes yields a relative extremum is where $f'(a)$ does not exist but $f(a)$ does exist. Consider the function $f(x) = x^{2/3}$ whose graph is shown in Figure 3. From the graph we see a relative minimum at

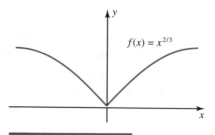

$f(x) = x^{2/3}$

FIGURE 3

$x = 0$, and $f'(x) = \frac{2}{3}x^{-1/3}$ does not exist at $x = 0$. Thus a relative extremum can also occur at a value where $f'(x)$ does not exist.

Recall that the critical values of $f(x)$ are the values $x = a$ in the domain of $f(x)$ where $f'(a) = 0$ or where $f'(a)$ does not exist. This shows that to locate relative extrema, we need to find the critical values of a function.

But a word of caution is in order. It is not true that a relative extremum occurs at every critical value. Consider the function $f(x) = x^3$ whose graph is in Figure 4. To find the critical value(s), set $f'(x) = 3x^2$ equal to 0. This yields the only critical value $x = 0$, since the derivative exists at all values of x. The graph shows that the function has no relative extremum at $x = 0$; in fact the graph has no relative extremum anywhere.

Let us summarize these results.

> A **relative extremum** of the function $f(x)$ will occur at a **critical value,** that is, a value $x = a$ in the domain such that $f'(a) = 0$ or $f'(a)$ does not exist. Not all critical values yield relative extrema.

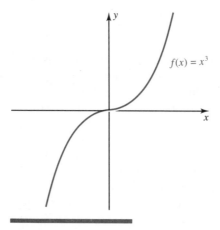

$f(x) = x^3$

FIGURE 4

The First Derivative Test

The relative maxima and minima of $f(x)$ are computed by finding the critical values and then applying a test to see whether the extremum is (1) a relative maximum, (2) a relative minimum, or (3) neither a maximum nor a minimum. This test involves the derivative of $f(x)$ and uses the properties of $f(x)$ studied in the previous section.

Consider again $f(x) = x^3 - 6x^2 + 9x$. As shown in Figure 2 the graph has a relative maximum at $x = 1$ and a relative minimum at $x = 3$. Also in Figure 2, note that the graph is increasing to the left of $x = 1$ and decreasing to the right of $x = 1$. This is why the critical value $x = 1$ is a relative maximum. Similarly the critical value $x = 3$ is a relative minimum because the graph is decreasing to the left and increasing to the right of $x = 3$.

Recall from the previous section that the intervals where the function is increasing and the intervals where it is decreasing are found by finding where the derivative is greater than 0 and less than 0. This applies to all functions. It is called the **first derivative test.**

The First Derivative Test

To find the relative extrema of $f(x)$

1. Find the critical values of $f(x)$ and the intervals where $f(x)$ is increasing and where $f(x)$ is decreasing
2. If $x = a$ is a critical value of $f(x)$, then
   i. $f(a)$ is a relative maximum if $f'(x) > 0$ to the left of a and $f'(x) < 0$ to the right of a.
   ii. $f(a)$ is a relative minimum if $f'(x) < 0$ to the left of a and $f'(x) > 0$ to the right of a.
   iii. $f(a)$ is not a relative extremum if
      (a) $f'(x) < 0$ to the left and to the right of a.
      (b) $f'(x) > 0$ to the left and to the right of a.

This means that for a critical value $x = a$, $f(a)$ is a relative maximum if $f'(x)$ changes from positive to the left of $x = a$ to negative to the right of $x = a$; $f(a)$ is a relative minimum if $f'(x)$ changes from negative to the left of $x = a$ to positive to the right of $x = a$; $f(a)$ is not a relative extremum if $f'(x)$ does not change sign.

EXAMPLE 2

Problem

Find the relative extrema of $f(x) = (x - 2)^2(x - 5)$. Sketch the graph of the function.

Solution We follow the steps in the first derivative test.

1. Find $f'(x)$.

$$f'(x) = 2(x - 2)(x - 5) + (x - 2)^2 = (x - 2)(2x - 10 + x - 2)$$

$$= 3(x - 2)(x - 4)$$

Solve $f'(x) = 3(x - 2)(x - 4) = 0$. This yields $x = 2$ and $x = 4$. These are the only critical values.

    Solve $f'(x) > 0$ and $f'(x) < 0$. The intervals are $(-\infty, 2)$, $(2, 4)$, and $(4, \infty)$. Choose test points, say, 0, 3, and 5. Since $f'(0) = 24$, $f'(3) = -3$, and $f'(5) = 9$, $f(x)$ is increasing in $(-\infty, 2)$ and $(4, \infty)$, while $f(x)$ is decreasing in $(2, 4)$. Record these facts in a chart as follows.

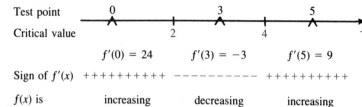

Test point   0   3   5

Critical value   2   4

                $f'(0) = 24$ $f'(3) = -3$ $f'(5) = 9$

Sign of $f'(x)$ ++++++++++   −−−−−−−−−−   ++++++++++

$f(x)$ is   increasing   decreasing   increasing

2. $f(2) = 0$ is a relative maximum because $f(x)$ is increasing to the left of $x = 2$ and decreasing to the right of $x = 2$.
$f(4) = -4$ is a relative minimum because $f(x)$ is decreasing to the left of $x = 4$ and increasing to the right of $x = 4$.

From this information and by plotting the points $(2, 0)$ and $(4, -4)$, we can sketch the graph as shown in Figure 5.

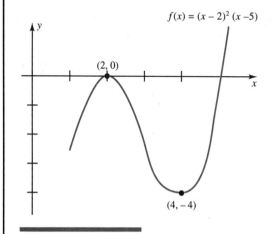

FIGURE 5

    Example 3 applies the first derivative test to a fourth-degree polynomial. In general, the derivative of a fourth-degree polynomial is a third-degree polynomial.

It is not always easy to factor a third-degree polynomial but the ones given in this text should be fairly straightforward.

EXAMPLE 3

Problem

Find the relative extrema, and sketch the graph of the function

$$f(x) = x^4 - 8x^3 + 16x^2$$

Solution

1. Find $f'(x)$.

$$f'(x) = 4x^3 - 24x^2 + 32x = 4x(x^2 - 6x + 8) = 4x(x - 2)(x - 4)$$

Solve $f'(x) = 0$. This yields $x = 0, 2, 4$. These are the only critical values.
   Solve $f'(x) > 0$ and $f'(x) < 0$. The intervals are $(-\infty, 0)$, $(0, 2)$, $(2, 4)$, and $(4, \infty)$. Choose test points, say, $-1, 1, 3$, and 5. Evaluate $f'(x)$ at each value and record the results in the following chart.

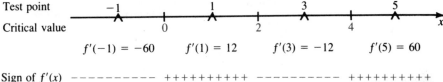

Test point	−1	1	3	5
Critical value	0	2	4	
	$f'(-1) = -60$	$f'(1) = 12$	$f'(3) = -12$	$f'(5) = 60$
Sign of $f'(x)$	− − − − − − − − − −	+ + + + + + + + + +	− − − − − − − − − −	+ + + + + + + + + +
$f(x)$ is	decreasing	increasing	decreasing	increasing

2. $f(0) = 0$ is a relative minimum because f is decreasing to the left and increasing to the right of $x = 0$.
   $f(2) = 16$ is a relative maximum because f is increasing to the left and decreasing to the right of $x = 2$.
   $f(4) = 0$ is a relative minimum because f is decreasing to the left and increasing to the right of $x = 4$.

From this information and by plotting the points $(0, 0)$, $(2, 16)$, and $(4, 0)$, we can sketch the graph as shown in Figure 6.

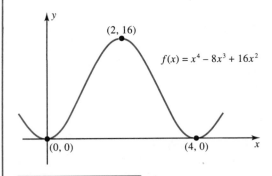

$(2, 16)$

$f(x) = x^4 - 8x^3 + 16x^2$

$(0, 0)$ $(4, 0)$

FIGURE 6

The next example shows how to identify when a function has a critical value that does not correspond to an extremum and when a function has an extremum at a point where the derivative does not exist.

EXAMPLE 4

Problem

Find the relative extrema, and sketch the graph of each function:
(a) $g(x) = 2 + x^3$, (b) $f(x) = x - 3x^{1/3}$.

Solution

(a) Compute

$$g'(x) = 3x^2$$

1. Solve $g'(x) = 0$. This yields $x = 0$. Thus the only critical value is $x = 0$. Find the intervals where $g(x)$ is increasing and where $g(x)$ is decreasing by arranging the necessary information in the following chart.

Test point	-1		1	
Critical value		0		x
	$f'(-1) = 3$		$f'(1) = 3$	
Sign of g'	$+++++++++++$		$++++++++++$	
$g(x)$ is	increasing		increasing	

2. $g(0) = 2$ is neither a relative maximum nor a relative minimum because $g(x)$ is increasing to the left of $x = 0$ and to the right of $x = 0$.

From this information and by plotting the point $(0, 2)$, we can sketch the graph as shown in Figure 7.

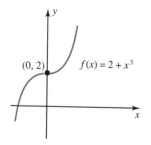

(0, 2) $f(x) = 2 + x^3$

FIGURE 7

(b) Compute

$$f'(x) = 1 - x^{-2/3} = 1 - \frac{1}{x^{2/3}}$$

1. Solve $f'(x) = 0$. This yields $x = \pm 1$. Also, $f'(x)$ does not exist when $x = 0$, but $f(0) = 0$. Thus $x = 0$ is a critical value. The function may change

from increasing to decreasing at $x = 0$, so it is included in the analysis. The critical values are $x = -1$, $x = 0$, and $x = 1$. Find the intervals where $f(x)$ is increasing and where $f(x)$ is decreasing by arranging the necessary information in the following chart.

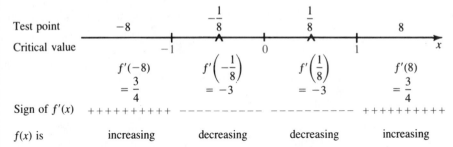

Test point	-8	$-\dfrac{1}{8}$	$\dfrac{1}{8}$	8
Critical value	-1	0	1	
	$f'(-8)$ $= \dfrac{3}{4}$	$f'\left(-\dfrac{1}{8}\right)$ $= -3$	$f'\left(\dfrac{1}{8}\right)$ $= -3$	$f'(8)$ $= \dfrac{3}{4}$
Sign of $f'(x)$	$+++++++++++$	$-----------$	$-----------$	$+++++++++++$
$f(x)$ is	increasing	decreasing	decreasing	increasing

2. $f(-1) = 2$ is a relative maximum because f is increasing to the left and decreasing to the right of $x = -1$.
   $f(0) = 0$ is neither a relative maximum nor a relative minimum because f is decreasing to the left and to the right of $x = 0$.
   $f(1) = -2$ is a relative minimum because f is decreasing to the left and increasing to the right of $x = 1$.

From this information and by plotting the points $(-1, 2)$, $(0, 0)$, and $(1, -2)$, we can sketch the graph as shown in Figure 8. The graph has a vertical tangent at $(0, 0)$, implying that the derivative does not exist at $x = 0$.

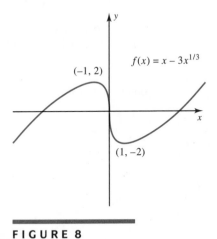

$$f(x) = x - 3x^{1/3}$$

FIGURE 8

An Application: Maximizing Profit

The ultimate goal of most businesses is to maximize profit. A fundamental principle of economics is that a firm will maximize its profit when marginal revenue equals marginal cost.* The first derivative test explains why this is so. The profit $P(x)$ that a firm makes when it sells x units of goods is defined as the difference between the revenue $R(x)$ and cost $C(x)$ functions.

$$P(x) = R(x) - C(x)$$

To maximize $P(x)$, solve $P'(x) = 0$. From the definition of $P(x)$ we have

$$P'(x) = R'(x) - C'(x)$$

Solve $P'(x) = 0$. This yields $R'(x) - C'(x) = 0$. Rearranging the latter equation gives

$$R'(x) = C'(x)$$

This states that marginal revenue $R'(x)$ equals marginal cost $C'(x)$. That is, to maximize profit, find the value of x where marginal revenue equals marginal cost.

     The next example is an economic model of a grocery chain.† The model uses the first derivative test to maximize the profit function of a particular product.

EXAMPLE 5

A firm determines that the revenue and cost functions for the volume of sales x for a particular product are

$$R(x) = 20x - 0.1x^2, \qquad x \geq 0$$
$$C(x) = 50 + 9x + 0.1x^2, \qquad x \geq 0$$

Problem

(a) Find the profit function $P(x)$ and the volume of sales x that maximizes profit.
(b) Show that the value of x that maximizes $P(x)$ in part (a) is also the solution of the equation $R'(x) = C'(x)$.

Solution (a) From the definition of the profit function

$$\begin{aligned} P(x) &= R(x) - C(x) \\ &= 20x - 0.1x^2 - (50 + 9x + 0.1x^2) \\ &= -50 + 11x - 0.2x^2 \end{aligned}$$

Solve $P'(x) = 0$.

$$P'(x) = 11 - 0.4x = 0$$

$$x = \frac{11}{0.4} = 27.5$$

The first derivative test shows that $P(x)$ has a relative maximum of $P(27.5) = 101.25$. The graph of $P(x)$ is given in Figure 9.

---

*Paul A. Samuelson, *Economics*, New York, McGraw-Hill Book Co., 1986, p. 491.
†Marilu Hurt McCarty, *Managerial Economics*, Glenview, IL, Scott, Foresman and Co., 1986, pp. 28–29.

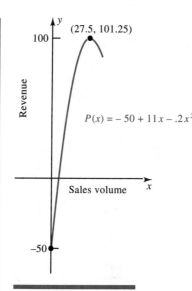

$P(x) = -50 + 11x - .2x^2$

FIGURE 9 *Source:* Marilu Hurt McCarty, *Managerial Economics,* Glenview, IL, Scott, Foresman and Co., 1986, pp. 28–29.

(b) Set the marginal revenue equal to the marginal cost.

$$R'(x) = C'(x)$$
$$20 - 0.2x = 9 + 0.2x$$
$$-0.4x = -11$$
$$x = 27.5$$

This states that maximum profit occurs at that volume of sales where marginal revenue equals marginal cost.

EXERCISE SET 4–2

In Problems 1 to 30 find the relative extrema and sketch the graph of the functions.

1. $f(x) = x^2 - 4$

2. $f(x) = 2x^2 - 5$

3. $f(x) = 2 - 3x^2$

4. $f(x) = 3 - 4x^2$

5. $f(x) = x^2 - 4x$

6. $f(x) = 3x^2 - 6x$

7. $f(x) = x^2(x + 1)$

8. $f(x) = x^2(x - 3)$

9. $f(x) = x^3 - 6x^2 - 1$

10. $f(x) = x^3 - 9x^2 + 4$

11. $f(x) = x^3 + 3x^2 - 9x$

12. $f(x) = 2x^3 + 3x^2 - 12x + 1$

13. $f(x) = x^3 - 3x^2 - 9x$

14. $f(x) = 2x^3 + 3x^2 - 36x + 3$

15. $f(x) = x^4 - 2x^2$

16. $f(x) = -x^4 + 6x^2$

17. $f(x) = -x^4 + 8x^2$

18. $f(x) = -x^4 + 12x^2 + 1$

19. $f(x) = x + 9/x$

20. $f(x) = 4x + 25/x$

21. $f(x) = (x^2 + 12)/x$

22. $f(x) = (4x^2 + 9)/x$

23. $f(x) = x^3(x - 4)$

24. $f(x) = x^3(8 - x)$

25. $f(x) = x^2(x^2 - 2)$

26. $f(x) = x^2(4 - x^2)$

27. $f(x) = x^2(x^2 + 2)$

28. $f(x) = x^2(4 + x^2)$

29. $f(x) = (x - 1)/(x + 1)$

30. $f(x) = (x + 3)/(x - 5)$

In Problems 31 to 34 find the relative extrema, and sketch the graph of the function from the given information. Assume $f(x)$ is continuous for all real numbers.

31. Critical values

    Sign of $f'(x)$

    $f(0) = 0$, $f(3) = 2$

32. Critical values

    Sign of $f'(x)$

    $f(-2) = 1$, $f(0) = 2$

33. Critical values

    Sign of $f'(x)$

    $f(-1) = 1$, $f(2) = 3$, $f(5) = 4$

34. Critical values

    Sign of $f'(x)$

    $f(2) = -1$, $f(3) = 2$, $f(4) = 4$

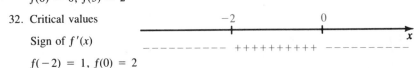

35. The profit $P(x)$ in thousands of dollars for selling x television sets in hundreds of sets is

    $$P(x) = x^2(x - 8)^2 \qquad \text{for } 0 < x < 6$$

    Find the number of television sets that maximizes profit.

36. Psychologists define learning rate $L(x)$ to be the percent of learning that an individual exhibits in x minutes while solving an intricate puzzle. For a particular puzzle the average $L(x)$ is given by

    $$L(x) = 100x^2 - 20x^3 + x^4 \qquad \text{for } 0 < x < 12$$

    Find where $L(x)$ is a maximum.

In Problems 37 to 44 find the relative extrema, and sketch the graph of the functions.

37. $f(x) = x^3(x^2 - 15)$ 38. $f(x) = x^2(10 - x^3)$

39. $f(x) = (x^2 - 1)(x^2 - 4)$

40. $f(x) = (x^2 - 3)(9 - x^2)$

41. $f(x) = (x - 5)^2(x + 1)^2$

42. $f(x) = (x + 6)^2(x - 4)^2$

43. $f(x) = (x - 5)^{2/3}$ 44. $f(x) = (x + 6)^{2/3}$

Referenced Exercise Set 4–2

1. Nursing is the largest cost item in most hospitals. In a paper* examining decision support models for budgeting nursing work-force requirements in a hospital, a key function is the cost function $C(x)$, where x is the number of nursing hours per week, measured in 1000 hours, and $C(x)$ is the cost of nursing, measured in 100,000 dollars. Several models are developed and applied to data from a large metropolitan hospital in the Sunbelt. For one model the cost function can be described by

   $$C(x) = 0.115x^2 - 3.046x + 28.686$$

   Use the first derivative test to find the value of x that minimizes C.

2. A national sample of married couples was surveyed to untangle the complex relationship between a person's actual pay and the feeling of being underpaid.† The data show that as income rises, people shift from comparing their income to that which is needed to "get along" to that which is required to "get ahead." One illustrative function measures the actual pay, x, of husbands versus the average amount, $f(x)$, they felt that

*Edward P. Kao and Maurice Queyranne, "Budgeting Costs of Nursing in a Hospital," *Management Science*, Vol. 31, No. 5, pp. 608–621.

†John Mirowsky, "The Psycho-Economics of Feeling Underpaid: Distributive Justice and the Earnings of Husbands and Wives," *American Journal of Sociology*, Vol. 92, No. 6, 1987, pp. 1404–1434.

they were underpaid. The data are approximated by the function

$$f(x) = 0.018x^2 - 0.757x + 9.047$$

This means that, on the average, husbands surveyed earning salary x (in thousands of dollars) felt they were underpaid by $f(x)$ thousands of dollars.

(a) Find the salary where the underpayment is a minimum. The authors refer to this salary as the "optimum pay."

(b) Find the amount of underpayment at the optimum pay.

(c) Graph the function and use it to describe what the authors refer to as a shift in the feelings of an income needed to "get along" versus to "get ahead."

3. Biological cells that are active electrically must function with great precision, making their motion very complicated. A. L. Hodgkin and A. F. Huxley won Nobel Prizes for being the first scientists to construct a model of the squid giant axon, the first such cell to be investigated satisfactorily.‡ An article describing their work states that the model cannot include the extrema of the function $f(x) = 3x - x^3$. What points are thus excluded?

Cumulative Exercise Set 4–2

In Problems 1 to 3, a function $f(x)$ is given.

(a) Find the intervals where $f(x)$ is increasing.

(b) Find the point(s) where $f(x)$ has a horizontal tangent line.

(c) Sketch the graph of the function.

1. $f(x) = 4 - x^2$

2. $f(x) = x^2(1 - x)$

3. $f(x) = \dfrac{x - 4}{x}$

In Problems 4 and 5 use the information given to sketch the graph of $f(x)$, assuming that $f(x)$ is continuous for all real numbers x.

4. Values where $f'(x) = 0$

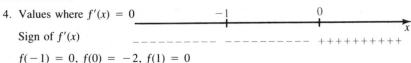

$f(-1) = 0$, $f(0) = -2$, $f(1) = 0$

5. Critical values

$f(-4) = 5$, $f(-1) = 3$, $f(3) = 5$

In Problems 6 and 7 find the relative extrema and sketch the graph of $f(x)$.

6. $f(x) = 2 - 8x + x^4$ 7. $f(x) = 8x^2 - x^4$

8. Let $f(x)$ be the function whose graph is shown.

(a) What are the critical values of $f(x)$?

(b) For what intervals is $f'(x) \leq 0$?

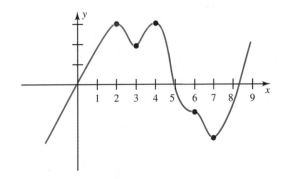

‡Jane Cronin, "Electrically Active Cells and Singular Perturbation Theory," *The Mathematical Intelligencer*, Vol. 12, No. 4, 1990, pp. 57–64. The function in question is described on page 60.

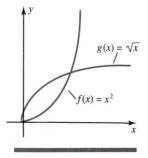

FIGURE 1

4–3 Concavity and the Second Derivative Test

Consider the graphs of the two functions in Figure 1. Each function is increasing in $(0, \infty)$, yet the *rate* of increase is different. The graph of $f(x) = x^2$ is increasing at a much greater rate than that of $g(x) = \sqrt{x}$. To illustrate this, let each function represent the position of a car moving in a straight line, so that at time x the car is at position y. Then the derivative of each function represents its velocity. Each function is increasing, but when x increases from 0 to 4, $f(x)$ moves from position $f(0) = 0$ to $f(4) = 16$, while $g(x)$ moves from $g(0) = 0$ to $g(4) = 2$. This is because the derivative of $f(x)$ is increasing, while the derivative of $g(x)$ is decreasing. In this section we study the rate of increase of the derivative of a function. We learn that the second derivative of the function provides valuable information about the graph of the function.

Concavity

The difference between the two graphs in Figure 1 can be described geometrically by saying that the graph of $f(x) = x^2$ lies above its tangent line at each point in $(0, \infty)$, while the graph of $g(x) = \sqrt{x}$ lies below its tangent line at each point in $(0, \infty)$. This is because $f'(x) = 2x$ is increasing while $g'(x) = 1/(2\sqrt{x})$ is decreasing on $(0, \infty)$. In Figure 2 two graphs are given along with three tangent lines for each: the curve in (a) lies on or above each tangent line and it is said to "open up," while the curve in (b) lies on or below each tangent line and it is said to "open down."

DEFINITION

The graph of a function $y = f(x)$ is **concave up** *on the interval* (a, b) if $f'(x)$ is increasing on (a, b) and is **concave down** *on* (a, b) if $f'(x)$ is decreasing on (a, b).

A function is defined to be *concave up at* $x = a$ if the graph of the function is concave up in some interval containing a and is *concave down at* $x = a$ if the

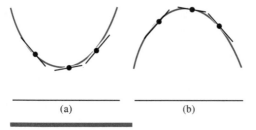

(a)   (b)

FIGURE 2

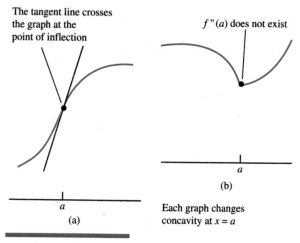

The tangent line crosses the graph at the point of inflection

$f''(a)$ does not exist

a

(b)

a

(a)

Each graph changes concavity at $x = a$

FIGURE 3

graph of the function is concave down in some interval containing *a*. Geometrically, a function is concave up at $x = a$ *if its graph close to a* lies above its tangent line at *a*, and it is concave down at $x = a$ if its graph close to *a* lies below its tangent line at *a*.

A point on the graph where the graph changes concavity is called a **point of inflection.** Thus at a point of inflection the graph changes from concave up to concave down, or vice versa. At a point of inflection the graph crosses its tangent line or $f''(x)$ does not exist. See Figure 3. Example 1 illustrates these definitions.

EXAMPLE 1

Problem

For the graph in Figure 4 locate the intervals where $f(x)$ is concave up and where $f(x)$ is concave down, and find the points of inflection.

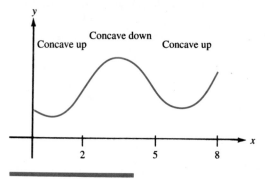

Concave up   Concave down   Concave up

FIGURE 4

Solution $f(x)$ is concave up in $(0, 2)$ and in $(5, 8)$, while it is concave down in $(2, 5)$. There are points of inflection at $x = 2$ and $x = 5$.

It is tempting to try to draw a connection between when a function is concave up and when the function is increasing. But a look at Figure 4 shows that a function can be both increasing and decreasing in an interval where it is concave up: $f(x)$ is decreasing in $(0, 1)$ and increasing in $(1, 2)$, yet $f(x)$ is concave up in $(0, 2)$. Similarly, a function can be increasing or decreasing in an interval where it is concave down. The chart lists the four possibilities.

$f(x)$ Is Increasing or Decreasing	$f(x)$ Is Concave Up or Concave Down	Graph of $f(x)$
A. Increasing	concave up	
B. Increasing	concave down	
C. Decreasing	concave up	
D. Decreasing	concave down	

In Figure 5 two functions are graphed along with three tangent lines for each function. The function in (a) is concave up and the function in (b) is concave down. Start at $x = a$ on the left of each graph and move to $x = b$; that is, let x increase from a to b. What can be said about the slopes of the tangent lines? In (a) the slopes of the tangent lines increase; they are negative on the left part of the graph,

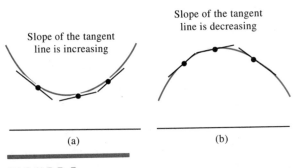

Slope of the tangent line is increasing

Slope of the tangent line is decreasing

(a)   (b)

FIGURE 5

then increase to 0 and then become positive. In (b) the slopes decrease as they start out positive, decrease to 0 and then become negative. This indicates that the derivative increases for a function that is concave up and the derivative decreases for a function that is concave down.

This statement can be coupled with the result from Section 4–1 about increasing functions and decreasing functions to yield a test for concavity. To derive the test, we combine the following two results. Statement I is from the preceding discussion and statement II is from Section 4–1.

I. $f(x)$ is concave up in (a, b) if $f'(x)$ is increasing on (a, b).

II. $f(x)$ is increasing in (a, b) if $f'(x) > 0$ on (a, b).

To tie the two together, replace $f(x)$ by $f'(x)$ in the second statement. We then get

III. $f'(x)$ is increasing in (a, b) if $f''(x) > 0$ on (a, b).

Combining I and III yields the test for concavity.

Procedure for Determining Intervals of Concavity

Assume that the first and second derivatives of $f(x)$ exist for all values of x in the interval (a, b).

1. $f(x)$ is concave up in (a, b) if $f''(x) > 0$ for all x in (a, b).
2. $f(x)$ is concave down in (a, b) if $f''(x) < 0$ for all x in (a, b).

EXAMPLE 2

Problem

Find the intervals of concavity and the point(s) of inflection of $f(x) = x^3 + 3x^2$.

Solution To apply the procedure, first calculate $f'(x)$ and $f''(x)$: $f'(x) = 3x^2 + 6x$ and $f''(x) = 6x + 6 = 6(x + 1)$. Thus $f''(-1) = 0$. Next find where $f''(x)$ is greater than and less than 0. Choose test points to the left and right of $x = -1$, say, $x = -2$ and $x = 0$. This gives $f''(-2) = 6(-1) = -6 < 0$ and $f''(0) = 6 > 0$. Record these facts in the following chart.

Test point

Possible point of inflection

Sign of $f''(x)$

f is   concave down   concave up

From the procedure $f(x)$ is concave up in $(-1, \infty)$ and concave down in $(-\infty, -1)$. Since $f(x)$ changes from concave down to concave up at $x = -1$, $f(x)$ has a point of inflection at $(-1, 2)$. See Figure 6 on the next page.

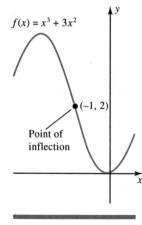

FIGURE 6

If $(a, f(a))$ is a point of inflection, $f''(x)$ changes from positive on one side of $x = a$ to negative on the other. For instance, in Example 2, $f(x)$ has a point of inflection at $(-1, 2)$ and $f''(x)$ is negative to the left of $x = -1$ and positive to the right of $x = -1$. What happens to $f''(x)$ at the point of inflection? In Example 2, $f''(-1) = 0$. But there is another possibility. In Figure 7 the graph of $f(x) = x^{1/3}$ is given. It has a point of inflection at $x = 0$ and $f''(0)$ does not exist. Thus there are two possibilities for $f''(x)$ at a point of inflection.

Second Derivative at Point of Inflection
If $f(x)$ has a point of inflection at $(a, f(a))$, then either $f''(a) = 0$ or $f''(a)$ does not exist.

It is very important to recognize what this result does not say. To find where a function has a point of inflection, we also need to compute the values where $f''(x)$

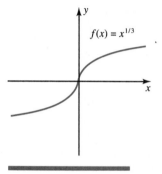

FIGURE 7

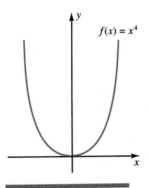

$f(x) = x^4$

FIGURE 8

$= 0$ and where $f''(x)$ does not exist. Recall that it is not enough to compute critical values when trying to locate relative extrema; more information is required about the behavior around the critical value, which is the essence of the first derivative test. So, too, to determine if a value where $f''(x) = 0$ or $f''(x)$ does not exist yields a point of inflection, we need to find where the function is concave up and where it is concave down. For example, $f(x) = x^4$ has $f''(x) = 12x^2$, and $f''(x) = 0$ when $x = 0$. But since $f''(x) > 0$ for all x to the left and to the right of $x = 0$, f is concave up both to the left and to the right of $x = 0$. Thus f does not have a point of inflection at $x = 0$ even though $f''(0) = 0$. See Figure 8.

The Second Derivative Test

The two graphs in Figure 9 illustrate the connection between concavity and relative extrema. Each graph has a horizontal tangent at $x = a$, so $f'(a) = 0$ and the function has a critical value at $x = a$. The graph in (a) is concave down and $f(x)$ has a relative maximum at $x = a$. The graph in (b) is concave up and $f(x)$ has a relative minimum at $x = a$. To summarize, let $x = a$ be a critical value of $f(x)$.

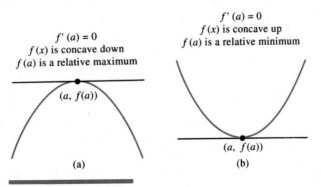

$f'(a) = 0$
$f(x)$ is concave down
$f(a)$ is a relative maximum

$(a, f(a))$

(a)

$f'(a) = 0$
$f(x)$ is concave up
$f(a)$ is a relative minimum

$(a, f(a))$

(b)

FIGURE 9

I. If $f(x)$ is concave down at $x = a$, then $f(x)$ has a relative maximum at $x = a$.

II. If $f(x)$ is concave up at $x = a$, then $f(x)$ has a relative minimum at $x = a$.

Therefore concavity can be used to determine whether a critical value yields a relative maximum or a relative minimum. Since the second derivative measures concavity, the second derivative can be used to determine whether a critical value is a relative maximum or a relative minimum. Statements I and II can be refined by using the test for concavity from the previous section.

Let $x = a$ be a critical value of $f(x)$.

III. If $f''(a) < 0$, then $f(x)$ is concave down at $x = a$, and thus $f(x)$ has a relative maximum at $x = a$.

IV. If $f''(a) > 0$, then $f(x)$ is concave up at $x = a$, and thus $f(x)$ has a relative minimum at $x = a$.

This is the essence of the second derivative test, which, like the first derivative test, determines whether a critical value is a relative maximum or a relative minimum.

The Second Derivative Test

Suppose $f(x)$ is differentiable and $f''(x)$ exists on an interval containing a. Let $x = a$ be a critical value.

1. $f(a)$ is a relative maximum if $f''(x) < 0$ at $x = a$.
2. $f(a)$ is a relative minimum if $f''(x) > 0$ at $x = a$.
3. If $f''(a) = 0$, the test gives no conclusion about a relative extremum at $x = a$; that is, $f(a)$ may be a relative maximum, a relative minimum, or neither.

EXAMPLE 3

Problem

Use the second derivative test to find the relative extrema for
(a) $f(x) = 5x^2 - 10x$, (b) $f(x) = x^3 - 3x^2$.

Solution (a) Compute $f'(x) = 10x - 10 = 10(x - 1)$. Solve $f'(x) = 0$. The only critical value is $x = 1$. To apply the second derivative test, we compute the second derivative, $f''(x) = 10$. This is a constant function, so

$$f''(1) = 10 > 0$$

Since $f''(1) > 0$, apply part 2 of the second derivative test and conclude that $f(x)$ has a relative minimum at $x = 1$. The relative minimum is $f(1) = -5$.
(b) Compute $f'(x) = 3x^2 - 6x$. Solve $f'(x) = 3x^2 - 6x = 3x(x - 2) = 0$. So $x = 0$ and $x = 2$. The critical values are $x = 0$ and $x = 2$. Compute $f''(x) = 6x - 6$. Substitute each critical value into f''. This yields

$$f''(0) = 0 - 6 = -6 < 0 \qquad f''(2) = 12 - 6 = 6 > 0$$

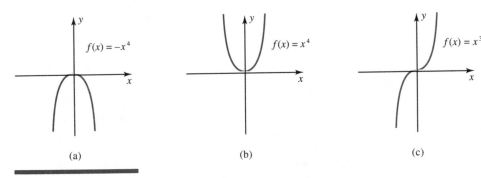

FIGURE 10

Part 1 of the second derivative test states that $f(x)$ has a relative maximum at $x = 0$ because $f''(0) < 0$; part 2 states that $f(x)$ has a relative minimum at $x = 2$ because $f''(2) > 0$. The relative maximum is $f(0) = 0$ and the relative minimum is $f(2) = -4$.

Part 3 of the second derivative test states that the test fails when $f''(a) = 0$ since no conclusion can be drawn about relative extrema. Figure 10 shows why this is the case. Even though each of the three functions graphed in Figure 10 has $f''(0) = 0$, their behavior at $x = 0$ is different. In (a), $f(x) = -x^4$ has a relative maximum at $x = 0$; in (b), $f(x) = x^4$ has a relative minimum at $x = 0$; in (c), $f(x) = x^3$ has neither a relative maximum nor a relative minimum because it has a point of inflection at $x = 0$. This means that if $f''(a) = 0$ for a critical value $x = a$, then the first derivative test must be applied to $x = a$. Also, if $f''(x)$ does not exist, the second derivative test cannot be used so the first derivative test must be applied.

Curve Sketching

The ideas presented thus far in this chapter can be combined to form a handy procedure for sketching graphs. The summary of these various techniques is a process called **curve sketching.**

Curve-Sketching Techniques

1. Compute $f'(x)$.
   a. Find the critical values by solving $f'(x) = 0$ and finding where $f'(x)$ does not exist.
   b. Find the intervals where $f(x)$ is increasing and where $f(x)$ is decreasing by solving $f'(x)$ greater than 0 and less than 0.
2. Compute $f''(x)$.
   a. Find the possible points of inflection by solving $f''(x) = 0$ and locating where $f''(x)$ does not exist.

b. Find the intervals where $f(x)$ is concave up and where $f(x)$ is concave down by setting $f''(x)$ greater than 0 and less than 0. Use this information to locate the points of inflection.

3. Use the first or second derivative test to locate the relative maxima and the relative minima.

4. Plot the relative extrema and the points of inflection. Keep in mind where the function is increasing, decreasing, concave up, and concave down, and plot several additional points that help to get an accurate sketch. For instance, plot the x- and y-intercepts if they exist and are easy to calculate; if the function is defined on a closed or half-open interval, plot the points corresponding to the end points.

5. Draw a sketch of the curve from this information.

Example 4 shows how to use the curve-sketching techniques by applying them to the functions in Example 3.

EXAMPLE 4

Problem

Sketch the graph of the functions (a) $f(x) = 5x^2 - 10x$, (b) $f(x) = x^3 - 3x^2$.

Solution We follow the steps in the procedure even though they do not have to be done precisely in the given order.

(a)

1. $f'(x) = 10x - 10 = 10(x - 1) = 0$ implies $x = 1$.
   a. Since $f'(1) = 0$, the only critical value is $x = 1$.
   b. $f'(x) > 0$ in $(1, \infty)$ and $f'(x) < 0$ in $(-\infty, 1)$, so $f(x)$ is increasing in $(1, \infty)$ and decreasing in $(-\infty, 1)$.

2. $f''(x) = 10$
   a. Since $f''(x)$ is always greater than 0, $f(x)$ is concave up in $(-\infty, \infty)$.
   b. $f''(x)$ is never equal to 0, so there is no point of inflection.

This information is recorded in the following chart.

Critical value		
Sign of $f'(x)$	$------------$	$+++++++++++$
$f(x)$ is	decreasing	increasing
Sign of $f''(x)$	$+++++++++++$	$+++++++++++$
$f(x)$ is		concave up

3. The second derivative test shows that $f(x)$ has a relative minimum at $x = 1$ because $f''(1) = 10 > 0$.

4. Plot $(1, -5)$, where $f(x)$ has its relative minimum. To find the y-intercept, let $x = 0$ and get $f(0) = 0$. Since $f(x) = 0$ at $x = 0$ and $x = 2$, plot the x-intercepts $(0, 0)$ and $(2, 0)$.

5. The graph is in Figure 11.

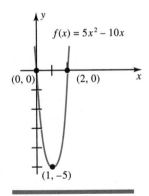

$f(x) = 5x^2 - 10x$

$(0, 0)$ $(2, 0)$

$(1, -5)$

FIGURE 11

(b)

1. $f'(x) = 3x^2 - 6x = 3x(x - 2)$
   a. Since $f'(0) = f'(2) = 0$, the critical values occur at $x = 0$ and $x = 2$.
   b. $f'(x) > 0$ in $(-\infty, 0)$ and $(2, \infty)$; $f'(x) < 0$ in $(0, 2)$. Thus $f(x)$ is increasing in $(-\infty, 0)$ and $(2, \infty)$ and $f(x)$ is decreasing in $(0, 2)$.
2. $f''(x) = 6x - 6 = 6(x - 1)$
   a. Since $f''(x) > 0$ in $(1, \infty)$ and $f''(x) < 0$ in $(-\infty, 1)$, $f(x)$ is concave up in $(1, \infty)$ and concave down in $(-\infty, 1)$.
   b. $f''(1) = 0$ and $f(x)$ changes concavity at $x = 1$, so there is a point of inflection at $(1, -2)$.

This information is recorded in the following chart.

Possible point of inflection		1	
Critical values	0		2
Sign of $f'(x)$	++++++++++	----------	++++++++++
$f(x)$ is	increasing	decreasing	increasing
Sign of $f''(x)$	------------------		+++++++++++++++++
$f(x)$ is	concave down		concave up

3. The second derivative test shows that $f(x)$ has a relative maximum at $x = 0$ because $f''(0) = -6 < 0$ and $f(x)$ has a relative minimum at $x = 2$ because $f''(2) = 6 > 0$.
4. Plot $(0, 0)$ and $(2, -4)$, where $f(x)$ has its relative extrema. Plot the point of inflection, $(1, -2)$. To find the y-intercept, let $x = 0$ and get $f(0) = 0$, but $(0, 0)$ is already plotted. Since $f(x) = 0$ at $x = 3$, plot the additional x-intercept $(3, 0)$.
5. The graph is in Figure 12.

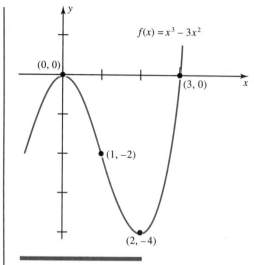

FIGURE 12

An Application: Public Subsidy

In some situations it is not satisfactory to analyze production decisions merely on the basis of cost-effectiveness or profit optimization. Certain not-for-profit institutions provide services to a community at prices below production costs. Total revenue from sales may be insufficient to pay total costs and a subsidy may be required. Examples of such institutions are public transit companies, low-cost higher education institutions, and cultural organizations. To illustrate production decisions when a public subsidy is required, economist Marilu McCarty used as an example a mass transit system in a congested urban area.* The cost $C(x)$ and revenue $R(x)$ functions were estimated for the daily demand x, where $C(x)$ and $R(x)$ are measured in thousands of dollars and x is the number of riders in thousands. If the system expected to make a profit, then it would seek to maximize $R(x) - C(x)$. But the system is designed to absorb a loss. The objective is then to minimize the *economic loss EL(x)*, defined by

$$EL(x) = C(x) - R(x)$$

The goal is to determine the demand x that minimizes $EL(x)$. This minimum value of $EL(x)$ is the subsidy paid by the government to the institution. The next example shows how to solve this problem for cost and revenue functions that are similar to those of the mass transit system. Referenced Exercise 2 presents the actual data.

---

*Marilu Hurt McCarty, *Managerial Economics*, Glenview, IL, Scott, Foresman and Co., 1986, pp. 479–486.

EXAMPLE 5

The cost and revenue functions for the mass transit system are

$$C(x) = 2000 + 140x - 30x^2 + x^3$$

$$R(x) = 20x - 9x^2$$

Problem

Determine the minimum economic loss.

Solution The economic loss function is defined by $EL(x) = C(x) - R(x)$. It is

$$EL(x) = 2000 + 120x - 21x^2 + x^3$$

To find the minimum, use the second derivative test. Compute $EL'(x)$ and $EL''(x)$.

$$EL'(x) = 120 - 42x + 3x^2 = 3(40 - 14x + x^2)$$

$$EL''(x) = -42 + 6x$$

To solve $EL'(x) = 0$, either factor the quadratic or use the quadratic formula.

$$EL'(x) = 3(x - 10)(x - 4) = 0$$

This implies that the critical values are $x = 4$ and $x = 10$. Substituting these values into $EL''(x)$ yields

$$EL''(4) = -18 \quad \text{and} \quad EL''(10) = 18$$

By using the second derivative test, we conclude that the relative minimum of $EL(x)$ occurs at $x = 10$. It is $EL(10) = 2100$. The function is graphed in Figure 13. Notice that there are functional values of $EL(x)$ that are smaller than $EL(10)$; in fact $EL(0) = 2000$. But these values of x are too small to be practical. That is, the transit system would not be properly serving the public if it accommodated so few riders. This means that an implicit assumption in the problem is that the transit system will service a certain minimum number of riders. In the next section we study problems that have natural boundaries like this.

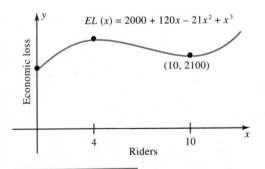

FIGURE 13 *Source:* Marilu Hurt McCarty, *Managerial Economics,* Glenview, IL, Scott, Foresman and Co., 1986, pp. 479–486.

EXERCISE SET 4–3

In Problems 1 to 6 find the largest intervals where the function is concave up and concave down. Find the points of inflection. Assume that the functions are defined on the whole real line.

1.

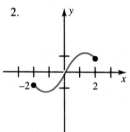

2.

3.

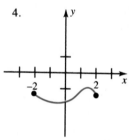

4.

5.

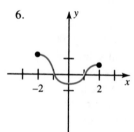

6.

In Problems 7 to 12 find the intervals where the function is concave up and concave down. Find any point of inflection.

7. $f(x) = x^2 - 4$ 8. $f(x) = 2x^2 - 5$

9. $f(x) = x^3 - 6x^2 - 2$ 10. $f(x) = x^3 - 9x^2$

11. $f(x) = x^3 + 3x^2 - 9x$

12. $f(x) = 2x^3 + 3x^2 - 12x + 1$

In Problems 13 to 18 apply the second derivative test to find the relative extrema.

13. $f(x) = x^3 - 3x^2 - 9x$

14. $f(x) = 2x^3 + 3x^2 - 36x + 3$

15. $f(x) = x^4 - 2x^2$ 16. $f(x) = -x^4 + 6x^2$

17. $f(x) = -x^4 + 8x^2$

18. $f(x) = -x^4 + 12x^2 + 1$

In Problems 19 to 36 use the curve-sketching techniques to graph the function.

19. $f(x) = x^3 + 6x^2$ 20. $f(x) = x^3 - x^2$

21. $f(x) = x^3 - 3x^2 - 9x$

22. $f(x) = 2x^3 + 3x^2 - 36x + 3$

23. $f(x) = x^4 - 2x^2$ 24. $f(x) = -x^4 + 6x^2$

25. $f(x) = 64x - 2x^4$ 26. $f(x) = 2x - 3x^{2/3}$

27. $f(x) = x + 9/x$ 28. $f(x) = 4x + 25/x$

29. $f(x) = (x^2 + 12)/x$ 30. $f(x) = (4x^2 + 9)/x$

31. $f(x) = x^2(x^2 - 2)$ 32. $f(x) = x^2(4 - x^2)$

33. $f(x) = x^2(x^2 + 2)$ 34. $f(x) = x^2(4 + x^2)$

35. $f(x) = (x - 1)/(x + 1)$

36. $f(x) = (x + 3)/(x - 5)$

In Problems 37 to 40 use the information given to sketch the graph of $f(x)$. Assume $f(x)$ is continuous for all real numbers.

37. Possible point
of inflection

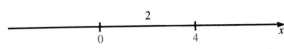

Critical values

Sign of $f'(x)$ $----------$ $++++++++++$ $----------$

$f(x)$ is   decreasing   increasing   decreasing

Sign of $f''(x)$ $+++++++++++++++$ $-------------$

$f(x)$ is   concave up   concave down

$f(0) = 0,\ f(2) = 1,\ f(4) = 2$

38. Possible point
of inflection

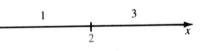

Critical values   · 0   2

Sign of $f'(x)$ $----------$ $++++++++++$ $----------$

$f(x)$ is   decreasing   increasing   decreasing

Sign of $f''(x)$ $+++++++++++++++$ $--------------$ $++++++$

$f(x)$ is   concave up   concave down   concave up

$f(0) = 0,\ f(1) = 1,\ f(2) = 2,\ f(3) = 1$

39. Possible points
of inflection

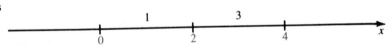

Critical values

Sign of $f'(x)$ $++++++++++$ $----------$ $++++++++++$ $----------$

$f(x)$ is   increasing   decreasing   increasing   decreasing

Sign of $f''(x)$ $-------------$ $++++++++++++++++$ $-------------$

$f(x)$ is   concave down   concave up   concave down

$f(0) = -1,\ f(1) = -2,\ f(2) = -3,\ f(3) = -1,\ f(4) = 0$

40. Possible points
of inflection

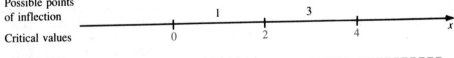

Critical values

Sign of $f''(x)$ $++++++++++$ $++++++++++$ $----------$ $----------$
$f(x)$ is   increasing   increasing   decreasing   decreasing

$----------$ $+++++$ $-----$ $-----$ $++++$ $----------$

concave   concave concave concave concave   concave
down   up   down   down   up   down

$f(0) = 0,\ f(1) = 2,\ f(2) = 4,\ f(3) = 2,\ f(4) = 1$

In Problems 41 to 45 use the curve-sketching techniques to graph the function.

41. $f(x) = x^3(x^2 - 15)$ 42. $f(x) = x^2(10 - x^3)$

43. $f(x) = (x^2 - 1)(x^2 - 4)$

44. $f(x) = (x + 6)^2(x - 4)^2$

45. $f(x) = (x + 6)^{2/3}$

In Problems 46 to 49 from the given figure find (a) the intervals where $f(x)$ is increasing, (b) the intervals where $f(x)$ is decreasing, (c) the sign of the derivative in each interval, and (d) the sign of the second derivative in each interval.

46.

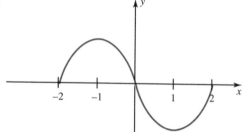

47.

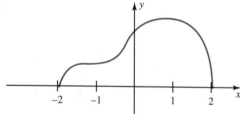

48.

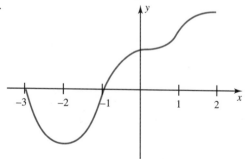

49.

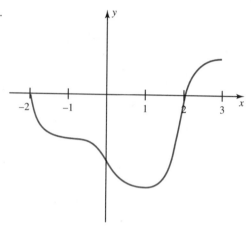

Referenced Exercise Set 4–3

1. In 1970 the U.S. government commissioned the Army to conduct a thorough study of the cost of converting to the metric system of measure. The methodology of the Army study could then be applied by various organizations, which would be similarly affected by the change. In the report* the cumulative costs of metrication for the U.S. Army are estimated by the function

   $$f(x) = -0.02x^3 + 0.12x^2 + 0.72x$$

   where x is measured in units of five-year periods starting with 1972 and $f(x)$ is measured in billions of dollars. For example, $x = 1$ represents 1977 (one five-year period after 1972) and $f(1) = 0.82$ means that the estimated cost of metrication for the Army by 1977 would be $0.82 billion. Find the year when the rate of increase of the budget would be a maximum.

2. For the economic loss function $EL(x)$ given in McCarty† for a mass transit system, find the value of x that minimizes $EL(x)$, assuming that the system must accommodate at least $x = 50$ thousand riders:

   $$EL(x) = 2000 + 72x - 1.1x^2 + 0.005x^3$$

3. Production models in business help a firm decide what is the optimum quantity to produce either to maximize price or profit or to minimize cost. In an article‡ studying agricultural production functions, with an emphasis

*Walter A. Lilius, "The Methodology of the Army Metric Study," *American Society of Quality Control—Technical Conference Transactions*, 1972, pp. 242–249.

†Marilu Hurt McCarty, *Managerial Economics*, Glenview, IL, Scott, Foresman and Co., 1986, pp. 479–486.

‡Jean-Paul Chavas et al., "Modeling Dynamic Agricultural Production Response: The Case of Swine Production," *American Journal of Agricultural Economics*, Vol. 67, 1987, pp. 636–646.

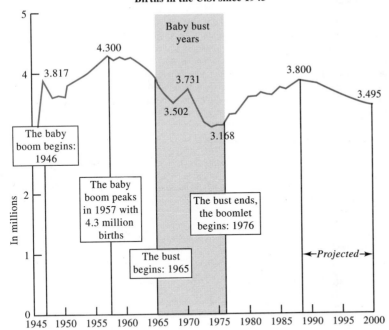

Births in the U.S. since 1945

4. The accompanying graph depicts the number of births in the U.S. from 1945 to 1989 based on figures supplied by American Demographics, Inc. (a) What kind of point on the curve is (1957, 4300)? (b) What kind of point on the curve is (1968, 3502)? (c) The figure describes the year 1965 as the beginning of the "bust." What kind of point on the curve is it?

on the swine industry, the following function relating the price P (dollars) to the weight x (lb) of an individual animal is given:

$$P(x) = 50.536 + 0.0988x - 0.00024x^2$$

Find the value of x that maximizes the price.

Cumulative Exercise Set 4–3

In Problems 1 to 3 find intervals where $f(x)$ is increasing and where $f(x)$ is decreasing. Find the point(s) where $f(x)$ has a horizontal tangent line. Graph the function.

1. $f(x) = x^2 - 12x + 1$

2. $f(x) = x^3 + 6x^2 - 18x$

3. $f(x) = (x^2 + 16)/x$

4. Use the information given to make a sketch of the graph of $f(x)$, assuming that $f(x)$ is continuous for all real numbers.

Values where $f'(x) = 0$ -2 3

Sign of $f'(x)$ $----------$ $++++++++++$ $----------$

$f(-3) = 1,\ f(-2) = 0,\ f(3) = 4,\ f(5) = 0$

In Problems 5 and 6 find the relative extrema and sketch the graph of the functions.

5. $f(x) = 2x^3 + 3x^2 - 1$

6. $f(x) = x + 25/x$

7. Find the relative extrema and sketch the graph of the function from the given information. Assume $f(x)$ is continuous for all real numbers.

Critical values -2 2

Sign of $f'(x)$ $--------- $ $+++++++++$ $----------$

$f(-2) = -1, f(0) = 1, f(2) = 3$

8. Find the intervals where the function $f(x) = 2x^3 - 3x^2$ is concave up and where it is concave down. Find any point of inflection.

9. Apply the second derivative test to find the relative extrema of $f(x) = 2x^3 - 3x^2 - 36x$.

In Problems 10 and 11 use the curve-sketching techniques to graph the function.

10. $f(x) = x^4 - 4x^3 - 2$

11. $f(x) = \dfrac{x^2 + 25}{x}$

12. Use the information given to sketch the graph of $f(x)$. Assume $f(x)$ is continuous for all real numbers.

Possible points
of inflection   1   3

Critical values -1 2   5 x

Sign of $f'(x)$ $++++++++++$ $++++++++++$ $----------$ $----------$

$f(x)$ is   increasing   increasing   decreasing   decreasing

Sign of $f''(x)$ $----------$ $+++++$ $-----$ $-----$ $+++++$ $----------$

$f(x)$ is   concave   concave concave concave concave concave
              down   up   down   down   up   down

$f(-1) = 0, f(1) = 2, f(2) = 4, f(3) = 2, f(5) = 1$

4—4 Absolute Extrema and Optimization Problems

A function may have a relative maximum or a relative minimum without being an absolute maximum or an absolute minimum. In fact most of the functions graphed in the previous sections have functional values that are larger or smaller than the relative extrema. In this section we cover absolute extrema. If $f(a)$ is a relative maximum, then $f(a)$ is greater than or equal to the other functional values $f(x)$ for all x restricted to an interval containing a. If $f(a)$ is an absolute maximum, then $f(a)$ is greater than or equal to all other functional values $f(x)$. A similar statement holds for relative minima and absolute minima.

There are several possibilities for absolute extrema. Some functions have neither an absolute maximum nor an absolute minimum. If the function is defined on a closed interval the absolute extremum can occur at an end point. And sometimes an absolute extremum occurs at a relative extremum.

Absolute Extrema

Consider the function $y = f(x)$ graphed in Figure 1. It has a relative minimum at $x = b$ and a relative maximum at $x = c$. But neither is an absolute extremum; $f(a)$ is greater than $f(c)$, so $f(c)$ is not an absolute maximum, and $f(d)$ is less than $f(b)$, so $f(b)$ is not an absolute minimum. In addition, $f(a)$ is greater than all other functional values, so it is the absolute maximum while $f(d)$ is the absolute minimum. Sometimes an absolute extremum occurs at an end point of an interval, as does $f(a)$ in Figure 1, and sometimes it does not, like $f(d)$. This leads to the following definition.

DEFINITION

Let $y = f(x)$ be defined on an interval I, where I can be either an open or closed interval. Suppose a is in I. Then

1. $f(a)$ is the **absolute maximum** *of* $f(x)$ *on* I if

   $f(a) \geq f(x)$ for all x in I

2. $f(a)$ is the **absolute minimum** *of* $f(x)$ *on* I if

   $f(a) \leq f(x)$ for all x in I

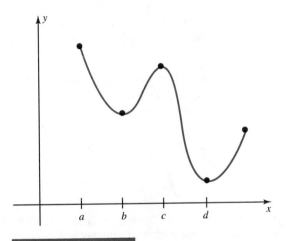

FIGURE 1

The graph in Figure 1 shows that an absolute extremum may occur at an end point of the interval. Of course, this can happen only when the end point is included in the interval, that is when it is closed or half-open (which means one end point is in the interval and the other is not). If the interval is open, then the absolute extrema cannot occur at an end point because the end points are not part of the interval. In fact the function may not have an absolute maximum or an absolute minimum. On the other hand, if the interval is closed, the end points are candidates for the absolute extrema. If an absolute extremum does not occur at an end point, then it occurs at an interior relative extremum. In more advanced courses the following theorem is proved.

Theorem

If $f(x)$ is continuous on the closed interval $[a, b]$, then $f(x)$ has an absolute maximum and an absolute minimum on $[a, b]$.

A nonvertical straight line drawn on an open interval is an example of a function that has neither an absolute maximum nor an absolute minimum on an open interval. If the end points are included, however, meaning the interval is closed, the absolute extrema of a nonvertical straight line occur at the end points.

The procedure to compute absolute extrema is the same as that for finding relative extrema, except that the end points must also be considered. That is, absolute extrema occur either at points that are relative extrema or at end points.

The Procedure for Finding Absolute Extrema for $f(x)$ Defined on a Closed Interval $[a, b]$

1. Calculate the critical values, $c_1, c_2, \ldots, c_n$ that lie in the interval $[a, b]$.
2. Calculate $f(a)$, $f(b)$, $f(c_1)$, $f(c_2)$, $\ldots$, $f(c_n)$.
3. The absolute maximum is the largest value of the numbers in step 2, and the absolute minimum is the smallest value of these numbers.

Example 1 shows how to compute absolute extrema once the relative extrema are known. It uses a function that was defined in Example 2 of Section 4–2 where the relative extrema were computed.

EXAMPLE 1

Let the function $f(x) = (x - 2)^2(x - 5)$ be defined on the interval $[\frac{3}{2}, 6]$.

Problem

Find the absolute extrema of $f(x)$.

Solution In Example 2 of Section 4–2, $f(x)$ was found to have a relative maximum of $f(2) = 0$ and a relative minimum of $f(4) = -4$. See Figure 2 for

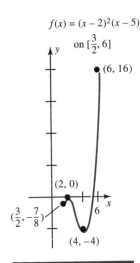

$f(x) = (x - 2)^2(x - 5)$

on $[\frac{3}{2}, 6]$

(6, 16)

(2, 0)

$(\frac{3}{2}, -\frac{7}{8})$

(4, –4)

FIGURE 2

the graph of the function. To see whether the relative extrema are absolute extrema, check the functional values of the end points. The end points occur at $x = \frac{3}{2}$ and $x = 6$. Their functional values are $f(\frac{3}{2}) = -\frac{7}{8}$ and $f(6) = 16$. To find the absolute extrema, find the largest and smallest functional values of the critical values and the end points. We arrange these four functional values in the following table:

	Critical Values		End Points	
x	2	4	$\frac{3}{2}$	6
$f(x)$	0	–4	$-\frac{7}{8}$	16

The largest of the four numbers for $f(x)$ is 16 and the smallest is -4. Therefore the absolute maximum is $f(6) = 16$ and the absolute minimum is $f(4) = -4$. The absolute maximum occurs at the right-hand end-point, while the absolute minimum occurs at an interior relative minimum.

Optimization Problems

Many applications of calculus depend on maximizing or minimizing a certain quantity. Often the variables in such problems have natural boundaries. For example, if x represents the number of items produced by a firm, x must be greater than or equal to 0. If t represents time, measured in hours, during an 8-hour shift, then t is restricted to the interval $[0, 8]$. If an absolute extremum is sought, then the end points must be checked.

An optimization problem is ordinarily a word problem, so it must first be translated into mathematics. Then the relative extrema are computed as well as the functional values of the end points. The extrema are then interpreted in terms of the original problem. Let us demonstrate the method with an example that is made easy because the function is given.

EXAMPLE 2

An object is thrown straight upward so that its height after t seconds is $f(t) = -16t^2 + 64t$, where f is measured in feet.

Problem

How high will the object travel and after how many seconds will it reach its maximum height?

Solution The object will reach its maximum height when $f(t)$ has its absolute maximum.

Are there natural end points for t? It makes no sense for t to be less than 0, and a negative value for $f(t)$ means the object is below ground level. Solving $f(t) < 0$ yields

$$-16t^2 + 64t = -16t(t - 4) < 0$$

This is true when $t < 0$ or when $t > 4$. Thus the natural restriction for $f(t)$ is $[0, 4]$.

There are no values of t where $f'(t)$ fails to exist, so to find the critical values, solve $f'(t) = 0$. Compute $f'(t) = -32t + 64$. Then $32t = 64$ and $t = 2$. The functional values of the end points are $f(0) = 0 = f(4)$. The absolute maximum occurs at the largest of the three values

$$f(0) = 0, \quad f(2) = 64, \quad f(4) = 0$$

Hence the object reaches its maximum height of $f(2) = 64$ at $t = 2$ seconds. The function is graphed in Figure 3.

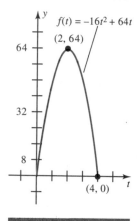

FIGURE 3

Example 3 deals with a familiar problem of building a new facility. Often constraints of cost, location, and size dictate the plans for the building. In this problem the facility is to be constructed utilizing an existing wall, thus allowing a larger area with a fixed cost. The cost is fixed because there is a given amount of fencing to be used.

EXAMPLE 3

A department store wants to build a rectangular outdoor storage facility on the rear wall of its building. An iron fence will enclose the facility on three sides, and the wall of the building will make the fourth side. (See Figure 4.)

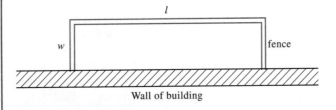

FIGURE 4

Problem

Find the dimensions of the storage facility with the largest area that can be built using 200 feet (ft) of fencing.

Solution From Figure 4 we assign letters to the variable quantities. Let l represent the length and w the width of the storage facility. Then the area, A, is given by the formula

$$A = lw \tag{1}$$

The variable A is to be maximized since we are asked for the largest storage facility. A must be expressed as a function of l or w alone, so there must be another relationship between w and l. The amount of fencing is fixed at 200 ft, and it is the sum of the three sides, whose lengths are w, w, and l. Therefore

$$200 = 2w + l$$

Solving for l yields

$$l = 200 - 2w \tag{2}$$

Substitute equation (2) into equation (1), which results in an expression of A as a function of w.

$$A(w) = w(200 - 2w) = 200w - 2w^2$$

To find the end points, note that it makes no sense for A to be negative. So set $A(w) \geq 0$. This yields

$$w(200 - 2w) \geq 0$$

This is satisfied for w in [0, 100]. Thus the end points are $w = 0$ and $w = 100$. To find the critical values, solve $A'(w) = 0$.

$$A'(w) = 200 - 4w = 0$$

The solution is $w = 50$. The end points are $w = 0$ and $w = 100$. The three values that are candidates for the absolute maximum are

$$A(50) = 5000$$
$$A(0) = 0$$
$$A(100) = 0$$

The largest value is $A(50) = 5000$. So the storage facility with the largest area has width 50 ft and length $l = 200 - 2w = 200 - 100 = 100$ ft.

In Example 3 the area of the new facility was maximized given that the cost, in terms of a given amount of fencing, was fixed. Another common type of problem is to increase production while keeping other variables fixed, such as cost, labor, and size of production facility. In Example 4 an orchard owner wishes to increase production while keeping the size of the orchard fixed. The object is to maximize the amount of apples produced given a fixed amount of land. The variable that the orchard owner can manipulate is the number of trees planted per acre. It then becomes the independent variable. The number of trees per acre determines how many apples are produced, which is therefore the dependent variable.

EXAMPLE 4

An apple orchard owner has determined that if 40 trees are planted per acre, the yield is 500 apples per tree. For each tree per acre that is added beyond 40, the yield is reduced by 10 apples per tree.

Problem

Find the number of trees per acre that should be planted to yield the maximum crop.

Solution Let x represent the number of trees planted per acre beyond 40. Let Y represent the total crop per acre. Then Y is the number of trees per acre times the number of apples per tree. For example, if $x = 0$, then the number of trees planted per acre is 40 and yield per tree is 500 apples, so $Y = 40 \cdot 500 = 20,000$ apples per acre.

The number of trees per acre is $40 + x$. The yield per tree diminishes by 10 for each tree planted beyond 40 so the yield per tree is $500 - 10x$. Hence the formula for $Y(x)$ is

$$\begin{array}{cc} \text{number} & \text{number of} \\ \text{of trees} & \text{apples per tree} \end{array}$$
$$Y(x) = (40 + x)\ (500 - 10x)$$
$$= 20,000 + 100x - 10x^2$$

We must find the value of x that maximizes $Y(x)$. First find the critical values. Solve $Y'(x) = 100 - 20x$ equal to 0. This yields

$$Y'(x) = 100 - 20x = 0$$

Thus $x = 5$ is the only critical value. The natural end points are $x = 0$ and $x = 50$. The three values to check are

$$Y(5) = (45)(450) = 20,250$$
$$Y(0) = 20,000$$
$$Y(50) = 0$$

Therefore the absolute maximum is $Y(5) = 20,250$. Hence the maximum yield occurs when 45 trees are planted per acre.

An Application: Dayton Trucking Company

In the mid-1970s the price of gasoline skyrocketed as OPEC (Organization of Petroleum Exporting Countries) forced up the price of crude oil. This caused great hardship for many industries, especially those whose transportation costs were significant. Winger gives an account of how one company, the Dayton Trucking Co., approached the problem.* The price of gasoline had become a critical variable in their cost evaluation for the first time. Truck performance was analyzed. They wanted to determine the speed that produced maximum fuel economy, that is, the speed at which miles per gallon (mpg) was a maximum. The data were summarized in the following function that measured fuel economy $F(x)$ in mpg for a truck operating at speed x, measured in miles per hour (mph).

$$F(x) = 0.15x - 0.0015x^2$$

The natural bounds for x were $x = 0$, because a negative value for x makes no sense, and $x = 55$, because the government had just imposed the 55 mph speed limit. The next example solves the problem.

EXAMPLE 5

Problem

Find the absolute maximum of the fuel economy function

$$F(x) = 0.15x - 0.0015x^2$$

where the domain of the function is $[0, 55]$.

Solution Find the critical values by solving $F'(x) = 0$.

$$F'(x) = 0.15 - 0.003x = 0$$
$$x = 50$$

Compare the three values $F(0) = 0$, $F(50) = 3.75$, and $F(55) = 3.7125$. The absolute maximum is $F(50) = 3.75$ mpg. Therefore the speed of 50 mph produced the greatest fuel economy.

*Bernard J. Winger, *Cases in Management Economics*, Columbus, OH, Grid Publishing, 1979, pp. 78–81.

EXERCISE SET 4–4

In Problems 1 to 14 find the absolute extrema, if they exist, for the function defined on the given interval.

1. $f(x) = x^2 + 2x + 1$ on $[-4, 4]$

2. $f(x) = x^2 - 2x + 1$ on $[-4, 4]$

3. $f(x) = 5 - 4x - x^2$ on $[-4, 6]$

4. $f(x) = 8x - 2x^2$ on $[-4, 6]$

5. $f(x) = x^3 + 3x^2 + 1$ on $[-2, 2]$

6. $f(x) = x^3 - 6x^2 + 1$ on $[-1, 5]$

7. $f(x) = x^3 + 3x^2 + 1$ on $[-3, -1]$

8. $f(x) = x^3 - 6x^2 + 1$ on $[-2, 2]$

9. $f(x) = x^4 - 2x^2 + 1$ on $[-2, 2]$

10. $f(x) = x^4 + 4x^3 + 1$ on $[-4, 2]$

11. $f(x) = (x - 2)^2(x - 1)$ on $[-1, 4]$

12. $f(x) = (x + 1)^2(x - 1)$ on $[0, 4]$

13. $f(x) = (x - 3)^2(x - 1)^2$ on $[-1, 4]$

14. $f(x) = (x + 1)^2(x - 1)^2$ on $[0, 4]$

15. An object is thrown straight upward so that its height after t seconds is $f(t) = -16t^2 + 128t$, where f is measured in feet. How high will the object travel and after how many seconds will it reach its maximum height?

16. An object is thrown straight upward so that its height after t seconds is $f(t) = -16t^2 + 256t + 6$, where f is measured in feet. How high will the object travel and after how many seconds will it reach its maximum height?

17. A family wants to build a rectangular deck on the back of their house. It will be enclosed on three sides by wooden fencing with the back of the house being the fourth side. If 40 feet of fencing is to be used, find the dimensions of the deck with the largest area that can be built.

18. An apple orchard owner has determined that if 30 trees are planted per acre, the yield is 600 apples per tree. For each tree per acre that is added beyond 30, the yield is reduced by 15 apples per tree. Find the number of trees per acre that should be planted to yield the maximum crop.

19. Divide the number 300 into two parts whose sum is 300 such that the product of one part and the square of the other is a maximum.

20. Divide the number 120 into two parts whose sum is 120 such that the product of one part and the cube of the other is a maximum.

21. A farmer wants to enclose with a fence a rectangular field that is adjacent to a straight river, where no fencing is required along the river. What are the dimensions of the field if 4000 feet of fencing is to be used and the area of the field is to be a maximum?

22. A manufacturing company offers the following discount schedule of prices: $20 for orders of 500 or fewer, with the cost being reduced 2 cents for each unit above 500. Find the order that will maximize the company's revenue.

23. A person is on an island 10 miles from a straight shore. The person wishes to get to a point 12 miles up the coast. If the person can row at the rate of 2 mph (miles per hour) and walk at the rate of 4 mph, to what point on the shore should the person row in order to minimize the time?

24. A manufacturing company makes an open box from a 5- by 6-inch rectangular piece of cardboard by cutting out equal squares at each corner and then folding up the sides. What are the dimensions of the box with maximum volume?

25. A manufacturing company wants to package its product in a rectangular box with a square base and a volume of 32 in.^{3}. The cost of the material used for the top and bottom is 3 cents per square inch (in.^{2}), while the cost of the material used for the sides is 6 cents per in.^{2}. What are the dimensions of the box with the minimum cost?

26. A manufacturing company wants to package its product in a closed can in the shape of a right circular cylinder. The company has determined that the volume of the can must be 16 in.^{3}. What are the dimensions of the can with the minimum surface area?

27. A manufacturing company wants to package its product in a rectangular box with a square base and a volume of 32 in.^{3}. The cost of the material used for the top is 1 cent per in.^{2}, the cost of the material used for the bottom is 3 cents per in.^{2}, and the cost of the

material used for the sides is 2 cents per in.^{2}. What are the dimensions of the box with the minimum cost?

28. At 1:00 P.M. plane A is 650 miles due west of plane B. If plane A flies at 100 mph due south and plane B flies at 150 mph due west, when will they be nearest each other and how near will they be?

29. U.S. postal regulations state that packages that are sent parcel post must have the sum of the length plus the girth (perimeter of a cross section) no more than 84 in. What are the dimensions of the rectangular package with square ends with the greatest volume that can be sent by parcel post?

30. U.S. postal regulations state that packages that are sent parcel post must have the sum of the length plus the girth (perimeter of a cross section) no more than 84 in. What are the dimensions of the cylindrical package with the greatest volume that can be sent by parcel post?

31. A "Norman window" consists of a rectangle with a semicircular area on top. See the accompanying figure. Find the width of the rectangle if the perimeter of the window is 12 feet and the area is as large as possible.

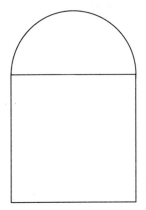

32. A long rectangular sheet of aluminum is to be made into a gutter by turning up equal sides the length of the strip at right angles. How many inches should be turned up on each side to maximize the volume of the gutter?

33. A rectangular pen is to be fenced in, and then an additional fence will be constructed in the center of the pen to make two identical rectangular enclosures. If the area is to be 60 square yards, find the dimensions of the pen with the minimum amount of fencing.

34. Show that among all rectangles having a fixed perimeter, the square has the maximum area.

35. Show that among all rectangles having a fixed area, the square has the minimum perimeter.

36. A principle in economics states that maximum profit is obtained when the marginal revenue, which is the derivative of the revenue function, equals the marginal cost, which is the derivative of the cost function. Explain this principle if $P(x)$ is the profit function, $R(x)$ is the revenue function, and $C(x)$ is the cost function when x goods are produced.

37. A box with no top and a square bottom is to have a volume of 10 in.^{3}. Find the dimensions of the box of least cost if the material of which the sides are to be made costs one-third per square inch as much as the material of which the bottom is to be made.

Referenced Exercise Set 4–4

1. In a study* to increase the productivity of the average personal automobile, Purdue University's Interdisciplinary Engineering Studies researched the feasibility of a plan for members of an enterprise to own and operate jointly a fleet of vehicles. In one model 26 families were monitored by having them keep diaries of when they would use a shared vehicle. By keeping records of cost and availability of the shared vehicles, the researchers generated the following function that relates the number of vehicles, x, to the cost per family per week, $f(x)$:

$$f(x) = 11.8x^3 - 138.2x + 200$$

The function is valid in the interval $[0, 4]$. Find the value of x that minimizes the cost per family per week.

2. There is extensive literature on the optimal governmental regulation of open-access resources, such as fisheries and watersheds. In a study† of market behavior and exploitation of fisheries an important function is the total revenue R produced that depends on the fishing effort x, measured by the number of days fished

*Jeffery K. Cochran and F. T. Sparrow, "Optimal Management of a Shared Fleet with Peak Demands," *Applications of Management Science*, Vol. 4, 1985, pp. 81–105.

†Lee G. Anderson and Dwight R. Lee, "Optimal Governing Instrument, Operation Level, and Enforcement in Natural Resource Regulation: The Case of the Fishery," *American Journal of Agricultural Economics*, Vol. 68, 1986, pp. 678–690.

by standard boats. One model considered assumes the cost $C(x)$ is a linear function. A typical pair of functions for $R(x)$ and $C(x)$ is

$$R(x) = 0.4x - 0.002x^2$$
and $\quad C(x) = 0.11x$

The natural bounds for these functions would be $x \geq 0$ on the left and the natural bound on the right would be that value of x such that $R(x) = C(x)$.

(a) Find the value of x that maximizes $R(x)$.
(b) Define the profit function $P(x) = R(x) - C(x)$ and find the value of x that maximizes $P(x)$.
(c) Graph $R(x)$ and $C(x)$ in the same coordinate system and state why the right-hand bound is a natural boundary for the problem.

3. In an article studying the trade-offs that research and development teams must make, an example from space research was analyzed.‡ NASA was trying to determine whether it was cost-effective to manufacture silicon cells for use in future applications. The price of silicon when the study was done (1981) was $65 per kilogram (kg). NASA set a goal of producing silicon at a price of $20 per kilogram. They could make silicon at this price, but its quality was not high enough. If they produced silicon at a higher price, what was the likelihood that it would be high quality? The research-

ers generated a function that measured the likelihood $P(x)$, measured in percent, that the silicon produced at price x would be high quality. The function is approximated by

$$P(x) = -0.0001x^3 + 0.0045x^2 - 0.25$$

There were two natural bounds for x. It made no sense to let x take on negative values and the price of $28 per kg was set as the highest price they would be willing to pay. So the domain of $P(x)$ is $[0, 28]$. Find the absolute maximum of $P(x)$ and graph the function.

4. This problem was suggested by John Dawson, our bird-loving friend from Penn State.§ In the autumn many people put up feeders for wild birds, thereby initiating the annual round of "squirrel wars." Seasoned veterans of the combat have learned to thwart the acrobatic rodents by suspending the feeders from wires in the shape of the letter **Y.** If the wire is suspended between two trees 10 feet apart such that there are two feet below the height at which the wires are attached to the trees, then the length L of the wire is given by the equation

$$L = x + \sqrt{10^2 + 4(2 - x)^2}$$

where x denotes the length of the tail of the **Y.** (See Figure 5.) What value of x will minimize the length of the wire?

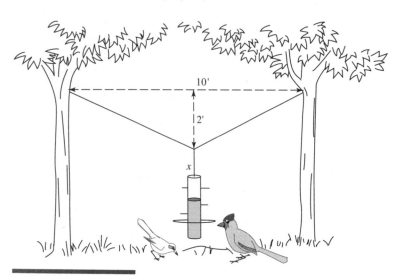

FIGURE 5

‡Virendra S. Sherlekar and Burton V. Dean, "Assessment of R & D Risk-Cost Trade-Offs," *Applications of Management Science,* Greenwich, CT, JAI Press, 1983, pp. 40–51.

§John W. Dawson, Jr., "Hanging a Bird Feeder: Food for Thought," *The College Mathematics Journal,* March, 1990, pp. 129–130.

Cumulative Exercise Set 4–4

In Problems 1 and 2, a function $f(x)$ is given.
(a) Find the intervals where $f(x)$ is increasing.
(b) Find the point(s) where $f(x)$ has a horizontal tangent line.
(c) Sketch the graph of the function.

1. $f(x) = x^3 - 4x^2 - 2$

2. $f(x) = \dfrac{x}{x - 4}$

In Problems 3 to 5 use the information given to sketch the graph of $f(x)$, assuming that $f(x)$ is continuous for all real numbers x.

3. Values where $f'(x) = 0$

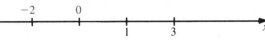

Sign of $f'(x)$ $----$ $++++$ $----$ 0000 $++++$

$f(-2) = -2$, $f(0) = 2$, $f(1) = 1$, $f(4) = 3$

4. Critical values

Sign of $f'(x)$ $+++++++++$ $-------$ $+++++++$ $-----$

$f(1) = 2$, $f(3) = 0$

5. Critical values

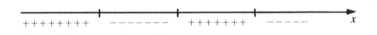

Sign of $f'(x)$ $----------$ $+++++++$ $++++++++$ $+++++$

0, $f(0) = 0$

In Problems 6 to 9 find the relative extrema and sketch the graph of $f(x)$.

6. $f(x) = 2 + 4x^3 - x^4$

7. $f(x) = x^5 - x$

8. $f(x) = x(x - 4)^2$

9. $f(x) = (x - 1)^3(x^2 - 1)$

10. Find the absolute extrema for $f(x) = (x^2 - 1)^2$ on the interval $[-2, 3]$.

11. What number exceeds its square by the maximum amount?

12. Let $f(x)$ be the function whose graph is shown.
   (a) List the critical values, if there are any.
   (b) List the relative extrema, if there are any.
   (c) List the points of inflection, if there are any.
   (d) List the vertical asymptotes, if there are any.
   (e) List the horizontal asymptotes, if there are any.
   (f) For what interval(s) is $f' > 0$?
   (g) For what interval(s) is $f(x)$ concave upward?

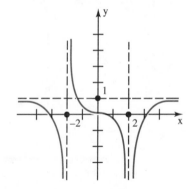

4—5 Implicit Differentiation and Related Rates

Thus far the relationships between the variables have been presented as **explicit functions,** meaning that an equation in the form $y = f(x)$ has been given. Two examples are

$$y = x^2 - 4 \qquad y = (x^3 + 4x)^{1/2}$$

However, often in applications equations do not express y as an explicit function of x. For example, the equation

$$x^2 + y^2 = 1$$

whose graph is a circle of radius 1, is not of the form $y = f(x)$. In this section we study equations that are not expressed as functions. They are called **implicit functions.** We also compute their derivatives. Applications of these types of derivatives, called *related rates,* are also studied.

Implicit Functions

The equation $x^2 + y^2 = 1$ does not express y as a function of x directly because, for every value of x in the interval $(-1, 1)$, there are two corresponding values of y that satisfy the equation. For instance, if $x = 0$, then y can be either 1 or -1; that is, the points $(0, 1)$ and $(0, -1)$ are on the graph. Geometrically, Figure 1 shows that any vertical line $x = a$ for $-1 < a < 1$ intersects the graph in more than one point, so the graph is not a function.

We say that equations such as $x^2 + y^2 = 1$ define y *implicitly* as a function of x. Intuitively this means that even though the equation does not express y in terms of x directly, it is theoretically possible to solve for y. However, this is not always an easy task. For example, given the equation $xy = 1$, it is easy to solve for y to get $y = 1/x$; but it is much more difficult to solve for y given the equation $y^5 - y^3 + xy = 1$. It may be possible to express the equation as several formulas.

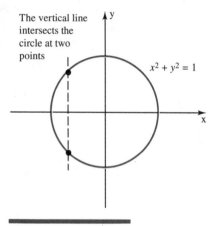

The vertical line intersects the circle at two points

$x^2 + y^2 = 1$

FIGURE 1

For example, $x^2 + y^2 = 1$ can be expressed as the two functions $y = \sqrt{1 - x^2}$ and $y = -\sqrt{1 - x^2}$.

Implicit Differentiation

If an equation expresses y as an implicit function of x, it is possible to compute the derivative of y with respect to x without having to solve for y directly. This technique is called **implicit differentiation.** It assumes that the equation can be expressed in the form $y = f(x)$, and then the derivative of each term of the equation is computed, keeping in mind that y is a function of x. For example, to find y' given the equation $x^2 + y^2 = 1$, find the derivative of each term separately. The derivative of the first term, x^2, is $2x$. However, the derivative of the second term, y^2, is not merely $2y$. The chain rule must be used to compute the derivative of y^2, because it is assumed that $y^2 = [f(x)]^2$. Therefore, using the chain rule, the derivative of y^2 is $2yy'$. The derivative of the constant term, 1, is 0. Putting these steps together and solving for y' yields

$$D_x(x^2 + y^2) = D_x 1$$
$$2x + 2yy' = 0$$
$$2yy' = -2x$$
$$y' = -\frac{2x}{2y}$$
$$y' = -\frac{x}{y}$$

The equation for the derivative contains both x and y, which is usually the case when computing a derivative implicitly. To evaluate the derivative at a particular point both the x and y coordinates must be substituted into the equation.

The important idea to remember when computing implicit derivatives is that the derivative of any term containing y will require the use of the chain rule so that it will have y' as part of its derivative. Also, any term containing a product, such as x^2y^3, will require the product rule as well as the chain rule. For instance, suppose an implicit function contains the term x^2y^3, then the derivative of this term is

$$D_x(x^2y^3) = x^2 D_x(y^3) + y^3 D_x(x^2)$$
$$= x^2 \cdot 3y^2 y' + y^3 \cdot 2x$$
$$= 3x^2 y^2 y' + 2xy^3$$

Example 1 illustrates how to compute the derivative of an implicit function and how to evaluate the derivative at a point that satisfies the implicit function.

EXAMPLE 1

Problem

Consider the equation

$$x^3 + x^2y - y^4 = 1$$

(a) Find y' using implicit differentiation.
(b) Evaluate y' at $(1, 1)$.

Solution (a) Find the derivative of each term.

$$D_x(x^3 + x^2y - y^4) = D_x(1)$$
$$D_x(x^3) + D_x(x^2y) - D_x(y^4) = 0$$
$$D_x(x^3) + x^2D_x(y) + yD_x(x^2) - D_x(y^4) = 0$$
$$3x^2 + x^2 \cdot y' + y \cdot 2x - 4y^3y' = 0$$
$$3x^2 + x^2y' + 2xy - 4y^3y' = 0$$

Gather terms containing y'.

$$x^2y' - 4y^3y' = -3x^2 - 2xy$$
$$(x^2 - 4y^3)y' = -3x^2 - 2xy$$
$$y' = (-3x^2 - 2xy)/(x^2 - 4y^3)$$

(b) Substitute $x = 1$ and $y = 1$ into the equation for y' in (a).

$$y' = \frac{-3 - 2}{1 - 4} = \frac{-5}{-3} = \frac{5}{3}$$

Often in applications the two variables in an implicit function both depend on a third variable. For example, suppose x and y are related by the equation $x^2 + y^2 = 2500$ where x and y represent distances, and each is changing with respect to time. If t represents time, then x and y are functions of t. By differentiating the equation relating x and y, we get a relationship in the two derivatives $D_t x$ and $D_t y$. The chain rule must be used on each term, including x^2. That is, the derivative of x^2 with respect to t is not merely $2x$ but $2xx'$, where x' is $D_t x$. Example 2 shows how to carry this out.

EXAMPLE 2

Let x and y be related by the equation

$$x^2 + y^2 = 2500$$

Suppose x and y change with respect to time, t.

Problem

Find an equation relating $D_t x$ and $D_t y$. Then evaluate the equation for $x = 40$, $y = 30$, and $D_t y = -4$.

Solution Think of the equation as an implicit function of t, where both x and y are implicit functions of t. Differentiate the equation with respect to t by differentiating each term with respect to t. This yields

$$D_t(x^2 + y^2) = D_t 2500$$
$$2x \cdot D_t x + 2y \cdot D_t y = 0$$
$$2x D_t x = -2y D_t y$$
$$D_t x = \frac{-y D_t y}{x}$$

Evaluating the equation for $D_t x$ when $x = 40$, $y = 30$, and $D_t y = -4$ yields

$$D_t x = \frac{-30(-4)}{40} = 3$$

Related Rates

In Example 2 the variables x and y are functions of a third variable t. The rate of change of x is related to the rate of change of y by the formula derived in the example by differentiating the original implicit function with respect to t. Such a problem is called a **related rates** problem. The next example illustrates this type of application.

EXAMPLE 3

A tropical storm is 50 miles offshore, and its path is perpendicular to a straight shoreline. It is approaching the shore at the rate of 4 mph. Meteorologists studying the behavior of the storm from a van on the shore want to stay exactly 50 miles from the storm and remain on the shoreline. They start at the point on the shoreline in the path of the storm.

Problem

Determine a formula for the speed that the truck must maintain to remain 50 miles from the storm. Then find the speed of the truck when the storm is 40 miles from the shore.

Solution Usually the first step in a related rates problem is to make a sketch of the problem. From the sketch the variables can be labeled. See Figure 2. Let the

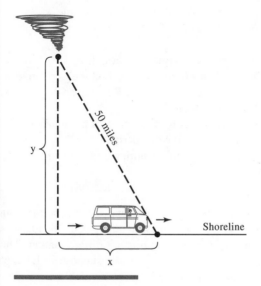

FIGURE 2

distance from the storm to the shore be represented by y and the distance that the van has traveled be represented by x. From the Pythagorean Theorem the relationship between x and y is

$$x^2 + y^2 = 50^2 = 2500$$

The problem asks for $D_t x$, given that $D_t y = -4$, which is negative because y is decreasing. The relationship between $D_t x$ and $D_t y$ is derived by differentiating the formula relating x and y with respect to t. This was done in Example 2. The relationship is

$$D_t x = \frac{-y D_t y}{x}$$

Since $D_t y = -4$, the formula becomes

$$D_t x = \frac{4y}{x}$$

When the storm is 40 mi from the shore, that is, $y = 40$, then from the Pythagorean Theorem $x = 30$. Therefore the speed of the van when $y = 40$ is

$$D_t x = \frac{4(40)}{30} = \frac{16}{3} \text{ mph}$$

Example 3 demonstrates the six steps used in solving related rates problems.

The Key Steps in Solving Related Rates Problems

1. Draw a sketch of the problem if possible.
2. Identify all quantities, both constants and variables, and give labels to the variables.
3. Find a formula relating the variables.
4. Using implicit differentiation, find the derivative with respect to time, t, of each side of the formula. Remember that each variable in the formula is a function of t.
5. The problem asks for the rate of a particular variable. Solve the formula derived in step 4 for this derivative.
6. Substitute the given quantities into the formula for the desired derivative.

It is important to perform steps 4 and 5 before doing the substitution called for in step 6. If the substitution is made beforehand, the variables will not appear in the formula when applying implicit differentiation. Thus the formula derived in step 4 will be incorrect. To illustrate, suppose in Example 3 the value $y = 40$ is substituted into the formula $x^2 + y^2 = 2500$ before the derivative is found. This would result in the formula $x^2 + 1600 = 2500$. Using implicit differentiation on this formula results in $2xx' = 0$, so $x' = 0$ whenever $x \neq 0$, which is certainly incorrect.

The next two examples illustrate how to apply the steps used to solve related rates problems.

EXAMPLE 4

A rock is thrown into a still pond and circular ripples move out. The radius of the disturbed region increases at the rate of 3 ft/sec.

Problem

Find the rate at which the surface area of the disturbed region is increasing when the farthest ripple is 20 feet (ft) from the place where the rock struck the pond.

Solution The sketch of the problem is one of concentric circles given in Figure 3, each circle representing a ripple. The farthest ripple is the largest circle. Let the radius of the farthest ripple be represented by r and the area of the disturbed region by A. The formula relating r and A is the area of a circle.

$$A = \pi r^2$$

It is given that $D_t r = 3$ and we are asked to find $D_t A$ when $r = 20$. Next use implicit differentiation to find $D_t A$ in terms of $D_t r$.

$$D_t A = 2\pi r D_t r$$

Substitute into this formula $r = 20$ and $D_t r = 3$.

$$D_t A = 2\pi(20)(3) = 120\pi \approx 376.99 \text{ ft}^2/\text{sec}$$

Thus when the radius is 20 ft the surface area of the disturbed region is increasing at the rate of about 377 ft²/sec.

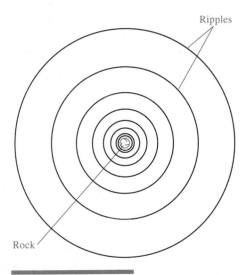

Ripples

Rock

FIGURE 3

EXAMPLE 5

An industrial filter in the shape of a right circular cone is 10 cm (centimeters) high and has a radius of 20 cm at its top. A solution is poured through the filter, and residue gathers in the filter at the rate of 2 cm³/min.

Problem

Find the rate at which the height of the residue is increasing when the height is 2 cm.

Solution The sketch is given in Figure 4. Let the height of the residue be represented by h, the radius by r, and the volume by V. The residue gathers in the shape of a cone. The formula for the volume of a cone is

$$V = \left(\frac{1}{3}\right)\pi r^2 h$$

It is given that $D_t V = 2$ and we are asked to find $D_t h$ when $h = 2$. It is possible to use implicit differentiation to find $D_t h$ in terms of $D_t V$ and $D_t r$, but the value of $D_t r$ is unknown. Since V is expressed as a function of two variables, r and h, and only one value of the two derivatives is given, it is necessary to express V as a function of h alone. The relationship between r and h is given by similar triangles. From Figure 4 the triangle of the cone and the triangle of the residue are similar. The ratios of corresponding sides of similar triangles are equal, which yields

$$\frac{r}{20} = \frac{h}{10}$$

Thus $r = 2h$. Substituting this into the formula for V yields

$$V = \left(\frac{1}{3}\right)\pi(2h)^2 h = \left(\frac{4}{3}\right)\pi h^3$$

Applying implicit differentiation to the formula yields

$$D_t V = \left(\frac{4}{3}\right)\pi \cdot 3h^2 D_t h$$
$$= 4\pi h^2 D_t h$$

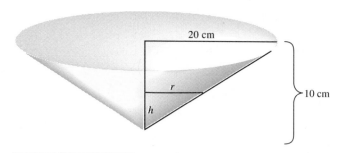

20 cm

r

h

10 cm

FIGURE 4

Solving for $D_t h$ and making the substitution $D_t V = 2$ gives

$$D_t h = \frac{2}{4\pi h^2}$$

This is the formula for $D_t h$. Next substitute the value $h = 2$ to find the rate at which the height of the residue is increasing when the height is 2 cm.

$$D_t h = \frac{2}{4\pi 2^2} = \frac{1}{8\pi} \approx 0.04 \text{ cm/min}$$

EXERCISE SET 4–5

In Problems 1 to 8 find $D_x y$ by using implicit differentiation.

1. $x^2 + 3y^2 = 4$

2. $x^2 - 5y^3 = 6$

3. $xy + x^2 y = 2$

4. $xy - x^2 y^2 = 4$

5. $x^3 + xy^3 - y^4 = 1$

6. $x^4 - x^2 y^3 + y^5 = 4$

7. $x^2 + xy^3 - y^{-1} = 1$

8. $x^4 - 2x^2 y + y^{-2} = 0$

In Problems 9 to 12 find $D_x y$ evaluated at the given point.

9. $x^2 + y^3 = 0;$ $(1, -1)$

10. $x^3 + y^2 = 2;$ $(1, -1)$

11. $3xy + x^2 y = 0;$ $(1, 0)$

12. $4xy - x^2 y^2 = 3;$ $(1, 1)$

In Problems 13 to 16 x and y are functions of t. Express $D_t y$ in terms of $D_t x$. Then find $D_t y$ when $x = 1$, $y = 1$, and $D_t x = 2$.

13. $x^2 + y^3 = 2$

14. $x^3 + 4y^3 = 5$

15. $3x - x^2 + y^2 = 3$

16. $x^{1/2} + 3y - y^{1/2} = 3$

17. A tropical storm is 100 miles offshore, and its path is perpendicular to a straight shoreline. It is approaching the shore at the rate of 6 mph. Meteorologists studying the behavior of the storm from a van on the shore want to stay exactly 100 miles from the storm and remain on the shoreline. Determine the speed of the truck when the storm is 60 miles from the shore.

18. A rock is thrown into a still pond and circular ripples move out, the radius of the disturbed region increasing at the rate of 2 ft/sec. Find the rate at which the surface area of the disturbed region is increasing when the farthest ripple is 30 ft from the place where the rock struck the pond.

19. A filter in the shape of an inverted cone is 20 cm high and has a radius of 30 cm at its top. A solution is poured through the filter, and residue gathers in the filter at the rate of 5 cm³/min. Find the rate at which the height of the residue is increasing when the height is 3 cm.

20. A firm determines that output y is related to input x, both measured in hundreds of units, by the formula $y = 3x^{1/3}$. If the demand is increasing at the rate of 20 units per month, at what rate must the input increase to meet this increase in demand if the current input is 2700 units?

21. A ladder 50 ft long is standing against a building. The base of the ladder is moved away from the building at the rate of 3 ft/sec. How fast is the top of the ladder moving down the wall when the base of the ladder is 40 ft from the foot of the building?

22. Water is being pumped into a cylindrical trough at the rate of 10 ft³/min. If the depth of the trough is 12 ft, the width of the top is 12 ft, and the length of the top is 20 ft, at what rate is the height of the water rising when the height is 6 ft?

23. A 6-ft tall person is walking away from a streetlight at the constant rate of 3 ft/sec. If the streetlight is 20 ft high, how fast is the length of the person's shadow increasing when the person is 30 ft from the streetlight?

24. A water tank is 30 meters (m) wide, 60 m long, and 10 m deep at one end and 15 m deep at the other end, with the depth increasing linearly. Water is pumped in at the rate of 5 m³/min. How fast is the water level rising when the water level is 3 m?

25. A spherical rubber balloon is being inflated with gas at a constant rate of 15 cm^3/sec. What is the rate of change of the radius when the volume is 45 cm^3?

26. Water is condensing on the surface of a spherical drop of water so that the water surface remains a sphere and the volume is increasing at the rate of 5 in.^{3}/min. At what rate are the surface area and the radius increasing when the radius is 10 in.?

27. A pulley, mounted on a dock, stands 10 ft above the water and a boat is being towed in toward the dock. The rope is attached to the boat at a point 2 ft above the water. If 0.5 ft of rope is being drawn in each second, at what rate is the boat moving when it is 6 ft from the dock?

28. An off-shore oil drill springs a leak. The oil slick spreads in a circular shape, and the radius of the slick is increasing at the rate of 30 ft/hr. Find the rate that the surface area of the oil slick is increasing when the radius is 100 ft.

29. A balloon is rising vertically at the rate of 15 ft/sec. An observer is standing on the ground 900 ft from the point where the balloon left the ground. Find the rate that the distance between the balloon and the observer is changing when the balloon is 1200 ft high.

30. Two roads intersect at right angles. A truck, traveling at the rate of 30 mph, reaches the intersection 5 minutes (min) before a car that is traveling at the rate of 40 mph on the other road. How fast is the distance between the truck and the car changing 5 min after the car reaches the intersection?

31. One airplane, traveling due north at 200 mph, passes a point 1 hour after another airplane, traveling due east at 150 mph, passes the same point. How fast is the distance between the airplanes increasing 1 hour after the first airplane passed the point?

32. A baseball player, running at the speed of 30 ft/sec, is running in a straight line from second to third base. A fielder, standing at a point 100 ft from third base on the left field foul line (which is at right angles to the path of the runner), throws the ball, at the speed of 150 ft/sec, when the runner is 50 ft from third base.

What is the rate of change of the distance between the ball and the runner 0.5 sec after the fielder throws the ball?

33. When helium expands, its pressure P is related to its volume V by the formula $P = kV^{-5/3}$. When the volume of a helium-filled balloon is 18 m^3 the pressure is 0.3 kg/m^2. What is the rate of change of the volume if the pressure is increasing at the rate of 0.05 N/m^2/sec?

34. Liquid is being drawn at the rate of 2 m^3/hr from a tank whose shape is a hemisphere with radius 10 m. How fast is the height of the liquid level changing 2 hours after the tank was full?

Referenced Exercise Set 4–5

1. In a paper* studying efficient glass-cutting techniques, various cost-saving measures were explored, including minimizing the amount of waste and the time used in the cutting procedure. Once a firm has an inventory of various sizes of virgin stock, the problem is to determine what sizes of glass should be cut from the different sheets of stock. Solve the following problem, which is representative of those considered in the paper. Two pieces of glass are to be produced by making two cuts in a sheet of stock. A vertical cut will produce one piece and a horizontal cut in the remaining piece of stock will produce the second piece. The piece left over is called the waste and it is to be made square. (See Figure 5.) If the perimeter of the stock plate is 20 in. and the total surface area of two pieces produced by the cuts is 20 in.^{2}, find the dimensions of the stock plate that minimizes the total length of the two cuts.

2. Population structure and regeneration patterns of trees are important aspects of forest management. In a paper studying size and age structure of two forested strands in the western Cascade Range in Oregon, Stewart studied four species of trees growing in various types of habitats.† In one part of the study a typical problem involved planting two types of trees, each planted on a different square plot of land. In the first plot seedlings were planted in rows every $1\frac{1}{3}$ meters, and in the other plot the seedlings were planted in rows every 4 meters. Find the minimum area of each plot that would be needed to plant 320 rows of trees.

*Adolf Diegel and Hans Vocker, "Optimal Dimensions of Virgin Stock in Cutting Glass to Order," *Decision Sciences*, Vol. 15, 1984, pp. 260–272.

†Glenn H. Stewart, "Population Dynamics of a Montane Conifer Forest, Western Cascade Range, Oregon, USA," *Ecology*, Vol. 67, 1986, pp. 534–544.

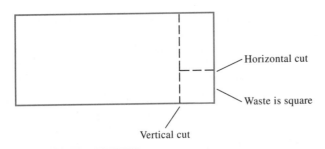

Horizontal cut

Waste is square

Vertical cut

FIGURE 5 *Source:* Adolf Diegel and Hans Vocker, "Optimal Dimensions of Virgin Stock in Cutting Glass to Order," *Decision Sciences,* Vol. 15, 1984, pp. 260–272.

Cumulative Exercise Set 4–5

In Problems 1 and 2 find intervals where $f(x)$ is increasing and where $f(x)$ is decreasing. Find the point(s) where $f(x)$ has a horizontal tangent line. Graph the function.

1. $f(x) = x^3 - 6x^2 - 1$

2. $f(x) = x^4 - 4x^3$

In Problems 3 and 4 find the relative extrema and sketch the graph of the function.

3. $f(x) = 2x^3 + 3x^2 - 12x$

4. $f(x) = 9x + 4/x$

5. Use the curve-sketching techniques to graph the function

$$f(x) = 3x^4 + 4x^3$$

6. Use the information given to sketch the graph of $f(x)$. Assume $f(x)$ is continuous for all real numbers.

Possible points of inflection		0		4
Critical values	-2		2	
Sign of $f'(x)$	$----------$	$+++++++++++$	$---------$	
$f(x)$ is	decreasing	increasing	decreasing	
Sign of $f''(x)$	$+++++++++++$	$----------$	$+++++++++++$	
$f(x)$ is	concave up	concave down	concave up	

$f(-2) = 0,\ f(0) = 1,\ f(2) = 2,\ f(4) = 1$

7. Find the absolute extrema, if they exist, for the function defined on the given interval.

$$f(x) = x^3 - 6x^2 + 1 \qquad \text{on } [-3, 2]$$

8. Divide the number 240 into two parts whose sum is 240 such that the product of one part and the cube of the other is a maximum.

9. A candy firm makes an open box from a 6-inch by 10-inch rectangular piece of cardboard by cutting out equal squares at each corner and then folding up the sides. What are the dimensions of the box with maximum volume?

10. Find $D_x y$ using implicit differentiation:
$$x^3 - 3x^2y + y^3 = 8$$

11. A firm determines that output y is related to input x, both measured in hundreds of units, by the formula $y = 110x^{1/4}$. If the demand is increasing at the rate of 10 units per month, at what rate must the input increase to meet this increase in demand if the current input is 100 units?

12. Two railroad tracks intersect at right angles. One train, traveling at the rate of 50 mph, reaches the intersection 10 minutes before a train that is traveling at the rate of 60 mph on the other track. How fast is the distance between the two trains changing 5 minutes after the second train reaches the intersection?

CASE STUDY   Minimizing Traffic Noise

The private car is by far the most popular means of transportation. However, for many people the car represents more than convenient transportation; their dreams and egos are front seat passengers. It is not surprising that the automobile population rate far outdistances its human counterpart. The role of highway transportation in our lives is profound. Indeed, a significant measure of a country's "development" is its highway system.

The Department of Transportation used to be an agency whose sole responsibility was the structure of roads. Most of our highways were built with static road design as the only critical issue. No concern was given to dynamic factors such as traffic flow, ramp control, merging, and congestion. But today traffic controllers must consider a multitude of factors to minimize traffic snarls and accidents. Estimating traffic demand and capacity has become as much an art as it is a science.

One of the most important considerations of highway operation is acceleration noise control. In the past the sole criterion for improving highway operations was to maximize volume throughout. Little consideration was given to the quality of highway conditions. For example, it is standard practice to reduce highway bottlenecks by ramp control techniques, such as using stoplights to control entrance to the roadway. But nowadays highway administrators give equal concern to the qualitative changes in traffic due to these controls. For instance, a particular control might have reduced overall travel time, but perhaps it created hazardous locations such as rapid decelerations far from the bottlenecks. Measurement of acceleration noise is a valuable tool to identify any hazardous locations. Noise control not only provides more comfortable travel, but it also helps maintain a safer and smoother operation.

One of the first and most extensive traffic flow research projects* was conducted by the Texas Highway Department and the U.S. Bureau of Public Roads in 1963, called the "Gulf Freeway Project." Various surveillance and control devices, including television and noise monitors, were installed along the Gulf Freeway in Houston. The freeway was divided into 56 sections, and each was monitored for several characteristics, including noise level.

Acceleration noise is defined as the disturbance of a vehicle's speed from a uniform speed. It is thus a measure of the deviation from a smooth ride and so it measures the smoothness of traffic. The frequency of violent acceleration and

*Donald R. Drew, *Traffic Flow Theory and Control*, New York, McGraw-Hill Book Co., 1968.

deceleration affects acceleration noise and reflects potentially hazardous conditions.

The primary goal of the Gulf Freeway study was to determine the means to ensure efficient traffic flow. This meant not only maximizing volume but also minimizing the threat of hazardous conditions by minimizing acceleration noise. What range of speeds would best satisfy both of these criteria?

Acceleration noise is a function of speed. Figure 1 is a representation of the relationship of acceleration noise, a, to the average speed of traffic, u, for 1 of the 56 sections of the freeway. The curve that represents the data was given as

$$a = 1.693 - 0.00284u^2 + 0.000045u^3$$

where a is measured in ft/sec^2 and u is measured in miles per hour (mph). This function is valid only in the interval [10, 60]. To see why 10 is a natural endpoint, refer to Figure 1: When $u = 0$, the vehicle is at rest, and $a = 0$. As the vehicle starts to accelerate quickly to merge, acceleration noise increases rapidly. It is represented by the dashed line. As the vehicle enters the traffic pattern, acceleration noise decreases.

The speed that yields minimum acceleration noise is found by calculating the derivative, a', and setting $a' = 0$. The derivative is computed as follows:

$$a' = -2 \cdot 0.00284u + 3 \cdot 0.000045u^2 = -0.00568u + 0.000135u^2$$
$$= u(-0.00568 + 0.000135u)$$

Setting $a' = 0$ yields

$$u(-0.00568 + 0.000135u) = 0$$

This means that either $u = 0$ or $-0.00568 + 0.000135u = 0$. But 0 is not in

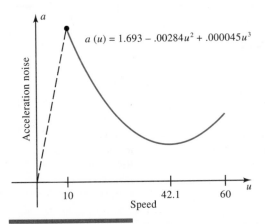

$$a(u) = 1.693 - .00284u^2 + .000045u^3$$

F I G U R E 1   *Source:* Donald R. Drew, *Traffic Flow Theory and Control,* New York, McGraw-Hill Book Co., 1968.

the domain of the function so the only critical value is the latter one. To find the critical value, we solve for u.

$$-0.00568 + 0.000135u = 0$$
$$0.000135u = 0.00568$$
$$u \simeq 42.1$$

The second derivative is $a'' = -0.00568 + 0.00027u$ and $a''(42.1) > 0$, so a has a minimum at $u = 42.1$. Thus the speed that yields the minimum amount of acceleration noise is 42.1 mph.

In another study the speed that yields the maximum flow of volume was found to be 35 mph. This allowed the highway administrators to establish a target speed interval of 35 to 42 mph. This ensured that while the freeway's average speed fluctuated within this interval, they were either maximizing volume, if the speed was close to 35 mph, or minimizing acceleration noise, if it was close to 42 mph.

Case Study Exercises

In Problems 1 to 4 the acceleration noise levels of several other sections of the Gulf Freeway are given. Find the speed at which the acceleration noise is a minimum.

1. $a = 1.572 - 0.00345u^2 + 0.000078u^3$
2. $a = 1.916 - 0.00402u^2 + 0.000088u^3$
3. $a = 2.006 - 0.00501u^2 + 0.000131u^3$
4. $a = 2.118 - 0.00298u^2 + 0.000073u^3$
5. If an acceleration noise test was done on a highway passing through a large city, would you expect the level of acceleration noise to be greater at 1:00 A.M. or at 5:00 P.M. and why?

CHAPTER REVIEW

Key Terms

4–1 Increasing and Decreasing Functions

Increasing Function	Critical Value
Decreasing Function	

4–2 Relative Extrema: The First Derivative Test

Extrema	Relative Minimum
Maximum	Relative Extremum
Minimum	Critical Value
Relative Maximum	First Derivative Test

4–3 Concavity and the Second Derivative Test

Concave Up	Point of Inflection
Concave Down	Curve Sketching

4–4   Absolute Extrema and Optimization Problems
Absolute Maximum   Absolute Extremum
Absolute Minimum

4–5   Implicit Differentiation and Related Rates
Explicit Function   Implicit Differentiation
Implicit Function   Related Rates

Summary of Important Concepts

First Derivative Test: If $f(a)$ is a critical value of $f(x)$, then

1. $f(a)$ is a relative maximum if $f(x)$ is increasing to the left of a and decreasing to the right of a;
2. $f(a)$ is a relative minimum if $f(x)$ is decreasing to the left of a and increasing to the right of a.

The Second Derivative Test: If a is a critical value of $f(x)$, then

1. $f(a)$ is a relative maximum if $f''(x) < 0$ at $x = a$;
2. $f(a)$ is a relative minimum if $f''(x) > 0$ at $x = a$;
3. the test fails if $f''(a) = 0$.

REVIEW PROBLEMS

In Problems 1 to 4 find intervals where $f(x)$ is increasing and decreasing. Find the point(s) where $f(x)$ has a horizontal tangent line. Sketch the graph of the function.

1. $f(x) = 9x^2 - 25$
2. $f(x) = x^2(4 - x)$
3. $f(x) = x/(3 - x)$
4. $f(x) = x^2/(4 - x)$

In Problems 5 to 8 find the relative extrema, and sketch the graph of the function.

5. $f(x) = x^2 + 6x - 10$
6. $f(x) = x^2(x - 4)$
7. $f(x) = x^2(4 - x^2)$
8. $f(x) = 3x^{2/3} - 2x$

In Problems 9 and 10 use the second derivative test to find the relative extrema.

9. $f(x) = 3x^4 - 4x^3$
10. $f(x) = x^4 - 2x^2$

In Problems 11 and 12 find the intervals where the function is concave up and concave down, and sketch the graph.

11. $f(x) = 2x^3 - x^4$
12. $f(x) = x^4 - 6x^2$

In Problems 13 to 15 find the absolute extrema, if they exist.

13. $f(x) = x^3 - 3x^2 + 4$ on $[-2, 2]$
14. $f(x) = (x + 1)^2(x - 1)^2$ on $[0, 4]$
15. $f(x) = x - x^{1/3}$ on $[-1, 8]$

16. An open box is to be constructed from a 10- by 12-inch rectangular piece of cardboard by cutting out equal squares at each corner and then folding up the sides. What are the dimensions of the box with maximum volume?

In Problems 17 and 18 find $D_x y$ by using implicit differentiation.

17. $x^2 + xy + y^2 = 1$
18. $x^3 - 4x^2y + y^{-2} = 1$

19. A water tank is 10 meters (m) wide, 40 m long, and 20 m deep at one end and 30 m deep at the other end, with the depth increasing linearly. Water is pumped in at the rate of 5 m^3/min. How fast is the water level rising when the water level is 3 m?

GRAPHICS CALCULATOR EXPLORATIONS

A function $f(x)$ with a formula that is a fourth-degree equation, called a quartic function, has for its derivative a third-degree equation, which can have as many as three roots. Thus $f(x)$ may have as many as three relative extrema. As an example, consider the function

$$f(x) = x^4 - 8x^2$$

Its derivative, $f'(x) = 4x^3 - 16x = 4x(x^2 - 4) = 4x(x + 2)(x - 2)$, has the three roots $x = -2, 0, 2$. Sketch the graph of this function on your calculator and see that there are three relative extrema: relative minima at $(-2, -16)$ and at $(2, -16)$, and a relative maximum at $(0,0)$. Not every function defined by a fourth-degree equation has three relative extrema. For example, the function $f(x) = x^4$ has only one relative extremum at $x = 0$.

In this Exploration we would like you to see how the cubic and quadratic terms (those terms with x^3 and x^2, respectively) affect the number and location of relative extrema of quartic functions. Consider the following family of functions, where a is some real number:

$$f(x) = x^4 - 4\left(1 + \frac{a}{3}\right)x^3 + 6ax^2 \tag{1}$$

We first look at two limiting cases of these functions, when $a = -3$ and when $a = 0$.

(a) Let $a = -3$ in equation (1). The corresponding function is

$$f(x) = x^4 - 18x^2$$

Graph this function and see that it has three relative extrema: relative minima at $x = -3$ and $x = 3$, and a relative maximum at $x = 0$.

Casio

1. Press GRAPH   ALPHA   x   x^y 4 − 18

   ALPHA   x   x^2 EXE

TI-81

1. Press Y =

2. Press X|T   ^ 4 − 18 X|T   ^ 2 ENTER

3. Press GRAPH

(b) Let $a = 0$ in equation (1). The corresponding function is

$$f(x) = x^4 - 4x^3$$

Graph this function and see that it has one relative extremum at $x = 3$. It also has a critical value at $x = 0$, which is a point of inflection.

Casio

1. Press GRAPH   ALPHA   x   x^y 4 − 4

   ALPHA   x   x^y 3 EXE

TI-81

1. Press $\boxed{\text{Y} =}$

2. Press $\boxed{\text{X}|\text{T}}$ $\boxed{\wedge}$ 4 $\boxed{-}$ 4 $\boxed{\text{X}|\text{T}}$ $\boxed{\wedge}$ 3 $\boxed{\text{ENTER}}$

3. Press $\boxed{\text{GRAPH}}$

Let us describe algebraically how as a increases from -3 to 0, the corresponding functions all have three relative extrema, but the function with $a = 0$ has only one. The function $f(x) = x^4 - 18x^2$ has three relative extrema because its derivative has three distinct factors $(x + 3, x, \text{and } x - 3)$, and as x increases past each zero of the derivative $(-3, 0, \text{and } 3)$ the factor changes sign so the derivative itself changes sign, and thus the function changes from increasing to decreasing or vice versa. The same behavior takes place for each function with a between -3 and 0. But for $a = 0$ the behavior is different.

For the function $f(x) = x^4 - 4x^3$ the derivative has two factors, x^2 and $x - 3$, and when x increases past $x = 0$ neither the derivative nor the function changes sign. This produces a point of inflection at $x = 0$.

Your graphics calculator can illustrate this same behavior geometrically. Successively graph several functions for the following values of a: -3, -2.5, -2, -1.5, -1, -0.5, -0.1, and -0.01. You will see that each function has three relative extrema. (For $a = -0.01$ you will probably need the zoom feature to see the second relative minimum.) Notice that as a gets closer to 0, one relative minimum gets closer to the relative maximum at $(0,0)$ until they appear to coincide when $a = 0$.

Further Explorations

1. Fill in the following table and graph each function.

a	$f(x) = x^4 - 4(1 + \frac{a}{3})x^3 + 6ax^2$	Relative extrema
-3	$x^4 - 18x^2$	$-3, 0, 3$
-2.5		
-2	$x^4 - \frac{4}{3}x^3 - 12x^2$	
-1.5		
-1		
-0.5		
-0.1		
-0.01		

In Problems 2 to 7 a function is given with a derivative that is not easily factored. Use the zoom feature to approximate the roots of the derivative and then use the first derivative test to sketch the graph of the function. Graph the function on your calculator to evaluate the accuracy of your graph.

2. $f(x) = x^3 + 3x^2 + x$
3. $f(x) = x^4 - x$
4. $f(x) = x^4 - 4x^3 + 4x$
5. $f(x) = x^4 - 2x^3 + 4x$
6. $f(x) = x^3 + 6x^2 + x$
7. $f(x) = x^4 + 2x$

Exponential and Logarithmic Functions

5

The Denver Fire Department used a variety of mathematical techniques to provide the best service at the lowest cost. (H. Armstrong Roberts)

Chapter Overview

Perhaps the most useful and universal functions in the application of calculus are exponential and logarithmic functions. They are used in such diverse areas as the growth of a population, the decay of radioactive material, the growth of investments, and the rate of learning.

The first section defines exponential functions and investigates some of their properties. Some graphs of simple types of exponential functions are sketched. The second section does the same for logarithmic functions. Sections 5–3 and 5–4 cover the derivatives of these functions and show how to use graphing techniques to draw the graphs of more complicated functions.

CASE STUDY PREVIEW

The Frustration of Dieting

Losing weight can be a trying experience. Excuses abound for our inability to diet successfully. This Case Study describes the mathematical foundation for weight control. It offers the frustrated dieter a bit of evidence for the case against short-term dieting.

CASE STUDY PREVIEW

Sears, Clothespins, and Unnatural Logarithms

Sears Roebuck and Co. started its mail-order business at the turn of the 20th century with the same kind of catalogs that are used today. These catalogs are still a major concern for Sears, but they also serve several purposes for economists. The Case Study uses logarithmic functions to explore the U.S. economy and lifestyle by examining automobiles, the Consumer Price Index, and clothespins.

CASE STUDY PREVIEW

Denver Fire Department Deployment Policy Analysis

In 1972 the mayor and fire chief of Denver agreed that a study of the fire department was in order in the face of mounting costs. The challenge was to minimize cost while keeping the same level of fire protection. A research team from the University of Colorado was commissioned to solve the problem. An exponential model was used to predict future trends by means of simulation methods.

5–1 Exponential Functions

In this section we introduce a new kind of function, called an exponential function, in which the exponent is a variable instead of a constant. An exponential function describes an object that "grows exponentially," meaning that it either expands or shrinks at an increasingly fast rate. We draw the graphs of exponential functions, study some of their properties, and define the exponential function involving the number e. The material is applied to an area of psychology that deals with learning curves.

Graphs

The graph of the quadratic function $f(x) = x^2$ is a parabola. It is quite different from the graph of the function $f(x) = 2^x$, which is called an exponential function because the exponent is a variable. Table 1 lists some points on the graph of $f(x) = 2^x$. These points are joined by the curve in Figure 1. In completing the table, remember that negative exponents refer to reciprocals, so that, for instance,

$$2^{-3} = \frac{1}{2^3} = \frac{1}{8}.$$

Table 1

x	-3	-2	-1	0	1	2	3
2^x	$\frac{1}{8}$	$\frac{1}{4}$	$\frac{1}{2}$	1	2	4	8

Table 1 lists only integral exponents, but exponential functions are defined for all real numbers. Rational exponents refer to roots. For instance, $2^{1/2} = \sqrt{2}$, $2^{5/3} = \sqrt[3]{2^5} = (\sqrt[3]{2})^5$, and $2^{-3/4} = \frac{1}{2^{3/4}} = \frac{1}{\sqrt[4]{2^3}} = \frac{1}{\sqrt[4]{8}}$. The meaning of an exponential function at an irrational exponent, such as $2^{\sqrt{2}}$, is beyond the scope of this book. However, it can be viewed intuitively from the graph in Figure 1 or derived from a calculator experiment by examining the limit of the numbers $2^{1.4}$, $2^{1.41}$, $2^{1.412}$, ...

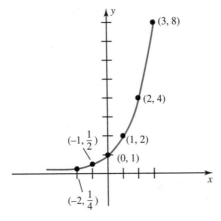

Graph of $f(x) = 2^x$

FIGURE 1

DEFINITION

If b is a positive number with $b \neq 1$, then a function $f(x) = b^x$ is called an **exponential function.** The number b is the **base** of $f(x)$.

The restriction $b \neq 1$ prevents the graph of $f(x) = b^x$ from being the horizontal line $f(x) = 1$. The restriction $b > 0$ ensures that the domain of an exponential function is the set of all real numbers, since otherwise there would be some values of x for which b^x is not a real number. For example, if $b = -1$ and $x = \frac{1}{2}$, then $(-1)^{1/2} = \sqrt{-1}$ is not a real number.

EXAMPLE 1

Problem

Sketch the graph of $f(x) = 3^x$.

Solution Table 2 lists some points on the graph, which is sketched in Figure 2.

Table 2

x	-3	-2	-1	0	1	2	3
3^x	$\frac{1}{27}$	$\frac{1}{9}$	$\frac{1}{3}$	1	3	9	27

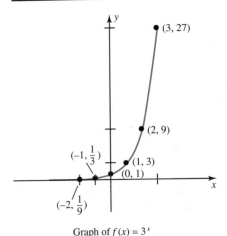

Graph of $f(x) = 3^x$.

FIGURE 2

The graphs of $f(x) = 2^x$ and $f(x) = 3^x$ are typical of the graphs of those exponential functions $f(x) = b^x$ with $b > 1$. The graphs pass through the point $(0, 1)$ and increase over the entire domain. As x gets larger and larger the graphs increase without bound, which is expressed compactly by writing $\lim\limits_{x \to \infty} b^x = \infty$. In addition, $\lim\limits_{x \to -\infty} b^x = 0$, so the x-axis is a horizontal asymptote of the graph of $f(x)$.

Example 2 examines a typical exponential function $f(x) = b^x$ for $0 < b < 1$.

EXAMPLE 2

Problem

Sketch the graph of $f(x) = \left(\dfrac{1}{2}\right)^x$.

Solution Write $\left(\dfrac{1}{2}\right)^x$ in the form $\left(\dfrac{1}{2}\right)^x = \dfrac{1}{2^x}$. Then construct Table 3. The graph is drawn in Figure 3.

Table 3

x	-3	-2	-1	0	1	2	3
$(\tfrac{1}{2})^x$	8	4	2	1	$\tfrac{1}{2}$	$\tfrac{1}{4}$	$\tfrac{1}{8}$

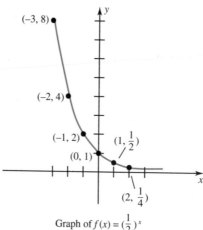

Graph of $f(x) = (\tfrac{1}{2})^x$

FIGURE 3

Properties

The graph of an exponential function $f(x) = b^x$ shows that if $b^r = b^s$, then $r = s$. This is the first of several properties of exponential functions that we list for reference. The other properties should be familiar from a previous study of algebra.

1. If $b^r = b^s$, then $r = s$
2. $b^r \cdot b^s = b^{r+s}$
3. $\dfrac{b^r}{b^s} = b^r \cdot b^{-s} = b^{r-s}$
4. $(b^r)^s = b^{rs}$

E X A M P L E   3

Problem

Find all values of x for which (a) $2^x = 64$, (b) $9^{(x+3)} = 3^{(3x-9)}$.

Solution (a) Since $2^6 = 64$, one value of x for which $2^x = 64$ is $x = 6$. According to property 1, this is the only value.
(b) Write $9^{(x+3)} = (3^2)^{(x+3)}$. By property 4, $(3^2)^{(x+3)} = 3^{2(x+3)}$. Thus $9^{(x+3)} = 3^{(2x+6)}$. It was given that $9^{(x+3)} = 3^{(3x-9)}$, so $3^{(2x+6)} = 3^{(3x-9)}$. By property 1, $2x + 6 = 3x - 9$. Therefore $x = 15$. This answer can be verified using the properties because $3^{(45-9)} = 3^{36} = (3^2)^{18} = 9^{18} = 9^{(15+3)}$.

The Exponential Function

Consider the function defined by

$$f(x) = \left(1 + \frac{1}{x}\right)^x$$

Then $f(1) = 2$ and $f(2) = \left(\dfrac{3}{2}\right)^2 = 2.25$. A calculator with a $\boxed{y^x}$ key was used to compute the additional values of $f(x)$ in Table 4. The numbers are rounded to five decimal places. Table 4 indicates that $f(x)$ approaches $2.71828 \ldots$ as x gets larger and larger. The exact number that $f(x)$ approaches is denoted by the letter e. An irrational number that occurs very often in applications, e is an important number in diverse branches of mathematics. It is singled out here because it is the most useful base for an exponential function.

D E F I N I T I O N

$$e = \lim_{x \to \infty} \left(1 + \frac{1}{x}\right)^x$$

Since e lies between 2 and 3, the graph of the function $f(x) = e^x$ is sandwiched between the graphs of $f(x) = 2^x$ and $f(x) = 3^x$, as shown in Figure 4. Although the number e might seem to be contrived, the function $f(x) = e^x$ occurs so often

Table 4

x	f(x)
1	$(2/1)^1 = 2$
2	$(3/2)^2 = 2.25$
3	$(4/3)^3 \approx 2.37037$
4	$(5/4)^4 \approx 2.44141$
5	$(6/5)^5 \approx 2.48832$
6	$(7/6)^6 \approx 2.52163$
7	$(8/7)^7 \approx 2.54650$
8	$(9/8)^8 \approx 2.56578$
9	$(10/9)^9 \approx 2.58117$
10	≈ 2.59374
100	≈ 2.70481
1,000	≈ 2.71692
10,000	≈ 2.71815
100,000	≈ 2.71827
1,000,000	≈ 2.71828
10,000,000	≈ 2.71828

that it is called *the* exponential function. A table in the Appendix lists the values of e^x for various values of x. Many calculators have an $\boxed{e^x}$ key. An approximation of e^k can be obtained by the sequence $\boxed{k}\,\boxed{e^x}$, although a color-coded key sometimes has to be pressed before tapping the $\boxed{e^x}$ key.

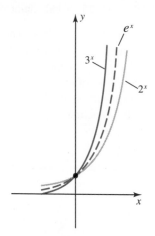

Graphs of three
exponential functions

FIGURE 4

DEFINITION

The function $f(x) = e^x$ is called **the exponential function.** The domain consists of all real numbers and the range is the set of positive numbers.

Example 2 examined the graph of the function $f(x) = \left(\dfrac{1}{2}\right)^x$, which is typical of all exponential functions of the form $f(x) = b^x$ for $0 < b < 1$. Example 4 examines the function $f(x) = e^{-x}$. It is important to remember that

$$e^{-x} = \left(\frac{1}{e}\right)^x = \frac{1}{e^x}$$

EXAMPLE 4

Problem

Sketch the graph of $f(x) = e^{-x}$.

Solution Construct Table 5 by using a calculator or by selecting the appropriate values for x from the table in the Appendix. The graph of $f(x) = e^{-x}$ is sketched by joining these points with the smooth curve shown in Figure 5.

Table 5

x	-2	-1	0	1	2
e^{-x}	7.38906	2.71828	1	0.36788	0.13534

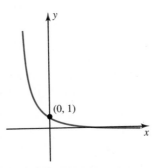

Graph of $f(x) = e^{-x}$

FIGURE 5

The graph of $f(x) = e^{-x}$ is similar to the graph of $f(x) = \left(\dfrac{1}{2}\right)^x$. To see this, compare Figure 5 with Figure 3. Notice that both curves pass through the point $(0, 1)$, decrease for all values of x, have the x-axis as a horizontal asymptote, and increase without bound as the values of x get smaller and smaller. The last two properties can be written $\lim\limits_{x \to \infty} e^{-x} = 0$ and $\lim\limits_{x \to -\infty} e^{-x} = \infty$.

Learning Curves

We have all experienced the range of emotions involved in learning a new skill, be it a physical activity like jumping rope or a mental discipline like differentiating functions. At first we are overwhelmed by the seeming complexity of the new task, but soon progress begins and improvement proceeds apace. However, later we reach a point at which further progress requires increasingly more time and dedication.

Psychologists introduced the name "learning curve" to deal with such situations in which a person's ability to learn a new skill is a function of the time devoted to learning it. A **learning curve** is the graph of an exponential function of the form $f(t) = a - be^{-ct}$, where a, b, and c are positive constants. Example 5 presents a typical setting for a learning curve.

EXAMPLE 5

The rate at which a person can learn to type is given by $f(t) = 70 - 60e^{-0.1t}$, where t is in weeks and $f(t)$ is in words per minute (wpm).

Problem

How many words per minute will the person ultimately be able to type?

Solution We derive the answer from the learning curve. The domain of $f(t)$ is restricted to $t \geq 0$ due to physical considerations. Without any training ($t = 0$), the person can type $f(0) = 70 - 60 = 10$ wpm. After one week the person can type $f(1) = 70 - 60e^{-0.1} = 70 - 54.3 = 15.7$ wpm. Table 6 lists additional time periods. (All wpm are rounded to one decimal place.)

The learning curve is the graph that is obtained from this table. It is sketched in Figure 6 and shows that $\lim\limits_{t \to \infty} f(t) = 70$, so as time goes by, the person's speed approaches 70 wpm.

Table 6

t	0	1	2	4	10	20	30	40
$f(t)$	10	15.7	20.9	29.8	47.9	61.9	67	68.9

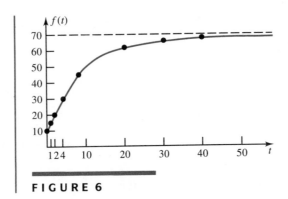

FIGURE 6

EXERCISE SET 5–1

In Problems 1 to 8 sketch the graph of the stated function by the plotting method.

1. $f(x) = 5^x$

2. $f(x) = 4^x$

3. $f(x) = 1.5^x$

4. $f(x) = 2.5^x$

5. $f(x) = (1/3)^x$

6. $f(x) = (1/4)^x$

7. $f(x) = (0.2)^x$

8. $f(x) = (0.1)^x$

In Problems 9 to 14 find all values of x that satisfy the given equality.

9. $2^x = 32$

10. $2^x = 128$

11. $5^x = 625$

12. $7^x = 16,807$

13. $9^{x+6} = 3^{20-2x}$

14. $9^{x+3} = 3^{x+10}$

In Problems 15 to 18 sketch the graph of the given exponential function.

15. $f(x) = e^{x+2}$

16. $f(x) = e^{x-2}$

17. $f(x) = e^{2-x}$

18. $f(x) = e^{-2x}$

In Problems 19 and 20 $f(t)$ represents the rate at which a person can learn to type, where t is in weeks and $f(t)$ is in words per minute (wpm). Find (a) the number of wpm that the person will be able to type after 50 weeks and (b) the number of wpm that the person will ultimately be able to type.

19. $f(t) = 60 - 55e^{-0.1t}$

20. $f(t) = 45 - 40e^{-0.01t}$

In Problems 21 and 22 $f(t)$ represents the number of bricks that an apprentice bricklayer can lay per hour, where t is in weeks. Find (a) the number of bricks that the apprentice will be able to lay per hour after 10 weeks and (b) the

number of bricks that the apprentice will ultimately be able to lay.

21. $f(t) = 28 - 27e^{-0.5t}$

22. $f(t) = 45 - 40e^{-0.2t}$

In Problems 23 to 30 sketch the graph of the stated function by the plotting method.

23. $f(x) = (1.1)^x$

24. $f(x) = (0.9)^x$

25. $f(x) = 6 - 3e^{-x}$

26. $f(x) = 15 - 20e^{-x}$

27. $f(x) = \dfrac{5}{1 + 3e^{-4x}}$

28. $f(x) = \dfrac{6}{1 + 2e^{-3x}}$

29. $f(x) = \dfrac{e^x - e^{-x}}{2}$

30. $f(x) = \dfrac{e^x + e^{-x}}{2}$

In Problems 31 to 34 find all values of x that satisfy the given equality.

31. $2^{2x+4} = 4^{8-x}$

32. $2^{2x-3} = 8^{x-4}$

33. $2^x = 6^x$

34. $3^x = 2^x$

35. Sketch the graphs of the exponential functions $f(x)$ and $g(x)$ on the same coordinate axis, where

$$f(x) = (e^x)^2 \quad \text{and} \quad g(x) = e^{(x^2)}$$

36. Sketch the graph of the exponential function

$$f(x) = e^{-x^2}$$

In Problems 37 and 38 let $f(x) = e^{cx}$, for some number c, and let $f(2) = 3$.

37. Use the law of exponents to find $f(8)$.

38. Use the law of exponents to find $f(-8)$.

39. If a thermometer is moved from inside a house where it reads 70°F, to outside the house where it is 10°F, the temperature of the thermometer will read

$$T(t) = 10 + 60e^{-0.46t}$$

where t is in minutes and $T(t)$ is in degrees Fahrenheit.
(a) What temperature does the thermometer read after 5 minutes?
(b) What temperature will the thermometer eventually reach?

40. If $1000 is deposited into an account that pays 5% interest compounded annually, the future amount is given by

$$A(t) = 1000(1.05)^t$$

where t is in years and $A(t)$ is in dollars. What is the future amount after 10 years?

41. It is estimated that the population of a certain city will be given by the equation

$$P(t) = \frac{5}{2 + e^{-0.1t}}$$

where t is in years after 1990 and $P(t)$ is in millions of inhabitants.
(a) Sketch the graph of $P(t)$.
(b) Is the population increasing or decreasing?
(c) What will happen to the population in the distant future?

42. Suppose a disease spreads according to the equation

$$E(t) = \frac{8}{2 + 5e^{-0.1t}}$$

where t is in days after the outbreak of the disease and $E(t)$ is in thousands of people.
(a) Sketch the graph of $E(t)$.
(b) How many people contracted the disease after 10 days?
(c) In the long run, how many people will contract the disease?

43. Let $f(x) = (-1)^x$. Plot the points on the graph of $f(x)$ for x an integer from 0 to 4. Can the entire graph be drawn by connecting these points with a smooth curve?

44. (a) Use a calculator to approximate this infinite sum to five decimal places.

$$1 + \frac{1}{1!} + \frac{1}{2!} + \frac{1}{3!} + \frac{1}{4!} + \cdots$$

(b) What number do you think this sum is equal to?

45. Let $f(x) = e^x$. Evaluate each limit by using a calculator to form a table in which one row contains values of h and the other row contains the corresponding difference quotient.

(a) $\lim\limits_{h \to 0} \dfrac{f(3 + h) - f(3)}{h}$

(b) $\lim\limits_{h \to 0} \dfrac{f(4 + h) - f(4)}{h}$

Referenced Exercise Set 5–1

1. A study of a game called a *lottery* derived a way to distinguish between different types of risk takers.* Consider a given lottery L whose utility is denoted by $U(L) = 0.817$. (*Utility* is a technical term whose meaning is not necessary to define here.) A decision maker is said to be *risk-averse* with respect to another lottery G if $U(G) < U(L)$, is *risk-neutral* if $U(G) = U(L)$, and is *risk-prone* if $U(G) > U(L)$. Let G be a lottery in which a player wins $2 90% of the time and loses $2 otherwise. The expected value is $E = (2)(.9) + (-2)(.1) = 1.6$. The utility of G is defined by

$$U(G) = (.9)f(2) + (.1)f(-2)$$

where

$$f(x) = \frac{1 - e^{-E(.1x + .5)}}{1 - e^{-E}}$$

Calculate $U(G)$, and determine whether the decision maker is risk-averse, risk-prone, or risk-neutral with respect to the lottery L.

2. An article on forecasting market development showed that two growth curves could be applied in a wide variety of settings.† A *logistic curve* is of the form

$$l(x) = \frac{a}{1 + ce^{-bx}}$$

*Joao L. Becker and Rakesh K. Sarin, "Lottery Dependent Utility," *Management Science*, November 1987, pp. 1367–1382.

†Nigel Meade, "Forecasting Using Growth Curves—An Adaptive Approach," *Journal of the Operational Research Society*, December 1985, pp. 1103–1115.

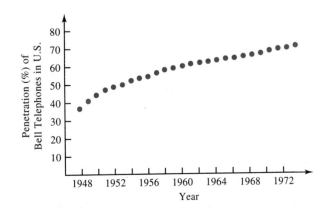

FIGURE 7

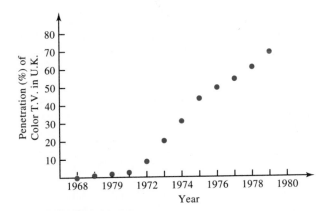

FIGURE 8

and a *Gompertz curve* is of the form

$$g(x) = ae^{-c(e^{-bx})}$$

where *a*, *b*, and *c* are positive numbers. Figure 7 shows the percentage of households in the United States that had Bell telephones in the period 1948–1972, while Figure 8 shows the percentage of households in the United Kingdom that had color television sets in the period 1968–1980. Which figure is modeled by which curve? (*Hint:* Set $a = b = c = 1$, and sketch the graphs of the resulting logistic curve and Gompertz curve.)

3. A study of the spread of automatic teller machines (ATMs) in the banking industry found that the proportion of banks that had adopted the innovation could be expressed as the logistic function

$$P(t) = \frac{1}{1 + e^{5.293 - 0.698t}}$$

where *t* denotes the number of years after the ATMs were first introduced in 1972.‡ Determine the proportion of banks which used ATMs in each listed year by calculating $P(t)$ for the appropriate value of *t*.
(a) 1975   (b) 1980   (c) 1988

‡Timothy H. Hannan and John M. McDowell, "Market Concentration and the Diffusion of New Technology in the Banking Industry," *The Review of Economics and Statistics*, November 1984, pp. 686–691.

4. A special report to the mathematical community entitled "A Challenge of Numbers" stated that "from entry into the ninth grade through receipt of the doctoral degree, a reduction of one-half per year of the number of students enrolled gives curiously accurate estimates of the numbers of students proceeding toward degrees."§ There were 3,600,000 ninth graders in 1976. How many of them earned a Ph.D. in mathematics by 1990?

§Bernard L. Madison, "A Challenge of Numbers," *Notices of the American Mathematical Society,* May–June 1990, pp. 547–554.

CASE STUDY   ## The Frustration of Dieting*

Who among us has not had to curtail eating habits in order to keep the body as trim as possible? It is true that some people are able to eat as much as they like, and as often as they desire, yet still maintain a slim figure. But others have to work hard to keep their weight down, whether for health or for vanity.

This Case Study examines dieting from a mathematical perspective. It draws one inescapable conclusion: there is a mathematical reason behind the fact that dieting is a long and arduous—in a word, frustrating—ordeal.

Equilibrium Weight   A person's weight depends primarily on two factors: energy intake and energy consumption (in calories per day). There are several other important factors, such as age, sex, and metabolic rate, but we restrict our consideration to the two energy factors. We also assume that each of the two main factors remains constant each day, meaning that a diet consists of the same number of calories per day and that the energy consumption is the same number of calories per pound per day.

Most people's energy consumption varies between 15 and 20 calories per pound per day. *We will assume that the energy consumption is 17.5 calories per pound per day.*

Let I denote the daily energy intake (in calories) and C denote the daily energy consumption (in calories per pound). A person who follows such a regimen will maintain a *weight equilibrium* of I/C pounds (lb).

$$w_{eq} = \frac{I}{C}$$

For instance, a person who eats 2625 calories a day will maintain a weight equilibrium of $2625/17.5 = 150$ lb. Since the energy consumption C remains constant, if I increases above 2625 calories, then the person will gain weight, while if I decreases below 2625 calories, then the person will lose weight.

The Weight Equation   One of the basic assumptions in physiology is that the rate of change of a person's weight w is directly proportional to the difference

*Adapted from Arthur C. Segal, "A Linear Diet Model," *College Mathematics Journal,* January 1987, pp. 44–45.

between the energy intake I and the energy consumption C. Since the rate of change is the derivative $\dfrac{dw}{dt}$, this assumption can be written as an equation.

$$\frac{dw}{dt} = k(I - C)$$

The most commonly used dietetic conversion factor is $k = 1/3500$ lb per calorie, so, since $C = 17.5w$

$$\frac{dw}{dt} = \frac{1}{3500}(I - 17.5w)$$

In Section 5–4 it will be possible to verify that the solution to this equation is

$$w(t) = w_{eq} + (w_0 - w_{eq})e^{-0.005t}$$

This equation is called the *weight equation,* where $w(t)$ represents the weight after t days and w_0 represents the initial weight. It is an important equation in many people's lives. Let us illustrate the weight equation for a hypothetical person named Rufus.

Rufus presently weighs 200 lb, so $w_0 = 200$. This means that his usual daily energy intake is $I = 200 \cdot 17.5 = 3500$ calories a day. He goes on a 3150 calories-per-day diet in order to lose weight. Now since $I = 3150$, the weight equilibrium will be

$$w_{eq} = \frac{3150}{17.5} = 180$$

Therefore, by adopting this diet, Rufus will eventually reach a weight of 180 lb and will maintain it.

How long will it take him to achieve this goal? To answer this question, set up the weight equation.

$$\begin{aligned} w(t) &= w_{eq} + (w_0 - w_{eq})e^{-0.005t} \\ &= 180 + (200 - 180)e^{-0.005t} \\ &= 180 + 20e^{-0.005t} \end{aligned}$$

Rufus' weight after $t = 7$ days is

$$w(7) = 180 + 20e^{-0.005 \cdot 7}$$

Use a calculator to evaluate this expression: $w(7) = 199.3$ (rounded to one decimal place). Thus Rufus will lose less than 1 lb during the first week of the diet. After 4 weeks Rufus will weigh about 197 lb because

$$w(28) = 180 + 20e^{-0.005 \cdot 28} \approx 197.4$$

Therefore after almost a month Rufus will have lost less than 3 lb. After a year, however, his weight will be $w(365) = 183.2$ lb, which is about 3 lb from his goal.

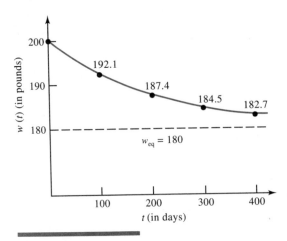

FIGURE 1

Does this mathematical model convince you of the difficulty involved in losing weight strictly by dieting? Poor Rufus stuck to his diet every day for a year, yet he was still 3 lb shy of his goal. Figure 1 shows the asymptotic nature of the weight equation. No wonder that losing weight is a long and arduous struggle!

Example 1 considers another person's valiant fight to remove extra pounds.

EXAMPLE 1

Tina weighs 130 lb. She wants to lose 10 lb.

Problem

(a) What should Tina's daily caloric intake be?
(b) If she maintains the diet for 100 days, how much weight will she lose?

Solution (a) Set $w_0 = 130$. Since Tina wants to lose 10 lb, her equilibrium weight is $w_{eq} = 120$. It is assumed that $C = 17.5$. Since $w_{eq} = I/C$, the daily caloric intake is $I = C \cdot w_{eq} = 17.5 \cdot 120 = 2100$. Therefore Tina must go on a 2100 calories-per-day diet.
(b) Tina's weight equation is

$$w(t) = 120 + 10e^{-0.005t}$$

If she adheres to this diet for $t = 100$ days, her weight will be

$$w(100) = 120 + 10e^{-0.5} = 120 + 10 \cdot 0.6 = 126$$

Since Tina will drop from 130 to 126 lb, she will end up losing 4 lb in 100 days.

Notice that Tina's weight can never reach exactly 120 lb on this diet because the weight equation $w(t) = 120 + 10e^{-0.005t}$ is always greater than

120. (Recall that e^x is positive for all values of x.) However, it is possible to determine when her weight will reach $w(t) = 121$ lb by solving the weight equation for t.

$$121 = 120 + 10e^{-0.005t}$$

The solution requires the logarithmic function, which will be discussed in the next section. By direct substitution into the weight equation, Tina will drop to 121 lb after 460 days. Now that's an ordeal!

Considering the cases of Rufus and Tina, is it any wonder that so many dieters give up in frustration?

Case Study Exercises

1. Rufus' initial weight is 200 pounds (lb). If he goes on a 2500 calories-per-day diet, how much weight will he lose in 100 days?
2. Rufus' initial weight is 200 lb. If he goes on a 2000 calories-per-day diet, how much weight will he lose in 100 days?
3. Tina weighs 120 lb and wants to lose 10 lb.
   (a) What should her daily caloric intake be?
   (b) If she maintains the diet for 100 days, how much weight will she lose?
4. Joan weighs 125 lb and wants to lose 15 lb.
   (a) What should her daily caloric intake be?
   (b) If she maintains the diet for 100 days, how much weight will she lose?
5. For Chris, $w_0 = 150$ and $I = 2275$. How much weight will Chris lose in 100 days?
6. Let your present weight be w_{eq}. Assume an average daily consumption of 17.5 calories per pound.
   (a) What is your daily energy intake I?
   (b) If you reduce I by 500 calories a day, what will your new equilibrium weight be?
7. (a) What is your weight equation for losing 10 lb?
   (b) What will you weigh in 100 days if you stick to the diet?

5–2 Logarithmic Functions

This section is organized like Section 5–1. First a new function, the natural logarithmic function, is introduced in terms of more familiar functions. Then its graph is drawn between the graphs of two similar functions and several of its properties are derived.

This approach is appropriate since the definition of the natural logarithmic function is based on the exponential function $f(x) = e^x$ and the natural logarithmic function is shown to be the inverse of $f(x)$. This inverse relationship is exploited in a discussion of the doubling time of an investment.

Graphs

Logarithmic functions are defined in terms of exponential functions: $y = \log_2 x$ if and only if y is the number such that $2^y = x$. Therefore $3 = \log_2 8$ because $2^3 = 8$. This leads to the definition of a *logarithmic function*.

$$f(x) = \log_2 x \qquad \text{if and only if} \qquad 2^{f(x)} = x$$

(Usually $\log_2 x$ is read "the logarithm of x to the base 2.") Notice that the logarithm of a number is an exponent. The definition yields $f(8) = 3$, so the point $(8, 3)$ lies on the graph of $f(x) = \log_2 x$. Additional points are listed in Table 1. The graph is drawn in Figure 1.

Table 1

x	1	2	4	8	16	32
$\log_2 x$	0	1	2	3	4	5

DEFINITION

Let b be a positive number with $b \neq 1$. The **logarithmic function with base b** is defined by

$$f(x) = \log_b x \qquad \text{if and only if} \qquad b^{f(x)} = x$$

The domain is the set of positive numbers and the range consists of all real numbers.

The next example illustrates that to find points on the graph of a logarithmic function $y = \log_b x$ it is usually easier to start with convenient values of y and to find the corresponding value of x using the equation $x = b^y$.

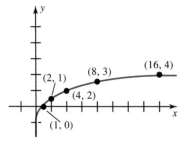

The graph of $f(x) = \log_2 x$

FIGURE 1

EXAMPLE 1

Problem

Sketch the graph of $f(x) = \log_{10} x$.

Solution By definition $y = \log_{10} x$ if $10^y = x$. Substitute some numbers for y and solve for x. For instance, if $y = 2$, then $x = 10^2 = 100$. Thus $(100, 2)$ is a point on the graph. Additional points are listed in the table. The graph is sketched in Figure 2.

x	0.01	0.1	1	10	100	1000
$\log_{10} x$	-2	-1	0	1	2	3

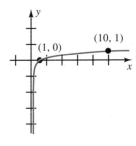

The graph of $f(x) = \log_{10} x$

FIGURE 2

Logarithms to the base 10 are called **common logarithms.** They are usually written $\log x$ instead of $\log_{10} x$. Common logarithms were invented about 1600 as an aid for astronomers in performing involved calculations. Scientific calculators, however, have rendered this use obsolete.

The graphs of $f(x) = \log_2 x$ and $f(x) = \log_{10} x$ are typical of the graphs of logarithmic functions $f(x) = \log_b x$ for $b > 1$. They pass through the point $(1, 0)$, they always increase, they are negative for $0 < x < 1$, and they are positive for $x > 1$.

The Natural Logarithm

Figure 3 shows the graphs of the functions $f(x) = 2^x$ and $f(x) = \log_2 x$ on the same coordinate system. The graphs are symmetric about the dashed line $y = x$, meaning that if one of them is flipped about this line, then it would land precisely on the other. As an example, the point $(3, 8)$ lies on the graph of $f(x) = 2^x$, the point $(8, 3)$ lies on the graph of $f(x) = \log_2 x$, and the distance from $(3, 8)$ to the dashed line is equal to the distance from $(8, 3)$ to the dashed line. When such a relation holds between two functions, each one is called the **inverse function** of the other. Thus $f(x) = \log_2 x$ is the inverse of $f(x) = 2^x$.

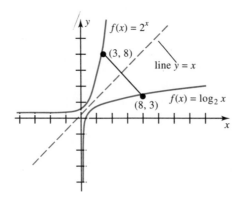

FIGURE 3

Figure 4 shows the graph of the exponential function $f(x) = e^x$ and its inverse, which is called the **natural logarithmic function.** The natural logarithm, which is denoted by $\ln x$ instead of $\log_e x$, is defined by

$$y = \ln x \qquad \text{if and only if} \qquad e^y = x$$

Notice that $\ln e = 1$ (since $e^1 = e$) and $\ln 1 = 0$ (since $e^0 = 1$). However, $\ln 0$ is not defined. The graph of $f(x) = \ln x$ is similar to the graph of $f(x) = \log_2 x$. It can be obtained from the graph of $f(x) = e^x$ as in Figure 4, or from Table 2. (In Table 2 the values of $\ln x$ are rounded to two decimal places. Each value of $\ln x$ in the table is an irrational number except for $x = 1$.) A more complete table can be found in the Appendix. Many calculators have $\boxed{\ln}$ and $\boxed{\log}$ keys. The key sequence $\boxed{.5}$ $\boxed{\ln}$ produces the value -0.69 in Table 2.

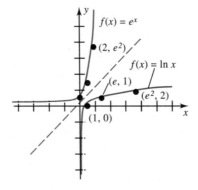

$\ln x$ is the inverse of e^x

FIGURE 4

Table 2

x	0.05	0.1	0.25	0.5	1	2	3	4
ln x	−3.00	−2.30	−1.39	−0.69	0	0.69	1.10	1.39

Example 2 sketches the graph of a function that is defined in terms of the natural logarithmic function. Its domain is the set of all real numbers and its graph is unlike Figures 1 and 2.

EXAMPLE 2

Problem

Sketch the graph of the function

$$f(x) = \ln(x^2 + 1)$$

Solution Construct a table using a calculator or the table of logarithmic functions in the Appendix. Plot the points and draw a smooth curve through them. This is carried out in Figure 5. Notice that $f(x)$ is defined for all real numbers.

x	−3	−2	−1	0	1	2	3
$\ln(x^2 + 1)$	2.30	1.61	0.69	0	0.69	1.61	2.30

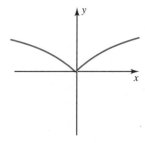

Graph of $f(x) = \ln(x^2 + 1)$

FIGURE 5

Properties

We derive several properties of logarithmic functions from the properties of exponential functions. Although these properties are stated in terms of the natural logarithmic function $f(x) = \ln x$, they hold for all logarithmic functions of the form $f(x) = \log_b x$ as well (for $b > 0$ and $b \neq 1$). In properties 4 to 7 assume that $r > 0$, $s > 0$, and k is any number.

1. $e^{\ln x} = x$ for $x > 0$
2. $\ln e^x = x$ for all x
3. $\ln 1 = 0$
4. If $\ln r = \ln s$, then $r = s$
5. $\ln rs = \ln r + \ln s$
6. $\ln \dfrac{r}{s} = \ln r - \ln s$
7. $\ln r^k = k \cdot \ln r$

The natural logarithm is defined by

$$y = \ln x \qquad \text{if and only if} \qquad e^y = x$$

Property 1 is obtained by substituting $\ln x$ for y in the right-hand equality. Similarly, substituting e^y for x in $y = \ln x$ yields $y = \ln e^y$. Writing this equality in the variable x gives property 2. Property 3 follows by substituting $x = 0$ in property 2. Property 4 can be inferred from the graph of $f(x) = \ln x$. Alternately, if $\ln r = \ln s$, then $e^{\ln r} = e^{\ln s}$. By property 1, $r = e^{\ln r}$ and $s = e^{\ln s}$, so $r = s$.

To prove property 5, use property 1 to write $r = e^{\ln r}$ and $s = e^{\ln s}$. Then $rs = e^{\ln r} \cdot e^{\ln s} = e^{\ln r + \ln s}$. By property 1, $rs = e^{\ln rs}$, so $e^{\ln rs} = e^{\ln r + \ln s}$. Property 5 then follows from property 1 of exponential functions. Properties 6 and 7 can be derived similarly.

Examples 3 and 4 illustrate these properties.

EXAMPLE 3

Problem

Write these expressions in simplified form: (a) $\ln(4x^3)$, (b) $\ln \left(\dfrac{6x^2y}{x + y} \right)$.

Solution (a) Using properties 4 and 6, we obtain

$$\ln(4x^3) = \ln(2^2 x^3) = \ln(2^2) + \ln(x^3) = (2 \ln 2) + (3 \ln x).$$

(b) Again by using properties 4 and 6, we have

$$\ln(6x^2y) = \ln 6 + 2 \ln x + \ln y.$$

Using property 5, we get

$$\ln \left(\frac{6x^2y}{x + y} \right) = \ln(6x^2y) - \ln(x + y)$$

$$= \ln 6 + 2 \ln x + \ln y - \ln(x + y)$$

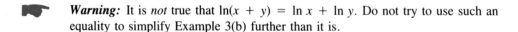

Warning: It is *not* true that $\ln(x + y) = \ln x + \ln y$. Do not try to use such an equality to simplify Example 3(b) further than it is.

EXAMPLE 4

Problem

Simplify (a) $\ln e^{(3x+4)}$, (b) $e^{\ln(3x+4)}$.

Solution (a) Rewrite property 2 in the variable v, so that $\ln e^v = v$. Now set $v = 3x + 4$. Then $\ln e^{(3x+4)} = 3x + 4$.
(b) Similarly rewrite property 1 in the variable v, so that $e^{\ln v} = v$. Once again set $v = 3x + 4$. Then $e^{\ln(3x+4)} = 3x + 4$.

The same reasoning used in Example 4 leads to a generalization of properties 1 and 2:

1'. $e^{\ln f(x)} = f(x)$ for $f(x) > 0$
2'. $\ln e^{f(x)} = f(x)$ for all $f(x)$

Doubling Time

One way to measure the effectiveness of an investment is by means of **doubling time,** which is the time required for the initial investment to double. By the Compound Interest Formula, if $1 is invested at an interest rate of $100r\%$ compounded annually (where r is written as a decimal), the formula for the amount after t years is given by

$$A(t) = (1 + r)^t$$

This is an exponential function. The doubling time is computed by solving the equation $A(t) = 2$ for t.

$$2 = (1 + r)^t$$

Since the logarithmic function is the inverse of the exponential function, take the natural logarithm of both sides.

$$\ln 2 = \ln(1 + r)^t$$
$$\ln 2 = t \cdot \ln(1 + r)$$
$$t = \frac{\ln 2}{\ln(1 + r)}$$

For instance, the doubling time of an 8% investment is 9 years because

$$t = \frac{\ln 2}{\ln 1.08} = \frac{0.69315}{0.07696} \approx 9$$

EXAMPLE 5

Problem

What is the doubling time for an investment that earns 16% compounded annually?

Solution By the Compound Interest Formula

$$A(t) = (1 + r)^t$$

Here $A(t) = 2$ and $r = 0.16$, so

$$2 = (1.16)^t$$
$$\ln 2 = t \cdot \ln(1.16)$$
$$t = \frac{\ln 2}{\ln 1.16} = \frac{0.69315}{0.14842} \approx 4.67017$$

Therefore the doubling time is about $4\frac{2}{3}$ years, or 4 years and 8 months.

EXERCISE SET 5–2

In Problems 1 to 4 sketch the graph of the stated function by the plotting method, using the definition of the logarithmic function.

1. $f(x) = \log_{10} 2x$

2. $f(x) = \log_{10} \frac{x}{2}$

3. $f(x) = \log_3 x$

4. $f(x) = \log_{20} x$

In Problems 5 to 8 sketch the graph of the stated function by the plotting method, using the table in the Appendix or a calculator.

5. $f(x) = \ln(3x)$

6. $f(x) = \ln(x/2)$

7. $f(x) = \ln(x^2 + 3)$

8. $f(x) = \ln(x^2 + 4)$

In Problems 9 to 12 write the given expression in simplified form.

9. $\ln(3x^4)$

10. $\ln(2x^5)$

11. $\ln\left(\frac{3xy^2}{x - y}\right)$

12. $\ln\left(\frac{x^3y^2}{x + y}\right)$

In Problems 13 to 16 simplify the expression.

13. (a) $\ln e^{3x}$ (b) $e^{\ln(5-x)}$

14. (a) $\ln e^{-2}$ (b) $e^{\ln 4}$

15. (a) $\ln e^5$ (b) $e^{\ln 0.5}$

16. (a) $\ln e^{3-8x}$ (b) $e^{\ln(2x-6)}$

In Problems 17 to 20 determine the doubling time for an investment that earns the stated interest compounded annually.

17. 6%   18. 10%   19. 20%   20. 5%

In Problems 21 to 26 sketch the graph of the stated function by the plotting method, using the table in the Appendix or a calculator.

21. $f(x) = \ln \frac{1}{x^2}$

22. $f(x) = \ln\left(\frac{1}{x^2 + 2}\right)$

23. $f(x) = \ln(2x + 3)$

24. $f(x) = \ln(x - 4)$

25. $f(x) = \ln |x|$

26. $f(x) = \ln \sqrt{x}$

In Problems 27 to 30 state the domain of the function in the problem stated above.

27. Problem 21   28. Problem 23

29. Problem 25   30. Problem 26

In Problems 31 to 34 write the given expression in simplified form.

31. $\ln \sqrt{xy}$

32. $\ln \sqrt{x + y}$

33. $\ln \dfrac{1}{\sqrt{x^3y}}$

34. $\ln \dfrac{1}{xy^2}$

In Problems 35 and 36 let $f(x) = \ln x$ and $g(x) = e^x$.

35. Compute $(f \circ g)(x)$.   36. Compute $(g \circ f)(x)$.

In Problems 37 to 44 solve for x.

37. $e^{2x} = 3$

38. $e^{5x} = 1$

39. $18 = e^{-x} - 2$

40. $5 = \dfrac{1}{1 - 2e^{-x}}$

41. $5 \ln x = 1$

42. $4 \ln(1/x) = 1$

43. $3^x = e^2$

44. $2^x = e^5$

In Problems 45 to 48 evaluate the expression without using a calculator.

45. (a) $\ln e$ (b) $e^{\ln 4}$

46. (a) $\ln 1$ (b) $e^{3 \ln 2}$

47. (a) $\ln \sqrt[3]{e}$ (b) $e^{4 \ln 3 - 3 \ln 4}$

48. (a) $\ln(1/\sqrt{e})$ (b) $e^{-\ln 2}$

49. Use a calculator to try to evaluate
    (a) ln 0   (b) ln − 2

50. Use a calculator to show that $\ln(2 + 3) \neq \ln 2 + \ln 3$.

51. If a thermometer is moved from inside a house where it reads 70°F, to outside the house where it is 10°F, the temperature of the thermometer will read

    $$T(t) = 10 + 60e^{-0.46t}$$

    where t is in minutes and $T(t)$ is in degrees Fahrenheit. If the thermometer is digital (meaning that it only displays whole degrees), when will the reading first reach 10°F?

52. Answer true or false.
    (a) $\ln(x^3 + 5x^2) = 3 \ln x + 10 \ln x$
    (b) $\ln(x^5) = (\ln x)^5$

In Problems 53 to 56 modify the argument that derived the doubling time formula to determine how long it will take an investment to increase to the given size at the given rate.

53. To double if it earns 10% compounded annually.

54. To triple if it earns 10% compounded annually.

55. To quadruple if it earns 5% compounded annually.

56. To increase by a factor of 2.5 if it earns 8% compounded annually.

57. Determine how long it will take for an investment that earns 8% interest compounded annually to triple.

58. (a) Prove property 6 of logarithmic functions.
    (b) Prove property 7 of logarithmic functions.

59. The weight equation for a 140-pound (lb) person who wants to lose 10 lb is

    $$f(t) = 130 + 10e^{-0.005t}$$

    where $f(t)$ is the weight (in lb) after being on the diet for t days. How long will it take the person to lose 5 lb? (This weight equation is derived in the Case Study in this chapter titled "The Frustration of Dieting.")

Referenced Exercise Set 5-2

Consider a job in which a number of workers produce a certain amount of products, say Boeing 707 jets. When the job is performed for the first time it cannot be expected to be completed in an acceptable amount of time. A certain period of time must elapse before the workers become familiar with the tasks involved and learn how to do them in a shorter amount of time. A *learning curve* is the graph of a function which measures the mastery of the job. If one hour is needed to produce the first unit, then the number of hours needed to produce the xth unit is given by

$$Y(x) = x^{\log L/\log 2}$$

where L is the learning rate of the particular job.* Data from the airframe manufacturing industry reveal that $L = 0.8$ is a typical learning rate. Assume such a value in Problems 1 and 2.

1. Compute the time (as a decimal part of an hour) needed to produce (a) the second unit, (b) the third unit, (c) the tenth unit, (d) the one-hundredth unit.

2. How many units have to be produced before a unit can be produced in each of these time periods?
   (a) 30 minutes (0.5 hour)
   (b) 6 minutes (0.1 hour)

Simple products and machine-paced processes were shown to exhibit learning rates between 0.9 and 1.0.†

3. Repeat Problem 1 for a learning rate of $L = 0.95$.

4. Repeat Problem 2 for a learning rate of $L = 0.95$.

In the "classical secretary problem" each applicant for a job is ranked by a manager according to some quality. As each applicant appears, her rank relative to those preceding her is recorded and the manager decides either to select or reject her. The object is to hire the best applicant, and the manager is rewarded accordingly. A Japanese specialist in operations research has solved this problem for various kinds of selections that the manager can make.‡

*Louis E. Yelle, "The Learning Curve: Historical Review and Comprehensive Survey," *Decision Sciences*, April 1979, pp. 302–328.

†Timothy L. Smunt, "A Comparison of Learning Curve Analysis and Moving Average Ratio Analysis for Detailed Operational Planning," *Decision Sciences*, Fall 1986, pp. 475–495.

‡Mitsushi Tamaki, "A Generalized Problem of Optimal Selection and Assignment," *Operations Research*, May–June 1986, pp. 486–493.

Problems 5 and 6 are taken from this study.

5. The reward for the model "without promotion" is $R(x) = x(1 - x)$, where x is a solution to the equation

$$2(1 - x) + \ln x = 0 \qquad 0 < x < 1$$

Solve the equation to one decimal place (by trial-and-error) and determine the reward.

6. The reward for the model "with promotion" is $R(x) = x\left(\dfrac{2}{\sqrt{e}} - x\right)$, where x is a solution to the equation

$$\sqrt{e}(1 + x) - \ln x = 3.5 \qquad 0 < x < 1$$

Solve the equation to one decimal place (by trial-and-error) and determine the reward.

7. In 1987 the world's population reached 5 billion.§ If it continues to grow 2% each year, in what year will it reach 10 billion?

8. Donald Knuth, arguably America's best-known computer scientist, has proved that the number of computer operations needed to alphabetize n words is about $n(\log_2 n)$.‖ About how many computer operations will it take to alphabetize 10,000 words?

Cumulative Exercise Set 5–2

In Problems 1 and 2 sketch the graph of $f(x)$ by the plotting method.

1. $f(x) = (0.2)^x$

2. $f(x) = \dfrac{2}{1 + 2e^{-3x}}$

3. Find all values of x that satisfy the equality $2^{x-2} = 8^{2x+1}$.

4. A disease spreads according to the equation

$$E(t) = \frac{5}{2 + 3e^{-0.1t}}$$

where t is in days after the outbreak of the disease and $E(t)$ is in thousands of people. (a) How many people contracted the disease after 10 days? (b) In the long run, how many people will contract the disease?

5. Write the expression in simplified form:

$$\ln\left(\frac{x^2 y^3}{x - y}\right)$$

6. Determine the doubling time for an investment that earns 12% interest compounded annually.

7. Solve for x: $6 \ln(1/x) = 1$.

8. Evaluate the expression without using a calculator:

$$e^{2 \ln 3 - 3 \ln 2}$$

CASE STUDY   ## Sears, Clothespins, and Unnatural Logarithms*

Consider the clothespin. No, not the one shown in Figure 1. That one is actually a Claes Oldenberg sculpture that stands in downtown Philadelphia. Instead, regard the usual clothespin, the one that has graced porches and backyards for over a century. It will serve as a central prop in our study of the American economy.

Sears catalogs also play an important role. Sears Roebuck and Co. began its mail-order operation at the turn of the 20th century based on these catalogs. The first ones were distributed in rural areas as a marketing strategy to sell items that were otherwise unavailable. Their overwhelming success spawned a vast

§Dan Shannon, "World Population at 5 Billion," *Los Angeles Times*, July 12, 1987.

‖Donald E. Knuth, *Sorting and Searching: The Art of Computer Programming III*, Addison-Wesley, Reading, MA, 1973.

*Adapted from Elliott W. Montroll, "On the Dynamics and Evolution of Some Sociotechnical Systems," *Bulletin of the American Mathematical Society*, January 1987, pp. 1–46.

F I G U R E 1 A photo of a Claes Oldenberg sculpture of a clothespin that stands in downtown Philadelphia.

nationwide network of over 800 stores which are woven into the lives of millions of people.

In 1987 Sears produced its first videocassette catalog to market some of its products. We will not attempt to predict whether this venture will be as successful in the 21st century as the standard catalogs have been in the 20th century. Instead, we will document several ways in which the standard catalogs have been used to study the American way of life over the past 100 years. We will see that they provide an economic index that measures the cost of living and indicates changes in the American life-style.

Cost of Living   On a given page in a Sears catalog you might see bicycles, VCRs, and—yes—clothespins. Since the clothespin is an invariant item, the 1988 model being indistinguishable from the 1900 model, its change in price reflects one kind of measure of the cost of living. Another measure results from examining bicycles. Since the models of the 1990s are considerably different from the models from 1900, the variation in price reflects an evolving technol-

Table 1

P	$\log_2 P$	$(\log_2 P - M)^2$
128.00	7	$2.94^2 = 8.6436$
39.40	5.3	$1.24^2 = 1.5376$
16.00	4	$-0.06^2 = 0.0036$
8.00	3	$-1.06^2 = 1.1236$
2.00	1	$-3.06^2 = 9.3636$
Sum	20.3	20.672

$$M = \frac{20.3}{5} = 4.06 \quad S = \sqrt{\frac{20.672}{5}} \approx 2.03$$

ogy and a varying public taste. VCRs, on the other hand, certainly were not featured in the 1900 catalog. Conversely, you are not likely to find buggy whips in the 1992 catalog. Yet in spite of such variations, one constant emerges. We describe it now.

Consider a page in a Sears catalog containing five items whose sales prices are \$128, \$39.40, \$16, \$8, and \$2. It is possible to compute the mean and standard deviation for the items on this page. However, instead of regarding each price P as a data entry, we use its logarithm to the base 2. For $P = \$128$ the data entry becomes $\log_2 P = \log_2 128 = 7$ because $2^7 = 128$. The remaining entries are listed in the second column of Table 1.

The mean of the $\log_2 P$ entries is $20.3/5 = 4.06$. We denote this mean by M, to distinguish it from the mean of the prices P, and refer to it as the *log-mean*. Similarly the standard deviation of the $\log_2 P$ entries is denoted by S and called the *log-standard deviation*. The third column of Table 1 reveals that $S = \sqrt{20.672/5} \approx 2.03$.

The values that we have obtained for this hypothetical page actually correspond to the log-mean M and the log-standard deviation S for all items that were listed in the 1975–1976 Sears catalog. Table 2 contains the values of M and S for selected annual Sears catalogs since 1900.

Table 2

Year	M	S
1900	0.150	2.43
1908	-0.023	2.29
1916	-0.068	2.38
1924–1925	0.422	2.32
1932–1933	0.691	1.91
1939–1940	0.627	2.62
1951–1952	1.785	2.34
1962	2.403	2.24
1975–1976	4.060	2.03

One fact stands out in Table 2. Although M varies considerably, S remains almost the same. For this reason S has been described as an "economic constant of motion" for marketing. For all Sears catalogs, which have ranged over various economic climates and two world wars, the standard deviation of the log-standard deviations is only 0.17. This reflects a remarkable uniformity and a very consistent marketing strategy.

Example 1 provides more practice in computing the log-standard deviation S.

EXAMPLE 1

A page of a Sears catalog lists five items at these prices: $78.25, $3.36, $1.68, $1.06, and $1.04.

Problem

Which annual Sears catalog has a log-mean and a log-standard deviation that are similar to the log-mean and the log-standard deviation of these five items?

Solution The prices P of the items are listed in the left-hand column of Table 3. The associated $\log_2$ prices are listed in column 2. Each number in the second column can be obtained by using the natural logarithm. For instance,

$$\log_2 78.25 = \frac{\ln 78.25}{\ln 2} = \frac{4.3599}{0.6931} = 6.29$$

The remaining calculations, which are shown at the bottom of Table 3, reveal that $M = 1.786$ and $S \approx 2.33$. These numbers approximate the values of M and S in the 1951–1952 Sears catalog.

Table 3

P	$\log_2 P$	$(\log_2 P - M)^2$
78.25	6.29	$4.50^2 \approx 20.2860$
3.36	1.75	$-0.04^2 \approx 0.0013$
1.68	0.75	$-1.04^2 \approx 1.0733$
1.06	0.08	$-1.71^2 \approx 2.9104$
1.04	0.06	$-1.73^2 \approx 2.9791$
Sum	8.93	27.2501

$$M = \frac{8.93}{5} = 1.786 \quad S = \sqrt{\frac{27.25}{5}} \approx 2.33$$

Unnatural Logarithms   The analysis of Sears catalogs has shown one use of a logarithm other than the natural logarithm. Another "unnatural" use involves the graphing of data. We show two ways in which such graphs arise.

In most graphs if the x-axis and the y-axis have the same units, then these units are marked uniformly along both axes. Sometimes it is more beneficial to scale one of the axes differently. Consider Figure 2, for instance. The y-axis is

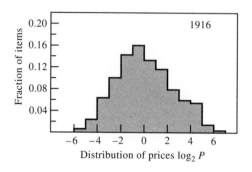

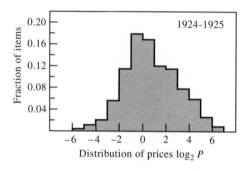

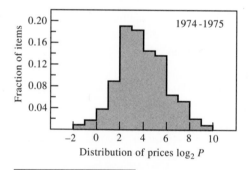

FIGURE 2 Histogram of distribution of prices in Sears Roebuck catalogs for years 1916, 1924–1925, and 1974–1975.

scaled uniformly, with each mark indicating 2% of the items in the 1916 Sears catalog. However, the units along the x-axis represent $\log_2 P$, where P is the price of an item. Therefore the interval from $x = 0$ to $x = 2$ represents prices in the range from $P = \$1$ to $P = \$4$ (since $\log_2 1 = 0$ and $\log_2 4 = 2$), while the interval from $x = 2$ to 4 represents prices in the range from $P = \$4$ to $P = \$16$.

Each histogram in Figure 2 is very close to a normal distribution. The same type of histogram results for every Sears catalog when $\log_2 P$ is used for the x-axis. In their marketing wisdom the Sears Roebuck Company has created catalogs whose price distribution year after year has maximized profit by maximizing a function associated with $\log_2 P$ (the so-called entropy function).

EXAMPLE 2

Problem

Use Figure 2 to determine which price range was listed most frequently in the 1916 catalog.

Solution The maximum interval of the histogram occurs from $x = -1$ to $x = 0$. The associated prices are $0.50 (since $\log_2 0.5 = -1$) and $1. Therefore items costing between $0.50 and $1 were listed most frequently. (They comprise about 17% of all listings.)

Figure 3 shows the graph of the log-mean M of the Sears catalogs. The years are marked uniformly along the x-axis and the values of $\log_2 M$ are marked along the y-axis. Therefore the point (1975, 4) lies on the graph because the average item in the 1975–1976 Sears catalog listed for about $M = $16 and $\log_2 16 = 4$.

The Sears catalogs list many diverse products, so the variation in the annual log-mean price represents an average over numerous technologies. The graph, called *the SR index*, has become a standard for comparison. For example, Figure 3 includes the average factory wholesale price to automobile dealers in

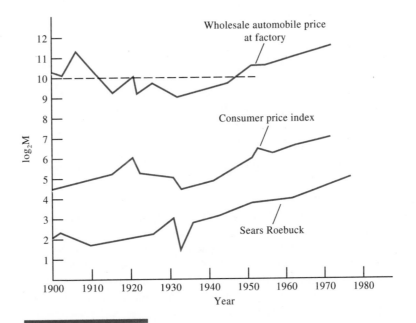

FIGURE 3 *Source:* From annual volumes, "Automobile Facts and Figures," American Automobile Association.

the United States. The graph is similar to the SR index after the initial drop in prices until about 1935. (The computer industry has also experienced such a period of decrease in prices since its inception.)

EXAMPLE 3

Problem

Use Figure 3 to approximate those years when the average factory wholesale price to automobile dealers was about $1000.

Solution The units along the y-axis of Figure 3 refer to $\log_2 M$, where M is the price to automobile dealers. Since $\log_2 1000 \approx 10$, we can locate the approximate years by drawing a horizontal line at $y = 10$. (It is the dashed line in Figure 3.) The approximate years are 1912, 1920, and 1945.

Case Study Exercises

1. Compute the log-mean and log-standard deviation for the five items whose sales prices are $128, $32, $8, $2, and $1.
2. Which annual Sears catalog (from Table 2) has a log-mean and a log-standard deviation that are similar to the log-mean and the log-standard deviation of these five items: $39.40, $32, $1.62, $1.41, and $1.41?
3. Use Figure 2 to determine which price range was listed most frequently in the 1924–1925 catalog.
4. Use Figure 2 to determine which price range was listed most frequently in the 1974–1975 catalog.

Problems 5 to 8 refer to Figure 3.

5. Approximate the average sales price of an item in the 1970 Sears catalog.
6. Approximate the average factory wholesale price to automobile dealers in 1970.
7. What was the Consumer Price Index in 1970?
8. Approximate the years in which the Consumer Price Index was about $64.

5–3 Differentiation of Logarithmic Functions

The definition of the derivative has been used for two purposes so far. First, in order to demonstrate its meaning, we applied it to compute the derivatives of several simple functions. Then it was used to derive the standard formulas that were applied to a wide range of functions. The derivative of the logarithmic function, however, cannot be derived by these techniques, so in this section we return to the definition. We begin by appealing to the definition to suggest a rule for the derivative of $f(x) = \ln x$. This rule will then be generalized, illustrated, and proved more rigorously.

The domain of the function $\ln x$ is $(0, \infty)$. In order to include all nonzero values of x in the domain, we often use the function $\ln |x|$. Since including absolute value signs complicates the formulas and problems, in the first part of the section we assume that the domain of each function is the largest possible set of real numbers. At the end of the section we handle functions like ln $|x|$.

The Derivative of ln x

Let us compute the derivative of $f(x) = \ln x$ at $x = a$ using the definition of the derivative. By definition

$$f'(a) = \lim_{h \to 0} \frac{f(a + h) - f(a)}{h}$$

$$= \lim_{h \to 0} \frac{\ln(a + h) - \ln(a)}{h}$$

In previous similar problems the next step is to simplify the difference quotient. However, this quotient does not simplify easily. At the end of the section a rigorous method for calculating this limit is given, but for now we evaluate the limit for two particular values of a and for various choices of h that approach 0. Using a calculator, we obtain the following table of values for $a = 5$ and $a = 8$. They give an indication of the rule for the derivative of $f(x) = \ln x$ (see Table 1).

For $a = 5$, as h approaches 0 the values of the quotient approach 0.2. This suggests that $f'(5) = 0.2$. It is instructive to write

$$f'(5) = \frac{1}{5}$$

In the same way, for $a = 8$, Table 1 suggests that $f'(8) = 0.125$. Once again, write

$$f'(8) = \frac{1}{8}$$

These two results suggest that $f'(a) = 1/a$. This is the rule for the derivative of $f(x) = \ln x$. The formal proof at the end of the section is based in part on this discussion.

Table 1

$a = 5$		$a = 8$	
h	$\dfrac{\ln(5 + h) - \ln 5}{h}$	h	$\dfrac{\ln(8 + h) - \ln 8}{h}$
0.1	0.198026	0.1	0.124225
0.01	0.199800	0.01	0.124922
0.001	0.199980	0.001	0.124992
0.0001	0.199998	0.0001	0.124999

The Derivative of the Logarithmic Function

$$D_x \ln x = \frac{1}{x}$$

EXAMPLE 1

$= 3x^2 - 5\left(\frac{1}{x}\right)$

Problem

Find the derivative of $f(x) = x^3 - 5 \ln x.$ $= 3x^2 - \frac{5}{x}$

Solution Use the rule for the derivative of a sum and the rule for a constant times a function.

$$D_x(x^3 - 5 \ln x) = D_x(x^3) - 5D_x(\ln x) = 3x^2 - 5\left(\frac{1}{x}\right) = 3x^2 - \frac{5}{x}$$

The Logarithmic Function Version of the Chain Rule

Most logarithmic functions found in applications are of the form $\ln u$ where $u = g(x)$ is a function of x. If $g'(x)$ exists, the chain rule can be used to find the derivative of $\ln u$. Recall that the chain rule states that if $y = f(g(x))$, then

$$y' = f'(g(x))g'(x)$$

To compute the derivative of $y = \ln u$, let $u = g(x)$ and $f(x) = \ln x$ in the formula. Then $y = f(g(x)) = f(u) = \ln u = \ln g(x)$. From the rule for the derivative of the logarithmic function, $f'(u) = f'(g(x)) = 1/u = 1/g(x)$. From the chain rule we get

$$y' = f'(g(x))g'(x) = \left(\frac{1}{g(x)}\right)g'(x) = \frac{g'(x)}{g(x)}$$

The form of the chain rule applied to the logarithmic function $f(x) = \ln g(x)$ is the formula

The Derivative of Logarithmic Functions

$$D_x \ln g(x) = \frac{g'(x)}{g(x)}$$

EXAMPLE 2

Problem

Find the derivative of (a) $f(x) = \ln(3x^2 + 5)$, (b) $f(x) = \ln(\ln x)$.

$\frac{6x +}{}$

$\left(\frac{6x}{3x^2+5}\right)$ $= \frac{1}{3x^2+5}$ $\frac{1}{\frac{1}{x}}$

$\frac{6x}{3x^2+5}$

Solution (a) Let $g(x) = 3x^2 + 5$. Then $g'(x) = 6x$ and we get

$$D_x \ln(3x^2 + 5) = \frac{g'(x)}{g(x)} = \frac{6x}{3x^2 + 5}$$

(b) Let $g(x) = \ln x$. Then $g'(x) = 1/x$ and we get

$$D_x \ln(\ln x) = \frac{g'(x)}{g(x)} = \frac{(1/x)}{\ln x} = \frac{1}{x \ln x}$$

The next example demonstrates how one derivative can require two versions of the chain rule, the generalized power rule and the logarithmic function version.

EXAMPLE 3

Problem

Find the derivative of

$$f(x) = [2 + \ln(1 + 3x)]^3$$

Solution First apply the generalized power rule. We get

$$f'(x) = 3[2 + \ln(1 + 3x)]^2[D_x(2 + \ln(1 + 3x))]$$

The derivative of $2 + \ln(1 + 3x)$ requires the derivative of a sum with $g(x) = 1 + 3x$. Then $g'(x) = 3$ and $D_x \ln(1 + 3x) = 3/(1 + 3x)$. Then

$$f'(x) = 3[2 + \ln(1 + 3x)]^2[D_x(2 + \ln(1 + 3x))]$$

$$= 3[2 + \ln(1 + 3x)]^2 \left[\frac{3}{1 + 3x} \right]$$

$$= \frac{9[2 + \ln(1 + 3x)]^2}{1 + 3x}$$

Graphing Logarithmic Functions

The curve-sketching techniques of the previous chapter can be used to graph logarithmic functions. The methods are illustrated in the next example.

EXAMPLE 4

Problem

Find the relative extrema and the points of inflection of the function $f(x) = x - \ln x$. Then draw the graph.

Solution Find the extrema via the second derivative test. Compute $f'(x)$ and $f''(x)$.

$$f'(x) = 1 - \frac{1}{x} \qquad f''(x) = \frac{1}{x^2}$$

Solve $f'(x) = 0$; $f'(x) = 1 - 1/x = 0$. Then $x = 1$, so the only critical value is $x = 1$. Although $f'(0)$ does not exist, $x = 0$ is not a critical value because 0 is not in the domain of $f(x)$. This is because ln x is not defined at $x = 0$.

Since $f''(1) = 1$, $f(x)$ has a relative minimum at $x = 1$. Since $f''(x) > 0$ for all x, $f(x)$ has no point of inflection. As x approaches 0, ln x approaches $-\infty$, and so $-$ln x approaches ∞, meaning that $f(x)$ has a vertical asymptote at $x = 0$. Note that $f(1) = 1 - \ln 1 = 1 - 0 = 1$. These facts can be used to draw the graph of $f(x)$ shown in Figure 1.

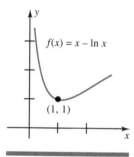

$f(x) = x - \ln x$

$(1, 1)$

FIGURE 1

Consider the function $y = \ln |x|$ whose domain is the set of all real numbers except $x = 0$. The definition of this function is

$$\ln |x| = \begin{cases} \ln x & \text{if } x > 0 \\ \ln(-x) & \text{if } x < 0 \end{cases}$$

If $x > 0$, then $y = \ln x$ and $y' = 1/x$. Suppose $x < 0$. Then $y = \ln(-x)$ and to find y' we use the rule for the derivative of logarithmic functions with $g(x) = -x$. The derivative of $\ln(-x)$ is $-1/-x = 1/x$. Thus the derivative of $\ln(-x)$ is equal to the derivative of ln x. A similar argument shows that the derivative of $\ln g(x)$ is the same as $\ln(-g(x))$.

$$D_x \ln |x| = \frac{1}{x}$$

$$D_x \ln |g(x)| = \frac{g'(x)}{g(x)}$$

EXAMPLE 5

Problem

Find the derivative of (a) $f(x) = \ln |3x - 2|$, (b) $f(x) = x \ln |1 + x^{-3}|$.

Solution (a) Let $g(x) = 3x - 2$. From the formula we get

$$f'(x) = \frac{3}{3x - 2}$$

(b) Let $g(x) = 1 + x^{-3}$ and apply the product rule.

$$f'(x) = xD_x \ln |1 + x^{-3}| + \ln |1 + x^{-3}|D_x x$$

$$= x \left[\frac{-3x^{-4}}{1 + x^{-3}} \right] + \ln |1 + x^{-3}|(1)$$

$$= \frac{-3x^{-3}}{1 + x^{-3}} + \ln |1 + x^{-3}|$$

$$= \frac{-3}{x^3 + 1} + \ln |1 + x^{-3}|$$

Proof of the Rule for the Derivative of the Logarithmic Function

Proof. If $f(x) = \ln x$, then

$$f'(x) = \lim_{h \to 0} \left[\frac{f(x + h) - f(x)}{h} \right] = \lim_{h \to 0} \left[\frac{\ln(x + h) - \ln x}{h} \right]$$

$$= \lim_{h \to 0} \left(\frac{1}{h} \right) \ln \left[\frac{x + h}{x} \right]$$

$$= \lim_{h \to 0} \ln \left[\frac{x + h}{x} \right]^{1/h}$$

We first deal with the case where $h > 0$. The case where $h < 0$ is handled in a similar way. To simplify this expression, let $m = x/h$. In this expression x is being held constant while h varies. As $h \to 0$, note that $m \to \infty$ and vice versa. In the last limit above write $(x + h)/x = 1 + h/x = 1 + 1/m$. The limit then becomes

$$f'(x) = \lim_{m \to \infty} \ln \left[1 + \frac{1}{m} \right]^{m/x} = \lim_{m \to \infty} \ln \left[\left(1 + \frac{1}{m} \right)^m \right]^{1/x}$$

$$= \lim_{m \to \infty} \frac{1}{x} \cdot \ln \left[1 + \frac{1}{m} \right]^m = \frac{1}{x} \cdot \lim_{m \to \infty} \ln \left[1 + \frac{1}{m} \right]^m$$

$$= \frac{1}{x} \cdot \ln \left[\lim_{m \to \infty} \left(1 + \frac{1}{m} \right)^m \right]$$

From the definition of e we have

$$\lim_{m \to \infty} \left(1 + \frac{1}{m} \right)^m = e$$

Therefore

$$f'(x) = \frac{1}{x} \cdot \ln \left[\lim_{m \to \infty} \left(1 + \frac{1}{m} \right)^m \right] = \frac{1}{x} \ln e = \frac{1}{x} \cdot 1 = \frac{1}{x}$$

Thus $D_x \ln x = 1/x$, as desired.

Elasticity

Economists define the **rate of growth** $Rf(x)$ of the function $f(x)$ as the ratio

$$Rf(x) = f'(x)/f(x)$$

It measures the relative rate of change of $f(x)$. For example, suppose an automobile manufacturer sells its car for \$20,000 and another sells its car for \$8000. Suppose both manufacturers increase the price, the first to \$21,000 and the second to \$8,500. So the first manufacturer's increase is \$1000 and the second manufacturer's increase is \$500. Which is the larger relative increase, meaning which is the larger percentage increase? Let the average price per car be represented by $f(x)$ for the first manufacturer and $g(x)$ for the second manufacturer. Let $x = 0$ represent the time when the original prices of \$20,000 and \$8000 were in effect. Then $f(0) = 20,000$ and $g(0) = 8000$. The rate of increase for each function is given by $f'(0) = 1000$ and $g'(0) = 500$. The rate-of-growth functions are

$$Rf(0) = f'(0)/f(0) = 1000/20,000 = 1/20 = .05$$
$$Rg(0) = g'(0)/g(0) = 500/8000 = 1/16 = .0625$$

This means that the relative rate of increase for the lower priced car is greater than that of the higher priced car, even though the actual size of the increase is only half that of the higher priced car.

  Another way to view the rate-of-growth function is to express it in terms of the derivative of the logarithm of $f(x)$:

$$Rf(x) = f'(x)/f(x) = D_x \ln f(x)$$

It is called the *logarithmic derivative of $f(x)$*. Economists use the rate-of-growth function to measure the relative change of various quantities, such as revenue and demand. In this way the relative changes of a product made by different companies can be compared. For instance, suppose two companies have demand curves $y = Q_1(p)$ and $y = Q_2(p)$, whose graphs are given in Figure 2. Because the graph of

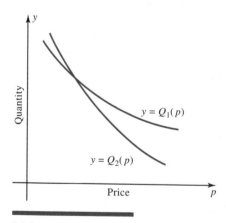

FIGURE 2

$Q_2(p)$ is steeper than the graph of $Q_1(p)$, a small change in the price p will result in a larger change in the quantity $Q_2(p)$ demanded versus the quantity $Q_1(p)$. From an intuitive point of view, economists refer to a steep demand curve such as the graph of $y = Q_2(p)$ as elastic while the graph of $y = Q_1(p)$ is called inelastic.

We now give a concrete definition of the **elasticity** $E(p)$ of a demand curve $y = Q(p)$. It is defined as the negative of the ratio of the rate of growth of $Q(p)$ to the rate of growth of p. That is:

$$E(p) = \frac{-(\text{rate of growth of } Q(p))}{\text{rate of growth of } p} = \frac{-D_p \ln Q(p)}{D_p \ln p} = \frac{-Q'(p)/Q(p)}{1/p} = \frac{-pQ'(p)}{Q(p)}$$

Elasticity is defined with a negative sign so that $E(p)$ is always positive. Notice that p and $Q(p)$ are always positive and $Q'(p)$ is always negative because $y = Q(p)$ is a decreasing function. If $E(p) > 1$ the demand curve is called **elastic** at p and if $E(p) < 1$ it is called **inelastic** at p.

EXAMPLE 6

The demand for oil is given by the formula

$$Q(p) = 60 - 2p$$

where p is the price per barrel and $Q(p)$ is the number of barrels demanded at price p in millions of barrels, for p between $5 and $30.

Problem

Find the values of p where $Q(p)$ is (a) elastic, (b) inelastic.

Solution First compute $Q'(p) = -2$. Then

$$E(p) = \frac{-p(-2)}{60 - 2p} = \frac{2p}{60 - 2p}$$

Determine the value of p where $E(p) = 1$ by solving

$$E(p) = \frac{2p}{60 - 2p} = 1$$

This simplifies to $2p = 60 - 2p$, and then $4p = 60$, which implies that $p = 15$. $E(p)$ is a continuous function that crosses the line $E = 1$ at $p = 15$. This means that $E(p)$ is either greater than 1 at each point in the interval to the left of 15, or it is less than 1 at each point in the interval. The same is true for the interval greater than 15. This means that we must determine whether $E(p)$ is greater than or less than 1 on the intervals $(5, 15)$ and $(15, 30)$. Choose a test point in each interval, say 10 and 20, respectively: $E(10) = 20/40 = 1/2$ and $E(20) = 40/20 = 2$.
(a) Since 10 is in the interval $(5, 15)$ and $E(10) < 1$, $E(p) < 1$ in $(5, 15)$ and so $Q(p)$ is inelastic there.
(b) Since 20 is in the interval $(15, 30)$ and $E(20) > 1$, $E(p) > 1$ in $(15, 30)$ and so $Q(p)$ is elastic there.

If the demand curve is inelastic, then an increase in p results in an increase in revenue while a decrease in p results in a decrease in revenue. This means that revenue is an increasing function in an interval where the demand is inelastic. If the demand curve is elastic, the opposite is true: An increase in p results in a decrease in revenue, meaning that revenue is decreasing in such an interval. Let us see why this is so using the demand curve in Example 6. If $Q(p) = 60 - 2p$, then the revenue $R(p)$, defined by (price $\times$ quantity), is

$$R(p) = pQ(p) = p(60 - 2p) = 60p - 2p^2$$

The question is to determine when $R(p)$ is increasing, so we compute $R'(p)$ and solve $R'(p) > 0$:

$$R'(p) = 60 - 4p > 0$$
$$4(15 - p) > 0$$

This is true when $p < 15$, meaning that $R(p)$ is increasing in $(5, 15)$. Therefore, from Example 6, in the interval $(5, 15)$, $E(p) < 1$, and thus $Q(p)$ is inelastic in this interval. The previous computation shows that $R(p)$ is increasing in this interval. In a similar way it can be shown that in the interval $(15, 30)$ the demand curve is elastic and $R(p)$ is decreasing.

EXERCISE SET 5–3

In Problems 1 to 24 find $f'(x)$.

1. $f(x) = \ln 6x$

2. $f(x) = \ln 9x$

3. $f(x) = \ln 3x + 5$

4. $f(x) = \ln 5x - 6$

5. $f(x) = \ln(x + 5)$

6. $f(x) = \ln(x - 7)$

7. $f(x) = \ln(x^2 + 3)$

8. $f(x) = \ln(3x^2 - 5)$

9. $f(x) = \ln(1 + 4x^2)$

10. $f(x) = \ln(3 - 5x^2)$

11. $f(x) = \ln(2x^3 + x^{-1})$

12. $f(x) = \ln(4 - x^{1/2} + x)$

13. $f(x) = x \ln 4x$

14. $f(x) = 3x \ln 7x$

15. $f(x) = x^{-3} \ln(x^2 - 1)$

16. $f(x) = 5x^{2/3} \ln\left(2x - \dfrac{3}{2}\right)$

17. $f(x) = (\ln 3x)/(2x - 5)$

18. $f(x) = \dfrac{x^5 - 11}{\ln 3x}$

19. $f(x) = \ln |3 - 5x|$

20. $f(x) = \ln |2x^3 + 1|$

21. $f(x) = \ln |4 - x^{5/2}|$

22. $f(x) = \ln |2x^3 + x|$

23. $f(x) = x \ln |7x|$

24. $f(x) = \dfrac{x^3}{\ln |x^2 - 1|}$

In Problems 25 to 28 find the relative extrema and sketch the graph.

25. $f(x) = x \ln x$

26. $f(x) = x \ln |x|$

27. $f(x) = \dfrac{\ln x}{x}$

28. $f(x) = x - \ln |x|$

In Problems 29 and 30 the demand function $Q(p)$ is defined in the given interval. Find where $Q(p)$ is (a) elastic and (b) inelastic.

29. $Q(p) = 80 - 5p$, $[5, 15]$

30. $Q(p) = 100 - 5p$, $[5, 30]$

In Problems 31 to 36 find $f'(x)$.

31. $f(x) = (x + \ln x)^2$

32. $f(x) = (4x + \ln 3x)^{-3}$

33. $f(x) = (\ln x + x^2)/\ln x$

34. $f(x) = x^2(x - \ln x)^2$

35. $f(x) = \ln(x^2 + \ln x)$

36. $f(x) = \ln(x + x \ln x)$

37. Sound intensity is measured in *decibels*. If $D(x)$ is the number of decibels when x is the power of the sound, measured in watts/cm^2, then the relationship between D and x is given by

$$D(x) = 10 \log_{10}(10^{16}x)$$

Find $D'(x)$.

38. The relationship between cost C and labor x is often given by a logarithmic equation. It is a form of a Cobb–Douglas function. Suppose a firm determines that

    $\ln C(x) = 8.3x^{-0.17}$

    Find $C'(x)$.

39. Economists define the *rate of growth* of a function $f(x)$ as the ratio $F(x)$ given by

    $$F(x) = \frac{f'(x)}{f(x)}$$

    Show that $F(x) = D_x(\ln f(x))$.

40. A manufacturer has a cost function

    $C(x) = 5000 + 600x - 0.3x^2$

    Find the rate of growth of $C(x)$ when $x = 100$.

41. Let $f(x) = \ln x$. Find a formula for $f^{(n)}(x)$ for n a positive integer by computing the first few derivatives of $f(x)$ and finding a pattern.

42. Find $f'(x)$ for $f(x) = \ln(\ln(\ln x))$.

43. Show that $f(x) = \ln x^2$ is not the same function as $g(x) = (\ln x)^2$ by first showing that their functional values are different for a particular value of x and then showing that their derivatives are different.

Referenced Exercise Set 5–3

1. The relationship between home range and body size of mammals is believed to be as important to carnivores as it is to herbivores. In an article studying this relationship, Lindstedt et al. use a formula that relates home range A, measured in hectares, and body mass m, measured in kilograms.* The relationship is

   $\ln A = 1 + \ln m$

   Express A in terms of m directly, and find $A'(1)$.

2. Accurate leaf-area measurements are critical for estimating fluxes of carbon, solar energy, and water in forests. In an article comparing various methods of estimating leaf areas in fir trees, Marshall and Waring show that the leaf-area index L, which is the projected surface area of foliage per unit ground area, is a function of the irradiance q below the canopy, which is the light level below the tops of the trees.† The relationship for a Douglas-fir forest is estimated as

   $L(q) = 2.5(2q - \ln q)$

   Find the value of q where $L(q)$ is a minimum.

Cumulative Exercise Set 5–3

1. Sketch the graphs of the functions $f(x) = 3^x$ and $g(x) = 3^{-x}$ on the same coordinate axis.

2. Sketch the graphs of the functions $f(x) = e^{x+1}$ and $g(x) = \ln(x + 1)$ on the same coordinate axis.

3. Find all values of x that satisfy the given equality.
   (a) $2^x = 1/4$
   (b) $3^{x-2} = 81$
   (c) $2^{2x-1} = 8^{x-1}$

4. A student's grade after t weeks of a 15-week semester is

   $g(t) = 100 - 60e^{-0.1t}$

   Determine the student's grade (a) at the beginning of the semester, (b) after one week, (c) after 10 weeks, and (d) at the end of the semester. (e) If the semester is extended indefinitely, what will the student's ultimate grade be?

5. Sketch the graphs of the functions $f(x) = \ln x$, $g(x) = \log_2 x$, and $h(x) = \log_3 x$ on the same coordinate axis.

6. Use the table in the Appendix or a calculator to sketch the graph of the function $f(x)$ by the plotting method, where

   $f(x) = e^{x/2} - (\ln x)^4$

   [The command for this function in the graphing module of the software program **Graph 2D/3D** is $\exp(x/2) - (\ln(x)) \triangleq 4$.]

7. Solve each equation for x.
   (a) $\ln e^{2x} - e^{\ln 6} = \ln 1$
   (b) $e^{\ln 3x} + 1 = \ln e^4$

*Stan L. Lindstedt et al., "Home Range and Body Size in Mammals," *Ecology*, Vol. 67, 1986, pp. 413–418.

†J. D. Marshall and R. H. Waring, "Comparison of Methods of Estimating Leaf-area Index in Old-growth Douglas Fir," *Ecology*, Vol. 67, 1986, pp. 413–418.

8. Suppose the weight equation for a 130-pound person who goes on a diet is given by

$$w(t) = 120 + 10e^{-0.005t}$$

where t is the time in days after the start of the diet. How many days will it take the person to drop to 126 pounds?

In Problems 9 to 11 find $f'(x)$.

9. $f(x) = \ln(4 - x^2)$

10. $f(x) = \ln(e^x) - \dfrac{1}{\ln x}$

11. $f(x) = x(1 - \ln x)^2$

12. The purpose of this problem is to develop a formula for the derivative of logarithmic functions to any base.
    (a) Let $y = \log_b x$. Fill in the blank: $b^y =$ ____.
    (b) Take the natural logarithm of both sides of the equation $b^y = x$ and solve for y.
    (c) Show that for any $b > 0$,

$$D_x \log_b x = \frac{1}{\ln b} \frac{1}{x}$$

5–4 Differentiation of Exponential Functions

Recall that the natural logarithm function and the exponential function are inverses of each other. This means that $\ln e^x = e^{\ln x} = x$. This property enables us to derive the derivatives of the functions $y = e^x$ and $y = e^{g(x)}$.

The Derivative of $y = e^x$

Because e^x and $\ln x$ are inverses of each other, the following property holds:

$$\ln e^x = x$$

Take the derivative of each side.

$$D_x(\ln e^x) = D_x(x)$$

On the left-hand side use the rule for the derivative of logarithmic functions, with $g(x) = e^x$.

$$\frac{D_x e^x}{e^x} = 1$$

Multiply by e^x to get

$$D_x e^x = e^x$$

The Derivative of the Exponential Function

$$D_x e^x = e^x$$

EXAMPLE 1

Problem

Find $f'(x)$ for (a) $f(x) = x^4 e^x - 5e^x$, (b) $f(x) = \dfrac{e^x}{x^2 + 5x}$.

Solution (a) Use the rule for the derivative of a sum, the rule for the derivative of a product, and the rule for a constant times a function.

$$D_x(x^4e^x - 5e^x) = D_x(x^4e^x) - 5D_x(e^x) = x^4D_x(e^x) + e^xD_x(x^4) - 5D_x(e^x)$$
$$= x^4e^x + e^x4x^3 - 5e^x = e^x(x^4 + 4x^3 - 5)$$

(b) Use the rule for the derivative of a quotient.

$$D_x\left(\frac{e^x}{x^2 + 5x}\right) = \frac{(x^2 + 5x)D_x(e^x) - e^xD_x(x^2 + 5x)}{(x^2 + 5x)^2}$$

$$= \frac{(x^2 + 5x)e^x - e^x(2x + 5)}{(x^2 + 5x)^2}$$

$$= \frac{e^x(x^2 + 5x - 2x - 5)}{(x^2 + 5x)^2} = \frac{e^x(x^2 + 3x - 5)}{(x^2 + 5x)^2}$$

The Exponential Function Version of the Chain Rule

The general exponential function is expressed in the form e^u where $u = g(x)$ is a function of x. If $g'(x)$ exists, the chain rule can be used to find the derivative of e^u. Recall that the chain rule states that if $y = f(g(x))$, then

$$y' = f'(g(x))g'(x)$$

To compute the derivative of $y = e^u$, let $u = g(x)$ and $f(x) = e^x$ in the formula. Then $y = f(g(x)) = f(u) = e^u = e^{g(x)}$. From the rule for the derivative of the exponential function $f'(u) = e^u$ or $f'(g(x)) = e^{g(x)}$. From the chain rule we get

$$y' = f'(g(x))g'(x) = e^{g(x)} \cdot g'(x) = g'(x)e^{g(x)}$$

The form of the chain rule applied to the exponential function

$$f(x) = e^{g(x)}$$

is the formula

Derivative of Exponential Functions

$$D_xe^{g(x)} = g'(x)e^{g(x)} \qquad \text{or} \qquad D_xe^u = e^uD_xu$$

Example 2 illustrates how to apply the exponential function version of the chain rule.

EXAMPLE 2

Problem

Find the derivative of (a) $f(x) = e^{2x}$, (b) $f(x) = e^{3x^2+7}$.

Solution (a) Let $g(x) = 2x$. Then $g'(x) = 2$ and

$$D_x e^{2x} = g'(x)e^{g(x)} = 2e^{2x}$$

(b) Let $g(x) = 3x^2 + 7$. Then $g'(x) = 6x$ and

$$D_x e^{(3x^2+7)} = g'(x)e^{g(x)} = 6xe^{3x^2+7}$$

The next example demonstrates how one derivative can require two versions of the chain rule, the generalized power rule, and the exponential function version.

EXAMPLE 3

Problem

Find the derivative of

$$f(x) = (x^{-1} + e^{5x+2})^3$$

Solution First apply the generalized power rule. We get

$$f'(x) = 3(x^{-1} + e^{5x+2})^2 \cdot D_x(x^{-1} + e^{5x+2})$$

The derivative of $x^{-1} + e^{5x+2}$ requires the derivative of a sum. Let $g(x) = 5x + 2$. Then $g'(x) = 5$ and $D_x e^{5x+2} = 5e^{5x+2}$. Thus

$$f'(x) = 3(x^{-1} + e^{5x+2})^2(-x^{-2} + 5e^{5x+2})$$

Graphing Exponential Functions

The curve-sketching techniques of the previous chapter can be used to graph exponential functions. The methods are illustrated in the next two examples.

EXAMPLE 4

Problem

Graph the function $f(x) = \dfrac{1}{1 + e^{-x}}$.

Solution To find $f'(x)$, write $f(x) = (1 + e^{-x})^{-1}$. Then

$$f'(x) = -(1 + e^{-x})^{-2}(-e^{-x}) = e^{-x}(1 + e^{-x})^{-2}$$

$$= \frac{1}{e^x(1 + e^{-x})^2} = \frac{1}{e^x(1 + 2e^{-x} + e^{-2x})} = \frac{1}{2 + e^x + e^{-x}}$$

The numerator is never 0, so there is no value of x such that $f'(x) = 0$. Each term in the denominator is positive so there is no value of x such that $f'(x)$ does not exist. Thus there is no relative extremum. To test for a point of inflection, we calculate $f''(x)$ using the quotient rule.

$$f''(x) = \frac{-(e^x - e^{-x})}{(2 + e^x + e^{-x})^2} = \frac{e^{-x} - e^x}{(2 + e^x + e^{-x})^2}$$

Solving $f''(x) = 0$ yields $e^{-x} - e^x = 0 = e^{-x}(1 - e^{2x})$. This is satisfied when $e^{2x} = 1$, which is true when $x = 0$. Therefore $f(x)$ has a possible point of inflection at $x = 0$. When $x < 0$, $f''(x) > 0$ since $e^{-x} > e^x$. When $x > 0$, $f''(x) < 0$. This means that $f(x)$ is concave up in $(-\infty, 0)$ and concave down in $(0, \infty)$. Thus $f(x)$ has a point of inflection at $x = 0$. Since $f'(x) > 0$ for all values of x, the graph is always increasing. As x gets very large e^{-x} approaches 0, so $f(x)$ approaches $1/(1 + 0) = 1$. Similarly as x approaches $-\infty$, $f(x)$ approaches 0. Putting this information together yields the graph in Figure 1.

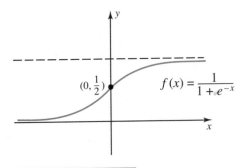

$$\left(0, \frac{1}{2}\right) \qquad f(x) = \frac{1}{1 + e^{-x}}$$

FIGURE 1

EXAMPLE 5

Problem

Graph the function $f(x) = \dfrac{1}{\sqrt{2\pi}} e^{-x^2/2}$.

Solution First find $f'(x)$.

$$f'(x) = \frac{1}{\sqrt{2\pi}} \left(\frac{-2x}{2}\right) e^{-x^2/2} = \left(\frac{-x}{\sqrt{2\pi}}\right) e^{-x^2/2}$$

The only value of x such that $f'(x) = 0$ is $x = 0$ since $e^{-x^2/2}$ is always positive. Thus the only critical value is $x = 0$. Applying the first derivative test shows that $f(x)$ is increasing in $(-\infty, 0)$ and decreasing in $(0, \infty)$ because $e^{-x^2/2}$ is always positive, and so $f'(x)$ is positive and negative when x is negative and positive. Therefore $f(x)$ has a relative maximum of $f(0) = 1/\sqrt{2\pi}$. The points of inflection are found by solving $f''(x) = 0$.

$$f''(x) = \left(\frac{x^2}{\sqrt{2\pi}}\right) e^{-x^2/2} - \frac{1}{\sqrt{2\pi}} e^{-x^2/2} = \left(\frac{1}{\sqrt{2\pi}}\right)(x^2 - 1)e^{-x^2/2}$$

The function has points of inflection at $x = -1$ and $x = 1$ and $f(x)$ is concave up in the intervals $(-\infty, -1)$ and $(1, \infty)$ and concave down in the interval $(-1, 1)$. The graph of $f(x)$ is given in Figure 2.

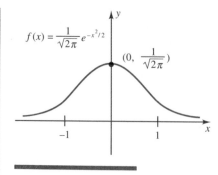

FIGURE 2

The function in Example 5 is the well-known "bell-shaped" curve that is used extensively in statistics. This function is known as the **standard normal probability density function.**

The Logistic Curve

Psychologists use various techniques to model the learning process. One type of model that uses exponential functions of the form $y(x) = a - be^{-ct}$, was introduced in Section 5–1. A more complex model of learning uses an exponential function of the form

$$y(x) = \frac{A}{1 + Be^{-kx}}$$

It is called the general form of the **logistic equation.** An example of a specific logistic equation is the function in Example 4, where $A = B = k = 1$. It is the function

$$f(x) = \frac{1}{1 + e^{-x}}$$

Its graph is given in Figure 1.

The first known application using the logistic equation was a study of world population made by the Belgian mathematician P. Verhulst. In 1840 Verhulst created a model of world population that was off by only 1% a century later.

One of the first applications of the logistic curve in the social sciences was made by Clark L. Hull.* He created a model of the learning process that has had a wide variety of uses in many disciplines.

The graph of a logistic equation is an S-shaped curve, as shown in Figure 3. When $x = 0$ the denominator is $1 + B$, so

$$y(0) = \frac{A}{1 + B}$$

*Clark L. Hull, *A Behavior System,* New Haven, CT, Yale University Press, 1952.

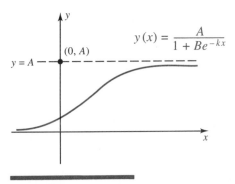

$$y(x) = \frac{A}{1 + Be^{-kx}}$$

FIGURE 3

As x increases without bound, the denominator approaches 1 since e^{-kx} approaches 0. This means that $y(x)$ approaches the value $A/1$ or simply A. It never reaches A, but as x increases, $y(x)$ gets arbitrarily close to A. When the logistic equation is applied to population growth, the constant A is the maximum level of the population that the environment can sustain.

In Hull's model of learning A represents the maximum amount of learning. For example, suppose a puzzle is given to a large group of individuals. Let $f(x)$ be the average amount of learning, measured in percent, per time x spent solving the puzzle. Then A is 1, or 100%. It represents total mastery of the puzzle. One such puzzle consists of arranging several different cubes with variously colored sides in a specific order. An individual's learning consists in discovering the various characteristics of the cubes. The puzzle is usually solved the first time merely by chance. Little actual learning has taken place. When the cubes are mixed up, it takes the person almost the same amount of time to solve it again. Once some information is acquired about the cubes, say their individual color schemes or which one must come first in the sequence, then learning takes place. Hull found that when solving such tasks, learning takes place very slowly at first. The rate of learning then rises rapidly. When mastery of the task is approached the rate of learning slows again. These are the characteristics of the logistic equation.

The maximum learning rate is the maximum of the derivative of the logistic equation, $y = f(x)$. It occurs at the point of inflection of $f(x)$. The next example looks at a logistic curve similar to one used by Hull.

EXAMPLE 6

A psychologist gives a puzzle to a large group of individuals. In the directions of the puzzle each subject was told 20% of the solution. On the average, after 10 minutes a subject had learned 50% of the solution.

Problem

Find the specific form of the logistic equation that measures the percent of learning $f(x)$ after x minutes of solving the puzzle.

Solution The maximum amount of learning is 100%, so $A = 1$. At $x = 0$, the time when the subject starts to solve the puzzle, 20% of the solution is known, so $f(0) = 0.2$. Therefore

$$f(0) = \frac{1}{1 + Be^0} = 0.2$$

Thus $1 = 0.2 + 0.2B$ and $0.8 = 0.2B$, so $B = 4$. Now solve the equation for k using the fact that $f(10) = 0.5$.

$$f(10) = \frac{1}{1 + 4e^{-10k}} = 0.5$$
$$1 = 0.5 + 2e^{-10k}$$
$$0.5 = 2e^{-10k}$$
$$0.25 = e^{-10k}$$
$$\ln(0.25) = \ln(e^{-10k}) = -10k$$
$$k = \frac{1.39}{10} = 0.139$$

Therefore the equation is

$$f(x) = \frac{1}{1 + 4e^{-0.139x}}$$

The logistic equation is also used in business to model a wide variety of situations. For example, Hannan and McDowell[†] studied how information spreads in an industry, with the banking industry as a specific example. They found that information spreads slowly at first, meaning that the proportion of firms using a new technological process remains small for a period of time. Then as firms using the process enjoy success with it the proportion of firms adopting it increases rapidly. To model the growth of automatic teller machines in the banking industry, they studied 89 local banks and developed the logistic equation

$$f(x) = \frac{1}{1 + 1.97e^{-0.33x}}$$

The logistic equation is used widely in marketing to predict future trends. Suppose a firm produces an innovative product that immediately captures a certain percentage of the market. The firm has allocated a specific amount of money to advertise the new product. The basic question is: How quickly should the advertising budget be spent? Past performance shows that the increased market share of innovative products often follows a logistic growth pattern. The market share increases slowly at first, then gains momentum, and then slows again (see Figure 4). Once the market share increase gains momentum, there is less need for advertising because information about the product will spread by word of mouth. In

[†]Timothy H. Hannan and John M. McDowell, "Market Concentration and the Diffusion of New Technology in the Banking Industry," *Review of Economics and Statistics,* Vol. 66, 1984, pp. 686–691.

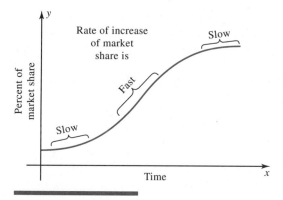

FIGURE 4

addition, as the market becomes saturated there are fewer potential buyers so the market share increase must slow. These factors imply that the advertising budget has the greatest effect in the early stages of the selling period. In fact, it should be entirely spent by the time the market share increase reaches a maximum. If $y = f(x)$ represents the percent of market share of the product at time x, then the rate of increase of market share is $f'(x)$. Thus the rate of increase of market share will attain a maximum when $f'(x)$ is a maximum, that is, when $f''(x) = 0$, or when $f(x)$ has a point of inflection. This means that the advertising budget should be spent by the time $x = a$ when $f(x)$ has a point of inflection (see Figure 5).

In practice, this is a dynamic process for the marketing department. As they spend advertising dollars, they monitor the product's market share. They compare the actual figures with their predictions, alter the predictions, and then make a decision on future advertising expenditures. If it appears that the product is close to its point of inflection on the predicted logistic curve, they may choose to spend more advertising dollars in hopes of altering the actual curve so that they reach the point of inflection sooner (see Figure 6).

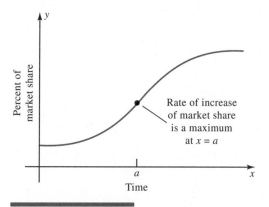

FIGURE 5

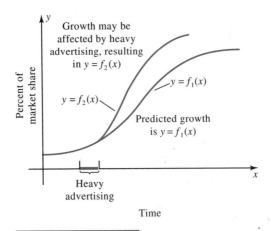

FIGURE 6

Thus in marketing, as in many other fields, the most important point on the curve is the point of inflection, which measures when $f(x)$ attains its maximum rate of growth. The next example computes the point of inflection of a logistic curve.

EXAMPLE 7

Consider the automatic teller machine logistic equation.

$$f(x) = \frac{1}{1 + 1.97e^{-0.33x}}$$

Problem

Find the point of inflection of $f(x)$ and interpret its significance from a marketing point of view.

Solution We compute $f'(x)$ and then $f''(x)$:

$$f'(x) = \frac{1.97(-0.33)e^{-0.33x}}{(1 + 1.97e^{-0.33x})^2} = \frac{-0.65e^{-0.33x}}{(1 + 1.97e^{-0.33x})^2}$$

To find $f''(x)$, let us make some substitutions that ease the computation. Let $a = -0.65$, $B = 1.97$ and $k = -0.33$. Then

$$f'(x) = \frac{ae^{kx}}{(1 + Be^{kx})^2}$$

$$f''(x) = \frac{a[ke^{kx}(1 + Be^{kx})^2 - 2e^{kx}(1 + Be^{kx})Bke^{kx}]}{(1 + Be^{kx})^4}$$

$$= \frac{ake^{kx}(1 + Be^{kx})[1 + Be^{kx} - 2Be^{kx}]}{(1 + Be^{kx})^4}$$

$$= \frac{ake^{kx}(1 - Be^{kx})}{(1 + Be^{kx})^3}$$

Now solve $f''(x) = 0$; $ake^{kx}(1 - Be^{kx}) = 0$ implies that $1 - Be^{kx} = 0$ and so $e^{kx} = 1/B$. Take the natural logarithm of each side and get $\ln e^{kx} = \ln(1/B) = kx$, since $\ln e^{kx} = kx$. Simplify $\ln(1/B) = \ln B^{-1} = -\ln B$. Therefore $x = \dfrac{-\ln B}{k}$. Substituting back for B and k yields

$$x = \frac{-\ln(1.97)}{-0.33} = \frac{0.68}{0.33} \approx 2.06$$

Thus the point of inflection occurs near $x = 2.06$. This means that the rate of increase of the automatic teller machines will be maximum just after two years. This means that advertising expenditures should be spent over two years. In practice, the marketing analysts will monitor the market share to see if the advertising or other factors will alter this predicted logistic growth.

EXERCISE SET 5–4

In Problems 1 to 24 find $f'(x)$.

1. $f(x) = e^{6x}$ $= 6e^{6x}$

2. $f(x) = e^{9x}$

3. $f(x) = e^{3x+5}$ $3e^{3x+5}$

4. $f(x) = e^{5x-6}$

5. $f(x) = 4e^{7x+5}$

6. $f(x) = 6e^{4-7x}$

7. $f(x) = e^{x^2+3}$

8. $f(x) = e^{3x^2-5}$

9. $f(x) = 2e^{1+4x^2}$

10. $f(x) = -3e^{3-5x^2}$

11. $f(x) = 4 - e^{2x^3+x-1}$

12. $f(x) = 3 + e^{4-x^{1/2}+x}$

13. $f(x) = xe^{4x}$

14. $f(x) = 3xe^{7x}$

15. $f(x) = x^{-3}e^{x^2-1}$

16. $f(x) = 5x^{2/3}e^{2x-3/2}$

17. $f(x) = (e^{3x})/(2x - 5)$

18. $f(x) = x^5 - 11/e^{3x}$

19. $f(x) = e^{5x} + e^{-5x}$

20. $f(x) = e^{3x} - 3e^{-3x}$

21. $f(x) = \dfrac{1}{2 + e^{-x}}$

22. $f(x) = \dfrac{1}{1 + e^{-2x}}$

23. $f(x) = \dfrac{1}{1 + e^x + e^{-2x}}$

24. $f(x) = \dfrac{2}{e^{2x} + e^{-2x}}$

In Problems 25 to 28 sketch the graph.

25. $f(x) = xe^x$

26. $f(x) = x + e^x$

27. $f(x) = \dfrac{1}{2 + e^{-2x}}$

28. $f(x) = \dfrac{4}{1 + e^{-3x}}$

In Problems 29 to 36 find $f'(x)$.

29. $f(x) = \ln(x + e^{4x})$

30. $f(x) = \ln(3 + xe^{2x})$

31. $f(x) = (\ln x + e^{5x})^2$

32. $f(x) = (2 \ln x + e^{3x})^{-3}$

33. $f(x) = \dfrac{e^{x^2} + x^2}{e^x}$

34. $f(x) = \dfrac{\ln x}{x - e^{3x}}$

35. $f(x) = e^{x^3} \ln x$

36. $f(x) = e^{3x} \ln(x + e^x)$

37. Let $f(x) = e^{2x}$. Find a formula for $f^{(n)}(x)$ for n a positive integer by computing the first few derivatives of $f(x)$ and finding a pattern.

38. Find $f'(x)$ for $f(x) = xe^x \ln x$.

In Problems 39 to 42 match the equation with the graphs in figures a to d on page 286.

39. $f(x) = \dfrac{1}{1 + e^{-x}}$

40. $f(x) = \dfrac{100}{1 + e^{-x}}$

41. $f(x) = \dfrac{1}{1 + 9e^{-x}}$

42. $f(x) = \dfrac{100}{1 + 9e^{-x}}$

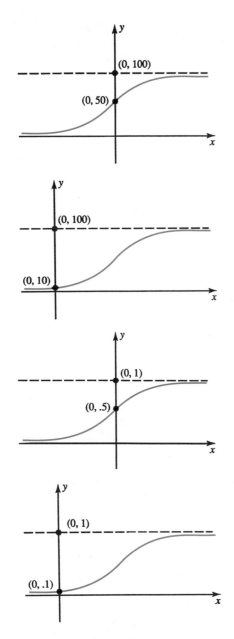

43. A psychologist gives a puzzle to a large group of individuals. In the directions of the puzzle each subject was told 10% of the solution. On the average, after 20 minutes a subject had learned 50% of the solution. Find the specific form of the logistic equation that measures the percent of learning $f(x)$ after x minutes of solving the puzzle.

44. A psychologist gives a puzzle to a large group of individuals. In the directions of the puzzle each subject was told 25% of the solution. On the average, after 15 minutes a subject had learned 50% of the solution. Find the specific form of the logistic equation that measures the percent of learning $f(x)$ after x minutes of solving the puzzle.

Referenced Exercise Set 5–4

1. In an article* studying the social cost of population growth, researchers used educational tax revenue as a measure of the cost of an expanding population. They focused on California's 1965 tax profile as a baseline for a community's ability to fund fully its educational system. The proportion of the tax paid by various age groups increased from age 16 to 65. Because many people retired at age 65, the proportion of tax paid per age decreased linearly from age 65 to age 75, after which it was always 0. The data are given by the formula $y = f(x)$ where $f(x)$ is the proportion of the total educational tax paid by those of age x. In the interval [16, 65] the data are given as an exponential function, and in the interval (65, 75] as a linear function.

$$f(x) = \begin{cases} 0.03e^{-0.0008(x-65)^2} & x \in [16, 65] \\ -0.003x + 0.225 & x \in (65, 75] \end{cases}$$

Graph $f(x)$ in [16, 75].

2. Thermal recovery of the skin after cooling is not only an important medical treatment but it is also a key indicator of normal versus pathological circulation. In a paper studying thermal recovery, Steketee and Van Der Hoek cooled the skin of the forehead of patients and measured the time it took to reach a steady-state temperature.† The temperature T of the forehead, measured in degrees centigrade, was found to be a function of time t, measured in minutes. For one group of patients, the functional relationship was found to be

$$T(t) = 35 + 2.1e^{-.017t}$$

The steady-state temperature was defined to be that value of t when $T'(t) = -.028$. Find the steady-state temperature for this group of patients.

*Norman R. Glass et al., "Human Ecology and Educational Crises," in *Is There an Optimum Level of Population?*, Fred Singer (ed.), New York, McGraw-Hill, 1971, pp. 205–218.

†J. Steketee, and M. J. Van Der Hoek, "Thermal Recovery of the Skin after Cooling," *Journal of Physics and Medical Biology*, Vol. 24, 1979, pp. 583–592.

3. The problem of finding an optimal advertising policy over a period of time is one of the most important questions in the field of marketing. In a paper studying oligopoly models, that is, business problems involving more than one firm, Teng and Thompson studied how firms market new products.‡ The price of the new product p, measured in dollars, was a function of time t, measured in months. In one market the new product sales price was found to obey the formula

$$p(t) = 100e^{.01t}$$

The object of the firm is to allow the price to increase to the point at which the increase is so great that it influences customers to purchase competitors' products. This point was reached when $p'(t) = 1.02$. Find this value of t.

Cumulative Exercise Set 5–4

1. Sketch the graph of $f(x) = e^{3-x}$.

2. If \$2000 is deposited into an account that pays 10% interest compounded annually, the future amount is given by

$$A(t) = 2000(1.1)^t$$

where t is in years and $A(t)$ is in dollars. What is the future amount after 5 years?

3. Solve for x: $e^{\ln(x+2)} = \ln e^{4x}$

4. Use a calculator to show that $\ln(4 - 3)$ is not equal to $\ln 4 - \ln 3$.

In Problems 5 to 12 find $f'(x)$.

5. $f(x) = x^2 \ln(1 - x^3)$ 6. $f(x) = (x + \ln 3x)^{-2}$

7. $f(x) = (\ln x)(x - \ln x)^2$

8. $f(x) = \ln(1 + 3x \ln x)$

9. $f(x) = e^{2 - 5x^2}$ 10. $f(x) = x^2 e^{3x^2}$

11. $f(x) = 2/(3 + e^{-5x})$

12. $f(x) = (3 \ln x + e^{5x})^{-2}$

‡Jinn-tsair Teng and Gerald L. Thompson, "Oligopoly Models for Optimal Advertising when Production Costs Obey a Learning Curve," *Management Science,* Vol. 29, 1983, pp. 1087–1104.

CASE STUDY   ## Denver Fire Department Deployment Policy Analysis*

In 1972 the mayor and the chief of the Fire Department of Denver, Colorado, agreed that a study of the fire department was in order in the face of mounting costs. They decided to apply modern operations analysis by contracting for a research study with faculty from the Department of Operations Research at the University of Colorado. The study was funded by HUD's Office of Policy Development and Research at a level of \$140,000. The challenge given the research team was: Can the fire department provide approximately the same level of fire suppression service at a lower cost?

The most critical cost was maintenance. Denver spent in excess of \$250,000 per year to staff each fire engine or ladder truck around the clock. In contrast, the cost of the vehicle itself was about \$60,000, and a fire house had a one-time cost of \$400,000. Thus the number of vehicles and their location accounted for the bulk of the city's expenditure for fire protection.

In the first phase of the analysis the research team used linear programming to study fire company locations. The objective function minimized cost and the constraints were a set of response time requirements. The latter included the assumptions that every vehicle was always available to respond to a fire incident, and that *alarm rates,* the number of incidents per time period, were less than or

*Adapted from D. E. Monarchi et al., "Simulation for Fire Department Deployment Policy Analysis," *Decision Sciences,* Vol. 8, 1977, pp. 211–227.

equal to a specific threshhold. The conclusion of the study was that the level of service would be unchanged with a reduction of five companies—from 44 to 39 fire stations. This reduction would take place by closing obsolete stations and building new houses at new locations.

The recommendations were to take place over a period of seven years. But strong opposition, especially from the fire fighters' union, prevented immediate implementation. The union contended that if alarm rates increased dramatically, suppression service would significantly decline because of the fewer number of stations.

The second challenge for the researchers was to provide evidence supporting their conclusion. They needed to produce realistic data depicting possible future trends of alarm rates. Their approach was to use a dynamic analysis that compared the service provided by the existing fire company locations versus the proposed configuration. The term "dynamic" means that every possible fire incident situation in the city would be considered. For each situation it would be determined whether a vehicle could respond. They first had to study suppression service with the current alarm rate and then determine the effect on service if the alarm rate increased. The important question was whether the new configuration of fire stations recommended by the research team would provide the same service as the current one, both for the current alarm rate and for possible increased rates in the future.

The specific method used for this dynamic analysis was a computer simulation model developed by the New York City–Rand Institute. By using simulation they could explore the effects of increasing alarm rates on fire department performance for each configuration.

The simulation model had three parts. It created an "incident generator," which developed 1000 separate types of incidents, reflecting what, when, and where fires occur and the equipment required to service them. A random number generator was used to simulate the occurrence of the incidents, depending on the probability of occurrence assigned to each incident. The simulator then monitored the equipment—it dispatched the necessary equipment, assigned a time that the equipment would expect to be in service, and then "returned" the vehicle to await the next call.

The third part of the model compared the performance of two fire station configurations, the current one versus the proposed one. The simulation was run for various alarm rates.

One of the key measures of performance is the occurrence of long response times. If a vehicle takes longer than 5 minutes to respond, its performance is poor. For a given average alarm rate the number of responses greater than 5 minutes (for the first vehicle to respond) was measured. The data for the new configuration are shown in Figure 1. As the average alarm rate increases, the number of long responses increases. For example, for the lowest alarm rate $t = 2.5$ (or 2.5 alarms per hour), $N = 3$, meaning there were 3 long responses out of 1000. For $t = 5$, $N = 14$, so when the alarm rate doubles from $t = 2.5$ to $t = 5$, N increases more than fourfold from 3 to 14.

A key point is that N increases at a fast rate from 2.5 to 5, but then the rate of increase of N slows dramatically. That is, while N increases as t

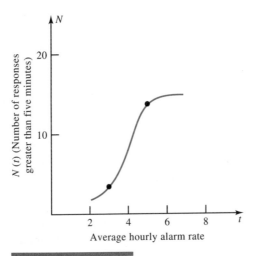

FIGURE 1

increases, the rate of increase, measured by the derivative, decreases when t approaches 5. The data suggest that the best type of function to model the behavior of $N(t)$ is a logistic curve, whose general equation is

$$N(t) = \frac{a}{b + e^{-kt}}$$

Analysis of the data from the simulation suggests that the logistic equation of best fit for the proposed configuration is

$$N(t) = \frac{0.375}{0.013 + e^{-0.87t}}$$

This function is used to determine when $N(t)$ starts to level off. At that point, with the new configuration, the fire department could maintain service even as the average alarm rate increases. That is, even though the alarm rate might increase, the number of long responses remains almost constant. The value of t where $N(t)$ starts to level off is the point where the derivative is a maximum, that is, at the point of inflection. It is found by solving $N''(t) = 0$. To compute this number, we use the quotient rule and the chain rule.

$$N'(t) = \frac{0.326e^{-0.87t}}{(0.013 + e^{-0.87t})^2}$$

$$N''(t) = \frac{-0.284e^{-0.87t}(0.013 - e^{-0.87t})}{(0.013 + e^{-0.87t})^3}$$

When we set $N''(t) = 0$ we get $0.013 = e^{-0.87t}$, and so $t = (\ln 0.013)/(-0.87) = 5.0$.

The simulation showed that $N(t)$ leveled off from $t = 5$ to 6, as in the logistic equation model, but when the average alarm rate was increased beyond 6, $N(t)$ began to rise appreciably. Hence the value $t = 5$ is critical. If Denver's average alarm rate ever reaches it, service can be maintained until t reaches 6, but by then the city would have to implement some kind of strategy, such as building new stations, to keep the same level of service. These relatively high alarm rates occur infrequently, however, as Denver's average fire company is available to respond from its quarters about 95 percent of the time.

When the simulation was run for the current configuration the results were almost identical. The only significant discrepancy occurred when t increased beyond 6, for then N increased at a far greater rate for the new configuration than for the current one. The research team recommended that if the new configuration was implemented, then the alarm rate would have to be closely monitored. If and when it reached $t = 5$, the city still had time until it reached $t = 6$ to take further steps to maintain service, like building new stations.

In summary, the research team used linear programming to determine that the fire company's cost would be minimized, subject to the present level of service being maintained, by closing five fire stations. The team used dynamic simulation at first to test the system to see if there were any hidden elements that would cause problems of service, such as poor response time for some combination of incidents. None was discovered. Then, when their conclusions were challenged the fact that the city had been modeled in a dynamic environment gave the team a "more believable" presentation. The predictions of the study had more credibility because the simulation effectively replicated the historical fire suppression data of the city.

CHAPTER REVIEW

Key Terms

5–1 Exponential Functions
Exponential Function
Base

e
Learning Curve

5–2 Logarithmic Functions
Logarithmic Function with Base b
Common Logarithms
Inverse Functions

Natural Logarithmic Function
Doubling Time

5–3 Differentiation of Logarithmic Functions
Rate of Growth
Elasticity

5–4 Differentiation of Exponential Functions
Standard Probability Density Function

Summary of Important Concepts

1. If $b^r = b^s$, then $r = s$
2. $b^r \cdot b^s = b^{r+s}$
3. $\dfrac{b^r}{b^s} = b^r \cdot b^{-s} = b^{r-s}$
4. $(b^r)^s = b^{rs}$
5. $e^{\ln x} = x$ for $x > 0$
6. $\ln e^x = x$ for all x
7. $\ln 1 = 0$
8. If $\ln r = \ln s$, then $r = s$
9. $\ln rs = \ln r + \ln s$
10. $\ln \dfrac{r}{s} = \ln r - \ln s$
11. $\ln r^k = k \cdot \ln r$

Formulas

$$e = \lim_{x \to \infty} \left(1 + \frac{1}{x} \right)^x$$

$$D_x \ln x = \frac{1}{x}$$

$$D_x \ln g(x) = \frac{g'(x)}{g(x)}$$

$$D_x e^x = e^x$$

$$D_x e^{g(x)} = g'(x) e^{g(x)}$$

REVIEW PROBLEMS

In Problems 1 to 5 sketch the graph of the function.

1. $f(x) = 2^{x+3}$

2. $f(x) = 5 + e^{-0.5x}$

3. $f(x) = \dfrac{1}{\sqrt{6}} e^{-x^2/2}$

4. $f(x) = \log_{1/e}(x - 3)$

5. $f(x) = \ln(4 - x^2)$

6. Evaluate (a) $\ln e^{-6.2}$, (b) $e^{\ln 0.5}$.

7. Solve for x.

$$10 = \frac{1}{e^{-x} - 2}$$

8. Evaluate

$$\lim_{h \to 0} \frac{\ln(10 + h) - \ln 10}{h}$$

9. Evaluate

$$\lim_{h \to 0} \frac{e^{2+h} - e^2}{h}$$

In Problems 10 to 12 find the derivative of the function.

10. $f(x) = 3 \ln x - \dfrac{1}{\ln x}$

11. $f(x) = [\ln(1 + 3x)^4]^5$

12. $f(x) = x \ln |x^2 - 1|$

13. Evaluate the derivative of the function $f(x) = \ln \sqrt{2x}$ in two ways: using the formula and using the properties of logarithmic functions.

14. Find the relative extrema and the points of inflection of the function $f(x) = \dfrac{\ln x}{x}$. Sketch the graph.

In Problems 15 and 16 find the derivative of the function.

15. $f(x) = x^2 e^x - e^{-x}$

16. $f(x) = (x^2 - e^{2x-1})^4$

17. Find the relative extrema and the points of inflection of the function $f(x) = e^x + e^{-x}$. Sketch the graph.

GRAPHICS CALCULATOR EXPLORATIONS

Your graphics calculator is an invaluable tool to develop your graphing facility. Use it not only to enter a formula and press "graph," but also to learn how to predict what the graph of a particular function looks like. This Exploration will help you predict what the graph of the more complicated function $y = 10 - e^{-0.5x}$ looks like by considering the relationships among graphs of simpler functions such as $y = e^x$, $y = e^{-x}$, and $y = -e^{-x}$.

Understanding how to graph families of curves will help take the mystery out of obtaining the graphs of complicated functions. For example, the graph of the function $y = 1 - e^{-x}$ can be obtained from the graph of the simpler function $y = e^x$ if you know a few additional facts about graphing.

Exploration 1

The function $y = 1 - e^{-x}$ can be obtained from $y = e^x$ in three steps:

(a) Insert a negative sign in the exponent to obtain e^{-x}.
(b) Multiply the resulting formula by -1 to obtain $-e^{-x}$.
(c) Add 1 to the formula to obtain $1 - e^{-x}$.

Graph the three functions $y = e^{-x}$, $y = -e^{-x}$, and $y = 1 - e^{-x}$ first on a piece of paper and then in the same viewing screen.

Solution (a) If you know what effect each of these algebraic procedures has on the corresponding graphs of the functions, it is possible to picture the graph of $y = 1 - e^{-x}$ from the graph of $y = e^x$. Your graphics calculator can help you understand these graphing techniques. In the same viewing screen, graph $y = e^x$ and then $y = e^{-x}$.

Casio

1. Press | **GRAPH** | **SHIFT** | e^x | **ALPHA** | **x** | **EXE** |

2. Press | **GRAPH** | **SHIFT** | e^x | **–** | **ALPHA** | **x** | **EXE** |

TI-81

1. Press | **Y =** |

2. Press | **2nd** | e^x | **X|T** | **ENTER** |

   | **2nd** | e^x | **(–)** | **X|T** | **ENTER** |

3. Press | **GRAPH** |

Each function is the mirror image in the y-axis of the other function. To test yourself, graph $y = e^{-2x}$. Picture the mirror image in the y-axis of this function. Now graph $y = e^{2x}$, its mirror image, and see how close your prediction is.

(b) The graphs of $y = e^{-x}$ and $y = -e^{-x}$ are mirror images in the x-axis of each other. Graph $y = e^{-x}$ and then predict what the mirror image in the x-axis looks like. Then graph $y = -e^{-x}$.

(c) The graphs of $y = -e^{-x}$ and $y = 1 - e^{-x}$ are the same except the latter is one unit above the former. The geometric affect of algebraically adding 1 to the formula raises the graph one unit.

(d) Consult the following display to graph the three functions.

Casio

1. Press | **GRAPH** | **SHIFT** | e^x | **(−)** | **ALPHA** | **x** | **EXE** |

2. Press | **GRAPH** | **(−)** | **SHIFT** | e^x | **(−)** | **ALPHA** | **x** | **EXE** |

3. Press | **GRAPH** | 1 | **−** | **SHIFT** | e^x | **(−)** | **ALPHA** | **x** | **EXE** |

TI-81

1. Press | **Y =** |

2. Press | **2nd** | e^x | **(−)** | **X|T** | **ENTER** |

   | **(−)** | **2nd** | **Y-VARS** | 1 | **ENTER** |

   | 1 | **−** | **2nd** | **Y-VARS** | 2 | **ENTER** |

3. Press | **GRAPH** |

Exploration 2

To graph the function $y = 100 - e^{-0.5x}$ break it up into simpler parts. Construct the graph starting with $y = e^x$, and then $y = e^{-x}$, $y = e^{-0.5x}$, $y = -e^{-0.5x}$, and finally $1 - e^{-0.5x}$. The only graphing technique not covered in Exploration 1 is multiplying the exponent by a number, which in this case involves multiplying the exponent $-x$ by 0.5 to get $-0.5x$ as the exponent. Graph the two functions, $y = e^{-x}$ and $y = e^{-0.5x}$ in the same viewing screen and you will see that the effect of multiplying the exponent by a number "stretches" the graph. Try to predict the shape of each of these functions and then graph them in the same viewing screen:

1. $y = e^{-x}$
2. $y = e^{-0.5x}$
3. $y = -e^{-0.5x}$
4. $y = 10 - e^{-0.5x}$

You may have to alter the range to view the last graph.

Casio

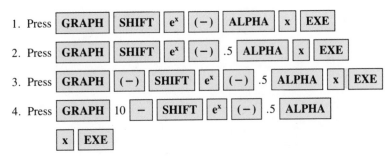

1. Press | **GRAPH** | **SHIFT** | e^x | **(−)** | **ALPHA** | **x** | **EXE** |

2. Press | **GRAPH** | **SHIFT** | e^x | **(−)** | .5 | **ALPHA** | **x** | **EXE** |

3. Press | **GRAPH** | **(−)** | **SHIFT** | e^x | **(−)** | .5 | **ALPHA** | **x** | **EXE** |

4. Press | **GRAPH** | 10 | **−** | **SHIFT** | e^x | **(−)** | .5 | **ALPHA** |

   | **x** | **EXE** |

TI-81

1. Press $\boxed{\text{Y} =}$

2. Press $\boxed{\text{2nd}}$ $\boxed{e^x}$ $\boxed{(-)}$ $\boxed{\text{X}|\text{T}}$ $\boxed{\text{ENTER}}$

   $\boxed{\text{2nd}}$ $\boxed{e^x}$ $\boxed{(-)}$ $\boxed{\cdot}$ 5 $\boxed{\text{X}|\text{T}}$ $\boxed{\text{ENTER}}$

   $\boxed{(-)}$ $\boxed{\text{2nd}}$ $\boxed{\text{Y-VARS}}$ 2 $\boxed{\text{ENTER}}$

   10 $\boxed{-}$ $\boxed{\text{2nd}}$ $\boxed{\text{Y-VARS}}$ 3 $\boxed{\text{ENTER}}$

3. Press $\boxed{\text{GRAPH}}$

Further Explorations

In Problems 1 to 5 sketch the graphs of the given functions in the same viewing screen and answer the accompanying question.

1. $y = 2^x$, $y = e^x$, $y = 3^x$. Where will the graph of $y = 4^x$ appear?
2. $y = 2^{-x}$, $y = e^{-x}$, $y = 3^{-x}$. Where will the graph of $y = 4^{-x}$ appear?
3. $y = e^x$, $y = e^{-x}$, $y = -e^{-x}$. Where will the graph of $y = 1 - e^{-x}$ appear?
4. $y = e^x$, $y = \ln x$. Where will the graph of $y = x$ appear?
5. $y = \ln x$, $y = \ln(x - 1)$, $y = -\ln(x - 1)$. Where will the graph of $y = 1 - \ln(x - 1)$ appear?
6. Graph the function $y = (1 + 1/x)^x$ and use the zoom feature and alterations of the range to show that $y = e$ is a horizontal asymptote. What does that say geometrically about the two functions?

The Integral

6

The case study connects calculus with fisheries. (H. Armstrong Roberts)

Chapter Overview

Calculus is divided into two branches, differential calculus, which was discussed in the previous four chapters, and integral calculus, which we introduce in this chapter. Just as the derivative of a function $f(x)$ is a new function derived from $f(x)$, so too the indefinite integral of $f(x)$ is a function computed from $f(x)$. The process of "finding the integral" of $f(x)$ is the reverse operation of differentiation. From a geometric point of view the derivative measures the slope of the tangent line, while the definite integral measures the area bounded by $f(x)$.

In the first section we study the reverse process of differentiation called antidifferentiation. A function whose derivative is equal to $f(x)$ is called an antiderivative of $f(x)$. Section 6–2 uses the limit process to calculate certain types of areas in the plane that are bounded by curves. Section 6–3 shows the close connection between antiderivatives and the areas studied in Section 6–2. The fourth section extends this concept of computing areas to more intricate regions.

CASE STUDY PREVIEW

What happens to a stable industry when a new, innovative production technique is introduced? How can the effect on the industry be measured? One method is called consumers' surplus, which is a measure of the gain to consumers as a result of the new mode of business. The Case Study centers on the effect on the Louisiana crawfish industry when fish farms using private ponds compete with standard fisheries using public watersheds, such as lakes and rivers. We see that the consumer gains via significant price decreases at the expense of the standard fisheries.

6–1 Antiderivatives

In this section several analogs of differentiation rules are presented. Each differentiation rule shows how to start with a function and then find its derivative. This section deals with the reverse operation: Start with the derivative and find a function that has it as its derivative. These new rules will be applied to the geometry of curves, the velocity of an object, and carbon-14 dating.

The differentiation rules considered in this section are the power rule, the exponential and logarithmic function rules, the sum of functions rule, and the constant multiple rule. Later sections will look at the analogs of various other differentiation rules.

Antiderivatives in Applications

There are many applications where the rate of change of a function is computed rather than computing the function directly. For instance, it is usually easier to compute the rate of increase of a population than the actual size of the population. In archaeology it is easier to estimate the rate of change of the aging process of items found at an ancient site, rather than computing the age of the site directly.

Suppose a firm computes that the marginal cost $MC(x)$ of a product is given by

$$MC(x) = x + 1$$

The marginal cost is the derivative of the cost function $C(x)$. Can $C(x)$ be computed from $MC(x)$? This is the type of question we answer in this section.

Antiderivatives

We know that the derivative of $f(x) = x^2$ is $2x$. Here we start with x^2 and compute $2x$ via the power rule. The reverse operation starts with the derivative, $2x$, and computes a function, x^2, whose derivative is $2x$. The process is called **antidifferentiation** and the function $f(x) = x^2$ is called an *antiderivative* of $2x$. In general antidifferentiation answers the question: Given a function $y = f(x)$, what is a function whose derivative is $f(x)$?

DEFINITION

Given a function $f(x)$, if $F'(x) = f(x)$, then $F(x)$ is an **antiderivative** of $f(x)$.

Example 1 illustrates not only the definition, but also why we said "an" instead of "the" antiderivative.

EXAMPLE 1

Problem

Verify that three antiderivatives of $f(x) = 3x^2$ are x^3, $x^3 + 1$, and $x^3 - 2$.

Solution The function x^3 is an antiderivative of $3x^2$ because $D_x x^3 = 3x^2$. The two other functions are also antiderivatives because $D_x(x^3 + 1) = D_x(x^3 - 2) = 3x^2$.

Example 1 uses the fact that the derivative of a constant function is 0. The converse of this is true and is a useful result. The proofs of this theorem and the next one are outlined in the exercises. Theorem 1 states that if a function has derivative 0, then it must be a constant function.

Theorem 1

If $F'(x) = 0$ for all x, then $F(x) = C$ for some real number C.

In Example 1 three functions are given whose derivatives are equal. Since the derivative measures the slope of the tangent line, the graphs of the three functions have the same slope at every point. See Figure 1.

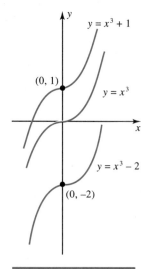

FIGURE 1

The three antiderivatives of $3x^2$ mentioned in Example 1 are similar because each is of the form x^3 plus a constant. In fact any function of the form $x^3 + C$, where C is a constant, is an antiderivative of $3x^2$.

In general if $F(x)$ is an antiderivative of $f(x)$, then $F(x) + C$ is also an antiderivative of $f(x)$. Are there other functions besides these that are antiderivatives? The next theorem shows that all antiderivatives of $f(x)$ are of the form $F(x) + C$. The constant C is called the **constant of integration.** It can be a positive number, a negative number, or zero.

Theorem 2

If $F(x)$ and $G(x)$ are both antiderivatives of $f(x)$, then there is a constant C such that

$$F(x) = G(x) + C$$

Another way to interpret this theorem is to say that any two antiderivatives of the same function differ by a constant.

The Indefinite Integral

Theorem 2 states that once one antiderivative $F(x)$ of $f(x)$ is known, then all antiderivatives are of the form $F(x) + C$. The **integral sign** $\int$ is used to express this.

$$\int f(x)\,dx = F(x) + C$$

This is called the general form of the antiderivative of $f(x)$. It states that the antiderivatives of $f(x)$ are the functions $F(x) + C$. The integral is similar to the derivative, $\dfrac{dy}{dx}$, in that it indicates an operation on the function. The meaning of the notation $\dfrac{dy}{dx}$ is to find the derivative, while the meaning of the notation $\int f(x)\,dx$ is to find all antiderivatives of $f(x)$. The term "dx" serves the same purpose in each notation. It indicates what the variable is. Thus $\int 2x\,dx$ is different from $\int 2x\,dt$. The expression $\int f(x)\,dx$ is called the **indefinite integral,** and $f(x)$ is called the **integrand.** To find the indefinite integral means to find all antiderivatives of the integrand.

Example 2 shows how to use this notation when finding the indefinite integral of another power function.

EXAMPLE 2

Problem

Find the indefinite integral $\displaystyle\int x^3\,dx$.

$\dfrac{x^4}{4} + C$

Solution Work backward as before. The function whose derivative is almost x^3 is x^4. But the derivative of x^4 is $4x^3$; that is, $D_x(x^4) = 4x^3$. Recall that the derivative of a constant times a function is the constant times the derivative. This means that the differentiation statement above can be divided by 4 to get

$$\frac{1}{4}D_x(x^4) = D_x\left(\frac{x^4}{4}\right) = x^3$$

Therefore

$$\int x^3\,dx = \frac{x^4}{4} + C$$

$$x^n = \frac{x^{n+1}}{n+1}$$

$$y^3 = \frac{x^{3+1}}{3+1} = \frac{x^4}{4} + C$$

The Power Rule for Antiderivatives

Example 2 illustrates how to compute the indefinite integral of all power functions x^n, except $n = -1$. To find the function whose derivative is x^n, start with x^{n+1}. The derivative of x^{n+1} has a factor of $n + 1$, so divide the function by that factor. Thus an antiderivative of x^n is $\dfrac{x^{n+1}}{n+1}$. This is the power rule for antiderivatives. It holds for all real numbers n except -1, because $n = -1$ produces division by 0.

The Power Rule

$$\int x^n \, dx = \frac{x^{n+1}}{n+1} + C \qquad n \neq -1$$

EXAMPLE 3

Problem

Find the indefinite integrals (a) $\int x^{-4} \, dx$, (b) $\int \sqrt{x} \, dx$, (c) $\int dx$.

Solution (a) Use the power rule with $n = -4$.

$$\int x^{-4} \, dx = \frac{x^{-4+1}}{-4+1} + C = \frac{x^{-3}}{-3} + C = -\frac{1}{3} x^{-3} + C$$

(b) Use the power rule with $n = 1/2$.

$$\int x^{1/2} \, dx = \frac{x^{(1/2)+1}}{1/2+1} + C = \frac{x^{3/2}}{3/2} + C = \frac{2x^{3/2}}{3} + C$$

(c) The integrand is understood to be $1 = x^0$. From the power rule with $n = 0$ we have

$$\int dx = \int x^0 \, dx = \frac{x^{0+1}}{0+1} + C = x + C$$

Exponential and Logarithmic Functions

The derivative rules for exponential and logarithmic functions can also be expressed as indefinite integrals. Recall the rules

$$D_x(\ln |x|) = \frac{1}{x} \qquad D_x e^{rx} = r e^{rx}$$

The power rule for antiderivatives excludes $n = -1$, so it does not give the antiderivatives of x^{-1}. However, since $D_x(\ln x) = 1/x = x^{-1}$, the general form of the antiderivative of x^{-1} is expressed in the logarithmic function rule.

Logarithmic Function Rule

Antiderivative of $f(x) = \dfrac{1}{x} = x^{-1}$:

$$\int x^{-1} \, dx = \ln |x| + C$$

From the formula $D_x(e^{rx}) = re^{rx}$ divide by r to get the exponential function rule.

> **Exponential Function Rule**
> Antiderivative of $f(x) = e^{rx}$:
>
> $$\int e^{rx}\, dx = \frac{e^{rx}}{r} + C$$

EXAMPLE 4

Problem

Find the indefinite integral of $f(x) = e^{5x}$.

Solution From the antiderivative rule of $f(x) = e^{rx}$ with $r = 5$ we have

$$\int e^{5x}\, dx = \frac{e^{5x}}{5} + C$$

Addition and Constant Rules

The first three antidifferentiation rules presented so far can be derived by working backward from their corresponding differentiation rules. In a similar way the formulas that state (a) "the derivative of a sum is the sum of the derivatives" and (b) "the derivative of a constant times a function is the constant times the derivative of the function" can be expressed as indefinite integrals.

> **The Sum Rule**
>
> $$\int [f(x) + g(x)]\, dx = \int f(x)\, dx + \int g(x)\, dx$$

> **The Constant Times a Function Rule**
>
> $$\int kf(x)\, dx = k \int f(x)\, dx \qquad \text{for any constant } k$$

A function with more than one term can be antidifferentiated term by term. It can be extended to a finite number of functions, so that

$$\int [f_1(x) + \cdots + f_n(x)]\, dx = \int f_1(x)\, dx + \cdots + \int f_n(x)\, dx$$

The constant times a function rule states that a constant can be moved through the integral sign in the same way that it can be moved through a derivative sign. The next two examples illustrate these rules.

EXAMPLE 5

Problem

Find the indefinite integrals (a) $\int (x^4 + e^{-x})\, dx$, (b) $\int \left(3x^{-5} - \dfrac{4}{x}\right) dx$.

Solution

(a) $\int (x^4 + e^{-x})\, dx = \int x^4\, dx + \int e^{-x}\, dx = \dfrac{x^{4+1}}{4+1} + \dfrac{e^{-x}}{-1} + C$

$$= \dfrac{x^5}{5} - e^{-x} + C$$

(b) $\int \left(3x^{-5} - \dfrac{4}{x}\right) dx = \int 3x^{-5}\, dx + \int \left(\dfrac{-4}{x}\right) dx$

$$= 3\int x^{-5}\, dx + (-4)\int \left(\dfrac{1}{x}\right) dx$$

$$= \dfrac{3x^{(-5)+1}}{-5+1} - 4\int \dfrac{1}{x}\, dx + C$$

$$= \dfrac{-3x^{-4}}{4} - 4\ln |x| + C$$

Sometimes algebraic manipulations can be used to express the integrand in a form that matches the antidifferentiation rules. The next example demonstrates one such manipulation.

EXAMPLE 6

Problem

Find the indefinite integral $\int \sqrt{x}(x^2 - 1)\, dx$.

Solution The integrand does not match any of those in the antidifferentiation rules, so an algebraic simplification is called for. Multiply through the parentheses to get

$$\int \sqrt{x}(x^2 - 1)\, dx = \int (\sqrt{x}\, x^2 - \sqrt{x})\, dx = \int (x^{5/2} - x^{1/2})\, dx$$

$$= \int (x^{5/2})\, dx - \int (x^{1/2})\, dx = \dfrac{x^{7/2}}{7/2} - \dfrac{x^{3/2}}{3/2} + C$$

$$= \dfrac{2x^{7/2}}{7} - \dfrac{2x^{3/2}}{3} + C$$

EXAMPLE 7

Problem

Find the function whose slope of the tangent line at x is $f'(x) = 3x^2 - 1$ and whose graph passes through the point $(0, 1)$.

Solution The first step is to find the indefinite integral of $f'(x)$. Use the fact that $f(x)$ is the antiderivative of $f'(x)$.

$$f(x) = \int f'(x)\, dx = \int (3x^2 - 1)\, dx = x^3 - x + C$$

This gives all functions whose slope is the given formula. To find which of these functions also passes through $(0, 1)$, substitute $f(0) = 1$ into the formula.

$$1 = f(0) = 0^3 - 0 + C = C$$

The function whose constant of integration is $C = 1$ is the solution, which is $f(x) = x^3 - x + 1$.

Recall that a production function $P(x)$ measures the production level P, or the output, of a firm using x amount of labor. The derivative of $P(x)$ is called marginal product, $MP(x)$. Antiderivatives can be used to compute the production function from the marginal product.

EXAMPLE 8

Problem

A firm determines that its marginal product $MP(x)$ is given by

$$MP(x) = x^2 - 10x + 20$$

where $MP(x)$ is measured in thousands of dollars and x in hundreds of workers. Compute the production function.

Solution Since $P'(x) = MP(x)$, we have

$$P(x) = \int MP(x)\, dx$$

$$= \int (x^2 - 10x + 20)\, dx = \frac{x^3}{3} - 5x^2 + 20x + C$$

Similar statements hold for marginal cost, marginal revenue, and marginal profit. If $R(x)$ is the revenue function of a firm, then $R'(x)$ is the marginal revenue. $R(x)$ can be computed from marginal revenue by the formula

$$R(x) = \int R'(x)\, dx$$

The Growth Equation

Many interesting applications of calculus lead to equations involving derivatives called **differential equations.** In fact whenever an antiderivative is computed we are solving a differential equation of the form $f'(x) = g(x)$, where $g(x)$ is given and $f(x)$ is unknown. A common and powerful type of differential equation is

$$f'(x) = kf(x)$$

for some constant k. It is called the **growth equation.** If k is positive, then k is called the *growth constant.* If k is negative, then k is called the *decay constant.* The next two examples illustrate an application from archaeology, where the growth equation plays a key role.

   The rate at which radioactive material changes to lead is proportional to the amount of the material present; that is, if $f(t)$ is the amount present, then $f'(t) = kf(t)$ for some constant k. Hence $f(t)$ is a function whose derivative is a constant times itself. The only type of function with that property is $f(t) = Ce^{kt}$. To check that $f(t)$ satisfies the differential equation, we compute $f'(t) = Cke^{kt}$, which is equal to $kf(t)$ as needed.

EXAMPLE 9

Living plant and animal tissues contain the radioactive substance carbon-14, which has a half-life of 5568 years (at the end of that time one-half of that amount remains). When the organism dies, no new carbon is received.

Problem

Find a formula for the amount of carbon-14 remaining after t years.

Solution From above, $f(t) = Ce^{kt}$. Since $f(0) = Ce^0 = C$, the amount present at time $t = 0$ is $C = f(0)$. When $t = 5568$, $f(0)/2$ is present, and thus

$$f(5568) = f(0)e^{5568k} = \frac{f(0)}{2}$$

Therefore $e^{5568k} = 1/2$, and so $\ln e^{5568k} = \ln 1/2$, which gives $5568k = \ln 1/2$ and $k = (1/5568)\ln 1/2$. Therefore

$$f(t) = Ce^{(1/5568)(\ln 1/2)t}$$

Example 10 illustrates one way that archaeologists use the function in Example 9 to date uncovered ruins.

EXAMPLE 10

Suppose an analysis of ashes found in an unearthed settlement reveals that 1/5 of the carbon-14 present in the original ashes has decomposed, so that 4/5 of $f(0)$ remains.

Problem

How old is the settlement?

Solution The task is to find t such that $f(t) = (4/5)f(0)$. Then

$$(4/5)f(0) = f(0)e^{kt} \qquad \text{implies} \qquad 4/5 = e^{kt}$$

Hence $\ln 4/5 = \ln e^{kt} = kt$. Thus

$$t = \left(\frac{1}{k}\right) \ln 4/5 = \frac{5568(\ln 4/5)}{\ln 1/2} = \frac{5568(\ln 4 - \ln 5)}{\ln 1 - \ln 2}$$

$$\approx \frac{5568(1.386 - 1.609)}{(0 - 0.693)}$$

$$\approx 1792$$

Thus the settlement is approximately 1792 years old.

EXERCISE SET 6–1

In Problems 1 to 24 find the antiderivative.

1. $\int (3x + 5)\, dx$

2. $\int (4x - 2)\, dx$

3. $\int (2x^4 - 7)\, dx$

4. $\int (3x^5 + 1)\, dx$

5. $\int (6x^{-2} + 5x)\, dx$

6. $\int (5x^{-3} - x^3 + 1)\, dx$

7. $\int (2x^{1/2} + 3x^{1/3} - 5x)\, dx$

8. $\int (5x^{1/5} - x^{2/3} + 7x)\, dx$

9. $\int (3x^{-1/3} + 7x^{-1/4} - 5)\, dx$

10. $\int (6x^{-1/6} - 3x^{-3/7} - 6x)\, dx$

11. $\int (x^{0.2} + 3x^{1.4})\, dx$

12. $\int (1.2x^{-1.6} - 4.1x^{-2.2})\, dx$

13. $\int (1/x^2 + 5/x^4)\, dx$

14. $\int (2/x^{0.5} - 4/x^{2.2})\, dx$

15. $\int (1/x^7 + 4/x^{1/4} + 4x^{1/4})\, dx$

16. $\int (3/x^{2.2} - 3/x^{2/5} - 2x^{2/5})\, dx$

17. $\int (3e^{2x} + 5/x)\, dx$

18. $\int (5e^{3x} - 2/x)\, dx$

19. $\int (x^3 + 2e^{-x} + x^{1/2} - 5/x)\, dx$

20. $\int (3e^{-4x} - 4x^{-3} - 15/x + x^{-1/2})\, dx$

21. $\int (5\sqrt{x} - 2e^{-4x} + 1/x)\, dx$

22. $\int (8/x + 2/\sqrt{7x} + 3e^{-2x})\, dx$

23. $\int (18/x + 2/\sqrt[3]{4x} - e^{7x})\, dx$

24. $\int (8/\sqrt[4]{3x} + 2/x - e^{-10x})\, dx$

In Problems 25 to 28 find the function whose slope of the tangent line to the curve $y = f(x)$ is given and whose graph passes through the given point.

25. $f'(x) = x^2 - 2$ at $(1, -1)$

26. $f'(x) = 2x^2 - 1$ at $(1, 1)$

27. $f'(x) = 4x^{-1} - 2x$ at $(2, -2)$

28. $f'(x) = 2x^{-2} - x$ at $(1, 1)$

In Problems 29 to 34 find $\int f(x)\, dx$ by using algebra to express the integrand in a form that fits the integration rules.

29. $\int (2x^4 - 7x)/x^2\, dx$ 30. $\int (3x^5 + 10x)/x^3\, dx$

31. $\int x^2(6x^{-2} + 5x)\, dx$

32. $\int x^3(5x^{-3} - x^3 + 1)\, dx$

33. $\int (2x^4 - 7x)^2\, dx$ 34. $\int (3x^5 + 10x)^2\, dx$

In Problems 35 and 36 find the position function $s(t)$ given the velocity function $v(t)$ and the initial position $s(0)$.

35. $v(t) = t^2 + 2t$, $s(0) = 0$

36. $v(t) = t^3 + t$, $s(1) = 1$

37. Find the age of a settlement where ashes found at the site are assumed to be 2/3 of the original amount; that is, 1/3 of the original amount has decomposed.

38. Do Problem 37 with the fraction 2/3 replaced by 3/4.

In Problems 39 to 42 determine the decay constant k so that the function $f(x) = Ce^{kx}$ is a solution to the differential equation $f'(x) = kf(x)$ where $f(x)$ is the amount of a radioactive substance remaining after x years and where the half-life of the substance is given.

39. The half-life is 1000 years.

40. The half-life is 100 years.

41. The half-life is 10 years.

42. The half-life is 1 year.

43. Give a reason for each step in the following proof of Theorem 1:

    Proof of Theorem 1: Suppose $F'(x) = 0$ for all x. Then the graph of $F(x)$ has slope 0 at every point. Thus the tangent line is horizontal at every point. This means that the graph of $F(x)$ is a horizontal line, since a horizontal line is the only graph with that property.

44. Give a reason for each step in the following proof of Theorem 2:

    Proof of Theorem 2: Suppose $F(x)$ and $G(x)$ are two antiderivatives of $f(x)$. We want to show that the function $F(x) - G(x)$ is a constant function, since then $F(x) - G(x) = C$ for some C and so $F(x) = G(x) + C$. Consider the function $H(x) = F(x) - G(x)$. Its derivative is

    $$H'(x) = F'(x) - G'(x) = f(x) - f(x) = 0$$

    Therefore $H(x)$ must be a constant function by Theorem 1. Hence $F(x) - G(x) = C$ and so $F(x) = G(x) + C$.

Referenced Exercise Set 6–1

1. In order to illustrate the application of dating methods in archaeology covered in his book, Joseph Michels discusses one of the archaeological expeditions he organized with students from the Pennsylvania State University.* The site was Sheep Rock Shelter in Huntingdon County, PA. For thousands of years, the shelter, located on the Juniata River, provided small platforms for various living mammals. Artifacts were found in the rock strata as deep as 14 feet below the present surface. In one level it was determined that of the original amount of carbon-14 contained in certain artifacts, 42% remained. Find the age of the artifacts.

In Problems 2 and 3 an article is quoted that discusses growth rates. Find the growth equation, $f'(x) = kf(x)$ that is satisfied by the function $f(x)$ given in the article.

2. In an article studying the causes and effects of wind damage to trees such as sugar maples, King investigates the factor influencing wind damage due to the rapid decline in wind speed beneath the canopy, or top, of the forest.† The relationship between the wind speed $V(h)$ and the height h of the wind is given by

   $$V(h) = 22e^{1.4h}$$

3. Because the bald eagle is an endangered species, the study of the growth rates of sibling eagles is very important to conservationists. In an article investigating the growth rates of nestlings, or broods, with one versus more than one young eagle, Bortolotti calculates the following growth equation that measures the difference in the size of two young eagles in the same nestling, born one day apart.‡ Let m be the difference in the mass, measured in percent, at time t, measured in days. The solution of the growth equation is

   $$m(t) = 38.1e^{-.068t}$$

*Joseph W. Michels, *Dating Methods in Archaeology,* New York, Seminar Press, 1973.

†David A. King, "Tree Form, Height Growth, and Susceptibility to Wind Damage in *Acer saccharum*," *Ecology,* Vol. 67, 1986, pp. 980–990.

‡Gary R. Bortolotti, "Evolution of Growth Rates in Eagles: Sibling Competition vs. Energy Considerations," *Ecology,* Vol. 67, 1986, pp. 182–194.

6–2 Area and the Definite Integral

One application of the definite integral is the measure of the area under a curve, which appears in many real-world problems. Areas of certain types of regions can be found using only geometry, such as squares, rectangles, and triangles. The classical Greeks even knew how to find the area of some regions bounded by specific curves. But the general problem of finding the area under an arbitrary continuous curve was not solved until about 1700 with the advent of calculus.

An Application from Economics

The area under a curve has a wide variety of uses in many disciplines, one of which is economics. One application analyzes the distribution of wealth and income in a society. For centuries economists, politicians, and social scientists have debated the question, "What constitutes a fair and equitable distribution of income and wealth?" Wealth is the value of the goods and property people own, while income is the gain derived from the use of human and material resources. For measuring inequalities in the distribution of income and wealth the most commonly used device is a graph called a **Lorentz diagram.** A Lorentz diagram illustrates what percentage of families received what percentage of a nation's total income. It is used to compare the income distribution of different countries as well as to evaluate an individual country's system.

Consider the Lorentz diagram for families in the United States in Figure 1, which is adapted from *Contemporary Economics* by Milton H. Spenser.* There are two curves in a Lorentz diagram. The lower one, called a *Lorentz curve,* measures the percentage of a nation's total income that is shared by those families with a certain income level. As an example of how to interpret the curve, consider $x = 40$ on the horizontal axis. It denotes those families with the 40% lowest family incomes. The vertical axis represents percentages of the total income of the United

*Milton H. Spenser, *Contemporary Economics*, New York, Worth Publishers, Inc., 1986, pp. 53–55.

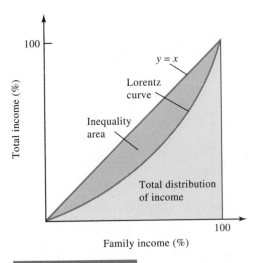

FIGURE 1 *Source:* Milton H. Spenser, *Contemporary Economics*, New York, Worth Publishers, Inc., 1986, pp. 53–55.

States. Corresponding to $x = 40$ on the Lorentz curve is $y = 16$, which means that the families with the lowest 40% of family incomes have 16% of the total income of the United States. The second curve in the diagram is $y = x$. It represents complete equality of distribution of income, because if income were distributed equally, families with the lowest x% of income would have $y = x$% of the total income. Economists call the area between these two curves the *inequality area* because it represents the extent of departure between an equal distribution of income and the actual distribution of income. The area of the region bounded above by the Lorentz curve and below by the x-axis is called the total distribution of income. The area of the region bounded above by $y = x$ and below by the x-axis is the area of complete equality of income distribution.

In this section we study how to compute the latter two areas. Each is the area of a region bounded above by a curve and below by the x-axis. Section 6–4 will show how to evaluate the area between two curves. Then we will compute the inequality income area for the United States.

Area Under a Curve

To find the area under a curve, we approximate the area by using rectangles whose area can be easily computed. The process entails fitting rectangles into the region. The first approximation is often not very close to the actual area. Better approximations are usually found by using more rectangles. The actual area is found by using a limit process.

We start by finding an initial approximation to the area under the curve $f(x) = x^2$ bounded by $x = 0$ to $x = 2$ and above the x-axis. Figure 2 shows this region. A rough approximation uses two rectangles as in Figure 3. Divide the interval $[0, 2]$ into two subintervals of equal width and choose the right-hand end point to calculate the height of each rectangle. Each width is 1 and the two heights are $f(1) = 1^2 = 1$ and $f(2) = 2^2 = 4$. Thus the sum of the areas of the two

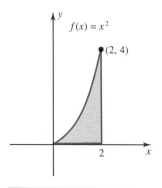

FIGURE 2

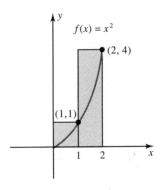

$f(x) = x^2$

(2, 4)

(1,1)

1   2   x

FIGURE 3

rectangles is

$$1 \cdot f(1) + 1 \cdot f(2) = 1 + 4 = 5$$

Figure 3 shows that this approximation is greater than the actual area, but it illustrates how to compute further approximations that are closer. To approximate the area of the region bounded by $y = f(x)$ (for now we must assume that $f(x) \geq 0$), the x-axis, $x = a$, and $x = b$, first divide the interval $[a, b]$ into n equal subintervals, each with width $(b - a)/n$. Evaluate $f(x)$ at the right-hand end point to find the height of each rectangle. Then the approximation is the sum of the areas of these rectangles. Example 1 illustrates this procedure.

EXAMPLE 1

Problem

Find the approximation of the area of the region bounded by $f(x) = x^2$, $x = 0$, $x = 2$, and the x-axis, using four subintervals.

Solution The width of each subinterval is $(2 - 0)/4 = 1/2$. The right-hand end points are 1/2, 1, 3/2, and 2. The approximation is the sum of the areas of the rectangles. (See Figure 4.)

$$\frac{1}{2} \cdot f(1/2) + \frac{1}{2} \cdot f(1) + \frac{1}{2} \cdot f(3/2) + \frac{1}{2} \cdot f(2)$$

$$= \frac{1}{2} \cdot \frac{1}{4} + \frac{1}{2} \cdot 1 + \frac{1}{2} \cdot \frac{9}{4} + \frac{1}{2} \cdot 4 = \frac{1}{2}\left(\frac{1}{4} + 1 + \frac{9}{4} + 4\right)$$

$$= \left(\frac{1}{2}\right)\left(\frac{30}{4}\right) = \frac{15}{4}$$

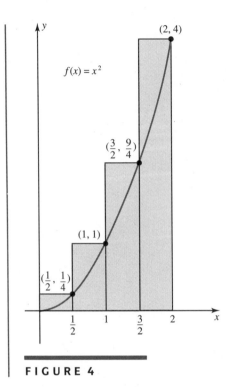

FIGURE 4

Area and the Definite Integral

From Figure 4 it is clear that the approximation of 15/4 is still greater than the area, but it is closer than the previous approximation of 5. To get even closer approximations, use more subintervals with smaller widths. In general the process of approximating the area entails dividing $[0, 2]$ into n subintervals and computing the sum of the areas of the rectangles. Then we let n get large to see if the process has a limiting value. Each subinterval has width $(2 - 0)/n = 2/n$. The right-hand end points are $2/n$, $4/n$, . . . , 2. For example, if $n = 8$, the end points are 1/4, 1/2, 3/4, 1, 5/4, 3/2, 7/4, 2. A computer was used to find the approximations to the area for the following values of n:

n	2	4	8	100	1000	10000	100000
area	5	3.75	3.188	2.707	2.671	2.667	2.6667

The areas of the approximations approach the number 8/3. In fact the larger the number of rectangles that are used, the closer the approximation is to 8/3. This means that as n increases without bound, the limit is equal to 8/3. This is expressed as

$$\lim_{n \to \infty} (\text{sum of the areas of the } n \text{ rectangles}) = \frac{8}{3}$$

Therefore the area of the region bounded by $f(x) = x^2$, the x-axis, $x = 0$, and $x = 2$ is equal to 8/3. This is what is meant by the **definite integral** of $f(x) = x^2$ from $x = 0$ to $x = 2$. It is written as

$$\int_0^2 x^2 \, dx = \frac{8}{3}$$

The function $f(x) = x^2$ is called the *integrand*. The number 2 is called the **upper limit of integration** and the number 0 is called the **lower limit of integration.**

For any continuous function $f(x)$, with $f(x) \geq 0$ in $[a, b]$, as the number n of subintervals of $[a, b]$ gets larger, the corresponding approximations get closer to the area under the curve. This can be seen from Figure 5. The limit of these approximations is the definite integral of $f(x)$ from a to b.

DEFINITION

If $f(x)$ is continuous on the interval $[a, b]$ and $[a, b]$ is divided into n subintervals whose right-hand end points are $x_1, x_2, \ldots , x_n$, then the *definite integral of $f(x)$ from a to b* is

$$\int_a^b f(x) \, dx = \lim_{n \to \infty} \frac{(b - a)}{n} [f(x_1) + \cdots + f(x_n)]$$

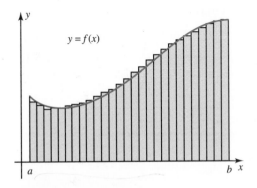

FIGURE 5

The next example shows how this limit behaves with a linear function as the integrand. We use the fact that

$$1 + 2 + \cdots + n = \frac{n(n + 1)}{2}$$

EXAMPLE 2

Problem

Find $\int_0^4 2x \, dx$.

Solution Divide $[0, 4]$ into n subintervals, each of width $4/n$. The right-hand end points are $4/n$, $8/n$, ..., 4. From the definition with $a = 0$ and $b = 4$ we have

$$\int_0^4 2x \, dx = \lim_{n \to \infty} \frac{(b - a)}{n} [f(x_1) + \cdots + f(x_n)]$$

$$= \lim_{n \to \infty} \frac{(4 - 0)}{n} \left[f\left(\frac{4}{n}\right) + f\left(\frac{8}{n}\right) + \cdots + f(4) \right]$$

$$= \lim_{n \to \infty} \frac{4}{n} \left[\frac{8}{n} + \frac{16}{n} + \cdots + 8 \right] = \lim_{n \to \infty} \frac{4}{n} \left(\frac{8}{n}\right) [1 + 2 + \cdots + n]$$

$$= \lim_{n \to \infty} \frac{32}{n^2} \left[\frac{n(n + 1)}{2} \right] = \lim_{n \to \infty} \frac{16(n + 1)}{n} = \lim_{n \to \infty} \left(16 + \frac{16}{n} \right) = 16$$

Thus $\int_0^4 2x \, dx = 16$.

The definite integral is defined in terms of a limit process. The definition does not stipulate that the function $f(x)$ be nonnegative in the interval, but when the definite integral is linked with the area under a curve this condition is important. When $f(x) \geq 0$ in $[a, b]$, then the area of the region bounded by $f(x)$, the x-axis, $x = a$, and $x = b$ is equal to the definite integral, $\int_a^b f(x) \, dx$. But if $f(x)$ is less than 0 in the interval, so that its graph is below the x-axis somewhere in the interval, then the two are not equal. The next section examines this situation.

DEFINITION

If $f(x) \geq 0$ in $[a, b]$, then the area bounded by $f(x)$, the x-axis, $x = a$, and $x = b$ is defined by

$$\int_a^b f(x) \, dx$$

A more general definition of the definite integral can be used that does not restrict the subintervals to have equal length. Also, the height of each rectangle can be computed by using any point in the subinterval, rather than the right-hand end point. The definition given here is equivalent to the less restrictive one and is easier to use. The graphics calculator exploration in Chapter 7 investigates this idea further and compares the different methods.

E X A M P L E   3

Problem

Compute $\int_0^1 (3x + 1)\, dx$.

Solution Divide $[0, 1]$ into n subintervals, each of width $1/n$. The right-hand end points are $1/n, 2/n, \ldots, 1$. From the definition with $a = 0$ and $b = 1$ we have

$$\int_0^1 (3x + 1)\, dx = \lim_{n \to \infty} \frac{(b - a)}{n} [f(x_1) + \cdots + f(x_n)]$$

$$= \lim_{n \to \infty} \frac{(1 - 0)}{n} \left[f\left(\frac{1}{n}\right) + f\left(\frac{2}{n}\right) + \cdots + f(1) \right]$$

$$= \lim_{n \to \infty} \frac{1}{n} \left[3\left(\frac{1}{n}\right) + 1 + 3\left(\frac{2}{n}\right) + 1 + \cdots + 3(1) + 1 \right]$$

The term "1" appears n times and we can factor 3 and $1/n$ from each of the remaining terms to get

$$= \lim_{n \to \infty} \frac{1}{n} \left[\frac{3}{n} (1 + 2 + \cdots + n) + n \right]$$

Now use the identity for $1 + 2 + \cdots + n$.

$$= \lim_{n \to \infty} \frac{1}{n} \left[\frac{3}{n} \frac{n(n + 1)}{2} + n \right] = \lim_{n \to \infty} \left[\frac{3(n + 1)}{2n} + \frac{n}{n} \right]$$

$$= \lim_{n \to \infty} \left[\frac{3}{2} \frac{(n + 1)}{n} + 1 \right] = \frac{3}{2} + 1 = \frac{5}{2}$$

Thus

$$\int_0^1 (3x + 1)\, dx = \frac{5}{2}$$

$\frac{3x^2}{2} + x = \frac{3}{2} + \frac{2}{2}$

$\frac{5}{2}$

Lorentz Diagrams Revisited

The area of complete equality of income distribution in a Lorentz diagram is the area of the region bounded by $y = x$ and the x-axis. It is the definite integral

$$\int_0^1 x\, dx$$

We use the definition to compute this area. Divide [0, 1] into n subintervals, each of width $1/n$. The right-hand end points are $1/n, 2/n, \ldots, 1$. With $a = 0$, $b = 1$, and $f(x) = x$, we have

$$\int_0^1 x \, dx = \lim_{n \to \infty} \frac{(b - a)}{n} [f(x_1) + \cdots + f(x_n)]$$

$$= \lim_{n \to \infty} \frac{(1 - 0)}{n} \left[f\left(\frac{1}{n}\right) + f\left(\frac{2}{n}\right) + \cdots + f(1) \right]$$

$$= \lim_{n \to \infty} \left(\frac{1}{n}\right) \left[\frac{1}{n} + \frac{2}{n} + \cdots + 1 \right]$$

Factoring $1/n$ from each term, we get

$$= \lim_{n \to \infty} \left(\frac{1}{n^2}\right) [1 + 2 + \cdots + n]$$

Now use the identity for $1 + 2 + \cdots + n$.

$$= \lim_{n \to \infty} \left(\frac{1}{n^2}\right) \left[\frac{n(n + 1)}{2} \right] = \lim_{n \to \infty} \frac{n + 1}{2n} = \frac{1}{2}$$

Thus

$$\int_0^1 x \, dx = \frac{1}{2}$$

This is the area of complete equality of income distribution. A country's actual income distribution is compared with it. For income in the United States the Lorentz curve given in Spenser can be approximated by the curve $y = x^2$. (See Figure 1.) The area of the region bounded by this curve and the x-axis is called the area of total income distribution. For this Lorentz curve it is

$$\int_0^1 x^2 \, dx$$

In the next section we will learn how to compute this area from the fundamental theorem of calculus.

A Few Concluding Remarks

The introduction of this section states that the ancient Greeks knew how to find the area under a curve for several specific curves. One such type of curve is a straight line, the graph of a linear function. This is because the area under a linear function can be divided into triangles and rectangles whose areas are easily computed. For instance, Example 3 shows that the definite integral of the linear function $f(x) = 3x + 1$ from $x = 0$ to $x = 1$ is 5/2. In the interval [0, 1] the function $f(x) = 3x + 1 > 0$, so the value of the definite integral should be equal to the area under the curve. From Figure 6 the area under the curve can be divided into a square with area 1 and a triangle with area $(1/2)(1)(3) = 3/2$. So the area under the curve

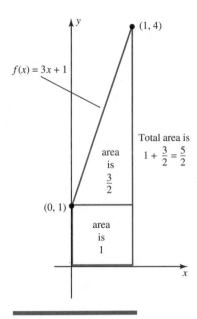

$(1, 4)$

$f(x) = 3x + 1$

Total area is

$1 + \dfrac{3}{2} = \dfrac{5}{2}$

area is $\dfrac{3}{2}$

$(0, 1)$

area is 1

x

FIGURE 6

is $1 + 3/2 = 5/2$, which is the same as the definite integral computed in the example.

In a similar way consider the Lorentz diagram in Figure 1: The area of total income distribution is the area of the region bounded by the linear function $y = x$. This region is a triangle and its area is $(1/2)(1)(1) = 1/2$, which is the same number computed from the definition. The definition of area is much more difficult to apply for nonlinear functions. The next section will give us an easier way to compute definite integrals.

The notation for the indefinite integral and the definite integral are similar and it is easy to confuse them. In the next section we will discover the close connection between the two concepts—basically the indefinite integral can be used to evaluate the definite integral. Remember that the definite integral $\displaystyle\int_a^b f(x)\,dx$ is a real number, whereas the indefinite integral $\int f(x)\,dx$, whose notation does not include limits of integration, is the set of functions consisting of all antiderivatives, $F(x) + C$ for $F(x)$ a particular antiderivative of $f(x)$.

EXERCISE SET 6–2

In Problems 1 to 6 let $f(x) = x^2$. For the given choice of n, use n rectangles to approximate the area under the curve $y = f(x)$ from $x = a$ to $x = b$.

1. $n = 2, \quad a = 1, \quad b = 2$
2. $n = 2, \quad a = 1, \quad b = 4$
3. $n = 4, \quad a = 1, \quad b = 2$
4. $n = 4, \quad a = 1, \quad b = 4$
5. $n = 6, \quad a = 0, \quad b = 3$
6. $n = 8, \quad a = 0, \quad b = 4$

In Problems 7 to 12 fill in the following table for the given choice of $f(x)$ by dividing $[0, 2]$ into n subintervals and computing the sum of the areas of the rectangles.

n	2	4	8	10
area				

7. $f(x) = 2x + 1$
8. $f(x) = 3x + 1$
9. $f(x) = x^2 + 1$
10. $f(x) = x^2 + 2$
11. $f(x) = 2x^2 + 2$
12. $f(x) = 2 + x^2$

In Problems 13 to 16 use the definition to find the definite integral.

13. $\int_0^4 3x \, dx$
14. $\int_0^8 2x \, dx$

15. $\int_0^4 (2x + 1) \, dx$
16. $\int_0^8 (3x + 1) \, dx$

In Problems 17 to 20 find the area under $f(x)$ from $x = a$ to $x = b$.

17. $f(x) = 2x + 1$, $a = 0$, $b = 2$
18. $f(x) = 3x - 1$, $a = 0$, $b = 1$
19. $f(x) = 3 - 2x$, $a = 0$, $b = 1$
20. $f(x) = 7 - 3x$, $a = 0$, $b = 2$

In Problems 21 to 24 use the definition to find the definite integral.

21. $\int_3^4 3x \, dx$
22. $\int_6^8 2x \, dx$

23. $\int_{-2}^0 (1 - 2x) \, dx$
24. $\int_7^8 (3x - 1) \, dx$

In Problems 25 to 32 use the definition to find the definite integral by using the identity

$$1^2 + 2^2 + \cdots + n^2 = \frac{n(n + 1)(2n + 1)}{6}$$

25. $\int_0^4 x^2 \, dx$
26. $\int_{-2}^2 2x^2 \, dx$

27. $\int_0^4 (2x^2 + 1) \, dx$
28. $\int_1^5 (3x^2 - 1) \, dx$

29. $\int_0^1 (x^2 + x) \, dx$
30. $\int_0^1 (3x^2 + x) \, dx$

31. $\int_1^3 \frac{1}{x} \, dx$
32. $\int_1^4 \frac{2}{x} \, dx$

33. Use the definition to show that $\int_a^b c \, dx = c(b - a)$ for the constant function $f(x) = c$.

34. Verify the result of Example 3 geometrically.

35. In Problem 23 verify that the definite integral is the area under the curve by finding the area geometrically.

Referenced Exercise Set 6–2

1. In an article studying the "well-being" of farm families, Kinsey* quotes one of the original articles† in economics that applied Lorentz curves to income inequality. The latter paper by Carlin and Reinsel constructed a Lorentz curve for the family income of farmers. It can be approximated by

$$y = \frac{x^2}{3} + \frac{2x}{3}$$

Graph this function using the facts that it passes through $(0, 0)$ and $(1, 1)$ and is similar in shape to $y = x^2$. Shade the inequality area.

2. In an article studying how farmland prices are determined, Brown and Brown discuss the possible options for a prospective seller of a tract of farmland.‡ They define $f(x)$ to be the likelihood that the seller will receive a bid for the land if the price is x thousand dollars. For a particular tract, $f(x) = 1/500$. They define the likelihood that the seller will receive a bid between \$1,900 and \$2,000 to be

$$\int_{1900}^{2000} \frac{1}{500} \, dx$$

Calculate this integral.

*Jean Kinsey, "Measuring the Well-being of Farm Households: Farm, Off-farm and In-kind Sources of Income: Discussion," *American Journal of Agricultural Economics*, Vol. 67, 1985, pp. 1105–1107.

†Thomas A. Carlin and Edward I. Reinsel, "Combining Income and Wealth—An Analysis of Farm Family 'Well-being'," *American Journal of Agricultural Economics*, Vol. 55, 1973, pp. 38–46.

‡Keith C. Brown and Deborah J. Brown, "Heterogenous Expectations and Farmland Prices," *American Journal of Agricultural Economics*, Vol. 66, 1984, pp. 164–169.

Cumulative Exercise Set 6–2

In Problems 1 to 3 find the antiderivative.

1. $\int \left(4x - \dfrac{6}{x^2} \right) dx$ 2. $\int (\sqrt{x} + 7)dx$

3. $\int \left(4e^x - \dfrac{3}{x} \right) dx$

4. Find the function whose slope of the tangent line to the curve is $f'(x) = 1 - 2x + 3x^2$ and whose graph passes through the point $(2, 7)$.

5. If the velocity function of an object is given by

$$v(t) = \frac{4t}{t^2 + 1}$$

what is the distance function $s(t)$ if $s(0) = 4$?

6. (a) Use four rectangles to approximate the area under the curve $y = x^2 + 1$ from $x = 0$ to $x = 1$.

   (b) Use the definition to find the area under the curve $y = x^2 + 1$ from $x = 0$ to $x = 1$.

7. Use four rectangles to approximate the area under the curve $y = e^x$ from $x = 0$ to $x = 1$.

8. Use the definition to evaluate the integral

$$\int_{-1}^{0} (-2x)dx$$

6–3   The Fundamental Theorem of Calculus

The definition of the definite integral involves a somewhat complicated limit process. It would be cumbersome if the definition were necessary to compute each definite integral. However, the **fundamental theorem of calculus** eases the burden by tying together the two branches of calculus, integral and differential calculus. This is done by using antiderivatives, or indefinite integrals, to calculate definite integrals.

The Fundamental Theorem of Calculus

We first present the theorem with an example. Its proof will follow.

The Fundamental Theorem of Calculus

Let $f(x)$ be continuous on the interval $[a, b]$ and let $F(x)$ be any antiderivative of $f(x)$. Then

$$\int_a^b f(x)\, dx = F(b) - F(a)$$

To represent the expression $F(b) - F(a)$ in a more compact form, we use the symbol $F(x) \Big|_a^b$. Thus the fundamental theorem states

$$\int_a^b f(x)\, dx = F(x) \Big|_a^b = F(b) - F(a)$$

Example 1 illustrates how to use the theorem.

EXAMPLE 1

Problem

Use the fundamental theorem of calculus to find the definite integrals

(a) $\int_2^4 2x \, dx$, (b) $\int_{-1}^2 (x^2 + 4x^3) \, dx$.

Solution (a) An antiderivative of $2x$ is x^2, so

$$\int_2^4 2x \, dx = x^2 \Big|_2^4 = 4^2 - 2^2 = 16 - 4 = 12$$

(b) An antiderivative of $x^2 + 4x^3$ is $x^3/3 + x^4$, so

$$\int_{-1}^2 (x^2 + 4x^3) \, dx = (x^3/3 + x^4) \Big|_{-1}^2$$

$$= (8/3 + 16) - (-1/3 + 1) = 18$$

The fundamental theorem states that any antiderivative can be used. To demonstrate what this means consider Example 1(a). The general antiderivative is $x^2 + C$. Using this antiderivative the computation is

$$\int_2^4 2x \, dx = x^2 + C \Big|_2^4 = 4^2 + C - (2^2 + C)$$

$$= (16 + C) - (4 + C) = 12$$

The constant drops out in the calculation, showing that any antiderivative can be used. Also, note that the theorem does not stipulate that $f(x) \geq 0$ in the interval as was required in calculating area in the previous section. Only when we use the definite integral to compute area will this requirement be important.

Verification of the Fundamental Theorem

The verification of the fundamental theorem of calculus provides insight into the connection between area and the definite integral. It starts by looking at areas under the graph of $f(x)$ and quickly shifts to the antiderivative.

Proof of the Fundamental Theorem of Calculus

Proof: Let $f(x) \geq 0$. The proof for $f(x) < 0$ is similar. Define the function $A(x)$ to be the area bounded by $f(x)$ and the x-axis between a and x, where $a \leq x \leq b$. We want to show that $A(x)$ is an antiderivative of $f(x)$, or that $A'(x) = f(x)$. Let h be a small positive number. Then $A(x + h)$ is the area from a to $x + h$. Since $A(x)$ is the area from a to x (Figure 1), $A(x + h) - A(x)$ is the area shaded in Figure 2. This area can be approximated by a rectangle of width h and height $f(x)$, whose area is $h \cdot f(x)$. Thus

$$A(x + h) - A(x) \simeq h \cdot f(x)$$

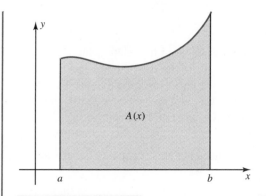

FIGURE 1

The approximation improves as h gets close to 0. Dividing by h yields

$$\frac{A(x + h) - A(x)}{h} \simeq f(x)$$

Take the limit of both sides as h approaches 0. Since x is not affected by h, the limit of the right-hand side is $f(x)$. Because the approximation improves as h gets close to 0, taking the limit means the approximation sign can be replaced by an equal sign. Thus the limit of the right-hand side is equal to $f(x)$. But the limit of the left-hand side is the definition of the derivative of $A(x)$. Therefore $A'(x) = f(x)$.

There is one more step in the proof. Since x represents any number between a and b, set $x = a$. Then $A(a) = 0$. Also, setting $x = b$ gives $A(b)$, which is the desired area. Since $A(x)$ is an antiderivative of $f(x)$, we have

$$\int_a^b f(x)\, dx = A(x) \Big|_a^b = A(b) - A(a) = A(b)$$

Therefore the desired area is the definite integral.

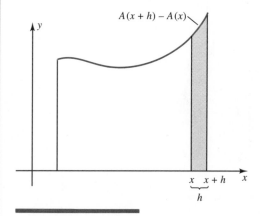

FIGURE 2

Area

Let us look at an example of the relationship between the definite integral and the area under a curve. Consider the function $f(x) = 2x$ in Example 1(a). Its graph is in Figure 3. The area under the curve bounded by the two lines $x = 2$ and $x = 4$ consists of two areas, the rectangle bounded above by the line $y = 4$ and the triangle above it. The area of the rectangle is $4 \cdot 2 = 8$. The area of the triangle is $1/2$ (base)(height) $= (1/2) \cdot 2 \cdot 4 = 4$. So the total area under $y = 2x$ from $x = 2$ to $x = 4$ is 12. Notice that this is equal to the definite integral $\int_2^4 2x \, dx$, computed in Example 1(a).

What is the area of the region between the x-axis and $f(x) = 2x$ from $x = -1$ to $x = 2$? It might seem that the answer is $\int_{-1}^2 2x \, dx$. However

$$\int_{-1}^2 2x \, dx = x^2 \Big|_{-1}^2 = 4 - 1 = 3$$

From Figure 4 the total area is the sum of the area from $x = -1$ to $x = 0$ and the area from $x = 0$ to $x = 2$; that is, $1 + 4 = 5$.

What went wrong? Why did the definite integral yield an incorrect answer? Because part of the graph is below the x-axis. An area problem must be divided into subproblems, those where the graph is above the x-axis and those where it is below. The definite integral from $x = -1$ to $x = 0$ gives the negative of the area.

$$\int_{-1}^0 2x \, dx = x^2 \Big|_{-1}^0 = 0 - (-1)^2 = -1$$

This leads to the next important property of definite integrals.

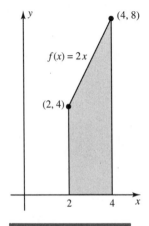

FIGURE 3

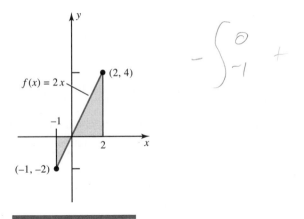

FIGURE 4

Areas Under the *x*-axis
If $f(x) \leq 0$ in $[a, b]$, then the area bounded by $f(x)$, the *x*-axis, $x = a$, and $x = b$ is given by

$$-\int_a^b f(x) \, dx$$

Therefore if $f(x) \geq 0$, the area and the definite integral are equal, and if $f(x) \leq 0$, the area is the negative of the definite integral. For example, to find the area bounded by $f(x) = 2x$, the *x*-axis, $x = -1$, and $x = 2$, we need to divide the region into two separate regions: the region from $x = -1$ to $x = 0$ where $f(x) \leq 0$ and the region from $x = 0$ to $x = 2$ where $f(x) \geq 0$. The area of the first region is then the negative of the definite integral from -1 to 0. The area of the second region is the definite integral from 0 to 2. The total area is then

$$-\int_{-1}^0 2x \, dx + \int_0^2 2x \, dx = -(-1) + 4 = 5$$

This means that if the graph of $y = f(x)$ is below the *x*-axis in the interval over which the area is to be found, then we must find where the function is positive and where it is negative. This is equivalent to evaluating the integral of $|f(x)|$ because if $f(x) < 0$ then $\int_a^b |f(x)| dx = \int_a^b - f(x) dx = -\int_a^b f(x) dx$. Thus

$$\int_a^b |f(x)| dx = \begin{cases} \int_a^b f(x) \, dx & \text{if } f(x) \geq 0 \text{ in } [a, b] \\ -\int_a^b f(x) \, dx & \text{if } f(x) \leq 0 \text{ in } [a, b] \end{cases}$$

For example, to find the area bounded by $f(x) = 2x$, the x-axis, $x = -1$, and $x = 2$, it is necessary to divide the area into two separate regions: the region from $x = -1$ to $x = 0$ where $f(x) \leq 0$, and the region from $x = 0$ to $x = 2$ where $f(x) \geq 0$. This means that

$$|f(x)| = |2x| = \begin{cases} -2x \text{ for } x \text{ in } [-1, 0] \\ 2x \text{ for } x \text{ in } [0, 2] \end{cases}$$

The area of the first region is then the negative of the definite integral from -1 to 0. The area of the second region is the definite integral from 0 to 2. The total area is then:

$$\int_{-1}^{2} |2x| \, dx = -\int_{-1}^{0} 2x \, dx + \int_{0}^{2} 2x \, dx = -(-1) + 4 = 5$$

Procedure for Finding the Area Under a Curve

To compute the area of the region bounded by $f(x)$, the x-axis, $x = a$, and $x = b$:

1. From the graph of $f(x)$ find the intervals where $f(x)$ is above the x-axis and where $f(x)$ is below the x-axis by locating the x-intercepts in $[a, b]$.
2. Compute the definite integral of $f(x)$ between each pair of x-intercepts.
3. If $f(x)$ is greater than 0 in an interval, the area is equal to the definite integral. If $f(x)$ is less than 0 in an interval, the area is the negative of the definite integral.
4. The total area is the sum of the areas in step 3.

The next example illustrates the procedure.

EXAMPLE 2

Problem

Find the area bounded by $f(x) = 2 - 2x$, the x-axis, $x = 0$, and $x = 4$.

Solution The graph is in Figure 5. The region consists of two triangles, one above the x-axis and one below it. To find the point of intersection of $f(x)$ and the x-axis, solve $f(x) = 0$; then $2 - 2x = 0$, so $x = 1$. From $x = 0$ to $x = 1$, $f(x)$ is above the x-axis, so the corresponding area is equal to the definite integral

$$\int_{0}^{1} (2 - 2x) \, dx = (2x - x^2) \Big|_{0}^{1} = (2 - 1) - 0 = 1$$

From $x = 1$ to $x = 4$ the graph is below the x-axis, so the area under the curve is equal to the negative of the definite integral.

$$-\int_{1}^{4} (2 - 2x) \, dx = -(2x - x^2) \Big|_{1}^{4} = -[(8 - 16) - (2 - 1)]$$

$$= -(-8 - 1) = -(-9) = 9$$

Therefore the total area is $1 + 9 = 10$.

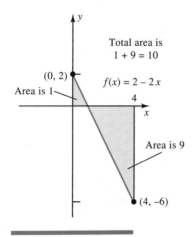

FIGURE 5

Properties of Definite Integrals

Some properties of antiderivatives or indefinite integrals were given in Section 6–1. Similar properties hold for definite integrals. We will list four of them.

Let $f(x)$ and $g(x)$ be functions and a, b, c, and k be real numbers. Then

1. $\displaystyle\int_a^b kf(x)\ dx = k\int_a^b f(x)\ dx$

2. $\displaystyle\int_a^b f(x)\ dx + \int_a^b g(x)\ dx = \int_a^b [f(x) + g(x)]\ dx$

3. $\displaystyle\int_a^a f(x)\ dx = 0$

4. $\displaystyle\int_a^c f(x)\ dx + \int_c^b f(x)\ dx = \int_a^b f(x)\ dx$ for $a \le c \le b$

The first two properties hold for indefinite integrals so they are also valid for definite integrals. Since the distance from a to a is 0, the third property states that the area under the curve between a and a is 0. The fourth property can be illustrated geometrically. The area bounded by the x-axis and $y = f(x)$ from $x = a$ to $x = b$ can be viewed as the sum of the area from $x = a$ to $x = c$ and the area from $x = c$ to $x = b$. This geometric argument works only for $f(x) > 0$, but a similar argument holds for the case when the graph of $f(x)$ is below the x-axis.

The next example shows how to use the properties to find more complicated definite integrals.

EXAMPLE 3

Problem

Find

$$\int_0^1 (4x^3 + 4x - 2e^x)\, dx$$

Solution

$$\int_0^1 (4x^3 + 4x - 2e^x)\, dx = \int_0^1 4x^3\, dx + \int_0^1 4x\, dx - \int_0^1 2e^x\, dx$$

$$= x^4 \Big|_0^1 + 2x^2 \Big|_0^1 - 2e^x \Big|_0^1$$

$$= 1^4 - 0^4 + 2(1^2) - 2(0^2) - 2e^1 + 2e^0$$

$$= 1 + 2 - 2e + 2 = 5 - 2e$$

Wages and Rent

A fundamental principle in economics is that "wages equal marginal product." This means that the employer will pay all employees doing similar work for similar pay, a wage equal to the marginal product of the last worker hired. This is because the last worker's value to the firm is that worker's marginal product, or the amount of increase in production the worker generates.

Along with this idea there are two additional principles. Consider the graph in Figure 6. Its curve is the marginal product function, which is equal to wages.

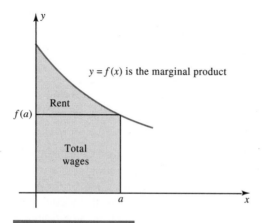

FIGURE 6

000   0·000 ✳

That is, $f(x)$ is the wage of each employee when x employees are hired. This curve is decreasing, meaning that as more workers are hired, the marginal product of the last worker decreases, and therefore the wage of the workers also decreases. If the number of employees is $x = a$, then the wage for each employee is $f(a)$. The total wages *TW* paid to all employees is the area bounded by the constant function $y = f(a)$, which is

$$TW = \int_0^a f(a)\ dx = f(a)x \ \Big|_0^a = f(a)(a - 0) = af(a)$$

In other words, the region defined by this integral is the rectangle whose length is a and whose height is $f(a)$, so its area is $af(a)$. The **total product** is the total area of the region bounded by $y = f(x)$ and the x-axis from $x = 0$ to $x = a$. Therefore total product *TP* is defined to be

$$TP = \int_0^a f(x)\ dx$$

There is another region in Figure 6. It is the region bounded by $y = f(x)$ and the horizontal line $y = f(a)$ from $x = 0$ to $x = a$. It represents a savings for the employer because it measures the amount of total product that is not paid out in wages. Economists call the area of this region the employer's *rent*.* From Figure 6 we see that rent is the total product minus total wages. That is

rent $= TP - TW$

When economists study the effects of a new innovation or product on an industry, the concept of rent provides a measure of how much the industry has gained from the innovation. See Referenced Exercise 1 and the Case Study for examples of this type of use of rent.

EXAMPLE 4

A firm determines that the marginal product $MP(x)$ for a level of labor x is

$$MP(x) = 0.03x^2 - 0.6x + 23$$

$MP(x)$ is defined on the interval $[0, 10]$ and is measured in thousands of dollars, while x is measured in hundreds of employees. The number of employees hired is $x = 10$ hundred.

Problem

Find (a) total wages *TW*, (b) total product *TP*, and (c) the employer's rent.

Solution The wage paid to each employee is the marginal product of the last employee hired, which is $MP(10) = 3 - 6 + 23 = 20$.
(a) Total wages is defined by

$$TW = \int_0^{10} MP(10)\ dx = \int_0^{10} 20\ dx = 20x \ \Big|_0^{10} = 200$$

*Paul A. Samuelson, *Economics*, 11th ed., New York, McGraw-Hill Book Co., 1980, pp. 465–467.

(b) The total product TP is defined by

$$TP = \int_0^{10} (0.03x^2 - 0.6x + 23)\, dx$$

$$= 0.01x^3 - 0.3x^2 + 23x) \Big|_0^{10} = 10 - 30 + 230 = 210$$

(c) The employer's rent is the area of the region in Figure 7. It is the area of the region under $y = MP(x)$ and above $y = 20$ from $x = 0$ to $x = 10$. This area is TP minus TW.

$$\text{rent} = TP - TW = 210 - 200 = 10$$

So the employer's rent is $10,000 when 1000 workers are hired at the salary of $20,000.

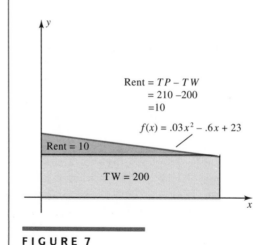

FIGURE 7

EXERCISE SET 6–3

In Problems 1 to 10 find the definite integral by using the fundamental theorem of calculus.

1. $\int_0^1 (3x + 5)\, dx$

2. $\int_1^3 (4x - 2)\, dx$

3. $\int_0^1 (x^2 + 1)\, dx$

4. $\int_{-2}^2 (3x^2 + 1)\, dx$

5. $\int_{-3}^2 (x^2 + 5x)\, dx$

6. $\int_1^2 (x^3 + x)\, dx$

7. $\int_0^8 (-x^2 + \sqrt[3]{x})\, dx$

8. $\int_1^4 (-x^2 + 1/\sqrt{x} + 7e^x)\, dx$

9. $\int_1^4 (-2x^{-3} + 6x^{-4} + 3\sqrt{x})\, dx$

10. $\int_1^4 (x^{-3} - 3x^{-2} - 3\sqrt{x})\, dx$

In Problems 11 and 12 find the definite integral by using geometric properties of the integrand. Verify by using the fundamental theorem of calculus.

11. $\int_{-2}^4 (x - 1)\, dx$

12. $\int_{-1}^4 (2 - x)\, dx$

In Problems 13 to 20 use the definite integral to find the area bounded by $f(x)$, the x-axis, $x = a$, and $x = b$.

13. $f(x) = x - 1$, $a = -3$, $b = 4$

14. $f(x) = 1 - x$, $a = -3$, $b = 3$

15. $f(x) = x^2 - 4$, $a = -2$, $b = 4$

16. $f(x) = 4 - x^2$, $a = -3$, $b = 2$

17. $f(x) = x^2 - x$, $a = 0$, $b = 2$

18. $f(x) = 4x - x^2$, $a = -1$, $b = 2$

19. $f(x) = x^3 - x$, $a = -1$, $b = 1$

20. $f(x) = x^3 - x^2$, $a = 0$, $b = 2$

In Problems 21 to 24 assume $\int_0^1 f(x)\, dx = 10$,

$\int_1^2 f(x)\, dx = 15$, and $\int_0^2 g(x)\, dx = -5$. Use the properties of definite integrals to evaluate the integral.

21. $\int_0^2 f(x)\, dx$

22. $\int_0^2 3f(x)\, dx$

23. $\int_0^2 [f(x) + g(x)]\, dx$

24. $\int_0^2 [2f(x) + 3g(x)]\, dx$

In Problems 25 to 30 find the definite integral.

25. $\int_0^1 (x^{0.2} + 3x^{1.4})\, dx$

26. $\int_0^1 (1.2x^{-1.6} - 4.1x^{-2.2})\, dx$

27. $\int_1^2 (3e^{2x} + 1/x)\, dx$

28. $\int_1^2 (5e^{3x} - 2/x)\, dx$

29. $\int_1^2 (x^{1.3} - 3e^{1.2x} + 1/2x)\, dx$

30. $\int_1^2 (-e^{1-3x} - e/x)\, dx$

In Problems 31 to 36 use the definite integral to find the area bounded by $f(x)$, the x-axis, $x = a$, and $x = b$.

31. $f(x) = x^3 - x$, $a = -2$, $b = 2$

32. $f(x) = x^3 - x^2$, $a = -1$, $b = 2$

33. $f(x) = x^3 - 2x^2$, $a = 0$, $b = 3$

34. $f(x) = x^3 - 4x$, $a = -3$, $b = 5$

35. $f(x) = 1 - 2/x$, $a = 1$, $b = 3$

36. $f(x) = 2 - 3/x$, $a = 1$, $b = 3$

In Problems 37 to 40 find $\int_1^4 f(x)\, dx$ from the given figure.

37.

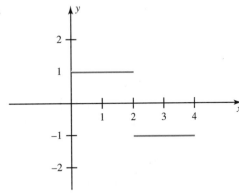

38.

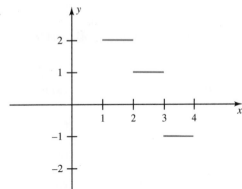

39.

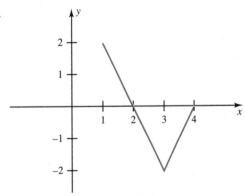

40.

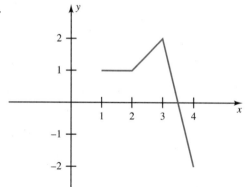

41. Prove the fundamental theorem for $f(x) < 0$.

42. What procedure would you use to find the area of the region bounded by $f(x) = x^2 - 4x + 3$ and the x-axis?

Referenced Exercise Set 6–3

1. Television is an example of an industry whose firms reap benefits from regulation. In an article studying the television industry, Fournier used rents to measure the effects of regulation on the industry.[*] The marginal cost function $MP(x)$ can be approximated by

$$MP(x) = 3x^2 - 60x + 800$$

where x is the average price of a show. Find the rent when $x = 10$. Here total product is income and total wages is total price.

2. In an article investigating the size of prey versus an optimal diet for carnivorous fish, Bence and Murdoch studied the relationship between the success rate $S(s)$ of the predator catching the prey versus the size s of the prey.[†] For one type of fish the relationship was calculated to be

$$S(s) = 1 + .861s - 91.57s^2 + 322.7s^3$$

The total success of attacks of prey between a mm (millimeters) and b mm is given by

$$\int_a^b S(s)ds$$

Find the total success for prey between 1 mm and 2 mm.

3. It is often difficult for economists to measure the benefits and costs of the uses of natural resources, such as federally owned lakes. In an article investigating ways to estimate the value of these resources, Seller et al. studied lakes in East Texas.[‡] They calculated the demand curve for Lake Livingston to be

$$V(x) = 10.04 - .12x$$

where $V(x)$ is the value of the land when the cost of a family visit is x dollars. One measure of the benefit of the resource involved computing the consumers' surplus, an economic measure defined in Section 6–4. It entails computing the integral of $V(x)$ from $x = 0$ to $x = 2$. Compute this integral.

Cumulative Exercise Set 6–3

In Problems 1 to 4 find the antiderivative.

1. $\int(3x^{-2} - x^4 + 5)dx$

2. $\int(x^{1/5} - x^{5/3} + 8x - 1)dx$

3. $\int(3e^{2x} - 5/x)dx$ 4. $\int(x^6 + 4x^2)/x^3\, dx$

5. Use four rectangles to approximate the area under the curve $y = f(x) = x^2$ from $x = 0$ to $x = 4$.

In Problems 6 and 7 use the definition to find the definite integral.

6. $\int_0^4 2x\, dx$ 7. $\int_6^8 3x\, dx$

8. Use the definition to find the definite integral by using the identity

$$1^2 + 2^2 + \cdots + n^2 = \frac{n(n+1)(2n+1)}{6}$$

$$\int_{-2}^2 x^2\, dx$$

In Problems 9 and 10 find the definite integral by using the fundamental theorem of calculus.

9. $\int_1^2 (x^3 + 2x - 1)dx$

[*]Gary M. Fournier, "Nonprice Competition and the Dissipation of Rents from Television Regulation," *Southern Economics Journal,* Vol. 51, 1985, pp. 754–765.

[†]James R. Bence and W. W. Murdoch, "Prey Size Aselection by the Mosquitofish: Relation to Optimal Diet Theory," *Ecology,* Vol. 67, 1986, pp. 324–336.

[‡]Christine Seller et al., "Validation of Empirical Measures of Welfare Change: A Comparison of Nonmarket Techniques," *Land Economics,* Vol. 61, 1985, pp. 156–175.

10. $\int_{1}^{4} (x^{-2} - 4x^{-2} - 2\sqrt{x})dx$

11. Assume $\int_{1}^{2} f(x)dx = 20$, $\int_{2}^{3} f(x)dx = -1$, and $\int_{1}^{3} g(x)dx = 5$. Use the properties of definite integrals

to evaluate

$$\int_{1}^{3} (f(x) + g(x))dx$$

12. Use the definite integral to find the area bounded by $f(x) = x^3 - x^2$, the x-axis, $x = -1$, and $x = 3$.

6–4 Area Bounded by Curves

Thus far we have considered regions bounded by a single curve and the x-axis. Often in applications a region is bounded both above and below by graphs of functions. In this section the concept of the area between a curve and the x-axis is generalized to include the area between two curves.

Areas Between Two Curves

Consider the area of the region bounded above by $f(x) = x + 2$ and below by $g(x) = x^2$. This region is graphed in Figure 1. View the region in two parts: The first part, region I, is the area below $f(x)$ from $x = -1$ to $x = 2$; the second part, region II, is the area below $g(x)$ from $x = -1$ to $x = 2$. Then the desired region is (region I–region II). (See Figure 2.) Its area is the difference between the two definite integrals.

$$\begin{aligned}
\text{(the desired area)} &= \text{region I} \quad - \text{region II} \\
&= (\text{area below } f(x)) - (\text{area below } g(x)) \\
&= \int_{-1}^{2} (x + 2)\, dx - \int_{-1}^{2} x^2\, dx \\
&= \left(\frac{x^2}{2} + 2x\right)\Bigg|_{-1}^{2} - \frac{x^3}{3}\Bigg|_{-1}^{2} = \frac{15}{2} - 3 = \frac{9}{2}
\end{aligned}$$

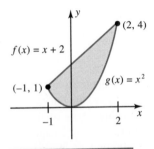

$f(x) = x + 2$

$(-1, 1)$

$(2, 4)$

$g(x) = x^2$

FIGURE 1

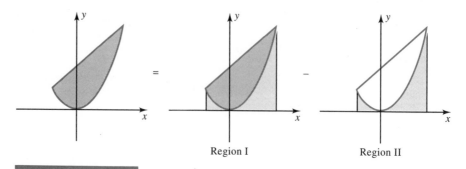

Region I   Region II

FIGURE 2

To compute this area, we can also use the third property from the previous section and get

$$= \int_{-1}^{2} (x + 2)\, dx - \int_{-1}^{2} x^2\, dx = \int_{-1}^{2} (x + 2 - x^2)\, dx$$

$$= \left(\frac{x^2}{2} + 2x - \frac{x^3}{3} \right) \Big|_{-1}^{2} = \frac{9}{2}$$

In general the area between two curves is the difference between the two definite integrals. We then use a property of definite integrals to simplify the integration.

The Area Bounded by Two Curves
If a region is bounded above by $f(x)$ and below by $g(x)$ between $x = a$ and $x = b$, then the area of the region is

$$\int_{a}^{b} [f(x) - g(x)]\, dx$$

Lorentz Diagrams

In Section 6–2 we defined a Lorentz diagram. The important part of the diagram is the area between the two curves $y = x$, the curve of complete equality of income distribution, and $y = x^2$, the Lorentz curve of the distribution of income in the United States. This area is called the area of inequality and measures the extent of departure between an equal distribution of income and the actual distribution of income. The area between these two curves is the difference between the area of the region bounded by $y = x$ and the area of the region bounded by $y = x^2$. We can now compute this area. It is

$$\text{area} = \int_0^1 x \, dx - \int_0^1 x^2 \, dx = \int_0^1 (x - x^2) \, dx$$

$$= \left(\frac{x^2}{2} - \frac{x^3}{3} \right) \Big|_0^1 = \frac{1}{2} - \frac{1}{3} = \frac{1}{6}$$

Economists compare this value with the inequality areas of other countries to determine which has a more equitable income distribution.

The formula for the area between two curves does not mention whether the functions are positive or negative. The only stipulation is that $f(x) \geq g(x)$ in $[a, b]$. The next example demonstrates what to do when two curves cross each other in the desired area.

EXAMPLE 1

Problem

Find the area of the region bounded by $f(x) = 3x^2 - 2x$, and $g(x) = x$ from $x = 0$ to $x = 2$.

Solution This region consists of two separate regions as shown in Figure 3. Find the points of intersection by solving $f(x) = g(x)$ to get $x = 0$ and $x = 1$.

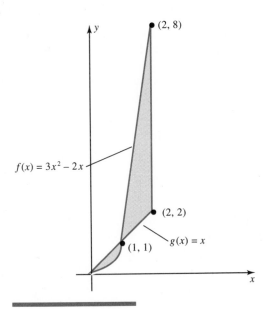

FIGURE 3

From 0 to 1, $g(x) \geq f(x)$, so we integrate $g(x) - f(x)$. From 1 to 2, $f(x) \geq g(x)$, so we integrate $f(x) - g(x)$. The area is the sum of the two integrals.

$$\text{area} = \int_0^1 [x - (3x^2 - 2x)] \, dx + \int_1^2 [(3x^2 - 2x) - x] \, dx$$

$$= \int_0^1 (3x - 3x^2) \, dx + \int_1^2 (3x^2 - 3x) \, dx$$

$$= \left(\frac{3x^2}{2} - x^3 \right) \Big|_0^1 + \left(x^3 - \frac{3x^2}{2} \right) \Big|_1^2$$

$$= \left(\frac{3}{2} - 1 \right) + \left[(8 - 6) - \left(1 - \frac{3}{2} \right) \right] = 3$$

End Points

In some problems the end points of the region are given and in others they are not. If the end points are not specified, they are found by computing the first coordinates of the points of intersection of the two curves. So the first step is to graph the region and locate the points of intersection. The next example shows how to do this.

EXAMPLE 2

Problem

Find the area of the region bounded by $y = x^3 - 2x$ and $y = x^2$.

Solution First sketch the curves. In order to refer to each curve, we name them. Let $f(x) = x^3 - 2x$ and $g(x) = x^2$. To find the points of intersection, solve $f(x) = g(x)$, so $x^3 - 2x = x^2$ or $x^3 - x^2 - 2x = 0$. Factoring yields $x(x - 2)(x + 1) = 0$. Thus the solutions are $x = -1, 0, 2$, and the curves intersect at $(-1, 1)$, $(0, 0)$, and $(2, 4)$. See Figure 4. From $x = -1$ to $x = 0$,

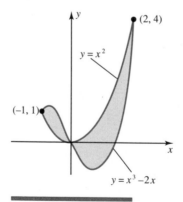

FIGURE 4

$f(x) \geq g(x)$, while from $x = 0$ to $x = 2$, $g(x) \geq f(x)$. Therefore the area of the region bounded by the two curves is

$$\int_{-1}^{0} [(x^3 - 2x) - x^2] \, dx + \int_{0}^{2} [x^2 - (x^3 - 2x)] \, dx$$

$$= \int_{-1}^{0} [x^3 - 2x - x^2] \, dx + \int_{0}^{2} [x^2 - x^3 + 2x] \, dx$$

$$= \left[\frac{x^4}{4} - x^2 - \frac{x^3}{3} \right] \Big|_{-1}^{0} + \left[\frac{x^3}{3} - \frac{x^4}{4} + x^2 \right] \Big|_{0}^{2}$$

$$= 0 - \left(\frac{1}{4} - 1 + \frac{1}{3} \right) + \frac{8}{3} - 4 + 4 - 0$$

$$= \frac{5}{12} + \frac{8}{3} = \frac{37}{12}$$

Procedure for Finding the Area of a Region Bounded by Two Curves

Find the area of a region bounded by two curves, $y = f(x)$ and $y = g(x)$, with the following steps:

1. Sketch the region. Find the points of intersection.
2. Find the end points, $x = a$ and $x = b$. If they are not given, then they are points of intersection.
3. Find the intervals where $f(x) \geq g(x)$.

   (i)   If $f(x) \geq g(x)$ in $[a, b]$, then the area is

$$\int_{a}^{b} [f(x) - g(x)] \, dx$$

   (ii) If $g(x) \geq f(x)$ in $[a, b]$, then the area is

$$\int_{a}^{b} [g(x) - f(x)] \, dx$$

   (iii) If there is a point c between a and b such that $f(x) \geq g(x)$ in $[a, c]$ and $g(x) \geq f(x)$ in $[c, b]$, then the area is

$$\int_{a}^{c} [f(x) - g(x)] \, dx + \int_{c}^{b} [g(x) - f(x)] \, dx$$

Consumers' Surplus and Producers' Surplus

A pair of economic concepts that are best explained by calculus are *consumers' surplus* and *producers' surplus*. Consider the demand function $y = D(x)$ and the

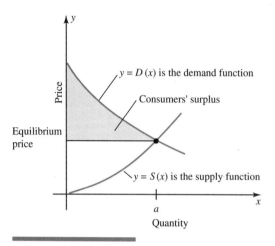

FIGURE 5

supply function $y = S(x)$, where x is the quantity of a product and y is the price. At price $D(x)$ consumers will demand x amount of the product. As x increases, $D(x)$ decreases, meaning that consumers will purchase more of the product at lower prices. This implies that $D(x)$ is a decreasing function. At price $S(x)$ producers are willing to supply x amount of the product. As x increases, $S(x)$ also increases because producers will be willing to supply more of the product at higher prices. Therefore $S(x)$ is an increasing function. In other words, $D(x)$ is the price that consumers are willing to pay for x amount of the product, and producers will supply x amount at price $S(x)$. In a free competitive market the actual selling price for the product will tend to be that price where supply equals demand, that is, where $S(x) = D(x)$. It is called the **equilibrium price,** and it is that quantity $x = a$ where $S(a) = D(a)$.

Figure 5 shows that the demand $D(x)$ is larger than $D(a)$ for values of x less than a. This means that some consumers would be willing to pay a higher price than the equilibrium price. In a certain sense these consumers have gained because the price they pay, the equilibrium price, is lower than what they would have been willing to pay. Economists refer to this gain as **consumers' surplus.** It is defined to be the area of the region bounded above by $y = D(x)$ and below by the constant function $y = D(a)$, from $x = 0$ to $x = a$. This area is

$$\text{consumers' surplus} = \int_0^a [D(x) - D(a)] \, dx$$

EXAMPLE 3

Suppose $D(x) = x^2 - 8x + 20$ for x in $[0, 4]$ and the equilibrium price occurs at $a = 2$.

Problem

Compute the consumers' surplus.

Solution The equilibrium price is $D(2) = 8$. The consumers' surplus is defined to be

$$\int_0^2 [(x^2 - 8x + 20) - 8] \, dx = \frac{x^3}{3} - 4x^2 + 12x \Big|_0^2 = \frac{8}{3} - 16 + 24 = \frac{32}{3}$$

Similarly the producers who were willing to supply goods at a price smaller than $S(a)$ gain because they would have made less revenue at the lower price. Economists call the sum of their gains the **producers' surplus.** It is defined to be the area of the region bounded above by $y = S(a)$ and below by $y = S(x)$ from $x = 0$ to $x = a$. It is

$$\text{producers' surplus} = \int_0^a [S(a) - S(x)] \, dx$$

EXAMPLE 4

Suppose $S(x) = x^2 + 2x$ for x in $[0, 4]$ and the equilibrium price occurs at $a = 2$.

Problem

Compute the producers' surplus.

Solution The equilibrium price is $S(2) = 8$. The producer's surplus is defined to be

$$\int_0^2 [8 - (x^2 + 2x)] \, dx = \left(8x - \frac{x^3}{3} - x^2 \right) \Big|_0^2 = 16 - \frac{8}{3} - 4 = \frac{28}{3}$$

Figure 6 shows the supply and demand curves in Examples 3 and 4 as well as the areas defining the consumers' surplus and producers' surplus.

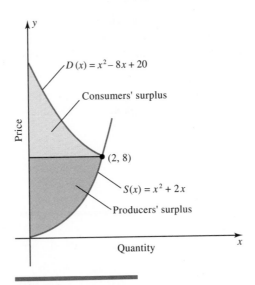

FIGURE 6

The concept of consumers' surplus measures the benefit consumers gain from the power of buying at low prices and not having to pay the higher price a monopolist may insist on. Another use of consumers' surplus is in measuring "social welfare." Sometimes economic value, revenue, or profit cannot be used to account for the value of a good. For instance, the economic value of air is negligible but its social welfare is great. The next example demonstrates how economists use consumers' surplus to help make a decision where the intangible quantity of utility is the only "profit."

EXAMPLE 5

A new community is debating whether to self-impose a tax to install street lights. The community has 3000 homes that will be taxed the same amount for the improvement. Contractors are willing to supply the lights according to the schedule

$$S(x) = x^2 + 9x$$

The community is willing to pay according to the schedule

$$D(x) = x^2 - 18x + 81$$

Price is measured in $100,000 units and x is measured in units of 100 street lights. To find the equilibrium price, solve $S(x) = D(x)$. This yields $a = 3$, so the equilibrium price is $S(3) = D(3) = 36$. Thus each homeowner must pay a tax of $3,600,000/3000 = $1,200 for the improvement. Is it worth it to the individual? Since there are no revenues from this project, the only profit is the increased utility achieved. Hence the consumers' surplus gives an indication of the worth of the project.

Problem

Compute the consumers' surplus.

Solution From the definition the consumers' surplus is

$$\int_0^3 [(x^2 - 18x + 81) - 36]\, dx = \frac{x^3}{3} - 9x^2 + 45x \bigg|_0^3 = 9 - 81 + 135 = 63$$

Thus the consumers' surplus is $6,300,000. Each homeowner gets a utility of $6,300,000/3000 = $2,100. Therefore the consumers' surplus is far greater than the amount spent on the improvement. The project is worthwhile from this point of view.

EXERCISE SET 6–4

In Problems 1 to 30 find the area between the curves.

1. $x = 0, \quad x = 2,$
   $f(x) = x^2 + 1, \quad g(x) = 0$

2. $x = 0, \quad x = 3,$
   $f(x) = x^2 + 4, \quad g(x) = 0$

3. $x = 0, \quad x = 2,$
   $f(x) = x^2 + 2x, \quad g(x) = 0$

4. $x = 0$, $x = 3$,
    $f(x) = x^2 + x$, $g(x) = 0$

5. $x = 0$, $x = 2$,
    $f(x) = x^3 + 3x$, $g(x) = 0$

6. $x = 0$, $x = 4$,
    $f(x) = x^3 + 6x$, $g(x) = 0$

7. $x = 0$, $x = 1$,
    $f(x) = x^2$, $g(x) = x$

8. $x = 0$, $x = 1$,
    $f(x) = x^2$, $g(x) = 2x$

9. $x = 0$, $x = 1$,
    $f(x) = x^2 + x$, $g(x) = 2x$

10. $x = 0$, $x = 1$,
    $f(x) = x^2 + 4x$, $g(x) = 5x$

11. $x = 0$, $x = 1$,
    $f(x) = 3x - x^2$, $g(x) = 2x$

12. $x = 0$, $x = 1$,
    $f(x) = 5x - x^2$, $g(x) = 6x$

13. $f(x) = x^2 + x$, $g(x) = 2x$

14. $f(x) = x^2 + 4x$, $g(x) = 5x$

15. $f(x) = 3x - x^2$, $g(x) = 2x$

16. $f(x) = 5x - x^2$, $g(x) = 6x$

17. $f(x) = x^2 - 2x$, $g(x) = 2x$

18. $f(x) = x^2 - 4x$, $g(x) = 5x$

19. $f(x) = 2 + x - x^2$,
    $g(x) = 2 - 3x$

20. $f(x) = 3 + 2x - x^2$,
    $g(x) = 3 - 4x$

21. $f(x) = x^3$, $g(x) = x^2$

22. $f(x) = x^3 - 4x$,
    $g(x) = x^2 - 4x$

23. $x = 0$, $x = 4$, $f(x) = x + 2$, $g(x) = x^2$

24. $x = 0$, $x = 6$, $f(x) = 3x + 2$,
    $g(x) = 2x^2$

25. $x = -1$, $x = 4$, $f(x) = 2x + 3$,
    $g(x) = x^2$

26. $x = -1$, $x = 5$, $f(x) = 3x + 4$,
    $g(x) = x^2 + x + 1$

27. $x = -2$, $x = 3$, $f(x) = 1 - 2x$,
    $g(x) = x^2 - 2x - 3$

28. $x = -3$, $x = 3$, $f(x) = 3x + 4$,
    $g(x) = x^2 + 3x$

29. $x = -2$, $x = 2$, $f(x) = 3x - x^2$,
    $g(x) = x^2 - x$

30. $x = 0$, $x = 4$, $f(x) = 4x - x^2$,
    $g(x) = 2x^2 - 5x$

In Problems 31 to 34 compute the consumers' surplus and the producers' surplus.

31. $D(x) = x^2 - 8x + 24$, $S(x) = x^2 + 4x$

32. $D(x) = x^2 - 12x + 48$, $S(x) = x^2 + 4x$

33. $D(x) = 48 - 10x - x^2$, $S(x) = x^2 + 10x$

34. $D(x) = 39 - 12x - x^2$, $S(x) = x^2 + 8x$

In Problems 35 to 40 find the area between the curves.

35. $x = 1$, $x = 2$, $f(x) = \dfrac{1}{x}$, $g(x) = 1$

36. $x = 2$, $x = 3$, $f(x) = \dfrac{1}{x}$, $g(x) = x$

37. $x = 1$, $x = 5$, $f(x) = \dfrac{4}{x}$, $g(x) = x$

38. $x = 0$, $x = 3$, $f(x) = \dfrac{2}{x + 1}$,
    $g(x) = x - 1$

39. $x = 0$, $x = 3$, $f(x) = \dfrac{5}{x + 2}$,
    $g(x) = x - 2$

40. $x = 1$, $x = 5$, $f(x) = \dfrac{8}{x}$, $g(x) = x^2$

Referenced Exercise Set 6–4

1. A new trash-to-steam plant is proposed for a community to reduce refuse. The community is willing to pay according to the schedule

$$D(x) = 1000 - 50x - x^2$$

Contractors are willing to build the plant according to the schedule

$$S(x) = x^2 + 30x$$

Compute the consumers' surplus and determine if the project is "profitable" in the sense that the consumers' surplus is greater than the cost.

2. In an article studying the adverse effects of ozone and other air pollutants on crop yields, Adams et al. assessed the benefits to agriculture from reductions in these pollutants.* The measurements used were consumers' surplus and producers' surplus. In one model, where a 10% reduction is assumed, the supply and demand curves are approximated by

$$S(x) = 0.03x^2 + 1.2x$$
$$D(x) = 0.03x^2 - 2.7x + 28.2$$

where $S(x)$ and $D(x)$ are measured in billions of dollars. The total benefit was the sum of the consumers' surplus and the producers' surplus. Compute the total benefit.

3. In many areas where soil is good but rainfall is poor, water irrigation projects are developed to support farm products. Often taxpayers are required to subsidize these projects. In 1978, California rice producers were given a subsidy in the form of reduced prices for water to irrigate the rice crop. In an article studying the effect of this subsidy on producers and consumers, Foster et al. used producers' surplus and consumers' surplus to measure the benefit of the subsidy to each group.† The demand curve $D(x)$ and the supply curve $S(x)$ at price x, measured in thousands of dollars, were approximated by

$$D(x) = 34 - 28x$$
$$S(x) = 30 + 2x + 16x^2$$

Compute the producers' surplus and the consumers' surplus.

4. In his book *Applied Mathematical Demography*, Nathan Keyfitz argues that there is an urgent need for haste in lowering the birth rate in less developed countries by showing the lag between the birth rates and death rates of such countries.‡ If the birth rate is represented by $b(t)$ and the death rate by $d(t)$, then the total lag $l(t)$ between the two rates over T years is defined by

$$l(t) = \int_0^T b(t)dt - \int_0^T d(t)dt$$

The birth rate $b(t)$ and the death rate $d(t)$ of one country are approximated by

$$b(t) = .01t^4 - .18t^2 + 1.3$$
$$d(t) = 1.2 - .06t$$

Compute $l(t)$ for $T = 10$.

Cumulative Exercise Set 6–4

In Problems 1 and 2 find the antiderivative.

1. $\int \dfrac{x}{3}\, dx$

2. $\int \dfrac{9}{x^2}\, dx$

3. Find the function whose slope of the tangent line to the curve is $f'(x) = 1/x$ and whose graph passes through the point $(1, 5)$.

4. If the velocity function of an object is given by $v(t) = -32t + 32$, what is the distance function $s(t)$ if $s(0) = 64$?

5. (a) Use four rectangles to approximate the area under the curve $y = 4 - x^2$ from $x = -1$ to $x = 1$.

   (b) Use the definition to find the area under the curve $y = 4 - x^2$ from $x = -1$ to $x = 1$.

6. Use the definition to evaluate the integral

$$\int_0^1 \dfrac{x}{2}\, dx$$

7. Find the area under the curve $f(x) = x^4$ from $x = -2$ to $x = 2$.

8. Use the fundamental theorem of calculus to evaluate the definite integral

$$\int_1^4 \left(\sqrt{x} - \dfrac{2}{x} \right) dx$$

9. Use the definite integral to find the area bounded by $f(x) = x^3$ and the x-axis, from $x = -2$ to $x = 2$.

In Problems 10 to 12 find the area between $f(x)$ and $g(x)$.

10. $f(x) = -x, g(x) = x^2$

11. $f(x) = x^3, g(x) = x$

12. $f(x) = 8x^3, g(x) = x^4 + 16x^2$

---

*R. M. Adams et al., "The Benefits of Pollution Control: The Case of Ozone and U.S. Agriculture," *American Journal of Agricultural Economics*, Vol. 68, 1986, pp. 886–893.

†William E. Foster et al., "Distribution Welfare Implications of an Irrigation Water Subsidy," *American Journal of Agricultural Economics*, Vol. 68, 1986, pp. 778–786.

‡Nathan Keyfitz, *Applied Mathematical Demography*, New York, Springer-Verlag, 1985, pp. 1–26.

CASE STUDY   ## Consumers' Surplus and Louisiana Fish Farming

Jim Larkin runs a fleet of small fishing boats in the Atchafalaya River basin in southern Louisiana. Jim's brother Larry owns a small farm nearby. Two brothers with two different careers. But Larry's innovative use of his ponds had a devastating effect on Jim's livelihood in the late 1970s.

The bulk of Jim's catch consists of shellfish, such as shrimp, crabs, and crawfish. Each type of shellfish has a distinct season, depending on its own natural characteristics. For instance, crawfish produce only one brood per year, and young crawfish grow rapidly, doubling in weight each month. Sexual maturity is reached from March to July, and the cycle is repeated. The size of the crawfish population is determined by water levels and temperature. When surface water of their habitat disappears in the fall, crawfish burrow into the ground. In addition, natural predators are reduced in this period. The lower the water level is in the fall, the higher will be the population the next year. This means there will be an abundance of crawfish as the water level rises from January to March. Production of good catches requires that the commercial fishing fleet be in synchronization with water changes and crawfish life cycles.

Rearing crawfish in a small pond is an entirely different story. As early as 1950 a few farmers were harvesting crawfish in ponds, but it was a haphazard effort. The pond-reared crawfish catch amounted to less than 1% of the total crawfish market. By 1980 the total crawfish market had increased from about 1 million pounds in 1950 to about 30 million pounds, a 3000% increase. But here's the catch—the pond production of crawfish accounted for a whopping 85% of the total harvest. Fish farming had become big business at the expense of the commercial fleet. The pond producers could circumvent the environmental barriers facing the commercial fleet. For instance, they could maintain the water level in the ponds year-round as well as supply a constant source of food for the fish.

The innovative aquaculture techniques of the pond producers have had a remarkable impact on the crawfish market, especially on the commercial fleet. But what effect does this have on the public? To what extent has pond-reared crawfish benefited consumers?

Economist Frederick W. Bell describes this phenomenon in an article studying the effects of new and innovative production techniques in the marketplace.* Economists use consumers' surplus to measure the benefit of such techniques to the public. One obvious benefit is that there are more crawfish to consume. As demand increases, the price of crawfish falls. Bell uses 1978 as a representative year in his study. He estimates the demand function in 1978 to be

$$D(x) = -0.0055x + 0.39$$

where x is the quantity of crawfish harvested in millions of pounds and $D(x)$ is dollars per pound. The demand function is graphed in Figure 1. Bell estimates

---

*Frederick W. Bell, "Competition from Fish Farming in Influencing Rent Dissipation: The Crawfish Industry," *American Journal of Agricultural Economics*, Vol. 68, 1986, pp. 95–101.

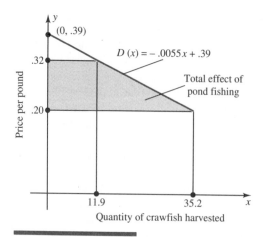

FIGURE 1

that without pond-reared crawfish, only 11.9 pounds would be produced at a price of \$0.32. Thus $D(11.9) = 0.32$. With the addition of pond-reared crawfish, 35.2 lb would be harvested at a price of \$0.20. Thus $D(35.2) = 0.20$. The total effect of pond-reared crawfish is measured by the difference in the two values of consumers' surplus, with and without pond-reared crawfish.

$$\begin{pmatrix} \text{total effect} \\ \text{of pond fish} \end{pmatrix} = \begin{pmatrix} \text{consumers' surplus} \\ \text{with pond fish} \end{pmatrix} - \begin{pmatrix} \text{consumers' surplus} \\ \text{without pond fish} \end{pmatrix}$$

$$= \int_0^{35.2} (-0.0055x + 0.39)\,dx - 35.2(0.20)$$

$$\quad - \left[\int_0^{11.9} (-0.0055x + 0.39)\,dx - 11.9(0.32) \right]$$

$$= (-0.00275x^2 + 0.39x) \Big|_0^{35.2} - 7.04$$

$$\quad - \left[(-0.00275x^2 + 0.39x) \Big|_0^{11.9} - 3.81 \right]$$

$$= -0.00275(35.2)^2 + 0.39(35.2) - 7.04$$

$$\quad - [-0.00275(11.9)^2 + 0.39(11.9) - 3.81]$$

$$= -3.41 + 13.72 - 7.04 + 0.39 - 4.64 + 3.81$$

$$= 2.83$$

This shows that pond-reared fishing greatly increases the benefit to consumers. The fish farmers have gained income as well as expanding the market for crawfish and increasing demand. On the other hand, commercial fishermen like Jim Larkin have experienced a decline in revenue.

The brothers seldom share crawfish stories.

Case Study Exercises

In Problems 1 to 4 compute the total effect of pond fishing for the given demand function and the quantities of crawfish harvested without pond fishing and those harvested with pond fishing.

1. $D(x) = -0.0055x + 0.30$,
   $x = 12$ and $x = 40$

2. $D(x) = -0.0055x + 0.30$,
   $x = 15$ and $x = 30$

3. $D(x) = -0.011x + 0.40$,
   $x = 10$ and $x = 50$

4. $D(x) = -0.011x + 0.40$,
   $x = 20$ and $x = 40$

In this article Bell computes the marginal cost function to be

$$MC(x) = \frac{7.26}{(13.7 - x)^2}$$

He defines "social welfare loss" by the formula

$$\int_a^b MC(x)\, dx - P(b - a)$$

where a and b are the quantities of crawfish harvested and P is the price per pound. In Problems 5 and 6 compute the social welfare loss for the given values taken from Bell.

5. $a = 7.71$, $b = 10.2$, $P = 0.20$
6. $a = 9.2$, $b = 11.9$, $P = 0.32$

CHAPTER REVIEW

Key Terms

6–1 Antiderivatives

Antidifferentiation
Antiderivative
Constant of Integration
Integral Sign

Indefinite Integral
Integrand
Differential Equation
Growth Equation

6–2 Area and the Definite Integral

Lorentz Diagram
Definite Integral

Limits of Integration
Area under a Curve

6–3 The Fundamental Theorem of Calculus

Fundamental Theorem of Calculus

6–4 Area Bounded by Curves

Area Bounded by Two Curves
Equilibrium Price

Consumers' Surplus
Producers' Surplus

Summary of Important Concepts

Formulas

Growth equation $\quad f'(x) = kf(x)$

The fundamental theorem of calculus $\quad \displaystyle\int_a^b f(x)\,dx = F(b) - F(a)$

REVIEW PROBLEMS

In Problems 1 and 2 find the antiderivative.

1. $\displaystyle\int (5x^{-1/5} - 2x^{-3/5} - 3x)\,dx$

2. $\displaystyle\int \frac{x^7 + 3x}{x^3}\,dx$

3. Find the function whose slope of the tangent line is $y' = x^{-2} - x$ and whose graph passes through $(1, 0)$.

4. Use the definition to find the definite integral.

$$\int_0^4 (2x + 1)\,dx$$

5. Find the area under $f(x)$ from $x = 0$ to $x = 2$ where $f(x) = 7 - 3x$.

6. Use the definition to find the definite integral

$$\int_0^2 x^2\,dx$$

by using the identity

$$1^2 + 2^2 + \cdots + n^2 = \frac{n(n + 1)(2n + 1)}{6}$$

7. Find the definite integral by using the fundamental theorem of calculus.

$$\int_1^4 \left(-x^2 + \frac{1}{\sqrt{x}} + 7e^x\right)\,dx$$

8. Use the definite integral to find the area bounded by $f(x) = 4 - x^2$, the x-axis, $x = -4$, and $x = 2$.

9. Find the definite integral.

$$\int_1^2 \left(5e^{3x} - \frac{2}{x}\right)\,dx$$

In Problems 10 to 12 find the area between the curves.

10. $x = 0, \quad x = 1, \quad f(x) = x^2 + 4x,$
    $g(x) = 5x$

11. $f(x) = 5x - x^2, \quad g(x) = 6x$

12. $x = 0, \quad x = 6, \quad f(x) = 3x + 2,$
    $g(x) = 2x^2$

GRAPHICS CALCULATOR EXPLORATIONS

The concepts of consumers' surplus and producers' surplus are key economic indicators, but they also provide an excellent means of studying the concept of area bounded by two curves. The supply and demand curves of a particular commodity determine the consumers' surplus, the theoretical amount consumers save at the given equilibrium price; and the producers' surplus, the theoretical amount producers save at the given equilibrium price. Both consumers' and producers' surplus are areas between curves. In this Exploration we will investigate how these areas are affected by a change in one of the curves. In each Exploration we will ask you to make a conjecture on whether a change in the curves causes the areas to increase or decrease by using your graphics calculator. Then we will indicate how to verify the conjecture algebraically.

Exploration 1

Consider the demand function $y = 4 - x$ and the supply function $y = 3x$. The equilibrium price is computed by solving $4 - x = 3x$, which yields $x = 1$, the equilibrium price. Now graph each function and view the areas that define the consumers' surplus and the producers' surplus. How will a change in the supply function affect each area? If we keep the same demand function and select a supply function with a smaller slope, will the areas both increase, both decrease, one increase and the other decrease, or is it impossible to tell? Select the new supply function $y = x$. The new equilibrium price is the solution of $4 - x = x$, that is, $x = 2$. Now graph the new supply function in the same viewing screen as the first one.

Casio

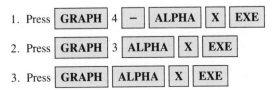

1. Press GRAPH 4 − ALPHA X EXE

2. Press GRAPH 3 ALPHA X EXE

3. Press GRAPH ALPHA X EXE

TI-81

1. Press Y =

2. Press 4 − X | T ENTER

  3 X | T ENTER

  X | T ENTER

3. Press GRAPH

The TI-80 has a command that shades the area of the consumers' surplus. For the functions given here we will first set the values for the range that produce the most impressive display.

1. Press RANGE . Enter new values for the RANGE variables:

  Xmin = (−)2   Ymin = (−)2   Xres = 1
  Xmax = 6   Ymax = 6
  Xscl = 1   Yscl = 1

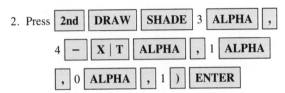

2. Press 2nd DRAW SHADE 3 ALPHA ,

  4 − X | T ALPHA , 1 ALPHA

  , 0 ALPHA , 1) ENTER

Compare the respective areas and try the following conjectures. If the answers are not immediate, try graphing more demand and supply functions. *Conjecture 1:* If the demand curve is held constant and the supply curve is changed to a curve with a smaller slope

(a) Does the equilibrium price increase or decrease?
(b) Does the consumers' surplus increase or decrease?
(c) Does the producers' surplus increase or decrease?

Exploration 2

Consider the demand function $y = 4 - x$ and the supply function $y = x$. As above, the equilibrium price is $x = 2$. Clear the screen and graph the functions to view the consumers' and producers' surplus. Now consider the new demand curve $y = 4 - 3x$ which is a more downward sloping curve than the first demand curve $y = 4 - x$. The new equilibrium price is $x = 1$, arrived at by solving $x = 4 - 3x$. Now graph the new demand function and view the areas that define the consumers' surplus and the producers' surplus.

Casio

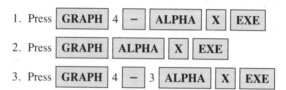

1. Press GRAPH 4 − ALPHA X EXE

2. Press GRAPH ALPHA X EXE

3. Press GRAPH 4 − 3 ALPHA X EXE

TI-81

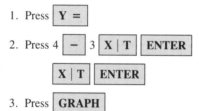

1. Press Y =

2. Press 4 − 3 X | T ENTER

   X | T ENTER

3. Press GRAPH

How will a change in the demand function affect each area? If we keep the same supply function and select a new demand function with a smaller slope, will the areas both increase, both decrease, one increase and the other decrease, or is it impossible to tell? Compare the respective areas and try the following conjecture.

Conjecture 2: If the supply curve is held constant and the demand curve is changed to curve with a smaller slope

(a) Does the equilibrium price increase or decrease?
(b) Does the consumers' surplus increase or decrease?
(c) Does the producers' surplus increase or decrease?

Exploration 3

Now try to justify your answers to the conjectures algebraically.

Conjecture 1: Keep constant the demand curve $y = 4 - x$. Start with the first supply curve $y = 3x$ and compute the consumers' surplus (CS) and producers' surplus (PS); they are CS = 0.5 and PS = 1.5. Compute each area for the new supply function $y = x$; they are CS = 2 and PS = 2. From Exploration 1 we see the equilibrium price increased from $x = 1$ to $x = 2$ and from the latter computation the consumers' surplus increased and the producers' surplus increased.

Conjecture 2: Keep constant the supply curve $y = x$. Start with the first demand curve $y = 4 - x$ and compute CS and PS; they are CS = 2 and PS = 2. Compute each area for the new demand function $y = 4 - 3x$; they are CS = 1.5 and PS = 0.5. From Exploration 1 we see the equilibrium price decreased from $x = 2$ to $x = 1$ and from the latter computation both areas decreased.

Further Explorations

In Problems 1 to 4 write a program that uses the given number of rectangles n to approximate the area under the curve of an arbitrary function $y = f(x)$ from $x = a$ to $x = b$. Then run the program for the function $f(x) = x^2$ from $x = 0$ to $x = 2$.

1. $n = 4$
2. $n = 6$
3. $n = 8$
4. $n = 10$
5. Write a program to compute the consumers' surplus and producers' surplus for the demand function $D(x) = x^2 - ax + b$ and supply function $S(x) = x^2 + cx$.

In Problems 6 to 8 run the program in Problem 5 for the given values of a, b, and c and graph the two functions in the same coordinate system.

6. $a = 6$, $b = 24$, $c = 2$
7. $a = 7$, $b = 20$, $c = 3$
8. $a = 2.2$, $b = 6.8$, $c = 1.2$

Techniques
of Integration

The social security system may not provide security in the future. (H. Armstrong Roberts)

Chapter Overview

The previous chapter showed how to compute the integral of various simple functions. This chapter shows how to integrate more complicated functions. There is no one method of integration that works for all functions; rather, there are various techniques that are applied on a trial-and-error basis.

The first section deals with the simplest technique. It uses a substitution of a variable to reduce the integrand to a familiar form whose integral can be found by a method that was demonstrated earlier. Section 7–2 considers more difficult functions. It relies on the chain rule. The third section explains how to apply a table of integrals. The table gives several formulas involving a diverse number of integrands. Sometimes an exact integrand cannot be found. In this case it is often preferable to approximate the definite integral by means of one of several techniques. This is called numerical integration, which is covered in Section 7–4.

CASE STUDY PREVIEW

Many people mistakenly believe that the taxes they pay to the social security system each year are put into a savings account to be held until their retirement. In reality, the money paid by the work force goes directly to fund those who are retired. Those in the age range of 20 to 64 support those retired over 64. The plan works well when there is a large work force to support a much smaller retired segment of the population.

The Case Study demonstrates that the birthrate is starting to lag well behind the death rate. And the gap is growing. The demographic consequences show that either the system will prove to be a great burden on the work force or the benefits paid to retirees will be greatly reduced.

7–1 Substitution

The fundamental theorem of calculus states that antiderivatives can be used to compute integrals. Thus far we have integrated only relatively simple functions. Just as there are various rules for finding derivatives that depend on the type of function, so too there are several methods for finding indefinite integrals of different types of functions. In this section we discuss the method of integration called **substitution.** To apply this method the integrand must be in a particular form. Determining whether a function fits the description is part of the technique.

Integration by Substitution

Let us start with an example. By the chain rule for derivatives

$$D_x(x^2 + 1)^4 = 4(x^2 + 1)^3 2x = 8x(x^2 + 1)^3$$

The corresponding indefinite integral is

$$\int 8x(x^2 + 1)^3 \, dx = (x^2 + 1)^4 + C$$

Finding this integral was straightforward because we knew $D_x(x^2 + 1)^4$ beforehand. The method of substitution allows us to perform the integration of such functions without knowing the corresponding derivative beforehand. **Integration by substitution** entails expressing the integrand such as the one above as the product of two functions, one of which is the derivative of the other, or at least almost its derivative. In the integrand above, the function $f(x) = x^2 + 1$ is part of the integrand, and its derivative, $f'(x) = 2x$, is also present. The integrand consists of $f(x)$ raised to the third power times $f'(x)$, with the factor 4 left over. Viewed this way, it is in the proper form for substitution. We write

$$\int 8x(x^2 + 1)^3 \, dx = \int 4 \cdot (x^2 + 1)^3 \cdot 2x \, dx$$

Make the substitution $u = x^2 + 1$. Then $\dfrac{du}{dx} = 2x$. Now formally multiply through by dx to get $du = 2x \, dx$. Then the integrand is written

$$\int 8x(x^2 + 1)^3 \, dx = \int 4 \cdot u^3 \cdot du$$

The new integrand is expressed in terms of the new variable u and is easily integrated.

$$\int 4u^3 \, du = u^4 + C$$

Making the reverse substitution for u in terms of x yields the same integral computed earlier.

$$\int 8x(x^2 + 1)^3 \, dx = (x^2 + 1)^4 + C$$

There are two key steps in applying substitution in this type of problem. First, it must be possible to express the integrand as the product of one function raised to a power and another function that is the derivative of the first function. The entire integrand, with the sole exception of a constant left over, must be described this way. The next step is to let the new variable u equal this function and then compute du, the *differential of u*, which is defined as follows:

DEFINITION

If $u = g(x)$, then the **differential of u** is $du = g'(x) \, dx$.

The differential actually has a deeper meaning and a wider application than is used in this text. Since we use differentials only in this chapter to explain techniques of integration, we will not expand on this definition.

   Once the substitution is made, the integrand is expressed in terms of u and constants. Next compute the integral in terms of u if possible. Finally, substitute back for x to find the integral. The result can always be checked by differentiating the result to see if it equals the integrand. Let us look at an example to illustrate the procedure.

EXAMPLE 1

Problem

Compute $\int x(1 - x^2)^5 \, dx$.

Solution Let $u = 1 - x^2$. Then $du = -2x \, dx$. The differential is not present exactly, but it varies only by a constant, -2. In this case the differential can be divided by the constant to get $(-1/2) \, du = x \, dx$. Now the substitution can be made that expresses the integrand entirely in terms of u.

$$\int x(1 - x^2)^5 \, dx = \int (1 - x^2)^5 \, x \, dx = \int u^5 \left(-\frac{1}{2}\right) du = -\frac{1}{2} \int u^5 \, du$$

$$= -\frac{1}{2} \frac{u^6}{6} + C = -\frac{1}{12}(1 - x^2)^6 + C$$

This result can be verified by computing the derivative of $(-1/12)(1 - x^2)^6 + C$ and checking that it equals the original integrand, $x(1 - x^2)^5$.

Notice how the constant $-1/2$ can be taken out of the integrand because of the rule that states that the integral of a constant times a function is the constant times the integral. This is why the expression for du can differ by a constant when deciding what substitution to make for u.

The next example shows how to use substitution to find definite integrals.

EXAMPLE 2

Problem

Find $\int_0^1 (x + 1)(x^2 + 2x + 1)^{-1/2} \, dx$

Solution First find the indefinite integral of $(x + 1)(x^2 + 2x + 1)^{-1/2}$. Let $u = x^2 + 2x + 1$. Then $du = (2x + 2) \, dx = 2(x + 1) \, dx$. The integrand contains only $(x + 1) \, dx$, so we must divide by the constant 2. Dividing by 2 yields $(1/2)du = (x + 1) \, dx$. By the method of substitution

$$\int (x + 1)(x^2 + 2x + 1)^{-1/2} \, dx = \int u^{-1/2} \frac{1}{2} \, du$$

$$= \frac{1}{2} \int u^{-1/2} \, du = \frac{1}{2} \frac{u^{1/2}}{1/2} + C = u^{1/2} + C = (x^2 + 2x + 1)^{1/2} + C$$

Now insert the limits of integration.

$$\int_0^1 (x + 1)(x^2 + 2x + 1)^{-1/2} \, dx = (x^2 + 2x + 1)^{1/2} \Big|_0^1 = 4^{1/2} - 1^{1/2} = 1$$

Sometimes the choice of substitution for u does not work. It might not express the entire integral in terms of u or the resulting integrand may not be easily integrated. Another substitution may be tried. Thus the method often involves trial and error. But not all integrals can be solved by substitution. For example, consider

$$\int (x^2 + 1)^{1/2} \, dx$$

The natural choice would be $u = x^2 + 1$. Then $du = 2x \, dx$. The factor of x is the problem. If the differential differed from the integrand by only a constant, then we could divide through by the constant. But here the differential and the integrand differ by the factor $2x$. If we try to divide through by x, then the left-hand side of the equation for the differential would have two variables, u and x. So the substitution would not entirely replace the variable x by the variable u. Whenever du varies from the integrand by a constant the substitution can be made.

In the examples presented so far the function in the integrand chosen for u has been raised to a power. If we let u be that function, then u' must also be present, at least up to a constant. The general formula governing this type of substitution problem can be expressed as follows, with $u = f(x)$ and $n \neq -1$:

$$\int [f(x)]^n f'(x) \, dx = \int u^n \, du = \frac{u^{n+1}}{n + 1} + C$$

$$= \frac{f(x)^{n+1}}{n + 1} + C, \qquad n \neq 1$$

There are other types of functions that can be integrated by substitution. The next type that we consider are certain exponential functions. If e is raised to the power $f(x)$ and $f'(x)$ is also present in the integrand, set $u = f(x)$ and find du to complete the substitution. The general formula, with $u = f(x)$, is

$$\int e^{f(x)} f'(x) \, dx = \int e^u \, du = e^u + C = e^{f(x)} + C$$

The formula can be verified using the definition of the antiderivative.

$$D_x(e^{f(x)} + C) = e^{f(x)} f'(x)$$

E X A M P L E   3

Problem

Find $\int x^2 e^{x^3} \, dx$.

Solution Let $u = x^3$. Then $du = 3x^2\ dx$ and $(1/3)\ du = x^2\ dx$. Thus

$$\int x^2 e^{x^3}\ dx = \int e^{x^3}(x^2\ dx) = \int e^u \frac{1}{3}\ du = \frac{1}{3}\ e^u + C = \frac{1}{3}\ e^{x^3} + C$$

The third form of the substitution method is derived from the derivative formula $D_x \ln |x| = 1/x = x^{-1}$. Thus $\int x^{-1}\ dx = \ln |x| + C$. If a function is raised to the power -1 and its derivative is present in the integrand, then the substitution can be made and the integral involves a logarithmic function. The same is true if the integrand is a rational function whose numerator is a constant multiple of the derivative of the denominator. This means that the denominator can be expressed in the numerator as a function raised to the power -1.

EXAMPLE 4

Problem

Find $\int x(x^2 - 4)^{-1}\ dx$.

Solution Let $u = x^2 - 4$. Then $du = 2x\ dx$ and $(1/2)\ du = x\ dx$. Thus

$$\int x(x^2 - 4)^{-1}\ dx = \int (x^2 - 4)^{-1}(x\ dx)$$

$$= \int u^{-1} \frac{1}{2}\ du$$

$$= \frac{1}{2} \int u^{-1}\ du = \frac{1}{2} \ln |u| + C$$

$$= \frac{1}{2} \ln |x^2 - 4| + C$$

The general formula, with $u = f(x)$, is

$$\int [f(x)]^{-1} f'(x)\ dx = \int \frac{f'(x)}{f(x)}\ dx$$

$$= \int u^{-1}\ du = \int \frac{du}{u}$$

$$= \ln |u| + C = \ln |f(x)| + C$$

The next example illustrates how to use the exponential and logarithmic forms of the substitution method of integration in which the choice for the substitution is not obvious.

EXAMPLE 5

Problem

Find (a) $\int \dfrac{e^{x^{-1}}}{x^2}\,dx$, (b) $\int \dfrac{x^{1/2}}{x^{3/2}-1}\,dx$.

Solution (a) Let $u = x^{-1}$. Then $du = -x^{-2}\,dx$ and $-du = x^{-2}\,dx$. Thus

$$\int \frac{e^{x^{-1}}}{x^2}\,dx = \int x^{-2}e^{x^{-1}}\,dx = -\int e^u\,du = -e^u + C = -e^{x^{-1}} + C$$

(b) Let $u = x^{3/2} - 1$. Then $du = (3/2)x^{1/2}\,dx$ and $(2/3)\,du = x^{1/2}\,dx$. Thus

$$\int \frac{x^{1/2}}{x^{3/2}-1}\,dx = \int \frac{1}{u}\left(\frac{2}{3}\right)\,du = \frac{2}{3}\int \frac{du}{u}$$

$$= \frac{2}{3}\ln|u| + C = \frac{2}{3}\ln|x^{3/2} - 1| + C$$

The method of substitution is versatile. The next example shows how to apply it in a case where the substitution is subtle.

EXAMPLE 6

Problem

Find $\displaystyle\int_0^1 \frac{x\,dx}{x+1}$.

Solution First find the antiderivative and then insert the limits of integration. Let $u = x + 1$. Then $x = u - 1$ and $du = dx$, so

$$\int \frac{x\,dx}{x+1} = \int \frac{u-1}{u}\,du = \int \left(1 - \frac{1}{u}\right)\,du = u - \ln|u| + C$$

$$= x + 1 - \ln|x + 1| + C$$

Insert the limits of integration.

$$\int_0^1 \frac{x\,dx}{x+1} = x + 1 - \ln|x + 1| \,\bigg|_0^1 = 2 - \ln 2 - (1 - \ln 1)$$

$$= 1 - \ln 2$$

An often troublesome aspect of substitution is the role of the differential, du. The differential is a vital ingredient in putting the integrand in proper form, but when the antiderivative is found it "disappears," in the sense that it merely indicates the variable of integration. That is, once the substitution is made the sole role of du is to label the variable, so it does not appear in the final solution.

EXERCISE SET 7–1

In Problems 1 to 26 find the antiderivative.

1. $\int 2x(x^2 + 5)^3 \, dx$

2. $\int 2x(x^2 - 2)^4 \, dx$

3. $\int 4x^3(x^4 - 7)^2 \, dx$

4. $\int 5x^4(x^5 + 1)^2 \, dx$

5. $\int x(3x^2 + 5)^3 \, dx$

6. $\int x(2x^2 - 3)^5 \, dx$

7. $\int x^3(3x^4 - 5)^3 \, dx$

8. $\int x^4(4x^5 + 1)^4 \, dx$

9. $\int x^5(3x^6 - 5)^{3/2} \, dx$

10. $\int x^3(5x^4 + 1)^{3/4} \, dx$

11. $\int 2x^3(x^4 - 7)^{-3} \, dx$

12. $\int 6x^4(x^5 + 10)^{-4} \, dx$

13. $\int x^{-3}(6x^{-2} + 5)^{-2} \, dx$

14. $\int x^{-4}(5x^{-3} - 1)^{-5} \, dx$

15. $\int (2x + 5)(x^2 + 5x)^{-3} \, dx$

16. $\int (2x - 2)(x^2 - 2x)^{1/4} \, dx$

17. $\int (x^3 - 1)(x^4 - 4x)^{1/2} \, dx$

18. $\int (x^4 + 2x)(x^5 + 5x^2)^{2/5} \, dx$

19. $\int (e^{2x} + 5e^{4x}) \, dx$

20. $\int (5e^{3x} - e^{-3x}) \, dx$

21. $\int (x^3 e^{x^4} + 2e^{-x}) \, dx$

22. $\int (x^{-3} e^{x^{-2}}) \, dx$

23. $\int \frac{2}{3x + 1} \, dx$

24. $\int \frac{x}{x^2 - 1} \, dx$

25. $\int x^2(x^3 - 1)^{-1} \, dx$

26. $\int (x^2 + 1)(x^3 + 3x)^{-1} \, dx$

In Problems 27 to 34 find the definite integral.

27. $\int_0^1 x(x^2 + 1)^2 \, dx$

28. $\int_0^1 x(x^2 - 2)^3 \, dx$

29. $\int_{-1}^1 x^3(x^4 + 1)^2 \, dx$

30. $\int_0^2 x^4(x^5 + 1)^4 \, dx$

31. $\int_0^3 x(x^2 + 1)^{1/2} \, dx$

32. $\int_0^1 x(1 - x^2)^{1/3} \, dx$

33. $\int_1^2 (x^3 + 1)(x^4 + 4x)^{-2} \, dx$

34. $\int_0^1 (x^4 + 5)(x^5 + 5x)^{-2} \, dx$

In Problems 35 to 46 find the antiderivative.

35. $\int (e^x + 1)e^x \, dx$

36. $\int (e^{2x+1} + 10) \, e^{2x} \, dx$

37. $\int \frac{\ln x}{x} \, dx$

38. $\int \frac{\ln x^2}{x} \, dx$

39. $\int x(x - 1)^2 \, dx$

40. $\int x(3x + 4)^4 \, dx$

41. $\int x(1 - 5x)^{1/2} \, dx$

42. $\int 5x(x + 4)^{-4} \, dx$

43. $\int x^3(x^2 - 1)^{2/3} \, dx$

44. $\int x^3(2 - 3x^2)^{-3} \, dx$

45. $\int \frac{x \ln(1 + x^2)}{1 + x^2} \, dx$

46. $\int \frac{e^x \ln(1 + e^x)}{1 + e^x} \, dx$

Referenced Exercise Set 7–1

1. In an article studying the possible environmental impact of uranium mining in northern Australia, Vardavas investigated how contaminants were transported by surface water and ground water.* This study is meant to predict human radiological exposure from contaminants from mine sites. One part of the study requires the calculation of the total water evaporation per day at Manton Dam. The evaporation rate $E(x)$ for a given wind speed x can be approximated by

$$E(x) = \frac{.5x + .8}{x + 1}$$

The total evaporation is then computed by finding the antiderivative of $E(x)$. Find this antiderivative.

---

*Ilias Mihail Vardavas, "Modelling the Seasonal Variation of Net All-wave Radiation Flux and Evaporation in a Tropical Wet-dry Region," *Ecological Modelling*, Vol. 39, 1987, pp. 247–268.

2. Talent searches identifying mathematically gifted high school youth have shown remarkably consistent sex differences. In an article studying the effect of a sex-linked gene that might facilitate high mathematical test score performance, Thomas considered data from the Iowa Test of Basic Skills and the SAT-mathematics test.† One part of the study investigated the properties of the function

$$f(x) = \frac{x}{x^2 + 1}$$

where $f(x)$ is a psychological measure, called the Mills ratio, and x is a percentile. Find the antiderivative of $f(x)$.

†Hoben Thomas, ''A Theory of High Mathematical Aptitude,'' *Journal of Mathematical Psychology,* Vol. 29, 1985, pp. 231–242.

7–2 Integration by Parts

In the previous section we mentioned that substitution is a trial-and-error method that does not work in all cases. In this section we consider another method of integration, integration by parts. To apply the technique, think of the integrand as formed by two distinct functions, or ''parts.''

Integration by Parts

The product rule for derivatives cannot be transformed directly into an integration formula, but it can be revised to give a powerful rule. Suppose u and v are functions of x. The product rule states that

$$D_x uv = u D_x v + v D_x u$$

Expressing this formula in terms of differentials yields

$$d(uv) = u \, dv + v \, du \quad .$$

It is now in a form that is more amenable for integration. But first solve for $u \, dv$.

$$u \, dv = d(uv) - v \, du$$

Taking the integral of both sides yields

$$\int u \, dv = \int d(uv) - \int v \, du = uv - \int v \, du$$

This is the formula for the technique of integration called **integration by parts.**

> **Integration by Parts**
> If u and v are differentiable functions, then
>
> $$\int u \, dv = uv - \int v \, du$$

The term "parts" refers to the way that the integrand is separated into two functions, u and dv. Thus there are two parts to the integrand. The underlying idea of the technique is that the integral $\int u \, dv$ is difficult, if not impossible, to calculate. The goal is to choose u and v so that the integral $\int v \, du$ is easier to calculate. The formula for integration by parts shows how to use $\int v \, du$ to compute $\int u \, dv$. The first example illustrates how to use the formula.

EXAMPLE 1

Problem

Compute $\int xe^x \, dx$

Solution Separate the integrand into u and dv as follows:

$$u = x \qquad dv = e^x \, dx$$

Then

$$du = dx \qquad v = \int dv = \int e^x \, dx = e^x$$

Since the constant of integration is usually only included in the final answer, it is not mentioned in the intermediate steps. From the formula

$$\int xe^x \, dx = \int u \, dv = uv - \int v \, du = xe^x - \int e^x \, dx = xe^x - e^x + C$$

It is important to realize that integration by parts is another trial-and-error method. For example, in Example 1 if the choice for u and dv is $u = e^x$ and $dv = x \, dx$, then $du = e^x \, dx$ and $v = \int x \, dx = x^2/2$. Then the formula yields $\int xe^x \, dx = x^2e^x/2 - \int x^2e^x/2 \, dx$. The latter integral is more complex than the original one. This choice of u and dv does not work. The object is to choose u and dv so that

1. It is possible to compute $v = \int dv$, and
2. It is possible to compute $\int v \, du$, or $\int v \, du$ is simpler than $\int u \, dv$.

The next example shows how to compute a seemingly simple integral by a judicious choice of u and dv.

EXAMPLE 2

Problem

Compute $\int \ln x \, dx$.

Solution Separate the integrand into u and dv as follows:

$$u = \ln x \qquad dv = dx$$

Then

$$du = \frac{1}{x} dx \qquad v = \int dv = \int dx = x$$

Substituting these expressions in the formula, we get

$$\int \ln x \, dx = \int u \, dv = uv - \int v \, du$$

$$= (\ln x)x - \int x\left(\frac{1}{x}\right) dx = x \ln x - \int dx$$

$$= x \ln x - x + C$$

Sometimes it is necessary to apply the formula more than once to compute the integral, as demonstrated in Example 3.

EXAMPLE 3

Problem

Compute $\int x^2 e^{3x} \, dx$.

Solution The idea is to eliminate the x^2 from the integrand. This is not possible in one application of the technique, but we can replace x^2 by x to simplify the integrand. Separate the integrand into u and dv as follows:

$$u = x^2 \qquad dv = e^{3x} \, dx$$

$$du = 2x \, dx \qquad v = \int dv = \int e^{3x} \, dx = \frac{1}{3} e^{3x}$$

$$\int x^2 e^{3x} \, dx = uv - \int v \, du = x^2 \frac{1}{3} e^{3x} - \int \frac{1}{3} e^{3x} 2x \, dx$$

$$= \frac{1}{3} x^2 e^{3x} - \frac{2}{3} \int x e^{3x} \, dx \tag{1}$$

The problem has been reduced to finding $\int x e^{3x} \, dx$. To compute the latter integral, use integration by parts again, this time with

$$u = x \qquad dv = e^{3x}$$

$$du = dx \qquad v = \frac{1}{3} e^{3x}$$

This is similar to Example 1. The formula yields

$$\int x e^{3x} \, dx = x \frac{1}{3} e^{3x} - \int \frac{1}{3} e^{3x} \, dx = \frac{1}{3} x e^{3x} - \frac{1}{9} e^{3x}$$

Now substitute this result into equation (1) to get the final answer.

$$\int x^2 e^{3x}\, dx = \frac{1}{3} x^2 e^{3x} - \frac{2}{3} \int xe^{3x}\, dx$$

$$= \frac{1}{3} x^2 e^{3x} - \frac{2}{3} \left[\frac{1}{3} xe^{3x} - \frac{1}{9} e^{3x} \right] + C$$

$$= \frac{1}{3} x^2 e^{3x} - \frac{2}{9} xe^{3x} + \frac{2}{27} e^{3x} + C$$

The next example shows how we sometimes need to employ more than one technique of integration in a problem. It uses substitution and integration by parts.

EXAMPLE 4

Problem

Compute $\int x(2x + 1)^{1/2}\, dx$.

Solution Choose u and dv as follows:

$$u = x \qquad dv = (2x + 1)^{1/2}\, dx$$

$$du = dx \qquad v = \int (2x + 1)^{1/2}\, dx$$

To compute v, we must use substitution. Let $w = 2x + 1$. Then $dw = 2\, dx$, $(1/2)dw = dx$, and so

$$v = \int (2x + 1)^{1/2}\, dx = \int w^{1/2} \left(\frac{1}{2} \right) dw = \frac{1}{2} \left(\frac{2}{3} \right) w^{3/2} = \frac{1}{3} (2x + 1)^{3/2}$$

Substituting into the formula for integration by parts yields

$$\int x(2x + 1)^{1/2}\, dx = \int u\, dv = uv - \int v\, du$$

$$= x \left(\frac{1}{3} \right) (2x + 1)^{3/2} - \int \frac{1}{3} (2x + 1)^{3/2}\, dx$$

$$= \frac{1}{3} x(2x + 1)^{3/2} - \frac{1}{3} \int (2x + 1)^{3/2}\, dx$$

Again, the substitution $w = 2x + 1$ is needed in the last integral. The final answer is

$$\int x(2x + 1)^{1/2}\, dx = \frac{1}{3} x(2x + 1)^{3/2} - \frac{1}{15} (2x + 1)^{5/2} + C$$

The integral in Example 4 could have been computed by the substitution $u = 2x + 1$. In fact substitution is the first method to try. This method is explored in

Exercise 41. We used integration by parts to illustrate the method and to show that the two methods can be used on the same problem.

Definite Integrals and Parts

To find a definite integral using integration by parts, first find the antiderivative and then insert the limits of integration. This yields the following formula:

$$\int_a^b u \, dv = uv \bigg|_a^b - \int_a^b v \, du$$

The next example shows that first we can calculate $\int v \, du$ and then insert the limits.

EXAMPLE 5

Problem

Compute $\int_0^4 x(2x + 1)^{1/2} \, dx$.

Solution First find the indefinite integral. This was done in Example 4. Then insert the limits of integration.

$$\int_0^4 x(2x + 1)^{1/2} \, dx = \left[\frac{1}{3} x(2x + 1)^{3/2} - \frac{1}{15} (2x + 1)^{5/2} \right] \bigg|_0^4$$

$$= \frac{1}{3} 4(9)^{3/2} - \frac{1}{15} (9)^{5/2} - 0 + \frac{1}{15}$$

$$= \frac{108}{3} - \frac{243}{15} + \frac{1}{15} = \frac{298}{15}$$

EXERCISE SET 7–2

In Problems 1 to 20 find the antiderivative.

1. $\int 2xe^x \, dx$

2. $\int 5xe^x \, dx$

3. $\int xe^{3x} \, dx$

4. $\int 5xe^{4x} \, dx$

5. $\int x(3x + 5)^{3/2} \, dx$

6. $\int x(2x - 3)^{5/3} \, dx$

7. $\int 3x(5x - 2)^{-3} \, dx$

8. $\int 2x(4x + 1)^{-4} \, dx$

9. $\int x \ln x \, dx$

10. $\int x \ln 5x \, dx$

11. $\int x^2 \ln x \, dx$

12. $\int x^4 \ln x \, dx$

13. $\int x^2 e^x \, dx$

14. $\int x^2 e^{5x-3} \, dx$

15. $\int (2x + 5)(x + 5)^{-3} \, dx$

16. $\int (x - 2)(4x - 3)^{-1/2} \, dx$

17. $\int x^2(x - 4)^{1/2} \, dx$

18. $\int x^2(1 - x)^{2/3} \, dx$

19. $\int x^2(1 - 3x)^{-2} \, dx$

20. $\int 4x^2(2 + 3x)^{-3} \, dx$

In Problems 21 to 26 find the definite integral.

21. $\int_0^1 xe^{2x} \, dx$

22. $\int_0^1 xe^{x+2} \, dx$

23. $\int_0^2 x(2x + 5)^{1/2} \, dx$

24. $\int_0^2 x(4x + 1)^{-1/2} \, dx$

25. $\int_0^2 x^2(x + 1)^{-4} \, dx$

26. $\int_0^2 x^2(2x + 3)^{-3} \, dx$

In Problems 27 to 34 find the antiderivative.

27. $\int x^2(x + 1)^{1/2} \, dx$

28. $\int x^2(1 - 3x)^{-1/2} \, dx$

29. $\int x^3(x + 1)^{1/2} \, dx$

30. $\int x^3(1 - 3x)^{-1/2} \, dx$

31. $\int x^3(x^2 + 1)^{1/2} \, dx$

32. $\int x^3(2 - 3x^2)^{-1/2} \, dx$

33. $\int x^3 e^{x^2} \, dx$ 34. $\int x^5 e^{x^3} \, dx$

In Problems 35 and 36 find the definite integral.

35. $\int_0^2 x^5(2x^2 + 1)^{1/2} \, dx$

36. $\int_0^1 x^5(x^2 + 3)^{-4} \, dx$

In Problems 37 to 40 find the antiderivative.

37. $\int (x + 1) \ln x \, dx$

38. $\int (x^2 + 1) \ln x \, dx$

39. $\int x^{-2} \ln x \, dx$

40. $\int x^{-3} \ln x \, dx$

41. Evaluate the integral in Example 4 using the substitution $u = 2x + 1$.

Referenced Exercise Set 7–2

1. In an article studying organic contaminants in natural environments, especially industrial and municipal water wastes, Chang and Rittmann focus attention on the contaminant activated carbon.* Let $q(r)$ be the surface concentration of activated carbon on a circular surface r units from the center of the concentration. The computation of the total concentration involves evaluating the definite integral of $r^2 q(r)$ from $r = 0$ to $r = R$, the radius of the circular region. Compute this integral for $q(r) = e^{-.1r}$ and $R = 1$ mm.

2. In business the term "mark-up" refers to the difference between the selling price to the customer and the cost the retailer paid for the item. The mark-up takes into account the retailer's cost to market and sell the item and the retailer's profit. In an article studying how mark-up varies during various phases of the business cycle, Goldstein applied his model to 20 manufacturing industries from 1949 to 1980.† The goal of each industry is to maximize profit subject to cost restrictions. Goldstein assumed that profit $P(t)$ varied with time t according to the relationship $P'(t) = e^{-rt}f(t)$, where r is the prevailing interest rate and $f(t)$ depended on selling price, market share, and competitors' prices. He defined total profit $P(t)$ as the antiderivative of $P'(t)$. Find $P(t)$ for $f(t) = t$.

Cumulative Exercise Set 7–2

In Problems 1 to 8 find the antiderivative.

1. $\int x^3(2x^4 + 3)^5 \, dx$

2. $\int x^{-2}(3x^{-1} - 1)^{-3} \, dx$

3. $\int (x^3 - 2x)(x^4 - 4x^2)^{2/3} \, dx$

4. $\int x^{-4} e^{x-3} \, dx$

5. $\int 3xe^{2x} \, dx$

6. $\int x(3x - 4)^{1/3} \, dx$

7. $\int x^5 \ln x \, dx$

8. $\int x^3(2 + x)^{1/2} \, dx$

*Ted Chang and Bruce E. Rittmann, "Mathematical Modelling of Biofilm on Activated Carbon," *Environmental Science and Technology*, Vol. 21, 1987, pp. 273–280.

†Jonathan Goldstein, "Mark-up Pricing over the Business Cycle: The Microfoundations of the Variable Mark-up," *Southern Economics Journal*, Vol. 53, 1987, pp. 233–246.

7–3 Tables of Integrals

Most of the integrands encountered thus far have required a direct application of one particular technique of integration. Often, however, an integration problem requires lengthy and cumbersome manipulation. For such problems a table of integrals is usually helpful. The table in the Appendix provides a list of antiderivatives that is intended to acquaint the student with such tables. More extensive tables can be found, such as in the *Mathematical Handbook of Formulas and Tables*, which contains more than 500 antiderivatives.

Using a table of integrals is usually straightforward, but there are a few pitfalls that need to be pointed out. Tables of integrals are organized by similarity of the integrands. The first few listings are generally the elementary formulas. These are usually followed by integrands involving $ax + b$ for constants a and b, which in turn are followed by integrands involving $ax^2 + b$, $(ax + b)^2$, and so forth.

Because the table is arranged in this way, it is easy to mistake the correct formula for a similar one. For example, consider the integral

$$\int \frac{dx}{x(3x + 2)}$$

Referring to the table, we find that formulas 12, 13, and 15 are all similar. Each could be mistaken for the proper formula, namely, formula 14 with $a = 3$ and $b = 2$. Formula 12 is incorrect because there is no x appearing in the denominator of the formula's integrand. Similarly, formulas 13 and 15 are incorrect because $(ax + b)^2$ appears in the integrand rather than $(ax + b)$.

The first example shows how to find an antiderivative when the integrand fits the formula directly. Later examples show how some integrands need to be expressed in a different form to apply a formula.

EXAMPLE 1

Problem

Find $\int \dfrac{dx}{x(3x + 2)^2}$.

Solution The integrand fits formula 15 with $a = 3$ and $b = 2$. Substituting these values into formula 15 yields

$$\int \frac{dx}{x(3x + 2)^2} = \frac{1}{2(3x + 2)} + \frac{1}{4} \ln \left| \frac{x}{3x + 2} \right| + C$$

Often some algebraic manipulation is necessary to see which formula is correct, as illustrated in the next example.

EXAMPLE 2

Problem

Find $\int \dfrac{dx}{1 - 4x^2}$.

Solution The integrand fits formula 6, but it cannot be applied immediately because the coefficient of x^2 in the problem is 4, while the coefficient of x^2 in the formula is 1. The coefficient must be factored out of the denominator and the integral sign because the coefficients of x^2 in the problem and the formula must agree. Use the fact that $1 - 4x^2 = 4(1/4 - x^2)$. Hence

$$\int \frac{dx}{1 - 4x^2} = \int \frac{dx}{4(1/4 - x^2)} = \frac{1}{4} \int \frac{dx}{1/4 - x^2}$$

Now formula 6 can be applied with $a = 1/2$.

$$\int \frac{dx}{1 - 4x^2} = \frac{1}{4} \int \frac{dx}{1/4 - x^2} = \frac{1}{4} \ln \left| \frac{1/2 + x}{1/2 - x} \right| + C$$

This formula holds only for $x^2 < a^2$; otherwise the antiderivative is not defined.

Most tables do not include integrals that can be simplified by a simple substitution. Sometimes a simple substitution of variable is required before applying the appropriate formula.

EXAMPLE 3

Problem

Find $\int x(x^4 + 4)^{1/2} \, dx$.

Solution The integrand is similar to formula 16, but it cannot be applied immediately because the problem contains x^4 instead of x^2 in the formula. The x outside the parentheses permits the substitution $u = x^2$. Then $du = 2x \, dx$ and so $(1/2)du = x \, dx$. Since $u^2 = x^4$, this substitution yields

$$\int x(x^4 + 4)^{1/2} \, dx = \frac{1}{2} \int (u^2 + 4)^{1/2} \, du$$

Now apply formula 16 with $a = 2$.

$$\int x(x^4 + 4)^{1/2} \, dx = \frac{1}{2} \int (u^2 + 4)^{1/2} \, du$$

$$= \frac{1}{2} \left[\frac{1}{2} u(u^2 + 4)^{1/2} + \frac{4}{2} \ln \left| u + (u^2 + 4)^{1/2} \right| \right] + C$$

$$= \frac{1}{4} x^2(x^4 + 4)^{1/2} + \ln \left| x^2 + (x^4 + 4)^{1/2} \right| + C$$

Formula 18 is a recursive formula, meaning that it might have to be applied several times to solve a problem.

EXAMPLE 4

Problem

Compute $\int x^3 e^{2x} \, dx$.

Solution Apply formula 18 with $n = 3$ and $a = 2$.

$$\int x^3 e^{2x}\, dx = \frac{1}{2} x^3 e^{2x} - \frac{3}{2} \int x^2 e^{2x}\, dx \tag{1}$$

To compute the integral on the right, again apply formula 18, this time with $n = 2$ and $a = 2$.

$$\int x^2 e^{2x}\, dx = \frac{1}{2} x^2 e^{2x} - \frac{2}{2} \int x e^{2x}\, dx \tag{2}$$

Once again apply formula 18, now with $n = 1$ and $a = 2$.

$$\int x e^{2x}\, dx = \frac{1}{2} x e^{2x} - \frac{1}{2} \int e^{2x}\, dx$$

$$= \frac{1}{2} x e^{2x} - \frac{1}{4} e^{2x} + C \tag{3}$$

Now put the pieces of the puzzle together. Substitute (3) into (2), and substitute the result into (1). The final answer is

$$\int x^3 e^{2x}\, dx = \frac{1}{2} x^3 e^{2x} - \frac{3}{2} \left[\frac{1}{2} x^2 e^{2x} - \frac{1}{2} x e^{2x} + \frac{1}{4} e^{2x} \right] + C$$

$$= \frac{1}{2} x^3 e^{2x} - \frac{3}{4} x^2 e^{2x} + \frac{3}{4} x e^{2x} - \frac{3}{8} e^{2x} + C$$

To evaluate a definite integral via the table of integrals, first find the indefinite integral and then evaluate the antiderivative at the limits of integration.

EXAMPLE 5

Problem

Find $\displaystyle\int_2^3 \frac{dx}{x(2x - 3)}$.

Solution The integrand fits formula 14 with $a = 2$ and $b = -3$. Substitute these values into the formula without the limits of integration.

$$\int \frac{dx}{x(2x - 3)} = \frac{1}{-3} \ln \left| \frac{x}{(2x - 3)} \right| + C$$

Inserting the limits of integration yields

$$\int_2^3 \frac{dx}{x(2x - 3)} = -\frac{1}{3} \ln \left| \frac{x}{(2x - 3)} \right| \Big|_2^3$$

$$= -\frac{1}{3} \ln \left| \frac{3}{3} \right| + \frac{1}{3} \ln \left| \frac{2}{1} \right| = \frac{1}{3} \ln 2$$

EXERCISE SET 7–3

In Problems 1 to 38 find the integral.

1. $\int \dfrac{dx}{x(3x + 2)^2}$

2. $\int \dfrac{dx}{x(5x + 6)^2}$

3. $\int \dfrac{dx}{x(3x - 4)^2}$

4. $\int \dfrac{dx}{x(5x - 7)^2}$

5. $\int \dfrac{dx}{1 - 9x^2}$

6. $\int \dfrac{dx}{4 - 9x^2}$

7. $\int \dfrac{dx}{16 - 9x^2}$

8. $\int \dfrac{dx}{4 - 25x^2}$

9. $\int \dfrac{dx}{36 - 4x^2}$

10. $\int \dfrac{dx}{25 - 4x^2}$

11. $\int x(x^4 + 9)^{1/2} \, dx$

12. $\int x(x^4 + 16)^{1/2} \, dx$

13. $\int x(4x^4 + 9)^{1/2} \, dx$

14. $\int x(9x^4 + 16)^{1/2} \, dx$

15. $\int 2x(4x^4 - 25)^{1/2} \, dx$

16. $\int 3x(9x^4 - 49)^{1/2} \, dx$

17. $\int_2^3 \dfrac{dx}{x(3x - 5)}$

18. $\int_2^3 \dfrac{dx}{x(6 - 5x)}$

19. $\int_1^2 \dfrac{dx}{x(3x + 2)^2}$

20. $\int_1^3 \dfrac{dx}{x(5x + 6)^2}$

21. $\int_0^1 x(x^4 + 1)^{1/2} \, dx$

22. $\int_0^1 x(x^4 + 4)^{1/2} \, dx$

23. $\int x^3 e^{3x} \, dx$

24. $\int x^4 e^{5x} \, dx$

25. $\int \dfrac{x \, dx}{(5x + 2)^2}$

26. $\int \dfrac{x \, dx}{(7x + 6)^2}$

27. $\int \dfrac{dx}{x(x - 4)}$

28. $\int \dfrac{x \, dx}{5x - 7}$

29. $\int \dfrac{dx}{x\sqrt{1 - 9x^2}}$

30. $\int \dfrac{dx}{\sqrt{4 + 9x^2}}$

31. $\int \dfrac{x \, dx}{\sqrt{16 + 9x^4}}$

32. $\int \dfrac{dx}{x\sqrt{4 - x^4}}$

33. $\int \dfrac{dx}{x\sqrt{9 + 4x^2}}$

34. $\int \dfrac{x \, dx}{49 - 4x^4}$

35. $\int \dfrac{x^2 \, dx}{9 - 4x^6}$

36. $\int \dfrac{x^3 \, dx}{9 - 4x^4}$

37. $\int x^5 \ln x^2 \, dx$

38. $\int x^5 e^{x^2} \, dx$

Referenced Exercise Set 7–3

1. The study of waiting situations, the so-called "queueing problems," usually assumes that people dislike having to wait. In an article studying the resulting psychological cost of waiting, Osuna defines the total stress of waiting $TH(t)$ as the integral of $H(t)$, the stress at time t.* Find the indefinite integral of $H(t)$ for

$$H(t) = \frac{t^2}{10 - t}$$

2. Purchasing land always involves risk, but the purchase of farmland usually involves a greater amount of risk than other types of land because there are so many additional factors. These include complex tax laws and government price supports that affect not only profit but land value. In an article studying farmland pricing strategies, Brown and Brown define the expected earnings per acre $E(a, b)$ of selling a piece of land for a price between $x = a$ and $x = b$ hundred dollars per acre to be

$$E(a, b) = \int_a^b f(x) \, dx$$

where $f(x)$ is the likelihood that the seller will get a bid of price x.† Find $E(1, 1.5)$ for

$$f(x) = \frac{x}{x + 1}$$

Cumulative Exercise Set 7–3

In Problems 1 to 8 find the antiderivative.

1. $\int \sqrt{1 + 5x} \, dx$

2. $\int \dfrac{6x \, dx}{\sqrt{x^2 + 4}}$

*Edgar Elias Osuna, "The Psychological Cost of Waiting," *Journal of Mathematical Psychology*, Vol. 29, 1985, pp. 82–106.

†Keith C. Brown and Deborah J. Brown, "Heterogenous Expectations and Farmland Prices," *American Journal of Agricultural Economics*, Vol. 66, 1984, pp. 164–169.

3. $\displaystyle\int (3 - x^2)^2 dx$

4. $\displaystyle\int \left(x - \frac{3}{x}\right) dx$

5. $\displaystyle\int xe^{-x} dx$

6. $\displaystyle\int x\sqrt{x - 2}\, dx$

7. $\displaystyle\int \frac{x\, dx}{2x - 2}$

8. $\displaystyle\int x^3 \ln x\, dx$

In Problems 9 and 10 find the definite integral.

9. $\displaystyle\int_2^8 \frac{dx}{x^2}$

10. $\displaystyle\int_0^4 x\sqrt{9 + x^2}\, dx$

In Problems 11 and 12 find the area between the curve $y = f(x)$ and the x-axis from $x = 0$ to $x = 2$.

11. $f(x) = x\sqrt{4 - x^2}$

12. $f(x) = xe^x$

7—4 Numerical Integration

The definite integral was defined by using approximations to the area under a curve that consisted of sums of areas of certain rectangles. A limit process was applied to define the definite integral. The fundamental theorem of calculus permits us to calculate integrals by using antiderivatives, thus avoiding the need to use approximations. But sometimes it is impractical or even impossible to find the correct antiderivative. In this case we need to approximate the area under the curve in order to approximate the definite integral.

There are many ways to approximate the definite integral. They are called **numerical integration** methods because they require several numerical calculations. With the advent of calculators and computers the tedium of these computations has been reduced significantly.

One method of approximation is to use rectangles, as is done when defining the definite integral. In this section we will present another method, called the **trapezoidal rule,** that usually produces a closer approximation than the method using merely rectangles. A third method, **Simpson's rule,** is described in the exercises.

An Example

Rectangles can yield good approximations to the area under a curve, but their upper leg is always horizontal. A trapezoid's upper leg can be chosen so that it usually comes closer to the curve. This is why trapezoids usually produce closer approximations to the area under a curve. In Figure 1 we see that the sum of the areas of the trapezoids is a better approximation to the area under the curve than the sum of the areas of the rectangles. Each trapezoid comes closer to the area under the curve.

Let us look at an example of how to approximate the area under a curve with trapezoids. Then we will develop the formula for approximating $\int_a^b f(x)\, dx$. Consider the area under $f(x) = x^2$ from $x = 0$ to $x = 4$. The first step is to divide the interval $[0, 4]$ into subintervals. We choose a convenient number, say, 4, and select the subintervals so that they have equal length. Then the end points of the subintervals are 0, 1, 2, 3, and 4. Each trapezoid has width 1. The heights of the trapezoids are $f(0) = 0$, $f(1) = 1$, $f(2) = 4$, $f(3) = 9$, and $f(4) = 16$. (See

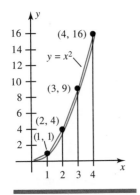

FIGURE 1

Figure 1.) The formula for the area of a trapezoid with heights h and k and width w is

$$\frac{w}{2}(h + k)$$

Let the area of the four trapezoids be A_1, A_2, A_3, and A_4. Then

$$A_1 = \frac{1}{2}[f(0) + f(1)] = \frac{1}{2}[0 + 1] = \frac{1}{2}$$

$$A_2 = \frac{1}{2}[f(1) + f(2)] = \frac{1}{2}[1 + 4] = \frac{5}{2}$$

$$A_3 = \frac{1}{2}[f(2) + f(3)] = \frac{1}{2}[4 + 9] = \frac{13}{2}$$

$$A_4 = \frac{1}{2}[f(3) + f(4)] = \frac{1}{2}[9 + 16] = \frac{25}{2}$$

The approximation is the sum of these areas. It is

$$A_1 + A_2 + A_3 + A_4 = \frac{1}{2} + \frac{5}{2} + \frac{13}{2} + \frac{25}{2} = \frac{44}{2} = 22$$

The actual area can be found using the fundamental theorem. It is

$$\int_0^4 x^2\, dx = \frac{x^3}{3}\Big|_0^4 = \frac{64}{3}$$

As a decimal rounded to two places the actual integral is 21.33. This compares favorably with the approximation via the trapezoidal rule with four subintervals, which is 22. A better approximation could be obtained by using more subintervals.

Trapezoidal Rule

We now present the formula for the trapezoidal rule. Assume that $f(x) \geq 0$ in the interval $[a, b]$ so that the area under the curve is equal to the definite integral. To approximate the definite integral $\int_a^b f(x)\, dx$, first divide the interval $[a, b]$ into n equal subintervals, each having width $(b - a)/n$. The elements of the subdivision are

$$a = x_0, x_1, \ldots, x_n = b$$

Observe that $x_k = a + k(b - a)/n$, for $k = 0, 1, \ldots, n$. The area A_k of a trapezoid whose base is $[x_{k-1}, x_k]$ and whose heights are $f(x_{k-1})$ and $f(x_k)$ is given by

$$A_k = \frac{b - a}{2n}[f(x_{k-1}) + f(x_k)]$$

To approximate the area under the curve, we sum all the areas of the trapezoids. Thus

$$\int_a^b f(x)\, dx \approx A_1 + A_2 + \cdots + A_n$$

$$= \frac{b-a}{2n}\,[f(x_0) + f(x_1)] + \frac{b-a}{2n}\,[f(x_1) + f(x_2)]$$

$$+ \cdots + \frac{b-a}{2n}\,[f(x_{n-1}) + f(x_n)]$$

The expression $(b-a)/2n$ can be factored from each term. Also, each term $f(x_k)$ occurs twice, except the first and last terms, $f(b)$ and $f(a)$. These simplifications result in the following formula, known as the *trapezoidal rule*.

Trapezoidal Rule

Suppose $f(x)$ is a continuous function on the interval $[a, b]$ and $f(x) \geq 0$ on $[a, b]$. To approximate the definite integral of $f(x)$ from a to b by trapezoids, divide $[a, b]$ into n equal subintervals whose end points are $a = x_0, x_1, \ldots, x_n = b$, then

$$\int_a^b f(x)\, dx \approx \frac{(b-a)}{2n}\,[f(a) + 2f(x_1)$$

$$+ 2f(x_2) + \cdots + 2f(x_{n-1}) + f(b)]$$

The first example shows that the formula yields the same result as the earlier computations.

EXAMPLE 1

Problem

Use the trapezoidal rule with $n = 4$ to approximate the integral $\int_0^4 x^2\, dx$.

Solution Each subinterval has width 1, and the heights of the trapezoids are $f(0) = 0$, $f(1) = 1$, $f(2) = 4$, $f(3) = 9$, and $f(4) = 16$. The formula yields

$$\int_0^4 x^2\, dx \approx \frac{4}{8}\,[f(0) + 2f(1) + 2f(2) + 2f(3) + f(4)]$$

$$= \frac{1}{2}\,[0 + 2 + 8 + 18 + 16] = \frac{44}{2} = 22$$

This agrees with earlier computations.

The next example shows that better approximations are usually obtained by increasing the number of subintervals.

EXAMPLE 2

Problem

Use the trapezoidal rule with $n = 8$ to approximate the integral $\int_0^4 x^2 \, dx$.

Solution Each subinterval has width $1/2$ since $b = 4$, $a = 0$, and $n = 8$. The heights of the trapezoids are $f(0) = 0$, $f(1/2) = 1/4$, $f(1) = 1$, $f(3/2) = 9/4$, $f(2) = 4$, $f(5/2) = 25/4$, $f(3) = 9$, $f(7/2) = 49/4$, and $f(4) = 16$. The formula yields

$$\int_0^4 x^2 \, dx \approx \frac{1}{4}\left[f(0) + 2f\left(\frac{1}{2}\right) + 2f(1) + \cdots + f(4) \right]$$

$$= \frac{1}{4}\left[0 + \frac{1}{2} + 2 + \frac{9}{2} + 8 + \frac{25}{2} + 18 + \frac{49}{2} + 16 \right]$$

$$= \frac{1}{4}\left(\frac{172}{2}\right) = \frac{43}{2} = 21.5$$

This is closer to the actual area of $21\frac{1}{3}$ than the earlier approximation using four subintervals. The approximation differs from the actual area by one-sixth. This is called the error of the approximation. As mentioned earlier, the error of the approximation can be further reduced by using more subintervals. In many applications of numerical integration the actual value of the integral cannot be easily computed, and it is often important to be able to compute a bound on the size of the error so that you can judge the accuracy of the approximation. More advanced books study techniques for estimating such errors.

The next two examples show how to use the trapezoidal rule when no antiderivative is attainable in a simple form.

EXAMPLE 3

Problem

Use the trapezoidal rule with $n = 4$ to approximate $\int_0^2 (x^3 + 1)^{1/2} \, dx$.

Solution Let $n = 4$, $a = 0$, and $b = 2$. Then

$$f(x_0) = f(0) = 1$$

$$f(x_1) = f\left(\frac{1}{2}\right) = \sqrt{\frac{9}{8}} \approx 1.06$$

$$f(x_2) = f(1) = \sqrt{2} \approx 1.41$$

$$f(x_3) = f\left(\frac{3}{2}\right) = \sqrt{\frac{35}{8}} \approx 2.09$$

$$f(x_4) = f(2) = \sqrt{9} = 3$$

Substituting these values into the formula yields

$$\int_0^2 (x^3 + 1)^{1/2} \, dx \approx \frac{1}{4}\left[f(0) + 2f\left(\frac{1}{2}\right) + 2f(1) + 2f\left(\frac{3}{2}\right) + f(2) \right]$$

$$= \frac{1}{4}(1 + 2.12 + 2.82 + 4.18 + 3) = \frac{1}{4}(13.12) = 3.28$$

EXAMPLE 4

Problem

Use the trapezoidal rule with $n = 5$ to approximate $\int_0^1 e^{-x^2} \, dx$.

Solution Let $n = 5$, $a = 0$, and $b = 1$. Then

$$f(x_0) = f(0) = 1$$

$$f(x_1) = f\left(\frac{1}{5}\right) = e^{-1/25} \approx 0.961$$

$$f(x_2) = f\left(\frac{2}{5}\right) = e^{-4/25} \approx 0.852$$

$$f(x_3) = f\left(\frac{3}{5}\right) = e^{-9/25} \approx 0.698$$

$$f(x_4) = f\left(\frac{4}{5}\right) = e^{-16/25} \approx 0.527$$

$$f(x_5) = f(1) = e^{-1} \approx 0.368$$

Substituting these values into the formula yields

$$\int_0^1 e^{-x^2} \, dx$$

$$\approx \frac{1}{10}\left[f(0) + 2f\left(\frac{1}{5}\right) + 2f\left(\frac{2}{5}\right) + 2f\left(\frac{3}{5}\right) + 2f\left(\frac{4}{5}\right) + f(1) \right]$$

$$= \frac{1}{10}(1 + 1.922 + 1.704 + 1.396 + 1.054 + 0.368)$$

$$= \frac{1}{10}(7.444) = 0.7444$$

The graph of the function $f(x) = e^{-x^2}$ that is the integrand in Example 4 is the familiar bell-shaped curve. It is particularly useful in various settings in probability, especially those dealing with normal distributions.

EXERCISE SET 7–4

In Problems 1 to 20 approximate $\int_a^b f(x)\,dx$ for the given choice of $f(x)$, a, and b, using the trapezoidal rule with n subintervals.

1. $f(x) = x^2$, $a = 0$, $b = 3$, $n = 3$

2. $f(x) = x^2$, $a = 0$, $b = 3$, $n = 6$

3. $f(x) = x^2$, $a = 0$, $b = 6$, $n = 6$

4. $f(x) = x^2$, $a = 0$, $b = 6$, $n = 12$

5. $f(x) = x^3$, $a = 0$, $b = 3$, $n = 3$

6. $f(x) = x^3$, $a = 0$, $b = 3$, $n = 6$

7. $f(x) = x^3$, $a = 0$, $b = 6$, $n = 6$

8. $f(x) = x^3$, $a = 0$, $b = 6$, $n = 12$

9. $f(x) = (x^3 + 1)^{1/2}$, $a = 0$, $b = 5$, $n = 5$

10. $f(x) = (x^3 + 1)^{1/2}$, $a = 0$, $b = 5$, $n = 10$

11. $f(x) = e^{-x^2}$, $a = 0$, $b = 5$, $n = 5$

12. $f(x) = e^{-x^2}$, $a = 0$, $b = 5$, $n = 10$

13. $f(x) = 1/(x + 1)$, $a = 0$, $b = 4$, $n = 4$

14. $f(x) = 1/(x + 1)$, $a = 0$, $b = 4$, $n = 8$

15. $f(x) = 1/(x + 1)$, $a = 0$, $b = 4$, $n = 6$

16. $f(x) = 1/(x + 1)$, $a = 0$, $b = 4$, $n = 10$

17. $f(x) = 1/x^2$, $a = 1$, $b = 5$, $n = 4$

18. $f(x) = 1/x^2$, $a = 1$, $b = 5$, $n = 8$

19. $f(x) = 1/x^2$, $a = 1$, $b = 3$, $n = 8$

20. $f(x) = 1/x^2$, $a = 1$, $b = 5$, $n = 10$

In Problems 21 to 24 approximate the definite integral using the trapezoidal rule with four subintervals.

21. $\int_0^2 (x^2 + x)^{1/2}\,dx$

22. $\int_0^2 (x^3 + x)^{1/3}\,dx$

23. $\int_0^4 (e^x + x^{1/3})\,dx$

24. $\int_1^5 (\ln x + x)^2\,dx$

In Problems 25 to 28 approximate the definite integral using the trapezoidal rule with eight subintervals.

25. $\int_0^8 (x^2 + 2x)^{1/2}\,dx$

26. $\int_0^4 (x^3 + x)^{1/4}\,dx$

27. $\int_0^4 (e^{-x} + 2x^{1/2})\,dx$

28. $\int_1^5 (x \ln x + x)^2\,dx$

Simpson's rule is another method of numerical integration that approximates sections of the curve by using segments of parabolas as opposed to line segments in the trapezoidal rule. Usually an approximation by Simpson's rule yields a closer approximation than the trapezoidal rule for the same number of subintervals. It is necessary to have an even number of subintervals for Simpson's rule. The formula for Simpson's rule follows:

Simpson's Rule

Suppose $f(x)$ is a continuous function on the interval $[a, b]$ and suppose $f(x) \geq 0$ on $[a, b]$. To approximate the definite integral of $f(x)$ from a to b by Simpson's rule, divide $[a, b]$ into n (an even number) equal subintervals whose end points are $x_0 = a, x_1, \ldots, x_n = b$, then

$$\int_a^b f(x)\,dx \approx \frac{(b - a)}{3n} [f(a) + 4f(x_1)$$
$$+ 2f(x_2) + 4f(x_3)$$
$$+ 2f(x_4) + 4f(x_5)$$
$$+ \cdots + 2f(x_{n-2})$$
$$+ 4f(x_{n-1}) + f(b)]$$

In Problems 29 to 32 approximate the definite integral by using Simpson's rule with four subintervals.

29. $\int_0^2 (x^2 + x)^{1/2}\,dx$

30. $\int_0^2 (x^3 + x)^{1/3}\,dx$

31. $\int_0^4 (e^x + x^{1/3})\,dx$

32. $\int_1^5 (\ln x + x)^2\,dx$

In Problems 33 and 34 approximate $\int_0^4 \dfrac{1}{x + 1}\,dx$ using Simpson's rule with n subintervals. Compare this answer with the approximations in Problems 13 and 14. Then compare both approximations with the actual indefinite integral computed by using the fundamental theorem of calculus.

33. $n = 4$

34. $n = 8$

In Problems 35 to 38 approximate $\int_0^4 \dfrac{1}{1 + x}\, dx$ using Simpson's rule with n subintervals for the given value of n. Compare the approximations with the trapezoidal rule and with the indefinite integral computed by using the fundamental theorem of calculus.

35. $n = 4$

36. $n = 8$

37. $n = 16$

38. $n = 2$

Referenced Exercise Set 7–4

1. To ensure that air quality satisfies set standards, air pollution control authorities must determine allowable emission rates for various types of pollutants. An important variable is the cost to treat the pollutants. The total treatment cost $C(x)$ is a function of the emission rate x. In an article investigating how pollution control authorities can establish an optimum scheme for calculating pollution cost functions, Hashimoto and Kimura studied the emission of nitrogen oxide in Tokyo City, where the primary pollutants were utility and industrial boilers and industrial internal combustion engines.* They were concerned with the case when the total cost could not be directly assessed, but the marginal cost $g(x)$ could be computed for various zones of the city. Then $C(x)$ is defined to be the integral of $g(x)$, which was computed by numerical integration. In one part of the study seven zones were used. The measurements are given in the table.

zone	1	2	3	4	5	6	7
$g(x)$	101	105	119	163	143	180	105

Take $g(0) = 100$ and approximate the total cost using the trapezoidal rule.

2. In 1962 Pierre Wenger discovered two persons who were contaminated with radium-226 and strontium-90. They agreed to have their bodily functions closely monitored for 10 years in order to study the long-range effects of contamination. In an article comparing the observed data with the predictions of various models, Wenger and Cosandey centered on the effects of contamination on excretion rates, where the units are in levels of radium-226 excreted per day.† They took periodic measurements, and they used numerical integration to compute the sum of the bodily excretion. Their measurements are summarized in the following table.

year	1962	1964	1966	1968	1970	1972
excretion rate	.91	.65	.28	.26	.20	.18

Approximate the sum of the bodily excretion using the trapezoidal rule.

3. In an article investigating the costs and benefits of sexual versus asexual reproduction in various plants in different environments and mating systems, Michaels and Bazzaz studied a perennial herb growing in Kickapoo State Park, Danville, Illinois.‡ In one part of the study the survival rate was given by $f(t) = 118e^{-.05t^2}$ for $t = 0$ to $t = 5$ months. Approximate the total number of survivors by integrating $f(t)$ from $t = 0$ to $t = 5$ using the trapezoidal rule with 5 intervals.

Cumulative Exercise Set 7–4

In Problems 1 to 8 find the antiderivative.

1. $\displaystyle\int (x^5 + 1)(x^6 + 6x)^{1/3}dx$

2. $\displaystyle\int x(2x - 1)^6 dx$

3. $\displaystyle\int x^{1/4} \ln x\, dx$

4. $\displaystyle\int x^5 e^{x^3} dx$

5. $\displaystyle\int \dfrac{dx}{x(2x + 3)^2}$

6. $\displaystyle\int \dfrac{dx}{9 - 4x^2}$

7. $\displaystyle\int x(25 + x^4)^{1/2}\, dx$

8. $\displaystyle\int \dfrac{x\, dx}{2x - 3}$

*Akihiro Hashimoto and Yuri Kimura, ''Determining the Optimal Scheme of Zoned Effluent Charges for the Control of Air Pollution,'' *Socio-Economic Planning Science*, Vol. 14, 1980, pp. 197–208.

†Pierre Wenger and Maurice Cosandey, ''Retention and Excretion of Radium-226 and Strontium-90 in Two Doubly Contaminated Persons,'' *Health Physics*, Vol. 31, 1976, pp. 225–229.

‡H. J. Michaels and F. A. Bazzaz, ''Resource Allocation and Demography of Sexual and Apomictic *Antennaria parlinii*,'' *Ecology*, Vol. 67, 1986, pp. 27–36.

In Problems 9 to 11 approximate $\int_a^b f(x)dx$ for the given choice of $f(x)$, a, and b, using the trapezoidal rule with n subintervals.

9. $f(x) = x^2$, $a = 0$, $b = 4$, $n = 4$

10. $f(x) = (x^3 + 1)^{1/2}$, $a = 0$, $b = 4$, $n = 4$

11. $f(x) = e^{-x^2}$, $a = 0$, $b = 6$, $n = 6$

12. Approximate the definite integral using the trapezoidal rule with 4 subintervals.

$$\int_0^2 (x^3 + x)^{1/2}dx$$

CASE STUDY   ## Demography and the Social Security System

Most people believe that the taxes they pay to the social security system each year are put into a savings account to be held until their retirement. Not so! The U.S. social security system is a pay-as-you-go plan, meaning that the money paid by the work force goes directly to fund those who are retired. In essence, the plan calls for those in the age range of 20 to 64 to support those over 64. The plan has worked well since its inception in 1935 because there has always been a large work force to support a much smaller retired segment of the population.

But times are changing. The present system faces grave consequences in the near future, primarily because both the birthrate and death rate are declining. This means that in a short while fewer people will be entering the work force while a larger number will be retiring. The latter is because of advances in health and medicine.

Demography   Demography answers many age-related questions about various populations, such as what is the effect of a lowered death rate on the retired segment of society, and what is the effect of abortions on the birthrate? The insurance industry was the first industry to use demography extensively, but now many diverse fields, including government, apply demography to a multitude of problems.

In his book on applied demography Nathan Keyfitz* uses calculus to give concise explanations of the theory of demography. As a specific application, Keyfitz describes the ramifications of demographic projections on the U.S. social security system.

Definitions of Life Table Functions   A life table for a given population is a table that contains the percentage of people who are expected to survive to a certain age. These percentages are computed from past records and mathematical projections. The probability of surviving from birth to age x is defined to be $l(x)$. Even though births and deaths occur in single increments, demographers assume $l(x)$ is a continuous function in order to use calculus for their projections. If the population is large, this is a valid assumption.

Consider those who have attained the age of t. The fraction of years survived by these people in the next n years is the "sum" of the probabilities of

---

*Nathan Keyfitz, *Applied Mathematical Demography*, 2nd ed., New York, Springer-Verlag, 1985.

their surviving at each of these ages from t to $t + n$. Since $l(x)$ is a continuous function, this sum is the integral

$$\int_{t}^{t+n} l(x)\, dx$$

The definite integral of $l(x)$ from 0 to infinity, or at least to the largest age that anyone lives, say, w, is the sum of all the probabilities, so it is 1.

Example 1 shows how numerical integration is used to compute the number of people in a segment of a population. The example studies the female population of Mexico in 1970. The following partial life table gives the values of $l(x)$ of the female population of Mexico in 1970.

Female Population, Mexico, 1970

Age	Fraction of Those Surviving to a Specific Age, $l(x)$
60	0.008939
62	0.008036
64	0.007071

E X A M P L E   1

Problem

Use the trapezoidal rule and the accompanying table to compute the fraction of females in Mexico between the ages of 60 and 64 in 1970.

Solution Find the integral of $l(x)$ from 60 to 64. The trapezoidal rule, with $n = 2$, states

$$\int_{60}^{64} l(x)\, dx \simeq \frac{64 - 60}{4}[l(60) + 2l(62) + l(64)] = l(60) + 2 \cdot l(62) + l(64)$$

$$= 0.008939 + 0.016072 + 0.007071 = 0.032082$$

Therefore the percentage of the female population in Mexico in 1970 between the ages of 60 and 64 was about 3.2%.

Pension Cost   Consider a pay-as-you-go pension plan in which each person over age 64 is paid the same amount, say, one unit. Also assume that the pension is funded by those aged 20 to 64 and that their salaries are the same amount, one unit. Suppose the total number of individuals in the population is P. Then the amount disbursed to the pensioners is P times the fraction of individuals over 64:

$$P \int_{64}^{w} l(x)\, dx$$

The amount of money paid into the fund will be a fraction of the workers' salaries. Let this fraction be g. Thus g is the premium paid by each worker. The amount of money paid into the fund by the workers is

$$gP \int_{20}^{64} l(x) \, dx$$

Since the fund is a pay-as-you-go plan the amount paid into the fund must equal the amount paid out to the pensioners. This means the two expressions must be equal. Setting them equal and solving for g yields

$$g = \frac{\displaystyle\int_{64}^{w} l(x) \, dx}{\displaystyle\int_{20}^{64} l(x) \, dx}$$

This means that the premium paid by the workers is a percentage of their salary equal to the ratio of the number of individuals over age 64 divided by the number of individuals aged 20 to 64.

The assumptions seem restrictive, but adapting the model to salaries and pension payments that vary with age and even within given ages, as well as other practical considerations—including accounting for some younger persons receiving pensions—complicates the formula for g without altering the principles of the argument. More realistic models have several additional terms that give minor adjustments to the computation of the premium. But the primary conclusion remains: The premium depends on the ratio of those over 64 to those between 20 and 64.

The formula for the premium g illustrates how demographic projections affect pension payments. The premium increases in two ways: if the numerator increases or if the denominator decreases. This means that if the number of pensioners increases, or if the number of individuals in the work force decreases, the premium paid by the workers increases.

Projections for the Social Security System   The U.S. social security system is a pay-as-you-go plan. The premium paid by the work force is dependent on the ratio of the number of pensioners to the number of individuals in the work force. This type of plan has demographic problems if this ratio has a significant increase. The premium paid by each worker will increase as well. The accompanying table shows the demographic projections of the ratio of pensioners to workers. The values in the table from 1990 on were computed by using numerical integration.

Significant pressure on the social security plan will come after the year 2000. The ratio will increase by more than 50% in the years from 2000 to 2025, from 0.20 to 0.31. The only thing that could avert the swing is a large number of births before the year 2000 that would increase the work force in the early 21st century. This seems unlikely.

Ratio of Individuals of Pensionable Age Versus Working Age

Year	Age 21–64	Age Over 64	Ratio Over 64/21–64
1950	85,944	12,397	0.1442
1960	92,181	16,675	0.1809
1970	103,939	20,085	0.1932
1980	122,115	24,523	0.2008
1990	137,500	28,933	0.2104
2000	148,589	30,600	0.2059
2025	146,645	45,715	0.3117
2050	147,635	45,805	0.3103

The social security system can be understood as a method of borrowing from future generations. When initiated, it was reasonable because there were no start-up funds to pay the retirees during its first few years. In the ensuing years the plan also was viable because the ratio of pensioners to the work force was a relatively small fraction, so that the premium paid by each worker was manageable. In addition, the premium was viewed as insurance for one's own retirement. But the ratio will increase dramatically soon so that either the system will prove to be a great burden on the work force or the benefits paid to retirees will be greatly reduced.

Case Study Exercises

In Problems 1 to 4 use the following table and the trapezoidal rule to find the integral of $l(x)$ from the given ages.

Female Population, Mexico, 1970

Age	Fraction of Those Surviving to a Specific Age, $l(x)$
60	0.008939
61	0.008570
62	0.008036
63	0.007512
64	0.007071

1. From age 60 to 62
2. From age 61 to 63
3. From age 60 to 63
4. From age 60 to 64

5. Compare the answer in Exercise 4 to Example 1. Which most likely gives the better approximation?

6. Give a reason why the percentage in the ratio of individuals of pensionable age versus working age increases so dramatically from the year 2000 to 2025.

CHAPTER REVIEW

Key Terms

7–1   Substitution

Substitution of a Variable   Differential of u
Integration by Substitution

7–2   Integration by Parts

Integration by Parts

7–4   Numerical Integration

Numerical Integration   Simpson's Rule
Trapezoidal Rule

Summary of Important Concepts

Formulas

If $u = g(x)$ then $du = g'(x)\, dx$

$$\int [f(x)]^n\, f'(x)\, dx = \frac{f(x)^{n+1}}{n+1} + C, \quad n \neq 1$$

$$\int e^{f(x)}\, f'(x)\, dx = e^{f(x)} + C$$

$$\int [f(x)]^{-1}\, f'(x)\, dx = \ln |f(x)| + C$$

$$\int u\, dv = uv - \int v\, du$$

REVIEW PROBLEMS

1. Find $\displaystyle\int (2x + 1)(x^2 + x)^{-3}\, dx$

2. Find $\displaystyle\int (e^{2x} + 3)^{-2} e^{2x}\, dx$

3. Find $\displaystyle\int_0^1 x(x^2 - 2)^3\, dx$

4. Find $\displaystyle\int 3xe^{4x}\, dx$

5. Find $\displaystyle\int x^2(1 - x)^{1/3}\, dx$

6. Find the definite integral $\displaystyle\int_0^2 x(4x + 1)^{-1/2}\, dx.$

In Problems 7 to 9 use the table of integrals to find the integral.

7. $\displaystyle\int \frac{dx}{x(5x - 7)^2}$

8. $\displaystyle\int x(9x^4 + 16)^{1/2}\, dx$

9. $\displaystyle\int \frac{dx}{14 - 9x^2}$

10. Use the trapezoidal rule with four subintervals to approximate $\displaystyle\int_0^4 x^3\, dx$.

11. Use the trapezoidal rule with six subintervals to approximate $\displaystyle\int_0^3 e^{-x^2}\, dx$.

GRAPHICS CALCULATOR EXPLORATIONS

When it is impossible or impractical to evaluate a definite integral by the fundamental theorem of calculus, numerical integration techniques can be used to compute an approximation of the integral. The drawback of these methods is the tedium of the many calculations required. Your graphics calculator can ease this burden significantly.

In this Exploration we present a program that will graph a function $f(x)$ that you input and will calculate the Riemann sum approximation of the definite integral $f(x)$ on the interval you input. You also will input the number of subintervals. Remember that you will get a better approximation when you choose a larger number of intervals.

Your manual has a program that will compute an approximation of the definite integral using Simpson's rule. If you run this program after the one below you can compare their respective computations.

In this program you must first set an appropriate range on the calculator screen. Then, depending on which calculator you have, you use the following steps to enter the function as the integrand:

Casio   Program location 0 must contain the expression that evaluates the function. For example, if you wanted to compute the integral of the exponential function, program location 0 would have to contain

$e^x \rightarrow Y$

Program location 1 will contain steps needed to graph the function, so in this case it will contain

Graph $Y = e^x$

TI-81   Enter the expression that evaluates the function to be integrated into Y_1.

When you execute the program you will be asked to input the values of the endpoints A and B and the number of subintervals N between A and B. Then you will be asked to choose whether you want to compute the heights of the rectangles in the approximation by taking the functional value of left endpoints, right endpoints, midpoints or random points within each subinterval. The variable W represents this choice and the prompt from the program is "POINT TYPE." You then enter 1, 2, 3, or 4 depending on whether you want to choose left endpoints, right endpoints, midpoints, or random points, respectively. For example, if you press 20 after the prompt "N" and you press 2 after the prompt "POINT TYPE" the Riemann sum will be computed over 20 subintervals using right endpoints. The graph will be shown on the screen until you press "EXE" (Casio) or "ENTER" (TI-81), and then the value of the approximation is displayed.

Casio

RSUMS
Cls
"A"?→A:"B"?→B
"N"?→N
"POINT TYPE"?→W
(B − A) ÷ N→H
Prog 1
1→J:0→R
Lbl 0
W = 1⇒Goto 1
W = 2⇒Goto 2
W = 3⇒Goto 3
W = 4⇒Goto 4
Lbl 1
A + (J − 1)H→X
X→P:X + H→Q
Goto 5
Lbl 2
A + JH→X:X→Q
X − H→P
Goto 5
Lbl 3
A + JH − H ÷ →X
X − H ÷ 2→P
X + H ÷ 2→Q
Goto 5
Lbl 4
A + (J − 1)H→X
X→P:X + H→
HRan# + X→X
Goto 5
Lbl 5:Prog 0
Y→F:F + R→R
Plot P,0
Plot P,F:Line
Plot Q,F:Line
Plot Q,0:Line
Isz J
J≤N⇒Goto 0
Line◢
"RS = ":HR→R◢

TI-81

PrgmS:RSUMS
ClrDraw
All-Off
Y_1-On
Disp "A"
Input A
Disp "B"
Input B
Disp "N"
Input N
Disp "POINT TYPE"
Input W
(B − A)/N→H
0→R
1→J
Lbl 0
If W = 1
Goto 1
If W = 2
Goto 2
If W = 3
Goto 3
If W = 4
Goto 4
Lbl 1
A + (J − 1)H→X
X→P
X + H→Q
Goto 5
Lbl 2
A + JH→X
X − H→P
X→Q
Goto 5
Lbl 3
A + JH − H/2→X
X − H/2→P
X + H/2→Q
Goto 5
A + (J − 1)H→X
X→P
X + H→Q
HRand + X→X
Goto 5
Lbl 5
Y_1→F
F + R→R
Line(P,O,P,F)

TI-81

Line(P,F,Q,F)
Line(Q,F,Q,O)
IS>(J,N)
Goto 0
Pause
HR→R
Disp "RS = "
Disp R

Further Explorations

1. Choose any function and compute the Riemann sum approximation with all four "point type" and compare results.
2. Choose a function whose integral can be easily computed using the fundamental theorem. Then execute this program using just a few subintervals and compare the results. How far off is the approximation? Then execute the program using more subintervals.
3. Compute the integral of a given function using the fundamental theorem and choose a very small number to represent the error of an approximation. Try to predict how many subintervals it will take to ensure that the Riemann sum approximation is within this error tolerance.
4. Enter the program in your calculator's manual that computes the numerical integration approximation via Simpson's rule. Choose a function and compare the results of this program versus the program given for the Riemann sum approximation.

Functions of Several Variables

8

The costs of high-tech industrial giants can be minimized by using functions of many variables. (H. Armstrong Roberts/Zefa)

379

Chapter Overview

Many applications of calculus use variable quantities that are functions of more than one variable. For example, profit is a function of cost, price, and sales volume. The total population of a species in a given environment depends on its rate of reproduction, its death rate, the availability of food, and the population of its predators.

In this chapter we study functions of more than one variable. The bulk of the material is on functions of two variables. Such functions are graphed in three dimensions so that it is possible to view them geometrically. In the first section we cover some examples of functions of several variables. The later sections show that many of the concepts that were developed for functions of one variable have analogous statements for functions of several variables.

In Section 8–2 we study the concept of rate of change of a function of several variables. It is an extension of the idea of the derivative of a function of one variable. Likewise in Section 8–3 the idea of relative extrema is covered. In Section 8–4 a constraint is added to the problem of finding relative extrema. These constraints, called Lagrange multipliers, have many applications to business and other disciplines.

The multivariable analog of the integral of a function of one variable is covered in the next two sections. We consider only functions of two variables, so Section 8–5 is called "Double Integrals." In this section we show how such integrals form a natural extension of the ideas developed for integrals of functions of a single variable. Section 8–6 applies the techniques of double integration to volumes of certain kinds of solid figures.

CASE STUDY PREVIEW

Jojoba oil is used for producing amides which, in turn, are an essential ingredient in many chemical processes. The procedure of extracting amides from jojoba oil is time-consuming and expensive. First the oil is refined and purified. Then other chemicals are added and the mixture is heated for several hours. Once the material becomes cakelike the amides can be extracted.

Researchers have tried various methods to maximize the amide yield while conserving cost to produce the cakelike material. The cost is a function of the two variables—temperature and time. The Case Study describes how the chemical industry uses this function of two variables to minimize the cost of producing amides.

8–1 Examples of Functions of Several Variables

Thus far we have considered only functions of one variable. Such functions are of the form $y = f(x)$ where x is the single independent variable. One value of the independent variable is substituted into the formula and a corresponding number is produced. For a function of two variables two numbers, given as an ordered pair, are substituted into a formula and another number is produced. In this section we consider examples of functions of two and more variables.

The Definition of a Function of Two Variables

An example of a function of two variables is

$$z = f(x, y) = 2x + 3y$$

The name of the function is f and z is called the **dependent variable.** The two variables x and y are **independent variables,** meaning that x and y can be chosen to be any numbers, while the corresponding value of z depends on which choice is made for x and y. For instance, let $x = 1$ and $y = -4$, then

$$z = f(1, -4) = 2(1) + 3(-4) = 2 - 12 = -10$$

Thus $z = -10$ when $x = 1$ and $y = -4$. This is succinctly stated by $f(1, -4) = -10$.

This leads to the definition of a function of two variables. The definition of functions of more than two variables is similar.

DEFINITION

The formula $z = f(x, y)$ defines f as a function of the two independent variables x and y if, for each ordered pair of real numbers (x, y) for which the formula is defined, there exists a unique corresponding value for the variable z.

The set of all ordered pairs of numbers for which $f(x, y)$ exists is called the *domain of f*. The set of all values of z that correspond to an ordered pair (x, y) is called the *range of f*. Usually the domain of a given function is not stated, and it is assumed that the domain is the largest set of ordered pairs for which the formula is defined. Occasionally, however, a restricted domain is given.

Sometimes the name of the function is omitted when only one function is being discussed. For example, the function above could have been given as simply $z = 2x + 3y$. If the dependent variable is not specifically mentioned, we assume its name is either z or the name of the function. That is, if the function above would have been given by the formula $f(x, y) = 2x + 3y$, where the dependent variable is not mentioned, it could be named either f or z.

EXAMPLE 1

Problem

If possible, evaluate the given functions at the points $(0, 0)$, $(-1, 2)$, and $(2, -3)$: (a) $f(x, y) = 4x - 5y + 2$, (b) $g(x, y) = 2x^2 + xy - \dfrac{1}{y + 3}$.

Solution (a) $f(0, 0) = 4(0) - 5(0) + 2 = 2$; $f(-1, 2) = 4(-1) - 5(2) + 2 = -12$; $f(2, -3) = 4(2) - 5(-3) + 2 = 8 + 15 + 2 = 25$.
(b) $g(0, 0) = 2(0)^2 + (0)(0) - 1/(0 + 3) = -1/3$; $g(-1, 2) = 2(-1)^2 +$

$(-1)(2) - 1/(2 + 3) = 2 - 2 - 1/5 = -1/5$; $g(2, -3) = 2(2)^2 +$ $(2)(-3) - 1/(-3 + 3) = 8 - 6 - 1/0$, which is not defined because $1/0$ is not defined, so $(2, -3)$ is not in the domain of g.

A function of three variables is defined similarly to a function of two variables. It differs only in that an ordered triple is substituted into a formula that has three variables. The next example gives an illustration of a function of three variables and one of four variables.

EXAMPLE 2

Define the functions

$$f(x, y, z) = x^2 - 4yz \qquad g(x, y, z, w) = xy + \frac{4z^2}{w}$$

Problem

(a) Find $f(2, 1, -1)$, (b) $g(1, 2, -3, 2)$.

Solution (a) $f(2, 1, -1) = (2)^2 - 4(1)(-1) = 4 + 4 = 8$.
(b) $g(1, 2, -3, 2) = (1)(2) + 4(-3)^2/2 = 2 + 18 = 20$.

Three-Dimensional Coordinate Systems

The graph of a function of two variables requires a three-dimensional coordinate system. This is because a point in the domain (a, b) is plotted in the xy-plane and then its functional value c is plotted on a third axis. Thus the point (a, b, c) is plotted in a three-dimensional system. Figure 1 demonstrates the usual convention for labeling the three axes, called the x-, y-, and z-axes. The z-axis is vertical, the

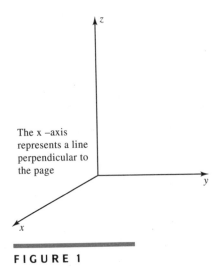

The x –axis represents a line perpendicular to the page

FIGURE 1

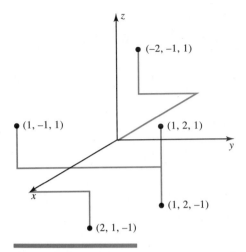

FIGURE 2

y-axis is horizontal, and the *x*-axis represents a line that is perpendicular to the plane of the paper. Since the figure is a two-dimensional depiction of a three-dimensional object, a perspective drawing is needed. That is why the *x*-axis is drawn at an angle. Only the positive half of each axis is drawn for clarity.

The point where the three axes intersect is called the *origin*. A point (a, b) in the *xy*-plane is given the three-dimensional coordinates $(a, b, 0)$. A point (a, b, c) is $|c|$ units above or below the point $(a, b, 0)$, depending on whether c is positive or negative. The coordinate system is divided into eight regions, called *octants*, four above the *xy*-plane and four below. Only one octant is named. It is the region where all three coordinates are positive, called the *first octant*. Several points in various octants are plotted in Figure 2.

Graphs of Functions of Two Variables

Since the domain of the function $z = f(x, y)$ is a set of ordered pairs (x, y), the graph of the function entails locating the points in the domain in the *xy*-plane. Then their corresponding functional values are plotted on the *z*-axis. Thus a point on the graph is an ordered triple (x, y, z). For example, consider the function $z = f(x, y) = 2x + 3y$. To locate the point on the graph corresponding to $(1, 2)$, first plot $(1, 2)$ in the *xy*-plane and locate $z = f(1, 2) = 8$ on the *z*-axis. Thus the point on the graph is $(1, 2, 8)$. This point is one unit out on the *x*-axis, two units to the right on the *y*-axis and eight units up on the *z*-axis. This point along with several others is plotted in Figure 3.

There is no easy method for graphing functions of two variables. The problem entails working algebraically with the formula to determine what type of two-dimensional figures arise when the graph is intersected with the graph of vertical and horizontal planes. These intersections are then pieced together to get a perspective figure. The planes used in the perspective drawings are the coordinate

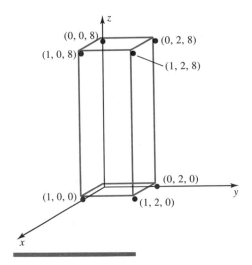

FIGURE 3

planes and those that are parallel to the coordinate planes. For instance, the set of points with z-coordinate 1 is the plane one unit above the xy-plane. Its equation is $z = 1$. Likewise

The graph of $x = a$ is a plane parallel to the yz-plane.
The graph of $y = b$ is a plane parallel to the xz-plane.
The graph of $z = c$ is a plane parallel to the xy-plane.

Figure 4 illustrates these planes.

The intersection of one of the coordinate planes with the graph of the function is called a **trace** of the graph. A **section** of a surface $z = f(x, y)$ is the intersection

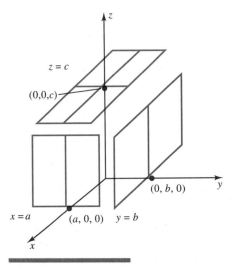

FIGURE 4

of the surface with a plane that is parallel to a coordinate plane. A section of the graph is found by solving simultaneously the two equations, the equation of the function and the equation of the plane. Thus a section is a two-dimensional figure in one of the planes $x = a$, $y = b$, or $z = c$.

The next example illustrates how to use sections to graph functions of two variables.

E X A M P L E 3

Problem

Graph the functions.
(a) $z = f(x, y) = 2x + 3y$, (b) $z = h(x, y) = x^2 + 4y^2$, (c) $z = g(x, y) = y^2 - x^2$.

Solution (a) The sections of the graph corresponding to $x = a$ are the straight lines $z = 2a + 3y$, each in the plane $x = a$, which is parallel to the yz-plane. The sections of the graph corresponding to $y = b$ are the straight lines $z = 2x + 3b$, each in the plane $y = b$, which is parallel to the xz-plane. The sections of the graph corresponding to $z = c$ are the straight lines $c = 2x + 3y$, each in the plane $z = c$, which is parallel to the xy-plane. The graph is the plane sketched in Figure 5.

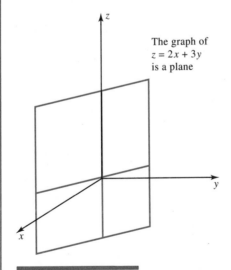

The graph of
$z = 2x + 3y$
is a plane

F I G U R E 5

(b) The trace in the yz-plane is found by letting $x = 0$ in the equation. It is the parabola $z = 4y^2$. The trace in the xz-plane is the parabola $z = x^2$. The sections of the graph corresponding to $x = a$ are the parabolas $z = a^2 + 4y^2$. The sections of the graph corresponding to $y = b$ are the parabolas $z = x^2 + 4b^2$. The sections of the graph corresponding to $z = c$ are the ellipses $c = x^2 + 4y^2$. The graph is an infinite bowl-shaped surface, sketched in Figure 6. It is called an elliptic paraboloid.

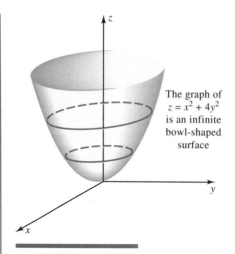

The graph of $z = x^2 + 4y^2$ is an infinite bowl-shaped surface

FIGURE 6

(c) The trace in the yz-plane is the parabola $z = y^2$ (see Figure 7a) and the trace in the xz-plane is the parabola $z = -x^2$ (see Figure 7b). The sections of the graph corresponding to $x = a$ are the parabolas $z = y^2 - a^2$. The sections of

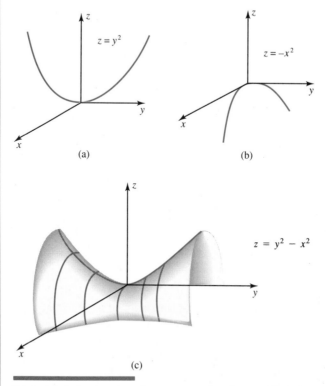

$z = y^2$

(a)

$z = -x^2$

(b)

$z = y^2 - x^2$

(c)

FIGURE 7   (a) The trace in the xy-plane is $z = y^2$. (b) The trace in the xz-plane is $z = -x^2$. (c) The graph of $z = y^2 - x^2$ is a saddle-shaped surface.

the graph corresponding to $y = b$ are the parabolas $z = b^2 - x^2$. The sections of the graph corresponding to $z = c$ are the hyperbolas $c = y^2 - x^2$. The graph is a saddle-shaped surface, sketched in Figure 7c. It is called a hyperbolic paraboloid.

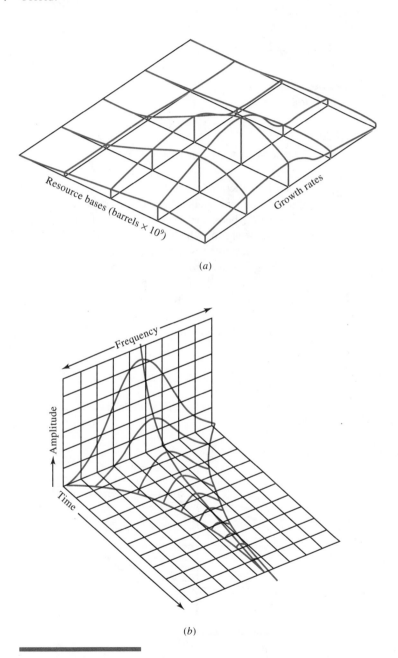

(a)

(b)

FIGURE 8

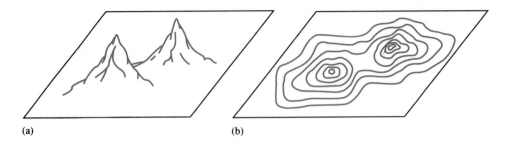

(a)   (b)

FIGURE 9

Applications of Sections

To see how sections are used to give a visual structure to a two-dimensional graph of a three-dimensional surface, consider the two graphs in Figure 8. The first graph appears in a book that investigates the future of the world supply of oil.* The second is taken from a book describing the structure of sound in music.† The surface in Figure 8a describes the relationship between the amount of oil in the ground (resources) and the growth in oil consumption. This affects the likelihood that the world supply of oil will exceed demand. This likelihood depends on the amount of resources available (already taken from the earth) and the growth rate of demand. The surface in Figure 8b describes the relationship among amplitude, frequency, and time of a musical note.

*Peter R. Odell and Kenneth E. Rosing, *The Future of Oil,* New York, Nichols Publishing, 1980, p. 182.
†Robert Erickson, *Sound Structure in Music,* Berkeley, CA, University of California Press, 1975, p. 62.

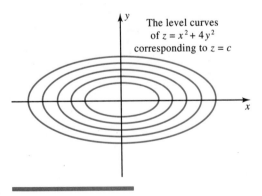

The level curves
of $z = x^2 + 4y^2$
corresponding to $z = c$

FIGURE 10

Level Curves

Another method used to describe the graph of a function of two variables is to sketch the sections $z = c$ in the xy-plane. Each graph is called a **level curve.** Level curves are commonly used on television weather reports and in contour maps, also called topographical maps. For example, Figure 9a illustrates a picture of a mountain range and Figure 9b depicts its corresponding topographical map. The level curves are the intersections of horizontal planes ($z = c$) and the earth, and thus describe the heights of the mountains.

For the function $z = x^2 + 4y^2$, whose graph is sketched in Figure 6, the level curves are the sections corresponding to $z = c$. They are the ellipses graphed in Figure 10.

EXERCISE SET 8–1

In Problems 1 to 4 find the functional value of $f(x, y) = 2x^2 - y^2$ at the given point.

1. (1, 1)
2. (−1, 2)
3. (0, −3)
4. (−4, 0)

In Problems 5 to 8 find the functional value of $f(x, y) = y(x^3 + 1)^{1/2} + xy^3$ at the given point.

5. (0, 2)
6. (0, −3)
7. (2, 3)
8. (2, −3)

In Problems 9 to 12 find the domain of the function.

9. $f(x, y) = \dfrac{y}{x - 2}$
10. $f(x, y) = \dfrac{2}{x(y^2 - 1)}$

11. $f(x, y) = \dfrac{y}{(x^3 + 1)^{1/2}}$
12. $f(x, y) = \dfrac{2}{x(y - 1)^{1/2}}$

In Problems 13 to 20 graph the functions, giving the traces and at least one section that is parallel to each coordinate plane.

13. $f(x, y) = 3x + 5y$
14. $f(x, y) = y - 4x$
15. $f(x, y) = 2 + 2x - 3y$
16. $f(x, y) = y - 2x - 3$
17. $f(x, y) = 4 + x^2 + 5y^2$
18. $f(x, y) = 1 + 2x^2 + y^2$
19. $f(x, y) = y^2 - 5x^2$
20. $f(x, y) = x^2 - 2y^2$

In Problems 21 to 26 graph the functions.

21. $f(x, y) = 1 - x^2 - y^2$
22. $f(x, y) = 2 - 2x^2 - y^2$
23. $f(x, y) = x^2 + 2x + y^2 + 4y$
24. $f(x, y) = 1 + x^2 + 2x + y^2$
25. $f(x, y) = 2x + x^2$
26. $f(x, y) = y^2 - 4y$

27. The equation of the sphere having a radius of one unit and its center at the origin is $x^2 + y^2 + z^2 = 1$. Is it possible to represent this sphere as a function?

28. Solve the equation $x^2 + y^2 + z^2 = 4$ for z and then keep only the positive square root. Graph the resulting function using the fact about spheres stated in Problem 27.

29. Give a geometrical test using lines perpendicular to the xy-plane to determine whether a given surface is a function $z = f(x, y)$. Recall the geometric test to determine whether a two-dimensional curve is a function.

30. A salesperson hires two employees. One earns $10 per hour and the second earns $12 per hour. Let the number of hours worked per week by the first employee be represented by x and the number of hours worked by the second employee be represented by y. If the salesperson's fixed cost is $300 per week, find a formula for the cost per week.

31. A manufacturing company makes three types of desks, small, medium, and large. If the profit is $40 for each small desk sold, $50 for each medium desk sold, and $60 for each large desk sold, find a formula for the profit by letting x represent the number of small desks sold, y the number of medium desks sold, and z the number of large desks sold.

In Problems 32 to 35 match the equation with the given graph.

32. $f(x, y) = x + y$

33. $f(x, y) = x^2$

34. $f(x, y) = x^2 + y^2$

35. $f(x, y) = x^2 - y^2$

(a)

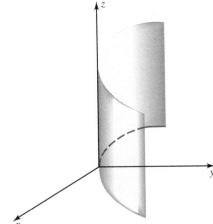

(b)

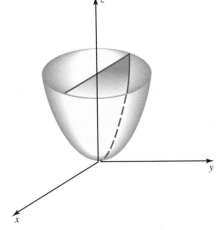

(c)

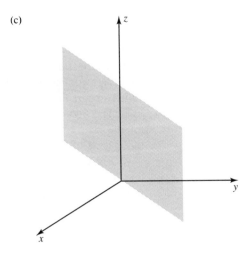

(d)

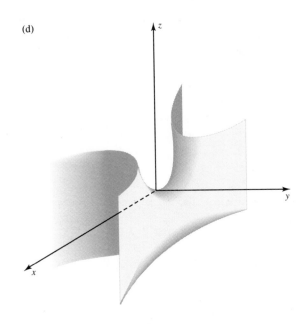

Referenced Exercise Set 8–1

1. Accurate forecasting of ambulance demand directly affects staffing, inventories, and assignment of units within the region served by the ambulance corps. In an article studying ambulance service to four counties in South Carolina, Baker and Fitzpatrick developed a model that accurately forecasted ambulance demand

while minimizing cost.* The cost function to be minimized was

$$C(x, y, z, w) = 103x + 109y + 116z + 117w$$

where the cost per call for an ambulance was $103 in Spartanburg County, $109 in Clarendon County, $116 in Horry County, and $117 in Florence County; and x, y, z, and w were the number of calls per day in each county, respectively. For a given week in 1983 in the study, the number of calls in each county was as follows: (a) on Monday, Spartanburg had 670 calls, Clarendon had 370 calls, Horry had 37 calls, Florence had 320 calls; (b) on Tuesday, Spartanburg had 690 calls, Clarendon had 300 calls, Horry had 38 calls, Florence had 290 calls; (c) on Wednesday, Spartanburg had 680 calls, Clarendon had 320 calls, Horry had 40 calls, Florence had 300 calls. Which day produced the least total cost?

2. Most large industries utilize research and development to improve their products and their marketing techniques. In an article evaluating the cost-effectiveness of research in the Florida citrus industry, Stranahan and Shonkwiler centered on the Florida frozen concentrate orange juice market.† The cost C was assumed to be a function of the output Y, the quantity of research R, and the price P. In one part of the study the cost was computed to be

$$C(Y, R, P) = 1.1Y^{.5}R^{.43}P^{-.009}$$

Compute $C(Y, R, P)$ when $Y = 10$ million gallons per day, $R = .2$ million dollars per day, and $P = $1 per gallon.

3. Many studies have been done on the harmful effects of large doses of radiation, such as the tragic accident at Chernobyl, U.S.S.R., in 1987. Johnson studied harmful effects of exposure to radiation at very low level dose rates which result from various environmental sources of radiation.‡ These dosages are called high natural radiation background. They occur from diverse sources such as sun rays and wristwatches. The article investigated data from the World Health Organization based on radiation reports from 45 countries. One part of the article studied cancer rates. The number of years required for detection y was found to be a function of the background radiation rate r, measured in rads per year, and the average cancer rate p, measured in cases per million persons per year. Thus

$$y(r, p) = \frac{kp}{r^2}$$

where k is a constant depending upon the country or region being studied. Compute $y(.88, .02)$ for $k = .04$.

---

*J. R. Baker and K. E. Fitzpatrick, "Determination of an Optimal Forecast Model for Ambulance Demand Using Goal Programming," *Journal of the Operations Research Society*, Vol. 37, 1986, pp. 1047–1059.

†H. A. Stranahan and J. S. Shonkwiler, "Evaluating the Returns to Postharvest Research in the Florida Citrus Processing Subsector," *American Journal of Agricultural Economics*, Vol. 68, 1986, pp. 88–94.

‡Richard E. Johnson, "Problems Involved in Detecting Increased Malignancy Rates in Areas of High Natural Radiation Background," *Health Physics*, Vol. 31, 1976, pp. 148–160.

8–2 Partial Derivatives

One of the most important concepts of calculus is rate of change. For a function of one variable, rate of change is measured by the derivative. For a function of two variables, rate of change can be measured in many ways so that the situation is more complicated. In this section we introduce the concept of a partial derivative, which is the three-dimensional analog of the derivative.

An Application from Business

Rate of change is used extensively in business. The demand function for a particular company or industry often is expressed as a function of more than one variable.

For instance, in an article studying the effect of the "health scare" on the cigarette industry, Bishop and Yoo consider the quantity demanded Q as a function of disposable income D and cigarette advertising expenditures A.* To model the industry's data, the authors use a Cobb–Douglas function of the form

$$Q(D, A) = 231.7D^{0.66}A^{0.07}$$

The article studies the effect on Q when D and A experience increases and decreases. Bishop and Yoo use several types of measures, each of which depends on the partial derivatives of Q. At the end of the section we compute these partial derivatives.

Partial Derivatives: A Geometric Description

The derivative of $y = f(x)$ measures how a change in x affects y. For a function of two variables $z = f(x, y)$ the partial derivatives measure how a change in x or in y affects z. Geometrically the derivative $f'(a)$ is the slope of the tangent line to the curve $y = f(x)$ at $x = a$. But a surface in three dimensions can have a tangent line in any direction, so there are an infinite number of tangent lines. (See Figure 1.) It suffices to study tangent lines in only two directions, the directions parallel to the x-axis and parallel to the y-axis. This is because the tangent lines in every other direction can be expressed in terms of these two.

Consider the direction parallel to the x-axis. How can we interpret the rate of change of $z = f(x, y)$ at the point (a, b) in this direction? The section of the graph in this direction is a two-dimensional curve obtained by taking the intersection of $z = f(x, y)$ with the plane $y = b$. The equation of this curve is $z = f(x, b)$.

*John A. Bishop and Jang H. Yoo, " 'Health Scare,' Excise Taxes and Advertising in the Cigarette Demand and Supply," *Southern Economics Journal*, Vol. 52, 1985, pp. 402–411.

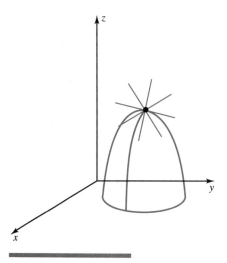

FIGURE 1

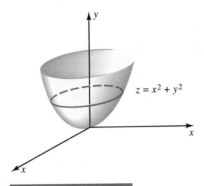

$z = x^2 + y^2$

FIGURE 2

Since this curve is two-dimensional, the slope of the tangent line at (a, b) is the derivative of the function of one variable, $z = f(x, b)$. This is called the partial derivative of $z = f(x, y)$ with respect to x. It is found by holding y constant and finding the derivative with respect to x.

For example, consider the function

$$z = f(x, y) = x^2 + y^2$$

Its graph, given in Figure 2, is an infinite bowl-shaped surface. To find the rate of change at $(1, 0)$ in the direction that is parallel to the x-axis, first find the intersection of $y = 0$ and the curve. It is $z = f(x, 0) = x^2 + 0 = x^2$. The derivative of this two-dimensional function is $2x$. This is expressed as

$$f_x(x, y) = 2x$$

It is called the *partial derivative of f with respect to x*. To get the slope of the tangent line at $(1, 0)$, substitute $x = 1$ into the formula for f_x and get $f_x(1, 0) = 2$. From Figure 3 we see that the function $z = f(x, 0)$ is a parabola in the xz plane whose tangent line at $(1, 0)$ is 2.

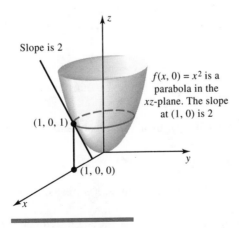

Slope is 2

$f(x, 0) = x^2$ is a parabola in the xz-plane. The slope at $(1, 0)$ is 2

$(1, 0, 1)$

$(1, 0, 0)$

FIGURE 3

Partial Derivatives: Formal Definition

The definition of the partial derivative uses the fact that if one variable is held constant the resulting expression is a function of one variable, and so its derivative can be computed. Thus a partial derivative of $f(x, y)$ is the derivative of a function of one variable that is obtained by holding constant one of the two variables of $f(x, y)$.

DEFINITION

The **partial derivative** of $f(x, y)$ with respect to x is defined as

$$f_x(x, y) = \lim_{h \to 0} \frac{f(x + h, y) - f(x, y)}{h}$$

The partial derivative of $f(x, y)$ with respect to y is defined as

$$f_y(x, y) = \lim_{h \to 0} \frac{f(x, y + h) - f(x, y)}{h}$$

There are several alternate notations for the partial derivatives. If it is clear that if f is a function of the two variables x and y, then sometimes $f_x(x, y)$ is expressed as simply f_x. Similarly f_y represents the partial derivative of f with respect to y. The notation for the partial derivative that is similar to the $\dfrac{dy}{dx}$ notation for the derivative is

$$\frac{\partial f}{\partial x} \quad \text{and} \quad \frac{\partial f}{\partial y}$$

This is read "the partial derivative of $f(x, y)$ with respect to x" and "the partial derivative of $f(x, y)$ with respect to y," respectively. Example 1 illustrates how to compute the partial derivatives of functions of two variables.

EXAMPLE 1

Problem

Compute $f_x(x, y)$ and $f_y(x, y)$ for the functions (a) $f(x, y) = x^3 + xy - y^{-1}$, (b) $f(x, y) = x^2 y^{-3} - e^{2xy} + 3$.

Solution (a) To find f_x, treat y as a constant and differentiate with respect to x.

$$f_x(x, y) = 3x^2 + y$$

Since y is held constant, the term y^{-1} is treated as a constant so that the partial derivative of it is 0.

To find f_y, treat x as a constant and differentiate with respect to y.

$$f_y(x, y) = x - (-1)y^{-2} = x + y^{-2}$$

(b) To find f_x, treat y as a constant and differentiate with respect to x.

$$f_x(x, y) = 2xy^{-3} - 2ye^{2xy}$$

To find f_y, treat x as a constant and differentiate with respect to y.

$$f_y(x, y) = -3x^2y^{-4} - 2xe^{2xy}$$

The next example demonstrates how to incorporate the chain rule with partial derivatives.

EXAMPLE 2

Problem

Consider the function $f(x, y) = (x^3 + xy^{-1})^4$. Compute (a) $f_x(x, y)$, (b) $f_y(x, y)$.

Solution (a) Use the extended power rule. Let $u = x^3 + xy^{-1}$ so that the partial derivative of u with respect to x is $3x^2 + y^{-1}$. Treat y as a constant and differentiate with respect to x.

$$f_x(x, y) = 4(x^3 + xy^{-1})^3(3x^2 + y^{-1})$$

(b) Again let $u = x^3 + xy^{-1}$. Then the partial derivative of u with respect to y is $-xy^{-2}$. Treat x as a constant and differentiate with respect to y.

$$f_y(x, y) = 4(x^3 + xy^{-1})^3(-xy^{-2})$$

If the partial derivative with respect to x is to be computed at a particular point (a, b), then the notation used is

$$f_x(a, b)$$

Likewise $f_y(a, b)$ represents the partial derivative with respect to y evaluated at the point (a, b). The next example illustrates this notation.

EXAMPLE 3

Problem

Consider the function $f(x, y) = 2x^3 + xy^2 - y$. Compute (a) $f_x(1, 3)$, (b) $f_y(-1, 2)$.

Solution (a) First find $f_x(x, y)$.

$$f_x(x, y) = 6x^2 + y^2$$

Next substitute the values $x = 1$ and $y = 3$.

$$f_x(1, 3) = 6(1)^2 + (3)^2 = 6 + 9 = 15$$

(b) Find $f_y(x, y)$.

$$f_y(x, y) = 2xy - 1$$

Next substitute the values $x = -1$ and $y = 2$.

$$f_y(-1, 2) = 2(-1)(2) - 1 = -4 - 1 = -5$$

Second-Order Partial Derivatives

Just as we can define higher-order derivatives of functions of one variable, so too we can define higher-order partial derivatives. The partial derivative of a partial derivative is called a second-order partial derivative, or simply a **second partial derivative.** The four second partial derivatives of $f(x, y)$ are

$$f_{xx}(x, y)$$
$$f_{xy}(x, y)$$
$$f_{yx}(x, y)$$
$$f_{yy}(x, y)$$

Just as with first-order partial derivatives, the shortened form of $f_{xx}(x, y)$ is f_{xx}. Also f_{xy}, f_{yx}, and f_{yy} are defined in a similar way. The functions f_{xy} and f_{yx} are called **mixed partial derivatives.** The notation for f_{xy} states that the partial derivative of $f(x, y)$ with respect to x is to be computed first, followed by computing the partial derivative of $f_x(x, y)$ with respect to y; f_{yx} calls for the reverse order of differentiation. For most functions that we will study these two functions are equal. The next example illustrates how to compute second partial derivatives.

EXAMPLE 4

Problem

Consider the function $f(x, y) = x^3y + xy^{-2} - ye^{3x}$. Compute (a) $f_{xx}(x, y)$, (b) $f_{xy}(x, y)$, (c) $f_{yx}(x, y)$, (d) $f_{yy}(x, y)$, (e) $f_{xx}(1, 3)$, (f) $f_{xy}(0, 1)$.

Solution First find $f_x(x, y)$ and $f_y(x, y)$.

$$f_x(x, y) = 3x^2y + y^{-2} - 3ye^{3x}$$
$$f_y(x, y) = x^3 - 2xy^{-3} - e^{3x}$$

(a) To find f_{xx}, compute the partial derivative of f_x with respect to x.

$$f_{xx}(x, y) = \frac{\partial}{\partial x} f_x(x, y) = \frac{\partial}{\partial x} (3x^2y + y^{-2} - 3ye^{3x}) = 6xy - 9ye^{3x}$$

(b) To find f_{xy} compute the partial derivative of f_x with respect to y.

$$f_{xy}(x, y) = \frac{\partial}{\partial y} f_x(x, y) = \frac{\partial}{\partial y} (3x^2y + y^{-2} - 3ye^{3x})$$

$$= 3x^2 - 2y^{-3} - 3e^{3x}$$

(c) To find f_{yx}, compute the partial derivative of f_y with respect to x.

$$f_{yx}(x, y) = \frac{\partial}{\partial x} f_y(x, y) = \frac{\partial}{\partial x} (x^3 - 2xy^{-3} - e^{3x})$$

$$= 3x^2 - 2y^{-3} - 3e^{3x}$$

Notice that $f_{xy} = f_{yx}$.

(d) To find f_{yy}, compute the partial derivative of f_y with respect to y.

$$f_{yy}(x, y) = \frac{\partial}{\partial y} f_y(x, y) = \frac{\partial}{\partial y} (x^3 - 2xy^{-3} - e^{3x}) = 6xy^{-4}$$

(e) To find $f_{xx}(1, 3)$, substitute the values $x = 1$ and $y = 3$ into $f_{xx}(x, y)$.

$$f_{xx}(1, 3) = 6(1)(3) - 9(3)e^{3(1)} = 18 - 27e^3$$

(f) To find $f_{xy}(0, 1)$, substitute the values $x = 0$ and $y = 1$ into $f_{xy}(x, y)$.

$$f_{xy}(0, 1) = 3(0)^2 - 2(1)^{-3} - 3e^{3(0)} = -2 - 3 = -5$$

The Cigarette Industry Revisited

The demand function computed by Bishop and Yoo for the cigarette industry was given in the beginning of the section as

$$Q(D, A) = 231.7D^{0.66}A^{0.07}$$

The analysis of the industry in the article uses several measures of the effect on Q by changes in D and A. These measures are based on the partial derivatives of Q. The investigators reached various conclusions concerning when it would be most advantageous for the industry to increase advertising in order to increase demand. The next example computes these partial derivatives.

EXAMPLE 5

Problem

Find (a) Q_D and (b) Q_A for the above demand function.

Solution (a) Treat A as a constant.

$$Q_D = 231.7(0.66)D^{0.66-1}A^{0.07} = 152.9D^{-0.34}A^{0.07}$$

(b) Treat D as a constant.

$$Q_A = 231.7(0.07)D^{0.66}A^{0.07-1} = 16.22D^{0.66}A^{-0.93}$$

EXERCISE SET 8–2

In Problems 1 to 14 find $f_x(x, y)$ and $f_y(x, y)$.

1. $f(x, y) = 4 + 3x^2 + 5y^{1/2}$

2. $f(x, y) = 1 + 2x^{2/3} + y^{-2}$

3. $f(x, y) = xy^2 - 5x^2$

4. $f(x, y) = x^2 - 2x^3y + y^3$

5. $f(x, y) = xy^2 - 3x^2y$

6. $f(x, y) = x^2y - 2x^2y^3$

7. $f(x, y) = x^{-2} + xy^{-2} - x^2y^3 + 4y$

8. $f(x, y) = x^{1/2} - x^{1/3}y + 3xy^{-3} + y$

9. $f(x, y) = xe^{2y} - \ln(2x + 3y)$

10. $f(x, y) = ye^{3x} + 4e^{(x-y)}$

11. $f(x, y) = (2x + 3y)^4$

12. $f(x, y) = (x^2 + 5y)^3$

13. $f(x, y) = (2x + 4xy)^{-2}$

14. $f(x, y) = (x^2y - 5x)^{1/3}$

In Problems 15 to 18 find $f_x(1, 0)$ and $f_y(2, 1)$.

15. $f(x, y) = 3xy^3 + 5x^{-1}y$

16. $f(x, y) = 4xy^4 - 6x^{-2}y + 3$

17. $f(x, y) = 2x^{1/2} - 3xy^3 + e^y$

18. $f(x, y) = x^3y + 4x^2y^2 + 6e^{2y}$

In Problems 19 to 24 find all second-order partial derivatives.

19. $f(x, y) = 3x^2 + 5y^{-2}$

20. $f(x, y) = 2x^{2/3} + y^{-1/2}$

21. $f(x, y) = 4 + 3x^2y + 5xe^{2y}$

22. $f(x, y) = 1 + 2x^{2/3}y + x \ln(1 + y^{-2})$

23. $f(x, y) = xy^2e^x + e^{2x}$

24. $f(x, y) = 2x^3ye^{3y}$

In Problems 25 and 26 find the values of x and y such that $f_x(x, y) = 0$ and $f_y(x, y) = 0$.

25. $f(x, y) = 1 - x^2 + y^2 + 2xy - 4y$

26. $f(x, y) = 2 - 2x^2 - y^2 + xy + 7x$

In Problems 27 to 30 find f_{xxx}, f_{xxy}, and f_{xyy}.

27. $f(x, y) = x^4 - x^3 + 2x + 2y^5 + 4y^2$

28. $f(x, y) = 1 + x^5 + x^{-1} + y^2$

29. $f(x, y) = 2xy + x^2y - xe^{2y}$

30. $f(x, y) = x^{-2} + xy^{-2} - \ln(y - 1)$

31. Give a definition of the partial derivative with respect to x of a function of three variables $w = f(x, y, z)$.

In Problems 32 and 33 find f_x and f_y by holding constant all variables except the one with which the partial derivative is being taken.

32. $f(x, y, z) = x^2 + x^3z - 2yz^3 + xyz$

33. $f(x, y, z) = 3x^{-2} + x^{1/3}z - x^2z^3 + xyz$

34. Find $f_{xyz}(1, 2, 3)$ for $f(x, y, z) = x^2 - yze^x - 4xyz^2$

35. Show that the function $f(x, y) = x^3 + 5x^2 + e^y$ satisfies the partial differential equation $f_{xy} = 0$.

36. Show that the function $f(x, y) = x^3 - 3xy^2$ satisfies the partial differential equation $f_{xx} + f_{yy} = 0$.

Referenced Exercise Set 8–2

1. In an article studying the economics of urban bus transportation Obeng* gathered data from 62 bus systems. The cost C was treated as a function of demand Q, fuel price F, the price of labor L, and the price of capital P. The data can be described by a Cobb–Douglas function of the form $C = 1.7Q^{1.1}F^{1.5}L^{0.36}P^{0.5}$. Find the partial derivatives of C.

2. In an article studying the optimal use of cobalt beams in treating cancer, Gantchew defines the dose distribution d along a straight line perpendicular to the central ray as a function of the width x of the ray and the length of penetration y as follows:†

$$d(x, y) = 0.93e^{[70x/(60 + y)]^{2.7}}$$

Compute d_x.

3. In their article on advertising in the cigarette industry, Bishop and Yoo also expressed price P as a function of disposable income D and cigarette advertising expenditures A.‡ In one part of the article the following function was defined:

$$P(D, A) = 0.125D^{0.56}A^{0.06}$$

Compute (a) P_D and (b) P_A.

Cumulative Exercise Set 8–2

1. Find the functional value of $f(x, y)$ at each point, where

$$f(x, y) = \frac{x^2 - y}{y}$$

(a) $(0.1, 0.0001)$ (b) $(0.1, -0.0001)$

2. Find the domain of the function in Problem 1.

*Kofi Obeng, "The Economics of Bus Transit Operation," *The Logistics and Transportation Review*, Vol. 20, 1986, pp. 45–63.

†M. G. Gantchew, "On the Application of a Simple Model of Cobalt-60 Beams," *Physics in Medicine and Biology*, Vol. 24, 1979, pp. 443–446.

‡John A. Bishop and Jang H. Yoo, " 'Health Scare,' Excise Taxes and Advertising Ban in the Cigarette Demand and Supply," *Southern Economics Journal*, Vol. 52, 1985, pp. 402–411.

In Problems 3 and 4 sketch the graph of the functions, giving the traces and one section that is parallel to each coordinate plane.

3. $f(x, y) = 4 - x - 2y$

4. $f(x, y) = 2 - 3x^2 - y^2$

In Problems 5 and 6 find $f_x(x, y)$ and $f_y(x, y)$.

5. $f(x, y) = \dfrac{x^2 - y}{y}$

6. $f(x, y) = e^{2xy + 3x}$

7. Find $f_x(0, 1)$ and $f_y(0, 1)$ for the function $f(x, y) = x + y + xy$.

8. Find all second-order partial derivatives of the function $f(x, y) = x - y - xy$.

8–3 Relative Extrema

Many applications of functions of more than one variable involve finding a value of a function that is a relative maximum or minimum. We concentrate on functions of two variables because they have a firm geometrical base. The concepts can be extended to functions of more than two variables. The ideas in this section are similar to corresponding results for functions of one variable. For instance, to compute relative extrema we use a theorem with first and second partial derivatives that is similar to the second derivative test for functions of one variable.

Definitions

The definition of a relative maximum and a relative minimum for a function of two variables is similar to the definitions for a function of one variable. The difference is that there are two independent variables instead of one. This means that $f(a, b)$ is a relative maximum if it is larger than all functional values of points (x, y) close to (a, b), whereby "close to" is meant within a circular region about (a, b). Geometrically, a relative maximum resembles a mountain peak or the top of a dome. A relative minimum resembles the bottom of a valley.

DEFINITION

A function $f(x, y)$ has a **relative maximum** at (a, b) if

$$f(a, b) \geq f(x, y)$$

for all (x, y) in some circular region around (a, b); $f(x, y)$ has a **relative minimum** at (a, b) if

$$f(a, b) \leq f(x, y)$$

for all (x, y) in some circular region around (a, b). The term **relative extremum** refers to either a relative maximum or a relative minimum.

EXAMPLE 1

Problem

Determine from the graphs given in Figure 1 if the function has a relative extremum at $(0, 0)$: (a) $f(x, y) = x^2 + 4y^2$, (b) $f(x, y) = y^2 - x^2$.

Solution (a) The graph in Figure 1a shows that all functional values $f(x, y) \geq 0$ for all (x, y), so that any circular region can be chosen to show that $f(x, y)$ has a relative maximum at $(0, 0)$.

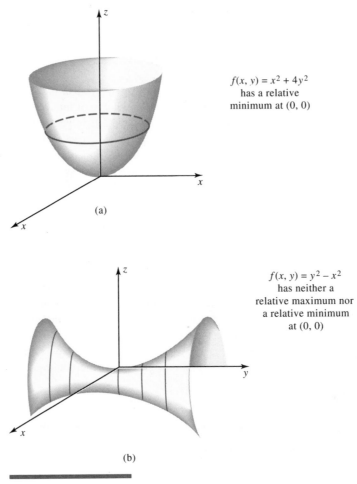

$f(x, y) = x^2 + 4y^2$
has a relative
minimum at $(0, 0)$

(a)

$f(x, y) = y^2 - x^2$
has neither a
relative maximum nor
a relative minimum
at $(0, 0)$

(b)

FIGURE 1

(b) Figure 1b shows that for points $(x, 0)$ on the x-axis, $f(x, 0) = -x^2 \leq 0$, and for points $(0, y)$ on the y-axis, $f(0, y) = y^2 \geq 0$. Any circular region containing $(0, 0)$ must contain points on each axis, so that some functional values will be greater than $f(0, 0) = 0$ and some will be smaller. Therefore $f(x, y)$ has neither a relative maximum nor a relative minimum at $(0, 0)$.

Critical Points

Recall that a function of one variable has a relative extremum at a point only if its derivative is 0 at the point, or if the derivative fails to exist. Figure 2 demonstrates that if a function of two variables has a relative maximum at (a, b), then the tangent lines to the graph at (a, b) must be horizontal, meaning that their slopes are 0. Since the partial derivative measures the slope of the tangent line, $f_x(a, b)$ and $f_y(a, b)$ must be 0. The same holds for relative minima. This is recorded in the following result.

> If $f(x, y)$ has a relative extremum at (a, b) and if both partial derivatives exist, then
>
> $$f_x(a, b) = 0 = f_y(a, b)$$

Extrema can also occur when a partial derivative fails to exist, but we will not consider this case.

A point (a, b) such that $f_x(a, b) = 0 = f_y(a, b)$ is called a **critical point.** The first step in locating relative extrema is to find the critical points. However, just as with functions of one variable, not all critical points correspond to relative extrema. For example, consider the function in Example 1(b), $f(x, y) = y^2 - x^2$. The partial derivatives are $f_x(x, y) = -2x$ and $f_y(x, y) = 2y$, so that each is 0 at $(0, 0)$, but the graph in Figure 1b shows that f has neither a relative maximum nor a relative minimum. This is an example of a **saddle point,** that is, a point where the partial derivatives are 0 but the function does not have a relative extremum at the point.

To find the critical points of a function, we need to solve $f_x = 0$ and $f_y = 0$ simultaneously, meaning that the same point (a, b) must satisfy both equations. This often requires that a system of two equations in two unknowns be solved, as illustrated in the next example.

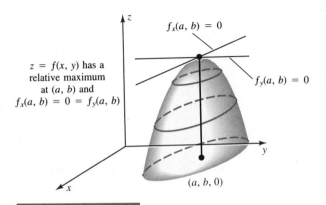

$f_x(a, b) = 0$

$f_y(a, b) = 0$

$z = f(x, y)$ has a relative maximum at (a, b) and $f_x(a, b) = 0 = f_y(a, b)$

$(a, b, 0)$

FIGURE 2

EXAMPLE 2

Problem

Find the critical points of the function

$$f(x, y) = x^2 - 4x + y^2 - 6y$$

Solution First compute the partial derivatives.

$$f_x(x, y) = 2x - 4$$
$$f_y(x, y) = 2y - 6$$

Set each partial derivative equal to 0.

$$f_x(x, y) = 2x - 4 = 0$$
$$f_y(x, y) = 2y - 6 = 0$$

This is a system of two equations in two unknowns. The first equation implies that $x = 2$ and the second implies $y = 3$. Thus $(2, 3)$ is the solution of the system of equations. Therefore $f_x(2, 3) = 0$ and $f_y(2, 3) = 0$, which means that $(2, 3)$ is the only critical point.

The Second Partial Derivative Test

According to the earlier result, if the function $f(x, y)$ in Example 2 has a relative extremum, it must occur at the critical value $(2, 3)$. But this does not mean that there is necessarily a relative extremum at $(2, 3)$; the point could be a saddle point. The following theorem shows how to determine whether critical points are relative extrema. It is referred to as the **second partial derivative test.**

Second Partial Derivative Test

Suppose $z = f(x, y)$ has first and second partial derivatives in a circular region around (a, b) where (a, b) is a critical point; that is

$$f_x(a, b) = 0 = f_y(a, b)$$

To simplify notation, let

$$A = f_{xx}(a, b), \quad B = f_{yy}(a, b), \quad \text{and} \quad C = f_{xy}(a, b)$$

Then

1. If $A > 0$ and $AB - C^2 > 0$, then $f(a, b)$ is a relative minimum.
2. If $A < 0$ and $AB - C^2 > 0$, then $f(a, b)$ is a relative maximum.
3. If $AB - C^2 < 0$, then $f(a, b)$ is not a relative extremum; it is a saddle point.
4. If $AB - C^2 = 0$, then the test yields no conclusion.

The next three examples show how to apply the test.

EXAMPLE 3

Problem

Find the relative extrema of the function $f(x, y) = x^2 - 4x + y^2 - 6y$.

Solution In Example 2 it was shown that the only critical point of $f(x, y)$ is $(2, 3)$. To apply the second partial derivative test, compute $f_{xx}(x, y) = 2$, $f_{yy}(x, y) = 2$, and $f_{xy}(x, y) = 0$. Then evaluate each at the critical point, which is immediate because each partial derivative is a constant.

$$A = f_{xx}(2, 3) = 2$$
$$B = f_{yy}(2, 3) = 2$$
$$C = f_{xy}(2, 3) = 0$$

Since $A > 0$ and $AB - C^2 = 2 \cdot 2 - 0 = 4 > 0$, part 1 of the test applies so that $f(2, 3) = -13$ is a relative minimum.

EXAMPLE 4

Problem

Find the relative extrema of the function $f(x, y) = x^2 + 4x - 2y^3 + 3y^2$.

Solution First find the critical points by setting the partial derivatives equal to 0.

$$f_x(x, y) = 2x + 4$$
$$f_y(x, y) = -6y^2 + 6y$$
$$f_x = 0 = 2x + 4 \tag{1}$$
$$f_y = 0 = -6y^2 + 6y \tag{2}$$

Equation 1 implies that $x = -2$. Equation 2 can be factored into

$$-6y(y - 1) = 0$$

This implies that $y = 0$ and $y = 1$. Thus there are two critical points, $(-2, 0)$ and $(-2, 1)$. We handle each critical point separately after the second partial derivatives are computed.

$$f_{xx}(x, y) = 2$$
$$f_{yy}(x, y) = -12y + 6$$
$$f_{xy}(x, y) = 0$$

(a) Consider the critical point $(-2, 0)$. Then

$$A = f_{xx}(-2, 0) = 2$$
$$B = f_{yy}(-2, 0) = 6$$
$$C = f_{xy}(-2, 0) = 0$$

Since $A > 0$ and $AB - C^2 = 2 \cdot 6 - 0 = 12 > 0$, part 1 of the test applies so that $f(-2, 0) = -4$ is a relative minimum.

(b) Consider the critical point $(-2, 1)$. Then

$$A = f_{xx}(-2, 1) = 2$$
$$B = f_{yy}(-2, 1) = -6$$
$$C = f_{xy}(-2, 1) = 0$$

Then $AB - C^2 = 2(-6) - 0 = -12 < 0$, and so part 3 of the test applies. Thus $f(x, y)$ has a saddle point at $(-2, 1)$.

Sometimes the expressions for f_x and f_y contain both x and y, which makes solving for the critical points a bit more complicated, as illustrated in the next example.

E X A M P L E   5

Problem

Find the relative extrema of the function $f(x, y) = x^2 - 5x - y^2 + xy$.

Solution First find the critical points by setting the partial derivatives equal to 0.

$$f_x(x, y) = 2x - 5 + y = 0$$
$$f_y(x, y) = x - 2y = 0$$

The second equation gives $x = 2y$. Substituting this equation into the equation $f_x = 0$ yields

$$2(2y) - 5 + y = 0$$
$$5y = 5$$
$$y = 1$$

Substituting $y = 1$ into $x = 2y$ yields $x = 2$. Therefore the critical point is $(2, 1)$. Next compute

$$A = f_{xx}(2, 1) = 2$$
$$B = f_{yy}(2, 1) = -2$$
$$C = f_{xy}(2, 1) = 1$$

Then $AB - C^2 = 2(-2) - 1 = -5 < 0$, and so part 3 of the test applies. Thus $f(x, y)$ has a saddle point at $(2, 1)$.

Part 4 of the test needs clarification. If $AB - C^2 = 0$, the second partial derivative test cannot determine whether the critical point is a relative extremum. It is necessary to use another test. For instance, all of the points close to the critical point can be tested. If they are all greater than $f(a, b)$, then $f(a, b)$ is a relative minimum; if they are all less than $f(a, b)$, then $f(a, b)$ is a relative maximum; if there are always some that are greater and some that are less than $f(a, b)$, then $f(x, y)$ has a saddle point at (a, b). This is usually a difficult task, however, and we will not pursue it.

EXERCISE SET 8–3

In Problems 1 to 6 determine whether the graph of the function has a relative extremum at $(0, 0)$.

1.

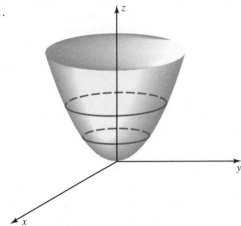

2.

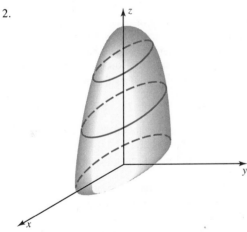

3.

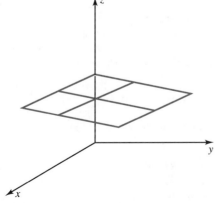

4.

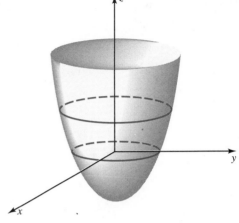

5.

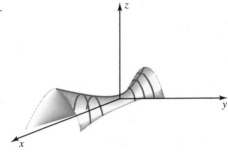

6.

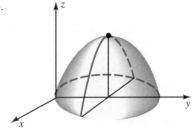

In Problems 7 to 12 find the critical points of the function.

7. $f(x, y) = x^2 + 6x + y^2 + 8y$

8. $f(x, y) = x^2 - 2x + y^2 + 6y$

9. $f(x, y) = 2x^2 + 4xy + y^2 - 6y + 1$

10. $f(x, y) = 3x^2 - 6xy + 2y^2 + 10y - 2$

11. $f(x, y) = x^2 - 2y^3 + 6y$

12. $f(x, y) = x^3 - 3x + 3y^2 + 12y$

In Problems 13 to 24 find the relative extrema and any saddle points of the function.

13. $f(x, y) = x^2 + 4x - 2y^3$

14. $f(x, y) = x^3 + 3y^2 + 12y$

15. $f(x, y) = x^2 - 8x + 2y^3 + 6y^2$

16. $f(x, y) = x^3 - 3x^2 + 3y^2 + 12y$

17. $f(x, y) = x^2 - 2y^3 + 6y$

18. $f(x, y) = x^3 - 3x + 3y^2 + 12y$

19. $f(x, y) = x^2 - 2x - 2xy + 2y^2 - 4y$

20. $f(x, y) = x^2 - 2xy + 3y^2 + 12y$

21. $f(x, y) = x - xy + y^3 - 3y^2$

22. $f(x, y) = x^2 + 4x - 2xy + 12y$

23. $f(x, y) = x^3 - 2x^2 - 2xy + y^2$

24. $f(x, y) = x^2 - 2xy - y^3 + 14y$

In Problems 25 to 30 find the relative extrema of the function.

25. $f(x, y) = x^3 - 2y^3 + 6y$

26. $f(x, y) = x^3 + 3y^3 - 9y$

27. $f(x, y) = x^3 - 6x^2 - 2y^3 + 24y$

28. $f(x, y) = x^3 + 3y^3 - 9y^2$

29. $f(x, y) = x^2 - 8x + 2y^3 + 6y^2 - 3y$

30. $f(x, y) = x^3 - 6x^2 - 8x + 3y^2 + 12y$

31. A firm utilizes x million dollars per year in labor cost and y million dollars per year in production cost. The cost function is

$$C(x, y) = 20 - 24x - 12y + x^2 + 3y^2$$

Find the amount that the firm should spend on labor and production each year to minimize cost.

32. A firm produces two types of radios, AM and AM–FM. If x thousand of AM radios and y thousand AM–FM radios are made per month, the revenue $R(x,y)$ and cost $C(x,y)$ functions are given as follows, where $R(x,y)$ and $C(x,y)$ are measured in millions of dollars:

$$R(x, y) = 5x + 6y$$
$$C(x, y) = 10 - 5x - 12y + x^2 + 2y^2$$

Find how many radios of each type should be produced to maximize profit.

33. A manufacturing firm wants to package its product in a rectangular box with no top. If the volume of the box is to be 64 cubic inches, find the dimensions that the box should be to use the least amount of material to construct it.

34. A firm wants to package its product in a rectangular box with no top and two intersecting partitions. If the volume of the box is to be 64 cubic inches, find the dimensions that the box should be to use the least amount of material to construct it.

Referenced Exercise Set 8–3

1. Extractive industries are those that extract minerals from the ground, such as metals or petroleum. In an article analyzing environmental controls on extractive industries, Stollery defines a welfare function W in terms of the quantity of minerals extracted x and the quantity of pollutants released y.[*] The welfare function measures the benefit to society based on the positive benefit due to extraction versus the negative benefit due to pollution. Stollery defines the function

$$W(x, y) = ax - bx^2 - cy^2 + d$$

where a, b, c, and d are constants depending on the industry and the method of extraction. Compute the values of x and y that maximize $W(x, y)$.

2. The design of an effective ramp is one of the most important aspects of highway design. When a car merges into traffic on a highway from a ramp, the car enters traffic in a gap between two vehicles. The distance between these two vehicles is called the gap size. Pignataro shows that the average gap size g is a function of the angle of approach x of the ramp, measured in degrees, and the length of the ramp y, measured in 100-foot sections.[†] The relationship is

$$g(x, y) = 5.5 - 0.83x + 0.04x^2 - 1.04y + 0.05y^2$$

Find the relative extrema of $g(x, y)$.

---

[*]Kenneth R. Stollery, "Environmental Controls in Extractive Industries," *Land Economics*, Vol. 61, 1985, pp. 136–144.

[†]Louis J. Pignataro, *Traffic Engineering: Theory and Practice*, Englewood Cliffs, N.J., Prentice-Hall, 1973, p. 169.

Cumulative Exercise Set 8–3

In Problems 1 to 4 graph the functions.

1. $f(x, y) = 2 + 3y - 4x$

2. $f(x, y) = -2 + x^2 + y^2$

3. $f(x, y) = x^2 - 4y^2$

4. $f(x, y) = 1 + x^2 + 2x + y^2$

In Problems 5 to 7 find $f_x(x, y)$ and $f_y(x, y)$.

5. $f(x, y) = x^3 - 2x^2y + y^4$

6. $f(x, y) = x^{-2}y + 6x^{1/2} + 9y^{2/3}$

7. $f(x, y) = (x^3 + 5y)^4$

8. Find all second-order partial derivatives of

$$f(x, y) = x^3 + 4xy^{1/2}$$

In Problems 9 to 12 find the relative extrema and any saddle points of the function.

9. $f(x, y) = x^3 - 3y^2 - 12y$

10. $f(x, y) = x^3 - 3x^2 - 3y^2 + 12y$

11. $f(x, y) = x^2 + 6x + 2xy - 12y$

12. $f(x, y) = x^3 + 3y^3 + 9y^2$

8–4 Lagrange Multipliers

The previous section dealt with finding the relative extrema of a function. This section also deals with optimizing functions, but the types of problems will have a little twist. Often in applications the optimum value of a function is subject to a **constraint.** For example, it might be necessary to find the maximum revenue subject to cost considerations. Another example might be to find the box with minimum surface area where the volume of the box is given.

An Application: Chemical Optimization at Monsanto

"Optimization" is a hallowed term in business. Sometimes a company's primary goal is to maximize profit, but at other times the objective is to minimize cost. As an example of the latter, consider the situation at Monsanto described by a consultant, Raymond Boykin, who was hired to evaluate Monsanto's chemical production methods.* In 1983 Monsanto opened a new chemical plant that caused capacity to exceed demand for many products. This prompted the company to hire a consultant to determine how to optimize production.

Boykin concentrated on one important product, maleic anhydride, which is used to make polyester resins. These resins are used in the manufacture of boat hulls, shower stalls, autobody parts, and counter tops. The chemical is produced by feeding raw material (butane) into a reactor under pressure. Production and cost are dependent on the raw material feed rate, the reactor velocity, and reactor pressure. Hence both production and cost are functions of three variables.

Because Monsanto is such a large producer of maleic anhydride, even small reductions in cost per pound would save large amounts of money. So Boykin recommended that Monsanto concentrate on minimizing cost once the amount of production was determined. This resulted in a constrained optimization problem: minimize cost subject to a certain level of production.

*Raymond F. Boykin, "Optimizing Chemical Production at Monsanto," *Interfaces,* Vol. 15, 1985, pp. 88–95.

In this section we study one method for solving such constrained optimization problems of functions of several variables. At the end of the section we will show how Monsanto solves the problem.

Lagrange Multipliers

Constrained optimization problems are often difficult to solve. An 18th-century mathematician, Joseph Louis Lagrange (1736–1813), discovered an ingenious method for solving these types of problems. Let us start with the general problem. Suppose $f(x, y)$ and $g(x, y)$ are functions of two variables. The problem is to find the values of x and y that either maximize or minimize the objective function $f(x, y)$ subject to the constraint $g(x, y) = 0$.

Lagrange's idea is to replace $f(x, y)$ with another function of three variables by introducing a new variable t called a Lagrange multiplier. The new function is

$$F(x, y, t) = f(x, y) + tg(x, y)$$

The variable t is always multiplied by the constraining function $g(x, y)$. The following theorem shows how to use Lagrange's method, called the technique of **Lagrange multipliers.** The proof of the theorem is beyond the scope of this text.

Theorem

If the function $f(x, y)$ has an extremum at (a, b), then there is a value of t, say, $t = c$, such that the partial derivatives of $F(x, y, t)$ each equal 0 at (a, b, c).

The theorem states that to find the extrema, first find the critical values of $F(x, y, t)$ by setting the partial derivatives equal to 0 and solving for x, y, and t. An extremum for $f(x, y)$ subject to $g(x, y) = 0$, if it exists, will be among these critical values. Example 1 illustrates how to apply the theorem.

EXAMPLE 1

Problem

Find the minimum value of $f(x, y) = x^2 + y^2$ subject to $g(x, y) = x + 2y - 10 = 0$.

Solution Form the function

$$F(x, y, t) = x^2 + y^2 + t(x + 2y - 10)$$

Find the partial derivatives of $F(x, y, t)$ and set each equal to 0.

$$F_x(x, y, t) = 2x + t = 0 \tag{1}$$

$$F_y(x, y, t) = 2y + 2t = 0 \tag{2}$$

$$F_t(x, y, t) = x + 2y - 10 = 0 \tag{3}$$

Solving (1) and (2) for t yields $t = -2x$ and $t = -y$. Thus $y = 2x$. Substituting $y = 2x$ into (3) yields

$$x + 2(2x) - 10 = 0$$
$$5x = 10$$
$$x = 2$$

Substitute $x = 2$ into (3) to get $y = 4$. Hence the minimum value of the function $f(x, y) = x^2 + y^2$, subject to $g(x, y) = x + 2y - 10 = 0$, occurs at the point $(2, 4)$. The minimum value is $f(2, 4) = 20$.

We have no easy means to verify that $f(2, 4)$ is actually a minimum rather than a maximum or a saddle point. Often it suffices to substitute into $f(x, y)$ several points close to $(2, 4)$ that also satisfy the constraint. Such points will produce functional values that are larger than 20. For instance, to select a point on $g(x, y) = 0$ close to $(2, 4)$, first choose a value of y that is close to 4, say, $y = 4.1$. Then the corresponding value of x is found by substituting $y = 4.1$ into $g(x, y) = 0$, which yields $x = 1.8$. Then $f(1.8, 4.1) = (1.8)^2 + (4.1)^2 = 20.05$.

The Geometric Interpretation

For functions of two variables the method of Lagrange multipliers has a geometric interpretation. To illustrate, graph both $f(x, y) = x^2 + y^2$ and $g(x, y) = x + 2y - 10 = 0$ in the same coordinate system. Figure 1 shows that $f(x, y)$ is an infinite bowl and $g(x, y) = 0$ is a plane. Their intersection is a two-dimensional curve, a parabola, whose minimum is at $(2, 4)$. Problem 34 will ask you to substitute algebraically $g(x, y) = x + 2y - 10 = 0$ into $f(x, y) = x^2 + y^2$. This produces the equation of the parabola in Figure 1; it is an equation in one variable whose derivative can be computed and whose minimum will occur at $(2, 4)$.

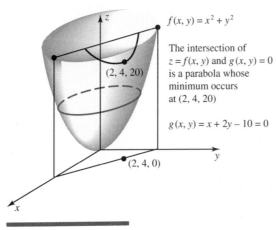

$f(x, y) = x^2 + y^2$

The intersection of $z = f(x, y)$ and $g(x, y) = 0$ is a parabola whose minimum occurs at $(2, 4, 20)$

$(2, 4, 20)$

$g(x, y) = x + 2y - 10 = 0$

$(2, 4, 0)$

FIGURE 1

The Method of Lagrange Multipliers

Before applying the method of Lagrange multipliers to various problems, let us summarize the steps in using the technique.

The Method of Lagrange Multipliers

In order to find the relative extrema of $f(x, y)$ subject to the constraint $g(x, y) = 0$, proceed as follows:

1. Construct the function

   $$F(x, y, t) = f(x, y) + tg(x, y)$$

2. Compute the partial derivatives of $F(x, y, t)$ and set each equal to 0.
3. Solve the system of equations formed in part 2.
4. The relative extrema of $f(x, y)$ subject to $g(x, y) = 0$ are among the solutions to the system of equations in step 3.

Example 2 illustrates how to apply the method of Lagrange multipliers to a problem where the system is a bit more complicated than the system in Example 1. This example also shows how to deal with a problem in which the constraint is expressed as an equation rather than as a function.

EXAMPLE 2

Problem

Find the relative maximum value of $f(x, y) = 8x - x^2 + 4y - y^2$ subject to $x + y = 10$.

Solution First express the constraint in the proper form, an expression in x and y, $g(x, y)$, set equal to 0. Subtract 10 from each side of the constraining equation to get

$$g(x, y) = x + y - 10 = 0$$

1. Construct the function

   $$F(x, y, t) = 8x - x^2 + 4y - y^2 + t(x + y - 10)$$

2. Compute the partial derivatives of $F(x, y, t)$.

   $$F_x(x, y, t) = 8 - 2x + t = 0 \tag{1}$$

   $$F_y(x, y, t) = 4 - 2y + t = 0 \tag{2}$$

   $$F_t(x, y, t) = x + y - 10 = 0 \tag{3}$$

3. Solving (1) and (2) for t yields

   $$t = 2x - 8 \quad \text{and} \quad t = 2y - 4$$

Thus $2x - 8 = 2y - 4$. This simplifies to the equation

$$x - y = 2 \tag{4}$$

Form the system of two equations consisting of (3) and (4).

$$x + y = 10$$
$$x - y = 2$$

Adding these equations yields $2x = 12$, or $x = 6$. Substituting this value into (4) gives $y = 4$.

4. The relative maximum of the value of $f(x, y) = 8x - x^2 + 4y - y^2$ subject to $x + y = 10$ must occur at (6, 4). Testing a few points close to (6, 4) verifies that $f(6, 4) = 12$ is a maximum.

The method of Lagrange multipliers can be applied to functions of more than two variables. Example 3 illustrates how to solve a particular type of word problem by applying Lagrange multipliers to a function of three variables.

EXAMPLE 3

Problem

Find three numbers whose sum is 60 and whose product is a maximum.

Solution First express the problem in a form where Lagrange multipliers can be applied. Let the three numbers be represented by x, y, and z. The problem calls for us to maximize

$$f(x, y, z) = xyz$$

subject to $x + y + z = 60$. Let $g(x, y, z) = x + y + z - 60 = 0$. Apply the steps in the method.

1. Construct the function

$$F(x, y, z, t) = xyz + t(x + y + z - 60)$$

2. Compute the partial derivatives of $F(x, y, z, t)$ and form the system of equations $F_x = 0$, $F_y = 0$, $F_z = 0$, $F_t = 0$.

$$F_x(x, y, z, t) = yz + t = 0 \tag{1}$$

$$F_y(x, y, z, t) = xz + t = 0 \tag{2}$$

$$F_z(x, y, z, t) = xy + t = 0 \tag{3}$$

$$F_t(x, y, z, t) = x + y + z - 60 = 0 \tag{4}$$

3. Solving (1), (2), and (3) for t and combining them yields

$$t = -yz = -xz = -xy \tag{5}$$

Assume that none of the numbers is 0, because then the product would be 0, which is not a maximum. Then (5) simplifies to $x = y = z$. Substituting these relationships into (4) yields $3x = 60$, so that $x = 20$. This means that $x = y = z = 20$ is the solution.

4. The relative maximum value of $f(x, y, z) = xyz$ subject to $x + y + z = 60$ must occur at $(20, 20, 20)$. Testing a few points close to $(20, 20, 20)$ verifies that the maximum value is $f(20, 20, 20) = 8000$.

It is not always easy to test the points close to the critical value. In this case suppose we considered points of the form $(20, 20 + h, 20 - h)$ for a small number h. These points are close to $(20, 20, 20)$ and satisfy the constraint because they sum to 60. Their product is $f(20, 20 + h, 20 - h) = 20(20 + h)(20 - h) = 20(400 - h^2) = 8000 - 20h^2$. This number is always less than 8000. This does not test all the points close to $(20, 20, 20)$, but it gives a good idea of what is required to prove that the maximum occurs at that point.

The Case Study deals with a common type of function in business called a Cobb–Douglas production function, which is of the form

$$f(x, y) = x^a y^b$$

where f is the production or output of a firm while x and y are the inputs, which could be raw goods or labor. The exponents a and b are constants. Example 4 shows how Lagrange multipliers can be used to solve an optimization problem containing a Cobb–Douglas production function.

EXAMPLE 4

The production capacity of a firm is given by

$$f(x, y) = x^{1/2} y^{1/2}$$

where x is the number of labor-hours available daily and y is the amount of capital invested in machinery. Labor costs are \$10 per labor-hour. Assuming that there are 8 working hours in a day and 260 working days in a year, and that the firm allocates \$5,000,000 in capital for wages and machinery, the constraining equation is

$$10(8)(260)x + y = 5,000,000$$

or

$$20,800x + y = 5,000,000$$

Problem

Find the values of x and y that maximize production capacity.

Solution Express the constraint in the form

$$g(x, y) = 20,800x + y - 5,000,000.$$

Apply the steps in the method.

1. Construct the function

$$F(x, y, t) = x^{1/2} y^{1/2} + t(20,800x + y - 5,000,000)$$

2. Compute the partial derivatives of $F(x, y, t)$ and form the system of equations $F_x = 0, F_y = 0, F_t = 0$.

$$F_x(x, y, t) = \frac{1}{2} x^{-1/2} y^{1/2} + 20,800t = 0 \tag{1}$$

$$F_y(x, y, t) = \frac{1}{2} x^{1/2} y^{-1/2} + t = 0 \tag{2}$$

$$F_t(x, y, t) = 20,800x + y - 5,000,000 = 0 \tag{3}$$

3. Solving (1) and (2) for t and combining them yields

$$-20,800t = \frac{1}{2} x^{-1/2} y^{1/2} = \frac{20,800}{2} x^{1/2} y^{-1/2}$$

Multiply the equation by $2x^{1/2} y^{1/2}$. This yields

$$y = 20,800x$$

Substituting this equation into (3) gives $2y = 5,000,000$, or $y = 2,500,000$. Then $x = 2,500,000/20,800 \approx 120.2$.

4. The maximum production occurs at $x \approx 120.2$, or about 120 labor-hours, and $y = \$2,500,000$.

Let us return to the optimization problem facing Monsanto, which was mentioned at the start of the section. The consultant determined that production P and cost C were functions of three variables. Let x represent the raw material feed rate, y the reactor velocity, and z the reactor pressure. The production function has the form $P = kxy/z$ and the cost function $C = ax + by + c/z$. The constants k, a, b, and c depend on which reactor is used. The object is to minimize cost subject to a given level of production. Boykin developed a computer program for Monsanto, which is run five times a day at each plant. The level of production is set and then the program determines the values of x, y, and z that minimize C. The next example illustrates how the program solves the problem.

EXAMPLE 5

Problem

Let $k = 1$, $a = 0.64$, $b = 4$, and $c = 1.6$ in the preceding production and cost functions. Set the production level at $P = 27$. Find the values of x, y, and z that minimize cost.

Solution The problem is to minimize

$$C = 0.64x + 4y + \frac{1.6}{z}$$

Subject to: $\quad \dfrac{xy}{z} - 27 = 0$

Define

$$C(x, y, z, t) = 0.64x + 4y + \frac{1.6}{z} + t\left(\frac{xy}{z} - 27\right)$$

Find the partial derivatives and form the system of equations.

$$C_x(x, y, z, t) = 0.64 + \frac{ty}{z} = 0$$

$$C_y(x, y, z, t) = 4 + \frac{tx}{z} = 0$$

$$C_z(x, y, z, t) = -1.6z^{-2} - txyz^{-2} = 0$$

$$C_t(x, y, z, t) = \frac{xy}{z} - 27 = 0$$

Solve for $-t$ in the first three equations.

$$-t = \frac{0.64z}{y} = \frac{4z}{x} = \frac{1.6}{xy}$$

$$y = 0.16x, \quad z = \frac{1}{0.4x}$$

Substituting these values into $C_t = 0$ gives

$$x(0.16x)(0.4x) - 27 = 0$$

$$x^3 = \frac{27}{0.064}$$

$$x = \frac{3}{0.4} = 7.5$$

Therefore the solution is $x = 7.5$, $y = 0.16(7.5) = 1.2$, $z = 1/3$.

The next example illustrates another application of Lagrange multipliers for solving real-world optimization problems.

EXAMPLE 6

An aircraft design engineer plans to construct a rectangular storage bin in the space between the rear wall of an aircraft cabin and the fuselage. In the design model of the bin, the floor of the bin is an extension of the floor of the cabin, which is represented by the xy-plane; one side wall of the bin is the rear wall of the cabin, represented by the xz-plane; and the fuselage is represented by the equation $z = 16 - x^2 - y^2$. (See Figure 2.)

Problem

Find the dimensions of the bin with greatest volume that can fit into the given space.

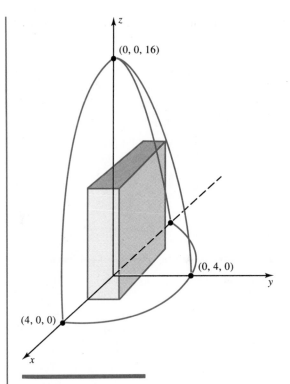

(0, 0, 16)

(0, 4, 0)

y

(4, 0, 0)

x

FIGURE 2

Solution The function to be maximized is the volume V of the bin. If we let the length, width, and height of the bin be x, y, and z, respectively, then $V = V(x, y, z) = xyz$. We must have all three variables, x, y, and z, greater than or equal to 0. The constraint is given by $z = 16 - x^2 - y^2$. The problem can be stated in a way to apply Lagrange multipliers as follows:

Maximize $V(x, y, z) = xyz$ subject to $z = 16 - x^2 - y^2$. Let $g(x, y, z) = 16 - x^2 - y^2 - z$. Apply the steps in the method.

1. Construct the function

   $F(x, y, z, t) = xyz + t(16 - x^2 - y^2 - z)$

2. Compute the partial derivatives of $F(x, y, z, t)$ and form the system of equations $F_x = 0$, $F_y = 0$, $F_z = 0$, $F_t = 0$:

   $F_x(x, y, z, t) = yz - 2xt = 0$ (1)

   $F_y(x, y, z, t) = xz - 2yt = 0$ (2)

   $F_z(x, y, z, t) = xy - t = 0$ (3)

   $F_t(x, y, z, t) = 16 - x^2 - y^2 - z = 0$ (4)

3. Solving (1), (2), and (3) for $2t$ and combining them yields

$$2t = yz/x = xz/y = 2xy \qquad (5)$$

Assume that none of the numbers is 0, because then the product would be 0, which is not a maximum. Then (5) simplifies to $x^2 = y^2 = z/2$. Substituting these relationships into (4) yields $4x^2 = 16$, or $x^2 = 4$, so $x = 2$. This means that the solution is $x = y = 2$ and $z = 8$.

4. The relative maximum value of $V(x, y, z) = xyz$ subject to $z = 16 - x^2 - y^2$ must occur at $(2, 2, 8)$. Testing a few points close to $(2, 2, 8)$ verifies that the maximum value is $f(2, 2, 8) = 32$.

EXERCISE SET 8–4

In Problems 1 to 18 use the method of Lagrange multipliers to find the extrema of the functions with the given constraints.

1. $f(x, y) = x^2 + y^2$
   subject to $g(x, y) = x + y - 1 = 0$

2. $f(x, y) = x^2 + 2y^2$
   subject to $g(x, y) = x + 2y - 3 = 0$

3. $f(x, y) = x^2 + 2x + y^2 - y$
   subject to $g(x, y) = x + 2y = 0$

4. $f(x, y) = x^2 + 4x + y^2 - 6y$
   subject to $g(x, y) = 3x + y - 1 = 0$

5. $f(x, y) = x^3 + 3x^2 + y^2 - 2y$
   subject to $g(x, y) = 3x + 2y + 3 = 0$

6. $f(x, y) = x^3 + 6x^2 + y^2 - 4y$
   subject to $g(x, y) = x + 3y + 1 = 0$

7. $f(x, y) = 2x^3 - 6x^2 + 2y^2 - 8y$
   subject to $x + 2y - 2 = 0$

8. $f(x, y) = 4x^3 + 6x^2 - y^2 - 2y$
   subject to $2x - 3y + 1 = 0$

9. $f(x, y) = 2x - x^2 + 2y^3 - 6y^2$
   subject to $x - 2y = 10$

10. $f(x, y) = 4x^3 + 6x^2 - y^2 - 2y$
    subject to $2x + y = 2$

11. $f(x, y) = xy$
    subject to $g(x, y) = 2x + y - 1 = 0$

12. $f(x, y) = 2xy$
    subject to $g(x, y) = x + 2y - 3 = 0$

13. $f(x, y) = xy + x^2$
    subject to $2x + 3y = 1$

14. $f(x, y) = xy - y^2$
    subject to $x - 2y = 0$

15. $f(x, y, z) = xyz$
    subject to $g(x, y, z) = 2x + y + z - 2 = 0$

16. $f(x, y, z) = 2xyz$
    subject to $g(x, y, z) = x + 2y - z - 4 = 0$

17. $f(x, y, z) = xyz$
    subject to $x + y + z = 300$

18. $f(x, y, z) = xyz$
    subject to $x + 2y + 3z = 600$

19. Find three numbers that sum to 30 such that their product is a maximum.

20. Find three numbers that sum to 120 such that their product is a maximum.

21. Maximize $f(x, y) = x^{1/2}y^{1/2}$
    subject to $100x + y = 5000$.

22. Maximize $f(x, y) = x^{1/2}y^{1/3}$
    subject to $200x + y = 8000$.

23. Find three numbers such that the sum of two of them and twice the third is equal to 600 and their product is a maximum.

24. Find three numbers such that the sum of two of them and twice the third is equal to 1200 and their product is a maximum.

25. Find the minimum distance from the line $3x + y = 5$ to the origin $(0, 0)$.

26. Find the minimum distance from the line $x - 2y = 4$ to the point $(1, 0)$.

In Problems 27 to 33 use the method of Lagrange multipliers to find the extrema of the functions with the given constraints.

27. $f(x, y) = xy^2$ subject to $2x - y = 1$

28. $f(x, y) = x^2y$ subject to $x + y = 3$

29. $f(x, y, z) = xyz^2$ subject to $2x + 3y + z = 2$

30. $f(x, y, z) = x^2yz$ subject to $x - 2y - 3z = 4$

31. $f(x, y, z) = xy^2 + yz^2$
    subject to $x + y + z = 2$

32. $f(x, y, z) = x^2yz + xz$
    subject to $x - y - z = 4$

33. $f(x, y, z) = xy^2 + xz^2$
    subject to $xy + yz = 2$

34. Substitute $g(x, y) = x + 2y - 10 = 0$ into $f(x, y) = x^2 + y^2$, and find the extremum of the resulting function.

35. A rectangular box with a top is to have a volume of 16 cubic feet. Find the dimensions that produce the least expensive box if the material for the sides of the box is \$2 per square foot and the material for the top and bottom is \$4 per square foot.

36. Find the dimensions of a rectangular box with a top that has the least surface area and a volume of 8000 cubic feet.

37. The postal service has a rule that rectangular packages cannot be sent with a specific rate if the length plus girth (the perimeter of the cross section taken perpendicular to the length) is more than 84 inches. Find the dimensions of the rectangular box with largest volume that can be sent with this rate.

38. A soft drink manufacturer sells its product in two markets, retail (in supermarkets) and commercial (in vending machines). The profit from the sale of x gallons of drink in the retail market and y gallons in the commercial market per week is given by

$$f(x, y) = 180x + 60y - 4x^2 - y^2 - xy$$

Find the number of gallons of drink that should be sold in each market per week to maximize profit.

39. A manufacturing company makes two types of suits: wool and polyester. The cost to manufacture each wool suit is \$100 and the cost to manufacture each polyester suit is \$50. If the manufacturer charges x dollars for each wool suit and y dollars for each polyester suit, the expected revenue per week is determined to be

$$R(x, y) = 140x + 80y - 10x^2 - 8y^2 + 4xy$$

(a) Find the cost function $C(x, y)$ assuming the fixed cost per week is \$1000.

(b) Determine the values of x and y that maximize the profit function

$$P(x, y) = R(x, y) - C(x, y)$$

40. A pharmaceutical company makes two types of pills: pain pills and cold pills. The expected revenue $R(x, y)$ and cost $C(x, y)$ functions are determined to be

$$R(x, y) = 60x + 95y - 2x^2 - 2y^2 + 4xy$$
$$C(x, y) = 10 + 20x + 15y$$

Find the profit function $P(x, y) = R(x, y) - C(x, y)$ and determine the values of x and y that maximize profit.

41. A food distributor markets white rice and brown rice. The expected revenue $R(x, y)$ and cost $C(x, y)$ functions are determined to be

$$R(x, y) = 190x + 100y - 20x^2 - 4y^2 - 8xy$$
$$C(x, y) = 15 + 30x + 20y$$

Find the profit function $P(x, y) = R(x, y) - C(x, y)$ and determine the values of x and y that maximize profit.

42. A manufacturing company makes two sizes of its product: large and small. The price at which the company will sell each product, p_l for the large size and p_s for the small size, depends upon the amount of items manufactured. If the company makes x large items and y small items, then it determines that $p_l = 92 - 5x$ and $p_s = 40 - 2y$. The cost function in the given time period is $C(x, y) = 50 + 4x + 8y + xy$. The revenue function $R(x, y)$ is given as $R(x, y) = xp_l + yp_s$. Then the profit function is $P(x, y) = R(x, y) - C(x, y)$. Determine the values of x and y that maximize profit.

43. Find the dimensions of the box with a top with maximum volume if the box is to be made from 216 square inches of material.

44. Find the minimum distance from the plane $6x - y - 2z = -3$ to the point $(1, 0, -1)$.

Referenced Exercise Set 8–4

1. In an article studying the economics of agricultural production, Chavas et al.* consider the swine industry in depth. The study uses constrained optimization to determine the maximum weight that an animal can attain subject to various limitations, such as cost of grain. Solve the following problem via Lagrange multipliers that is similar to the problems presented by Chavas: Find the values of x and y that maximize $z = 3x + 4y$ subject to $a = 2x^2 + 3y^2$ where a is the weight of the animal, x is the amount of feed, and y is the amount of veterinary care for the animal.

2. The Swedish roundwood market is a good example of a market in which both buyers and sellers are highly monopolized. This means that various types of simple bargaining models can be tested against the actual results in the marketplace. In one such article, Johannson and Lofgren assume that the owners wish to maximize profit P subject to a utility function u which measures the benefit to the owners when the price of roundwood is greater than the price determined by free competition.† The latter is sometimes possible because of the monopolistic power enjoyed by the sellers. Profit and utility are both functions of the price x and the output y. The two functions are

$$P(x, y) = f(x) - xy$$
$$u(x, y) = x[g(y) - a] - k$$

where $f(x)$ and $g(y)$ are measures of the monopolistic powers of the sellers and the buyers, respectively. Johannson and Lofgren then found conditions when $P(x, y)$ would be maximized subject to the constraint $u(x, y) = 0$ by applying the method of Lagrange multipliers to the function

$$L(x, y, t) = P(x, y) - tu(x, y)$$

Find the maximum of $L(x, y, t)$ for $f(x) = x^2$, $g(y) = y$, $a = 2$, and $k = 1$.

Cumulative Exercise Set 8–4

1. Consider the function

$$f(x, y) = \frac{x^2 - y}{y}$$

(a) Evaluate $f(1, 1)$.
(b) Find the domain of $f(x, y)$.

2. Sketch the graph of the function $f(x, y) = 9 - x^2 - y^2$, giving the traces and one section that is parallel to each coordinate plane.

3. Find $f_x(x, y)$ and $f_y(x, y)$ for the function

$$f(x, y) = \frac{xy}{x - y}$$

4. Find all second-order partial derivatives of the function

$$f(x, y) = x - x^2y + 2y^3$$

5. Determine whether the graph of the function $f(x, y) = x^2 - y^2$ has a relative extremum at the point $(0, 0)$. (See Figure 3.)

6. Find the critical points of the function $f(x, y) = 2x^3 - 6x - y^2$.

In Problems 7 and 8 find the relative extrema and any saddle points of the given function.

7. $f(x, y) = (x - 4)^2 + (y - 1)^2$

8. $f(x, y) = x^3 + 2y^3 + 3x^2 - 6y^2 - 45x - 48y$

In Problems 9 to 11 use the method of Lagrange multipliers to find extrema of the functions with the given constraints.

9. $f(x, y) = 6x + 20y + 3xy - 2x^2 - 4y^2$ subject to $x + y = 16$

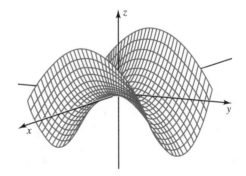

$$f(x, y) = x^2 - y^2$$

FIGURE 3

---

*Jean-Paul Chavas et al., "Modeling Dynamic Agricultural Production Response: The Case of Swine Production," *American Journal of Agricultural Economics*, Vol. 67, 1985, pp. 636–646.

†Per-Olov Johannson and Karl-Gustaf Lofgren, "A Bargaining Approach to the Modeling of the Swedish Roundwood Market," *Land Economics*, Vol. 61, 1985, pp. 65–75.

10. $f(x, y) = \sqrt{x^2 + y^2}$ subject to $2x + 5y = 14$

11. $f(x, y, z) = xy^2z^2$ subject to $x + y + z = 15$

12. Three sides of a box-shaped tree-house need painting: the front, one side, and the roof. The volume of the tree house is 2016 cubic feet. What dimensions of the tree house will minimize the cost of painting if it costs $9 per square foot to paint the face, $7 per square foot to paint the side, and $4 per square foot to paint the top?

8–5 Double Integrals

In the preceding sections of this chapter the concept of the derivative of a function of one variable was generalized to partial derivatives of functions of several variables. In this section we describe how to generalize the integral of a function of one variable to the integral of a function of two variables.

One of the fundamental notions of calculus is that the integral relies on the antiderivative, which is the reverse operation of finding the derivative. In a similar manner the definition of an integral of a function of two variables will also rely on the reverse operation of computing partial derivatives.

Partial Differentiation Reversed

We start with an example of reversing the process of partial differentiation. To find the partial derivative with respect to x of $f(x, y) = 6xy^2 - 5x^4y$, we treat y as a constant and differentiate with respect to x. To reverse the process, hold y constant and find the antiderivative with respect to x. This yields

$$\int (6xy^2 - 5x^4y)\, dx = 6\left(\frac{x^2}{2}\right)y^2 - 5\left(\frac{x^5}{5}\right)y + C(y) = 3x^2y^2 - x^5y + C(y)$$

The constant of integration $C(y)$ represents not just a constant, but an arbitrary function of y. The reason for this can be seen when checking the answer by computing the partial derivative with respect to x of $3x^2y^2 - x^5y + C(y)$ which is $6xy^2 - 5x^4y$. Notice that the partial derivative with respect to x of $C(y)$ is 0.

We write $\int f(x, y)\, dx$ for the operation of computing the partial antiderivative of $f(x, y)$ with respect to x, holding y constant. The first example further illustrates this operation.

EXAMPLE 1

Problem

Compute $\int (6xy^5 + 12x^{-2}y^3 - 7)\, dx$

Solution Hold y constant and find the antiderivative with respect to x.

$$\int (6xy^5 + 12x^{-2}y^3 - 7)\, dx = 6\left(\frac{x^2}{2}\right)y^5 + 12\left(\frac{x^{-1}}{-1}\right)y^3 - 7x + C(y)$$

$$= 3x^2y^5 - 12x^{-1}y^3 - 7x + C(y)$$

The generalization of antidifferentiation to functions of two variables can be used to compute definite integrals of the form

$$\int_a^b f(x, y)\, dx$$

The limits of integration are actually functions. Here they should be viewed as the constant functions $x = a$ and $x = b$.

EXAMPLE 2

Problem

Compute $\int_1^2 (6xy^5 + 12x^{-2}y^3 - 7)\, dx$

Solution The first step is to find the antiderivative with respect to x. This was done in Example 1. Just as with functions of one variable, any antiderivative can be used when computing definite integrals, so we take $C(y) = 0$. Then insert the limits of integration, remembering that they are $x = 1$ and $x = 2$.

$$\int_1^2 (6xy^5 + 12x^{-2}y^3 - 7)\, dx = (3x^2y^5 - 12x^{-1}y^3 - 7x)\,\Big|_{x=1}^{x=2}$$

$$= 3(4)y^5 - 12\left(\frac{1}{2}\right)y^3 - 7(2) - [3(1)y^5 - 12(1)y^3 - 7(1)]$$

$$= 12y^5 - 6y^3 - 14 - 3y^5 + 12y^3 + 7$$

$$= 9y^5 + 6y^3 - 7$$

Notice in Example 2 that the definite integral with respect to x produces a function of y. The variable x does not appear because a number is substituted for x whenever x occurs in the antiderivative. Thus the result of a single integral of $f(x, y)$ is a function of one variable. This function can then be integrated, producing a double integral. Example 3 illustrates this property.

EXAMPLE 3

Problem

Compute $\int_0^2 \left[\int_1^2 (6xy^5 + 12x^{-2}y^3 - 7)\, dx\right] dy$

Solution First work within the brackets. The definite integral in the brackets was computed in Example 2. Substitute its value, $9y^5 + 6y^3 - 7$, for the expression in the brackets to get

$$\int_0^2 \left[\int_1^2 (6xy^5 + 12x^{-2}y^3 - 7)\, dx\right] dy$$

$$= \int_0^2 (9y^5 + 6y^3 - 7)\, dy = \left(\frac{9y^6}{6} + \frac{6y^4}{4} - 7y\right)\Big|_{y=0}^{y=2}$$

$$= \frac{9(64)}{6} + \frac{6(16)}{4} - 7(2) = 96 + 24 - 14 = 106$$

Suppose the order of the integrals in Example 3 was reversed and integration with respect to y was carried out before integration with respect to x. Would the double integral be changed? To answer this question we must examine the processes used in Examples 1, 2, and 3 but with the order reversed. The computations will be carried out in Example 4.

EXAMPLE 4

Problem

Compute

$$\int_1^2 \left[\int_0^2 (6xy^5 + 12x^{-2}y^3 - 7) \, dy \right] dx$$

Solution Consider the integral inside the square brackets. The notation

$$\int f(x, y) \, dy$$

stands for the partial antiderivative of $f(x, y)$ with respect to y. It is evaluated by holding x constant and finding the antiderivative with respect to y.

$$\int (6xy^5 + 12x^{-2}y^3 - 7) \, dy = 6x\left(\frac{y^6}{6}\right) + 12x^{-2}\left(\frac{y^4}{4}\right) - 7y + C(x)$$

$$= xy^6 + 3x^{-2}y^4 - 7y + C(x)$$

Now insert the limits of integration and use the antiderivative with respect to y that was found above (note the $C(x)$ drops out):

$$\int_0^2 (6xy^5 + 12x^{-2}y^3 - 7) \, dy = (xy^6 + 3x^{-2}y^4 - 7y) \Big|_{y=0}^{y=2}$$

$$= x(64) + 3x^{-2}(16) - 7(2) - [x(0) + 3x^{-2}(0) - 7(0)]$$

$$= 64x + 48x^{-2} - 14$$

The definite integral with respect to y produces a function of x that can be integrated to obtain the desired double integral.

$$\int_1^2 \left[\int_0^2 (6xy^5 + 12x^{-2}y^3 - 7) \, dy \right] dx = \int_1^2 (64x + 48x^{-2} - 14) \, dx$$

$$= (32x^2 - 48x^{-1} - 14x) \Big|_{x=1}^{x=2}$$

$$= 32(4) - 48\left(\frac{1}{2}\right) - 14(2) - (32 - 48 - 14) = 76 - (-30) = 106$$

Order of Integrals

Compare the definite integrals in Examples 3 and 4. It is not a coincidence that they are equal. In fact, for most functions $f(x, y)$, if the order of integration is reversed then the double integrals will be the same, that is,

$$\int_a^b \left[\int_c^d f(x, y) \, dx \right] dy = \int_c^d \left[\int_a^b f(x, y) \, dy \right] dx$$

This means that the brackets are not needed, which in turn allows us to define the double integral of $f(x, y)$ without having to specify an order for the integrals.

The limits of integration define a rectangular region R in the plane. Here R is the set of points (x, y) satisfying $c \leq x \leq d$ and $a \leq y \leq b$. It is called the **region of integration** and is shown in Figure 1.

The equality of the two orders of integration means that either order of integration can be chosen. The integral is referred to as an **iterated integral** because it is computed by integrating twice. The display gives the formal definition.

DEFINITION

The **double integral** of $f(x, y)$, called the **integrand,** over the rectangular region of integration R, is expressed as

$$\int \int_R f(x, y) \, dx \, dy \qquad \text{or} \qquad \int \int_R f(x, y) \, dy \, dx$$

and is equal to

$$\int_a^b \left[\int_c^d f(x, y) \, dx \right] dy = \int_c^d \left[\int_a^b f(x, y) \, dy \right] dx$$

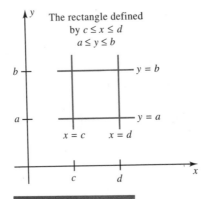

The rectangle defined
by $c \leq x \leq d$
$a \leq y \leq b$

$y = b$

$y = a$

$x = c \qquad x = d$

FIGURE 1

As an illustration, consider the region R in the display and let the integrand be the function $f(x, y) = C$ for C a positive number. The double integral of $f(x, y)$ over R is the volume of a box with the lengths of its sides $d - c$ and $b - a$, respectively, and with height C. Then

$$\int \int_R C \, dx \, dy = C[\text{area of } R] = C(d - c)(b - a)$$

The next example shows how to evaluate a double integral when the region is given.

EXAMPLE 5

Problem

Evaluate $\int \int_R x^2(1 - y^3) \, dx \, dy$ over the region R defined by $0 \le x \le 3$ and $-1 \le y \le 1$.

Solution Either order of integration can be selected. We choose to integrate first with respect to x and then with respect to y. The reverse order can be used as a check. The boundaries of the region are $x = 0$ to $x = 3$ and $y = -1$ to $y = 1$. These are the limits and the iterated integral is

$$\int_{-1}^{1} \int_{0}^{3} [x^2(1 - y^3)] \, dx \, dy = \int_{-1}^{1} \left[\left[\frac{x^3}{3} (1 - y^3) \right] \Big|_{x=0}^{x=3} \right] dy$$

$$= \int_{-1}^{1} 9(1 - y^3) \, dy = 9\left(y - \frac{y^4}{4} \right) \Big|_{y=-1}^{y=1}$$

$$= 9\left[\left(1 - \frac{1}{4} \right) - \left(-1 - \frac{1}{4} \right) \right] = 9(2) = 18$$

The final example in this section shows how to evaluate double integrals involving exponential functions.

EXAMPLE 6

Problem

Evaluate the double integral of the function $f(x, y) = ye^{xy}$ over the rectangle $1 \le x \le 2, 0 \le y \le 1$.

Solution We will integrate first with respect to the variable x. Thus our mission is to compute the double integral

$$\int_{0}^{1} \int_{1}^{2} ye^{xy} \, dx \, dy$$

First evaluate the inner integral by treating y as a constant

$$\int ye^{xy} \, dx = \int e^{xy} (y \, dx) = e^{xy} + C(y)$$

Inserting the limits of integration yields

$$\int_1^2 y e^{xy} \, dx = e^{xy} \Big|_{x=1}^{x=2} = e^{2y} - e^y$$

Next,

$$\int_0^1 (e^{2y} - e^y) \, dy = \frac{1}{2} \int_0^1 e^{2y}(2dy) - \int_0^1 e^y \, dy$$

$$= .5e^{2y} - e^y \Big|_{y=0}^{y=1}$$

$$= (.5e^2 - e) - (.5 - 1) = .5e^2 - e + .5 \approx 1.4762$$

In Example 6 integration was carried out first with respect to x. To see why this order is preferred, try integrating with respect to y first. We leave this task as an exercise.

EXERCISE SET 8–5

In Problems 1 to 4 compute the antiderivative.

1. $\int (7xy^6 + 6x^3y^{-2} - 5) \, dx$

2. $\int (5x^4y + 6x^2y^{1/2} - 5x + y) \, dx$

3. $\int (8xy^7 - 20x^{-4}y^4 - 3x - y) \, dy$

4. $\int (x^2y^3 - 12x^{1/4}y^{-2} + x - 2y + 3) \, dy$

In Problems 5 to 8 compute the definite integral.

5. $\int_1^2 (5x^4y + 2x^{-1}y^3 - 5) \, dx$

6. $\int_1^3 (4x^3y^2 + x^{-2} - 2y + 1) \, dx$

7. $\int_0^2 (5x^4y + 2x^{-1}y^3 - 5) \, dy$

8. $\int_1^3 (4x^3y^2 + x^{-2} - 2y + 1) \, dy$

In Problems 9 to 12 compute the double integral.

9. $\int_1^3 \left[\int_1^2 (3xy^2 - 12x^{-3}y^2 - 1) \, dx \right] dy$

10. $\int_1^2 \left[\int_{-1}^1 (4xy^3 + 6x^2y^{-1} - y) \, dx \right] dy$

11. $\int_1^2 \left[\int_2^3 (10x^{1/2}y^4 + 12x^{-1/2}y^3 - x) \, dy \right] dx$

12. $\int_0^2 \left[\int_1^3 (3x^{1/3}y^2 + 12x^{-1/3} + x - y) \, dy \right] dx$

In Problems 13 to 16 evaluate the double integral over the rectangular region R.

13. $\iint_R 4x^3(1 - y^2) \, dx \, dy$, where R is defined by

$0 \le x \le 2$ and $0 \le y \le 1$

14. $\iint_R x^2y(2 - 3y^2) \, dx \, dy$, where R is defined by

$0 \le x \le 2$ and $1 \le y \le 3$

15. $\iint_R (x^3y + x - 4y^3) \, dx \, dy$, where R is defined by

$1 \le x \le 2$ and $-1 \le y \le 1$

16. $\iint_R (x^{-2}y + x^{1/2} - 4y) \, dx \, dy$, where R is defined

by $1 \le x \le 2$ and $1 \le y \le 3$

In Problems 17 to 22 compute the antiderivative.

17. $\int x\sqrt{x^2 + 3y} \, dy$

18. $\int \frac{3 + 5y}{\sqrt{x}} \, dy$

19. $\displaystyle\int \frac{6x + 2y}{3x^2 + 2xy}\, dx$

20. $\displaystyle\int xe^{x^2 + 9y}\, dy$

21. $\displaystyle\int x\sqrt{x^2 + 3y}\, dx$

22. $\displaystyle\int \frac{10x}{\sqrt{3y + 5x^2}}\, dy$

In Problems 23 to 28 compute the definite integral. Notice that the corresponding indefinite integrals were computed in Problems 17 to 22.

23. $\displaystyle\int_4^5 x\sqrt{x^2 + 3y}\, dy$

24. $\displaystyle\int_2^7 \frac{3 + 5y}{\sqrt{x}}\, dy$

25. $\displaystyle\int_3^5 \frac{6x + 2y}{3x^2 + 2xy}\, dx$

26. $\displaystyle\int_1^6 xe^{x^2 + 9y}\, dy$

27. $\displaystyle\int_0^4 x\sqrt{x^2 + 3y}\, dx$

28. $\displaystyle\int_3^6 \frac{10x}{\sqrt{3y + 5x^2}}\, dy$

In Problems 29 to 32 compute the double integral.

29. $\displaystyle\int_1^3 \left[\int_1^2 \frac{1}{xy}\, dx \right] dy$

30. $\displaystyle\int_1^3 \left[\int_1^2 \left[\frac{x}{1 + y} \right] dy \right] dx$

31. $\displaystyle\int_0^1 \left[\int_0^2 [x(x^2 + y)^{1/2}]\, dx \right] dy$

32. $\displaystyle\int_0^2 \left[\int_1^2 [xy(x^2 + y^2)^{1/2}]\, dy \right] dx$

In Problems 33 to 36 evaluate the double integral over the rectangular region R.

33. $\displaystyle\iint_R ye^{x + y^2}\, dy\, dx$

where R is defined by $-1 \le x \le 1$ and $0 \le y \le 1$

34. $\displaystyle\iint_R \frac{y}{x^2} e^{y/x}\, dx\, dy$

where R is defined by $1 \le x \le 2$ and $0 \le y \le 1$

35. $\displaystyle\iint_R (3x^2 + y)\, dx\, dy$

where R is defined by $1 \le x \le 5$ and $0 \le y \le 2$

36. $\displaystyle\iint_R xe^{x^2 + y}\, dx\, dy$

where R is defined by $0 \le x \le 2$ and $2 \le y \le 3$

37. Evaluate each integral in parts (a) and (b).

(a) $\displaystyle\int_3^{11} \int_0^4 \sqrt{x}\sqrt{y} - 2\, dx\, dy$

(b) $\displaystyle\int_0^4 \int_3^{11} \sqrt{x}\sqrt{y} - 2\, dy\, dx$

(c) Compare parts (a) and (b). Explain your comparison.

38. Evaluate each integral in parts (a) and (b).

(a) $\displaystyle\int_{-1}^1 \int_0^2 (1 - 6x^2y)\, dx\, dy$

(b) $\displaystyle\int_0^2 \int_{-1}^1 (1 - 6x^2y)\, dy\, dx$

(c) Compare parts (a) and (b). Explain your comparison.

In Problems 39 and 40 compute the double integral carefully because one order of integration is easy while the other is not.

39. $\displaystyle\iint_R xe^{xy}\, dx\, dy$

where R is defined by $0 \le x \le 2$ and $1 \le y \le 3$

40. $\displaystyle\iint_R \frac{2x + 3x^2y}{1 + 2y + y^2}\, dx\, dy$

where R is defined by $0 \le x \le 1$ and $0 \le y \le 2$

41. Evalute the double integral in Example 6 in the reverse order by computing

$$\int_1^2 \int_0^1 ye^{xy}\, dy\, dx$$

In Problems 42 to 45 express the limits of integration using both orders of integration, given that a function is to be integrated over the region R in the accompanying graphs.

42.

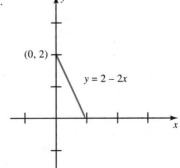

$(0, 2)$

$y = 2 - 2x$

43.

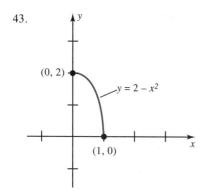

44.

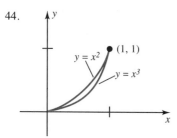

45.

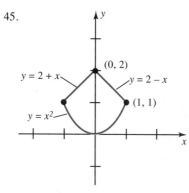

Referenced Exercise Set 8–5

1. In an article investigating the effects of urban renewal on rent increases on the land directly affected, Fishelson and Pines define a transportation cost function $C(x, H)$ that measures the dollars per year the average family spends on transportation when the family lives a distance x miles from the center of the urban renewal

site and H is the yearly housing expenditure of the family.* The authors compute the marginal cost $MC(x, H)$ to be

$$MC(x, H) = \frac{kx^2}{H}$$

where k is a constant depending on the particular city and the location of the urban renewal site in the city. They then calculate the total transportation cost using double integrals. Find the total cost from $x = 0$ to $x = 5$ miles, from $H = 1$ to $H = 10$.

2. Tax reform has become a priority in recent years. Many economists have called for tax revisions for farmers that would replace an income tax with a tax on purchases. In an article investigating this proposal, Chambers and Lopez define a cost of labor function $CL(p, t)$ in terms of profit or income p and income tax t.† They compute the marginal cost function $MCL(p, t)$ to be

$$MCL(p, t) = \frac{kp}{1 - t}$$

where k is a constant depending on the particular financial position of the farm. Then the total cost of labor is the double integral of this function. Find the total cost from $t = 0.25$ to $t = 0.30$ when $k = 1$ and p is held constant with $p = 10$ thousand dollars.

Cumulative Exercise Set 8–5

In Problems 1 and 2 draw the graph of the given function.

1. $f(x, y) = x^2 + 4y^2$

2. $f(x, y) = x^2 + 4x + y^2 - 2y$

3. Find $f_x(x, y)$ and $f_y(x, y)$.

$$f(x, y) = x^2 - 2x^3y + 6y^{1/3}$$

4. Find all second-order partial derivatives.

$$f(x, y) = 2x^3 - e^{2x}y^2$$

In Problems 5 and 6 find the relative extrema and any saddle points of the function.

5. $f(x, y) = x^3 - y^2 + 12y$

6. $f(x, y) = x^3 + y^3 + 6y^2$

*Gideon Fishelson and David Pines, ''Market Versus Social Valuation of Redevelopment Projects in an Urban Setting,'' *Socio-Economic Planning,* Vol. 18, 1984, pp. 419–424.

†Robert G. Chambers and Ramon E. Lopez, ''Tax Policies and the Financially Constrained Farm Household,'' *American Journal of Agricultural Economics,* Vol. 69, 1987, pp. 369–377.

In Problems 7 and 8 use the method of Lagrange multipliers to find the extrema of $f(x, y)$ with the given constraints.

7. $f(x, y) = 2xy$ subject to $3x + 2y - 1 = 0$

8. $f(x, y) = x^2y$ subject to $2x + y = 1$

In Problems 9 to 12 compute the double integral.

9. $\displaystyle\int_1^2 \left[\int_{-1}^1 (3xy^2 + 3x^2y - 1)dx\right] dy$

10. $\displaystyle\iint_R (x + 2xy)dx\, dy$, where R is defined by $0 \leq x \leq 1$ and $0 \leq y \leq 2$

11. $\displaystyle\iint_R (x^{-3}y + x^{1/2} - 2y)dx\, dy$, where R is defined by $1 \leq x \leq 2$ and $1 \leq y \leq 3$

12. $\displaystyle\iint_R x^2/(1 + y^2)dx\, dy$, where R is defined by $1 \leq x \leq 2$ and $1 \leq y \leq 2$

8–6 Volume

One measure of a figure in two-dimensional space is its area. We have seen that the definite integral provides a powerful tool for computing area.

The situation is similar when we consider figures in three-dimensional space. There a measure of a figure is its volume. We will see that the definite double integral introduced in the preceding section provides a powerful tool for computing the volume of a solid. First it will be necessary to evaluate double integrals over regions which are not necessarily rectangles.

Double Integrals over Other Regions

In the preceding section we mentioned that the limits of integration of a double integral over a rectangular region are actually functions. However, in that section there was no need to use the fact that they are actually linear functions of the form

$x = $ constant

or

$y = $ constant

In this section we will show that it is possible to compute double integrals in which the limits of the inner integral are nonlinear functions, provided that the limits of the outer integral are constant functions. Because this remark is so important we single it out. The first example will then illustrate how to compute double integrals with variable limits.

 Warning: In a double integral the limits of the outer integral must always be constant functions. The limits of the inner integral may be variable functions.

EXAMPLE 1

Problem

Compute

$$\int_0^1 \int_{x^2}^x 3xy^2 \, dy \, dx$$

Solution First calculate the inner integral.

$$\int_{x^2}^x 3xy^2 \, dy = xy^3 \Big|_{y \,=\, x^2}^{y \,=\, x} = xx^3 - x(x^2)^3 = x^4 - x^7$$

Then perform the outer integral of this function.

$$\int_0^1 \int_{x^2}^x 3xy^2 \, dy \, dx = \int_0^1 (x^4 - x^7) \, dx = \left(\frac{x^5}{5} - \frac{x^8}{8} \right)\Big|_0^1 = \frac{1}{5} - \frac{1}{8} = \frac{3}{40}$$

The double integral in Example 1 is defined over a region with variable limits. The shaded region in Figure 1 defines the limits of integration. It is bounded above by the line $y = x$ and below by the parabola $y = x^2$.

There are two different ways to view such a region, and each view corresponds to a different order of integration. Figure 2 shows the two views. In part (a) the

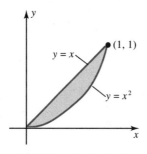

FIGURE 1

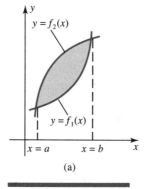 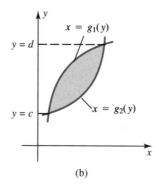

| (a) | (b) |

FIGURE 2

region is bounded above by a function of the form $y = f_2(x)$ and below by a function of the form $y = f_1(x)$. This view corresponds to integrating first with respect to y and then with respect to x. The functions $f_1(x)$ and $f_2(x)$ form the limits of the inner integral, while the constant functions $x = a$ and $x = b$ form the limits of the outer integral.

The view in part (b) corresponds to the reverse order of integration, first with respect to x and then with respect to y. Here the boundaries are viewed from a horizontal perspective. The region is bounded on the right by a function of the form $x = g_2(y)$ and on the left by a function of the form $x = g_1(y)$. These functions form the limits of the inner integral, while the constant functions $y = c$ and $y = d$ form the limits of the outer integral. Example 2 evaluates such an integral using both orders of integration.

EXAMPLE 2

Problem

Compute the integral $\displaystyle\iint_R 2xy\, dx\, dy$ over the region defined by $x^2 \le y \le 4$ and $0 \le x \le 2$.

Solution (a) Since the variable boundary of R is given directly as $y = x^2$, the more immediate order of integration is first with respect to y and then x. The limits of the inner integral are $y = x^2$ and $y = 4$, while the limits of the outer integral are $x = 0$ and $x = 2$. See Figure 3a. The iterated integral is

$$\int_0^2 \int_{x^2}^4 2xy\, dy\, dx = \int_0^2 \left[xy^2 \Big|_{y=x^2}^{y=4} \right] dx = \int_0^2 (16x - x^5)\, dx$$

$$= \left(8x^2 - \frac{x^6}{6} \right) \Bigg|_{x=0}^{x=2} = \left(32 - \frac{64}{6} \right) - 0 = \frac{64}{3}$$

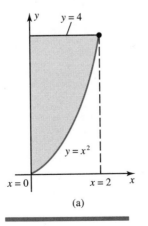

(a)

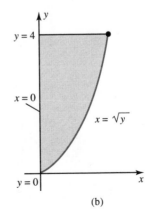

(b)

FIGURE 3

(b) To switch the order of integration, view the region as being bounded by functions of the form $x = g(y)$. (See Figure 3b.) From this perspective, the formula $y = x^2$ yields the function $x = \sqrt{y}$. For the inner integral, the upper limit is the function $x = \sqrt{y}$ and the lower limit is the function $x = 0$. The limits of the outer integral are $y = 0$ and $y = 4$. The iterated integral is

$$\int_0^4 \int_0^{\sqrt{y}} 2xy \, dx \, dy = \int_0^4 \left[x^2 y \Big|_{x=0}^{x=\sqrt{y}} \right] dy$$

$$= \int_0^4 y^2 \, dy = \frac{y^3}{3} \Big|_0^4 = \frac{64}{3}$$

Notice that the answers in parts (a) and (b) of Example 2 are equal. This is expected because we merely switched the order of integration.

Sometimes computing the integral using one order of integration is easier than the reverse order. Unfortunately there is no general rule to tell which order is better, so instead of worrying about how to start a problem, you are better off attempting one order and, if it does not seem to be working, then try the reverse order.

Volume

One way to compute the area under the graph of a function of one variable is to find the definite integral of the function within the given limits. The initial approach, however, was to form sums of rectangles and then take the limit of the sums.

In a similar way, the volume of a solid in three dimensions can be found by summing the volumes of boxes whose heights are functional values of $f(x, y)$ and whose length and width are small increments of dx and dy. One such box is shown in Figure 4. Taking the analogy with the case in two dimensions one step further,

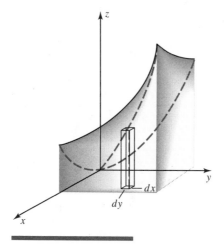

FIGURE 4

the volume of a solid under the graph of a function of two variables can be computed much more easily by finding the definite double integral between appropriate limits.

Look at Figure 5 for instance. It shows the graph of $f(x, y) = x^2 + y^2$ over the region R defined by $0 \le x \le 1$ and $0 \le y \le 2$. The volume of the solid bounded above by $f(x, y)$ and below by R can be found by computing the double integral of $f(x, y)$ over R.

DEFINITION

If $f(x, y) \ge 0$ for all points in a region R, then the **volume** of the solid bounded above by $f(x, y)$ and below by R is equal to

$$\iint_R f(x, y) \, dx \, dy$$

Figure 6 portrays this result geometrically. Examples 3 and 4 will show how to compute volumes of solids whose bases are rectangles.

EXAMPLE 3

Problem

Compute the volume of the solid bounded above by $f(x, y) = x^2 + y^2$ and below by the rectangle R defined by $0 \le x \le 1$ and $0 \le y \le 2$. This solid is graphed in Figure 5.

Solution The display states that the volume is the double integral of $f(x, y) = x^2 + y^2$ over R. This is given by

$$\int_0^2 \int_0^1 (x^2 + y^2) \, dx \, dy = \int_0^2 \left[\left(\frac{x^3}{3} + xy^2 \right) \Big|_{x=0}^{x=1} \right] dy$$

$$= \int_0^2 \left(\frac{1}{3} + y^2 \right) dy = \left(\frac{y}{3} + \frac{y^3}{3} \right) \Big|_{y=0}^{y=2}$$

$$= \frac{2}{3} + \frac{8}{3} = \frac{10}{3}$$

In the previous section we saw that the double integral does not depend on the order of the integration. Therefore one way to verify that the volume of the solid in Example 3 is 10/3 is to compute the double integral

$$\int_0^1 \int_0^2 (x^2 + y^2) \, dy \, dx$$

EXAMPLE 4

Problem

Compute the volume of the solid bounded above by $f(x, y) = e^{x+y}$ and below by the rectangle R defined by $0 \le x \le 1$ and $1 \le y \le 2$.

Solution

$$\int_1^2 \int_0^1 e^{x+y}\, dx\, dy = \int_1^2 \left[e^{x+y} \Big|_{x=0}^{x=1} \right] dy = \int_1^2 (e^{1+y} - e^y)\, dy$$

$$= e^{1+y} - e^y \Big|_{y=1}^{y=2} = (e^3 - e^2) - (e^2 - e)$$

$$= e^3 - 2e^2 + e \approx 8.0257.$$

The final example in this section shows how to find the volume of a solid whose base has some boundaries that are not constant functions.

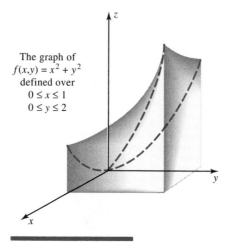

The graph of
$f(x,y) = x^2 + y^2$
defined over
$0 \leq x \leq 1$
$0 \leq y \leq 2$

FIGURE 5

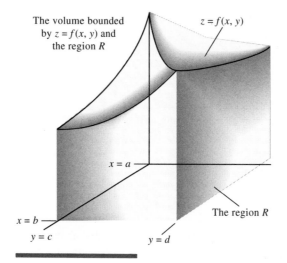

The volume bounded
by $z = f(x, y)$ and
the region R

$z = f(x, y)$

$x = a$

$x = b$

$y = c$

$y = d$

The region R

FIGURE 6

EXAMPLE 5

Problem

Compute the volume of the solid bounded above by the function $f(x, y) = 4 - x - y$ and below by the triangle R formed from the lines $x + y = 4$, $x = 0$, and $y = 0$.

Solution Two issues must be settled before the desired volume can be computed, the order of integration and the limits. We choose to integrate first with respect to y. Thus the limits of the inner integral are of the form $y = f(x)$. It can be seen from Figure 7 that the lower limit is $y = 0$ and the upper limit is $y = 4 - x$.

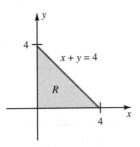

FIGURE 7

Recall the warning that the limits of the outer integral must be constant functions. Here they are $x = 0$ and $x = 4$. Thus the desired volume is

$$\int_0^4 \int_0^{4-x} (4 - x - y) \, dy \, dx = \int_0^4 \left[(4 - x)y - \frac{y^2}{2} \Big|_{y=0}^{y=4-x} \right] dx$$

$$= \int_0^4 \left[(4 - x)^2 - \frac{(4 - x)^2}{2} \right] dx$$

$$= \frac{1}{2} \int_0^4 (4 - x)^2 \, dx$$

$$= \frac{1}{2} \frac{(4 - x)^3}{-3} \Big|_{x=0}^{x=4} = \frac{-1}{6} (4 - x)^3 \Big|_{x=0}^{x=4}$$

$$= \frac{-1}{6} (0 - 4^3) = \frac{64}{6}$$

The solid whose volume was found in Example 5 is the triangular pyramid shown in Figure 8. It is reminiscent of the pyramids whose volumes were computed by the ancient Egyptians about 4000 years ago.

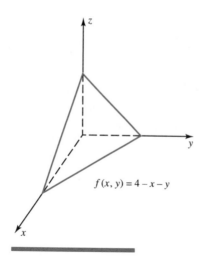

$f(x, y) = 4 - x - y$

FIGURE 8

EXERCISE SET 8–6

In Problems 1 to 6 evaluate the double integral.

1. $\int_0^1 \int_{x^2}^x 4xy^3 \, dy \, dx$

2. $\int_0^2 \int_{x^2}^{2x} 4x^2y^3 \, dy \, dx$

3. $\int_0^1 \int_{y^3}^1 xy^2 \, dx \, dy$

4. $\int_0^1 \int_{y^3}^y xy^4 \, dx \, dy$

5. $\int_0^2 \int_{-x^2}^{x^2} (x + y^2) \, dy \, dx$

6. $\int_0^2 \int_{1-x^2}^{x^2+1} (x - y) \, dy \, dx$

In Problems 7 to 12 evaluate the double integral over the nonrectangular region R.

7. $\iint_R 4xy \, dx \, dy$, where R is defined by

$x^2 \le y \le 1$ and $0 \le x \le 1$

8. $\iint_R (x + y) \, dx \, dy$, where R is defined by

$x^3 \le y \le 1$ and $0 \le x \le 1$

9. $\iint_R (2x + 3y) \, dx \, dy$, where R is defined by

$x^3 \le y \le x^2$ and $0 \le x \le 1$

10. $\iint_R (x - xy) \, dx \, dy$, where R is defined by

$x^4 \le y \le x^2$ and $0 \le x \le 1$

11. $\iint_R (x + 2y) \, dx \, dy$, where R is defined by

$x^2 \le y \le 2x$ and $0 \le x \le 2$

12. $\iint_R f(x, y) = x + 2y$, where R is defined by

$0 \le y \le 4$ and $(y/2) \le x \le \sqrt{y}$

In Problems 13 to 16 compute the volume of the solid bounded above by $f(x, y)$ and below by the rectangle R.

13. $f(x, y) = x^3y + x - 4y^3$, where R is defined by $1 \le x \le 3$ and $0 \le y \le 1$

14. $f(x, y) = x^{-2}y + x^{1/2} - 4$, where R is defined by $1 \le x \le 3$ and $1 \le y \le 2$

15. $f(x, y) = xe^{x^2+y}$, where R is defined by $0 \le x \le 2$ and $2 \le y \le 3$

16. $f(x, y) = y^2e^{2x+y^3}$, where R is defined by $2 \le x \le 3$ and $1 \le y \le 2$

In Problems 17 to 20 compute the volume of the pyramid bounded above by the given function $f(x, y)$ and below by the triangle R formed from the lines $x = 0$, $y = 0$, and the given line.

17. $f(x, y) = 4 - x - y$, where the line is $x + y = 2$

18. $f(x, y) = 4 - x - y$, where the line is $2x + y = 4$

19. $f(x, y) = 6 - x - y$, where the line is $x + y = 6$

20. $f(x, y) = 6 - x - y$, where the line is $3x + y = 6$

In Problems 21 to 24 evaluate the double integral.

21. $\displaystyle\int_0^4 \int_0^x \sqrt{xy}\, dy\, dx$

22. $\displaystyle\int_1^4 \int_0^7 (x + 4y)\, dx\, dy$

23. $\displaystyle\int_1^2 \int_y^{3y} \frac{1}{x}\, dx\, dy$ 24. $\displaystyle\int_0^1 \int_{2x}^{3x} e^{x+y}\, dx\, dy$

In Problems 25 to 28 evaluate the double integral over the region R.

25. $\displaystyle\iint_R (\sqrt{x} + y)\, dx\, dy$, where R is defined by

$0 \le y \le 1$ and $0 \le x \le 3y$

26. $\displaystyle\iint_R \sqrt{1 - x^2}\, dx\, dy$, where R is defined by

$0 \le y \le 1$ and $y \le x \le 1$

27. $\displaystyle\iint_R \frac{x}{1 + y^2}\, dx\, dy$, where R is defined by

$x^2 \le y \le 4$ and $0 \le x \le 2$

28. $\displaystyle\iint_R y e^{x^2}\, dx\, dy$, where R is defined by

$0 \le y \le 1$ and $y^2 \le x \le 1$

In Problems 29 to 32 compute the volume of the solid bounded above by $f(x, y)$ and below by the rectangle R.

29. $f(x, y) = \sqrt{x}$, where R is defined by $0 \le x \le 1$ and $0 \le y \le 4$

30. $f(x, y) = \sqrt{y}$, where R is defined by $0 \le x \le 4$ and $0 \le y \le 9$

31. $f(x, y) = e^{x+y}$, where R is defined by $0 \le x \le 2$ and $-1 \le y \le 1$

32. $f(x, y) = x\sqrt{x^2 + y}$, where R is defined by $0 \le x \le 1$ and $0 \le y \le 1$

In Problems 33 to 36 compute the volume of the solid bounded above by the given function $f(x, y)$ and below by the given region R.

33. $f(x, y) = x^2 + y^2$, R is the square formed from the lines $x = 0$, $x = 3$, $y = 0$, and $y = 3$.

34. $f(x, y) = x^2 + y^2$, R is the rectangle formed from the lines $x = 0$, $x = 3$, $y = 0$, and $y = 6$.

35. $f(x, y) = 3 - x - y$, R is the triangle formed from the lines $x = 1$, $y = 0$, and $y = x$.

36. $f(x, y) = 3 - x - y$, R is the triangle formed from the lines $x = 0$, $y = 0$, and $y = x$.

37. Verify the volume of the solid that was calculated in Example 3 by computing the double integral

$$\int_0^1 \int_0^2 (x^2 + y^2)\, dy\, dx$$

38. Compute the volume of the pyramid in Figure 8 by integrating first with respect to x.

Cumulative Exercise Set 8–6

1. Consider the function
$$f(x, y) = \frac{x^2 + 2xy + y^2}{x + y}$$
   (a) Evaluate $f(1, 1)$.
   (b) Evaluate $f(1, -1)$.
   (c) Find the domain of $f(x, y)$.

2. Sketch the graph of the function $f(x, y) = 4x^2 - y^2$, giving the traces and one section that is parallel to each coordinate plane.

3. Let $f(x, y) = 3x^4 - 4xy^3 - 2y$.
   (a) Find $f_x(x, y)$ and $f_y(x, y)$.
   (b) Find all second-order partial derivatives of $f(x, y)$.

4. Find the relative extrema and any saddle points of the function $f(x, y) = 12x - 2xy + 4y + y^2$.

5. Use the method of Lagrange multipliers to find extrema of the function $f(x, y) = x^2 + 3y^2 - 12xy$ subject to $x + y = 16$.

6. Find two positive numbers such that the sum of their squares is equal to 50 and their product is a maximum.

7. Compute the antiderivative.

$$\int_{-2}^2 \frac{x}{\sqrt{x^2 + y^2}}\, dx$$

In Problems 8 and 9 compute the double integral.

8. $\int_0^{1/3} \int_{-2}^0 (x + 3y + 2)^4 \, dx \, dy$

9. $\int_0^1 \int_0^{x^3} e^{y/x} \, dx \, dy$

10. Evaluate the double integral over the rectangular region R, where R is defined by $0 \le x \le 3$ and $-1 \le y \le 1$.

$\iint\limits_R (2x + 3y) \, dx \, dy$

11. Compute the volume of the solid bounded above by $f(x, y) = 1 + x^2 + y^2$ and below by the rectangle $0 \le x \le 1$ and $0 \le y \le 4$.

12. Compute the volume of the solid bounded above by $f(x, y) = 6x + 2y + 5$ and below by the square formed from the lines $x = 0$, $x = 2$, $y = 0$, and $y = 2$.

C A S E   S T U D Y

Minimum Cost Production in the Chemical Industry

Jojoba oil is used for producing amides, which, in turn, are an essential ingredient in many chemical processes. The procedure of extracting amides from jojoba oil is time-consuming and expensive. First the oil is refined and purified. Then other chemicals are added and the mixture is heated for several hours. Once the material becomes cakelike the amides can be extracted.

Researchers have tried various methods to maximize the amide yield while conserving cost to produce the cakelike material. The two most important variables are temperature and time. In an article* discussing various strategies for improving amide yield, researchers discovered that the relationship between the variables is best described by a Cobb–Douglas production function, which is a power function of the form

$$f(x, y) = ax^b y^c$$

Let $f(x, y)$ represent the amount of amide yield (as a percentage of oil weight) at temperature x (measured in degrees Celsius) and time y (measured in hours). Test results showed that proper production function is

$$f(x, y) = 1.3x^{0.5}y^{0.4}$$

This function is useful in planning production. Various constraints on cost are given and the corresponding production yield is computed. A particular firm would decide what the most reasonable limits on cost would be and then use this function to predict whether the amide yield would be great enough. If not, further study of the cost constraints would be required.

For example, cost considerations require that if x increases, then y must decrease. This means that if the processing time increases, then the temperature would have to decrease to keep the production cost the same. One such constraint is

$$x + 2y = 200$$

*A. Shani et al., "Synthesis of Jojobamide and Homojojobamide," *Journal of American Oil and Chemical Society,* March 1980, pp. 112–114.

This means that when $x = 100$, $y = 50$, and when $x = 110$, $y = 45$; so that the cost of using a temperature of $100°$ for 50 hours is equal to the cost of using a temperature of $110°$ for 45 hours. The next example shows how Lagrange multipliers are used to determine the maximum amide yield subject to the given cost constraint.

EXAMPLE

Problem

Find the maximum of $f(x, y) = 1.3x^{0.5}y^{0.4}$ subject to the constraint $x + 2y = 200$.

Solution Introduce the Lagrange multiplier t in the function

$$F(x, y, t) = 1.3x^{0.5}y^{0.4} + t(x + 2y - 200)$$

Find the partial derivatives and set them equal to 0.

$$F_x(x, y, t) = 1.3(0.5)x^{-0.5}y^{0.4} + t$$
$$= 0.65x^{-0.5}y^{0.4} + t = 0 \qquad (1)$$

$$F_y(x, y, t) = 1.3(0.4)x^{0.5}y^{-0.6} + 2t$$

$$= 0.52x^{0.5}y^{-0.6} + 2t = 0 \qquad (2)$$

$$F_t(x, y, t) = x + 2y - 200 = 0 \qquad (3)$$

Solve (1) and (2) for $-t$.

$$-t = 0.65x^{-0.5}y^{0.4} = 0.26x^{0.5}y^{-0.6}$$

Disregard $-t$ and multiply the resulting equation by $x^{0.5}y^{0.6}$.

$$0.65y = 0.26x$$

Hence $x = 2.5y$. Substitute this relationship into (3).

$$2.5y + 2y - 200 = 0$$
$$4.5y = 200$$
$$y \approx 44.4$$
$$x \approx 111$$

This means that with the given constraint the maximum amide yield is $f(111, 44.4) \approx 1.3(10.5)(4.56) \approx 62.2\%$ when the material is heated at $111°$ for 44.4 hours.

Case Study Exercises

In Problems 1 to 4 find the maximum amide yield for the given cost function and the production function.

$$f(x, y) = 1.3x^{0.5}y^{0.4}$$

1. $x + 2.4y = 200$ 2. $x + 1.8y = 200$

3. $2x + 4.5y = 440$ 4. $3x + 5.8y = 600$

In Problems 5 and 6 find the maximum amide yield for the given cost function and the given production function.

5. $x + 2.4y = 200,$
     $f(x, y) = 1.3x^{0.6}y^{0.3}$

6. $x + 1.8y = 200,$
     $f(x, y) = 1.7x^{0.55}y^{0.35}$

CASE STUDY   **Volumes: Old and New**

Numerous methods for calculating the volume of a solid figure have been developed over the past 4000 years. The search for these methods led to some curious episodes in mathematical history.

The story unfolds in ancient Egypt about 1900 B.C. A scroll from that time, called the Moscow Papyrus after the city where it has been preserved, contains a method for finding the volumes of pyramids. Colossal pyramids dominate the Egyptian landscape, so it is not particularly surprising that such knowledge was desirable. The Egyptians could calculate not only the volume of a pyramid but the volume of its frustums as well. (The "frustum" of a pyramid is the solid that remains after a portion of the pyramid has been lopped off its top. See Figure 1.) For a pyramid whose base is a $b \times b$ square, the modern formula for the volume of the frustum of height h with an $a \times a$ square top is

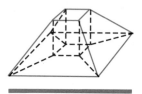

FIGURE 1

$$V = \frac{h(a^2 + ab + b^2)}{3}$$

This formula is truly remarkable! It stood as a high point for another 1500 years. Then along came Archimedes.

It is generally known that Isaac Newton and Gottfried Leibniz discovered calculus in the late 17th century. What is not so well known is that Archimedes beat them by almost 2000 years. In fact, his method for calculating volumes is very similar to the method we described in this chapter.

So why doesn't Archimedes receive proper credit? Mainly because few people could understand the advanced level of his thinking. Indeed, there were periods when few of his works were known.

The famous astronomer Johannes Kepler was one of the first people to go beyond Archimedes. His inspiration? The excellent wine produced by the local vineyards in 1612. As the court mathematician for Emperor Rudolph II, Kepler was charged with improving the crude methods then known for estimating the volume of wine casks. He succeeded admirably, and three years later he published his results in a book called *The Volume Measurement of Barrels*.

Kepler's results, and those of several contemporaries, were superseded by the work of Newton and Leibniz, whose discoveries in the 1680s offer an instance of simultaneous codiscovery in science. The whole episode was quite controversial and unpleasant at the time. It began when a plague that ravaged London forced classes to be suspended at Cambridge University on January 14, 1665, the day that Newton was awarded his bachelor of arts degree. Newton returned home and spent a remarkable year in which he made three brilliant discoveries: the universal law of gravitation, the nature of light, and calculus.

A year later, when Cambridge reopened, Newton returned to the University and published his findings. However, some results were criticized so harshly that the overly sensitive scientist vowed to refrain from further publication. Although he wrote three pamphlets on calculus over the next decade, he published them privately, so they were not generally available when Leibniz published similar results in a series of journal articles beginning in 1684.

Initially Newton and Leibniz corresponded and shared their results with one another. When a third party accused Leibniz of plagiarism, however, a battle erupted between the academic giants. The bitter feelings ran so deep that the third volume of Newton's monumental work *Principia Mathematica*, published after Leibniz's death, erased all mention of Leibniz's work cited in the first two editions.

The controversy centered around a trip that Leibniz had made to London in 1673. Newton's adherents claimed that Leibniz had seen a copy of Newton's manuscript at the time. Leibniz's adherents responded that Leibniz never saw it, and even if he had he would not have been able to understand the material at that stage of his development. Besides, they argued, Leibniz's approach was entirely different than Newton's. Today most historians agree that Newton discovered calculus ten years before Leibniz but that Leibniz made his discoveries independently. All agree that priority in formal publication goes to Leibniz.

Newton's approach to the calculus is difficult to follow. After all, his intention was to derive equations for universal laws of nature, not to develop a new branch of mathematics. Thus this text, like most calculus texts over the past 300 years, follows Leibniz, using his notation dx and $\int y\ dx$ and his terms *calculus differentialis* and *calculus integralis*. The main difference in today's approach is the fundamental role of limits, an approach popularized by Augustin Louis Cauchy in a brilliant series of textbooks in the 1820s.

The focus of this Case Study is volume, so this historical overview should mention one other person, Giuseppe Fubini. In 1907 he proved that the volume of a solid can be calculated by integrating double integrals separately and in either order. Our whole approach revolves around this result.

Two recent articles examined volume in entirely different settings—famous domes throughout the world and water in the Great Salt Lake. This Case Study calculates the volume of a frustum of a pyramid using integral calculus, then explores the articles.

Pyramids

We will compute the volume of a frustum of a pyramid with a base that forms a $b \times b$ square, a top that forms an $a \times a$ square, and an altitude of h for the special case $b = 10$, $a = 4$, and $h = 3$. Exercise 10 describes the general case.

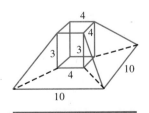

FIGURE 2

To find the volume of the frustum, first drill a rectangular solid from its middle as shown in Figure 2. Since the dimensions of this solid are $4 \times 4 \times 3$, its volume is 48. The entire frustum is composed of this solid and the surrounding region, which we decompose into eight pieces, as shown in Figure 3. Refer to Figure 4 to find the volume of each piece because it orients the piece in a way that suggests a method for calculating its volume. The volume of the frustum is $V = 48 + 8P$, where P is the volume of each piece.

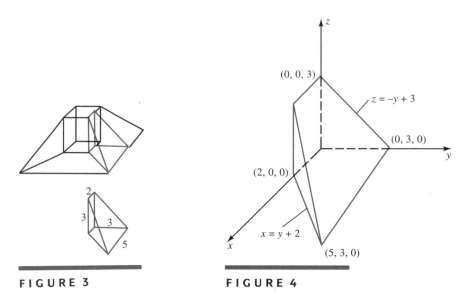

FIGURE 3   **FIGURE 4**

EXAMPLE 1

Problem

What is the volume P of the piece shown in Figure 4?

Solution The piece is bounded above by a function $f(x, y)$ and below by a region R that is bounded by the lines $x = 0$ and $x = y + 2$ (the inner limits) and the constant functions $y = 0$ and $y = 3$ (the outer limits). The equation for the function $f(x, y)$, derived from the line $z = -y + 3$ in the y-z plane, is $f(x, y) = -y + 3$. Thus the volume of each piece is

$$P = \int_0^3 \int_0^{y+2} (-y + 3)\, dx\, dy = \int_0^3 \left[(-y + 3)x \, \Big|_{x=0}^{x=y+2} \right] dy$$

$$= \int_0^3 (-y + 3)(y + 2)\, dy = \int_0^3 (6 + y - y^2)\, dy$$

$$= \frac{27}{2}$$

Since $V = 48 + 8P$ and $P = \frac{27}{2}$, the volume of the entire frustum is $V = 48 + 108 = 156$. This conclusion can be verified by substituting $a = 4$,

$b = 10$, and $h = 3$ into the Egyptian formula for the frustum of a pyramid:

$$V = \frac{h(a^2 + ab + b^2)}{3} = \frac{3(4^2 + 4 \cdot 10 + 10^2)}{3} = 16 + 40 + 100 = 156$$

One particular case of the Egyptian formula is worth noting. By setting $a = 0$ the formula yields the volume of the entire square pyramid:

$$V = \frac{hb^2}{3}$$

Of Domes and Lakes

A recent article calculated the volume of domes around the world.* Many famous domes are hemispheres and cylinders, figures whose volumes are known to students of solid geometry. But the twin towers atop the Benedictine Monastery at Einsiedeln, Switzerland, built between 1704 and 1719, are different. (See Figure 5.) They are built upon a square base of about 15 meters by 15 meters in a shape shown in Figure 6.

Example 2 computes the volume of each dome by double integration. The function $f(x,y)$ has been constructed to produce the dome's actual volume, but it is not the one that describes the figure.

*Anthony J. LoBello, "The Volumes and Centroids of Some Famous Domes," *Mathematics Magazine*, June 1988, pp. 164–170.

FIGURE 5

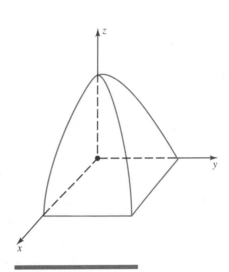

FIGURE 6

EXAMPLE 2

Problem

What is the volume of a dome that is bounded above by the function

$$f(x,y) = 8x - \frac{4}{15}y^2$$

and below by a square that measures 15 meters by 15 meters?

Solution The volume is

$$V = \int_0^{15} \int_0^{15} (8x - \frac{4}{15}y^2) \, dx \, dy = \int_0^{15} (4x^2 - \frac{4}{15}xy^2) \Big|_{x=0}^{x=15} \, dy$$

$$= \int_0^{15} (900 - 4y^2) \, dy = 900y - \frac{4y^3}{3} \Big|_0^{15} = 9000$$

Thus the volume of the dome is 9000 cubic meters.

The second article investigated the implications of a radical change that occurred in the Great Salt Lake.† The Utah Department of Natural Resources reported that from 1963 to 1986 the volume of water tripled while the top surface area doubled. The authors asked what kind of water basins could support such a phenomenon. Their methods for calculating the top surface areas and volumes are beyond the scope of this book, but we will examine one kind of basin they examined.

The basin shown in Figure 7 is a called a **paraboloid.** It is generated in the following manner. First draw the curve $z = ky^2$ in the y-z plane, where k is a constant. Then revolve the curve about the z-axis. The variable z becomes the

---

†Daniel Cass and Gerald Wildenberg, "Relations between Surface Area and Volume in Lakes," *The College Mathematics Journal,* November 1990, pp. 384–389.

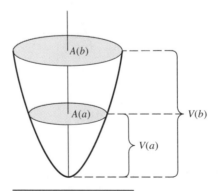

FIGURE 7

depth of the water in the parabolic basin. The top surface area and the volume of the paraboloid are given by the equations

$$A(z) = \pi k^2 z^4 \qquad V(z) = \frac{\pi k^2 z^5}{5}$$

Can the basin of the Great Salt Lake be a paraboloid? To answer this question, let a be the depth of the paraboloid in 1963 and b be the depth in 1986. Substituting $A(b) = 2A(a)$ into the formula for the area of the top surface yields $b^2 = 2a^2$. Substituting $V(b) = 3V(a)$ into the formula for the volume yields $b^5 = a^5$. It is impossible for the equations $b^2 = 2a^2$ and $b^5 = a^5$ to be satisfied simultaneously. Therefore the basin cannot be a paraboloid generated by $z = ky^2$.

Cass and Wildenberg showed that the basin could not be any paraboloid. Exercise 9 will show that the basin cannot be cone-shaped either. The authors also considered hemispheres, but they concluded "We were surprised to find that a hemispherical basin fails." Their work does demonstrate that numerous basins satisfy the given property, but none of them is as simple as a paraboloid, cone, or hemisphere.

Case Study Exercises

1. What is the volume P of the outer piece of the frustum of a pyramid that is bounded above by the plane $z = 6 - 3y$ and below by a region R that is bounded by the lines $x = 0$ and $x = y + 1$ and the constant functions $y = 0$ and $y = 2$?
2. What is the volume P of the outer piece of the frustum of a pyramid that is bounded above by the plane $z = 6 - 6y$ and below by a region R that is bounded by the lines $x = 0$ and $x = y + 2$ and the constant functions $y = 0$ and $y = 1$?
3. What is the volume of the frustum of the pyramid defined in Problem 1?
4. What is the volume of the frustum of the pyramid defined in Problem 2?
5. Problem 14 in the Moscow Papyrus describes a pyramid whose square base is 4 units on a side and whose altitude is 6 units. What is the volume of the pyramid?
6. What is the volume of a pyramid whose square base is 6 units on a side and whose altitude is 4 units?

In Problems 7 and 8 find the volume of a dome that is bounded above by the function

$$f(x,y) = 8x - \frac{4}{15}y^2$$

and below by a square that measures b meters by b meters for the given value of b.

7. $b = 30$
8. $b = 45$

9. The top surface area and the volume of a cone-shaped basin are given by the equations

$$A(z) = \pi k^2 z^2 \qquad V(z) = \frac{\pi k^2 z^3}{3}$$

Can the basin of the Great Salt Lake be cone-shaped?

10. The volume V of the frustum of a pyramid with a base $b \times b$, top $a \times a$, and altitude h is given by $v = a^2 h + 8P$, where P is the volume of each inner piece. The inner piece is bounded above by the function

$$f(x,y) = \frac{2h}{a-b}(y-h)$$

and below by a region that is bounded by the lines $x = 0$ and $x = y + a/2$ and the constant functions $y = 0$ and $y = (b-a)/2$. Use integration to confirm the Egyptian formula for V.

11. Evaluate the integral in Example 2 in this order:

$$\int_0^{15} \int_0^{15} \left(8x - \frac{4}{15}y^2 \right) dy \, dx$$

CHAPTER REVIEW

Key Terms

8–1   Examples of Functions of Several Variables
Dependent Variable

Independent Variable

Function of Two Variables

Trace

Section

Level Curves

8–2   Partial Derivatives
Partial Derivative

Second Partial Derivative

Mixed Partial Derivative

8–3   Relative Extrema
Relative Maximum

Relative Minimum

Relative Extremum

Critical Point

Saddle Point

Second Partial Derivative Test

8–4   Lagrange Multipliers
Constraint

Lagrange Multipliers

8–5   Double Integrals
Region of Integration

Iterated Integral

Double Integral

Integrand

8–6   Volume
Volume

REVIEW PROBLEMS

1. Find the functional value of $f(x, y) = 3x^2 - 4y^2$ at the point $(1, -1)$.

2. Find the domain of the function $f(x, y) = \dfrac{xy}{x^2 - 1}$.

3. Graph the function $f(x, y) = 1 + x^2 + 4y^2$, giving the traces and at least one section parallel to each coordinate plane.

4. Graph the function $f(x, y) = x^2 + 4x + y^2 + 2y$.

5. Find $f_x(x, y)$ and $f_y(x, y)$, where $f(x, y) = xy^3 - (2x + 3y)^4$.

6. Find $f_x(1, 0)$ and $f_y(2, 1)$ for $f(x, y) = 2xy^3 + 3x^{-1}y$.

7. Find all second-order partial derivatives for $f(x, y) = 3x^{2/5} + x^2y^{-2}$.

8. Find the values of x and y such that $f_x(x, y) = 0$ and $f_y(x, y) = 0$ for $f(x, y) = 1 - x^2 + y^2 + 2xy - 4y$.

9. Find the critical points of the function $f(x, y) = 3x^2 - 12xy + y^2 + 10y - 2$.

10. Find the relative extrema and any saddle points of the function $f(x, y) = x - xy + y^3 - 3y^2$.

11. Find the relative extrema of the function $f(x, y) = x^3 - 6x^2 - 2y^3 + 24y$.

12. A firm utilizes x million dollars per year in labor cost and y million dollars per year in production cost. The cost function is

$$C(x, y) = 100 + 24x + 12y + 2x^2 + 3y^2$$

Find the amount that the firm should spend on labor and production each year to minimize cost.

In Problems 13 and 14 use the method of Lagrange multipliers to find the extrema of the function with the given constraints.

13. $f(x, y, z) = xy$ subject to $3x + y - 5 = 0$

14. $f(x, y, z) = xyz$ subject to $x + y + 2z - 1 = 0$

15. Find three numbers that sum to 45 such that their product is a maximum.

16. Maximize $f(x, y) = x^{1/2}y^{1/3}$ subject to $200x + y = 8000$

17. Compute the antiderivative

$$\int (14xy^6 + 6x^3y^{-2} - y + x - 1)\, dx$$

18. Compute the definite integral.

$$\int_1^2 (10x^4y + 2x^{-3}y^3 - 1)\, dx$$

19. Compute the double integral.

$$\int_1^3 \left[\int_1^2 (6xy^2 - 24x^{-3}y^2 - 1)\, dx \right] dy$$

20. Evaluate the double integral over R:

$$\iint_R 4x^3(1 - y^2)\, dx\, dy, \text{ where } R \text{ is defined by}$$

$$0 \le x \le 1 \quad \text{and} \quad 0 \le y \le 1$$

21. Evaluate the double integral over R:

$$\iint_R 4xy\, dx\, dy, \text{ where } R \text{ is defined by}$$

$$x^2 \le y \le 1 \quad \text{and} \quad 0 \le x \le 1$$

▦ GRAPHICS CALCULATOR EXPLORATIONS

Most people find it challenging to depict a three-dimensional graph on a two-dimensional piece of paper. It is especially difficult to draw a saddle point. One way to help view three-dimensional graphs in two dimensions is by using level curves. Your graphics calculator is particularly useful in finding the relative extrema of functions of two variables by locating the level curves of the relative extremum. The level curves resemble a concentric set of circles surrounding the extremum. This Exploration will investigate how to picture relative extrema this way.

Example 1

Consider the relatively simple function $f(x, y) = x^2 + y^2$. Its graph has a relative minimum at $(0,0,0)$. If we let $z = x^2 + y^2$, we get level curves of the function by letting z take on specific values. For example, if we let $z = 1$ the equation becomes

$$1 = x^2 + y^2$$

The graph of this equation is a circle of radius 1 in the xy-plane. In a three-dimensional coordinate system its graph is a cylinder extending up and down from the latter circle in the xy-plane. The level curve is the intersection of this cylinder with the xy-plane, which is this circle. To graph this level curve using your calculator requires a bit of algebra. We can only graph an expression of the form "$y =$" so we solve the equation for y. This yields

$$y = \pm\sqrt{1 - x^2}$$

To graph these two equations we need two graph statements on the calculator. That is, first graph $y = \sqrt{1 - x^2}$, and then graph $y = -\sqrt{1 - x^2}$.

Casio

1. Press   GRAPH   √   (1   −   ALPHA   x x^2)   EXE

2. Press   GRAPH   (−)   √   (1   −   ALPHA   x x^2)   EXE

TI-81

1. Press   Y =

2. Press   2nd   √   (1   −   X | T x^2)   ENTER   (−)   2nd   Y-VARS   ENTER   ENTER

3. Press   GRAPH

This is the level curve corresponding to $z = 1$. To graph another level curve choose another value for z, for example $z = 2$. Then solve the equation $2 = x^2 + y^2$ for y. This yields

$$y = \pm\sqrt{2 - x^2}$$

As before, graph $y = \sqrt{2 - x^2}$, and then graph $y = -\sqrt{2 - x^2}$.

Casio

1. Press   GRAPH   √   (2   −   ALPHA   x x^2)   EXE

2. Press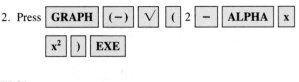

TI-81

1. Press $\boxed{\text{Y} =}$

2. Press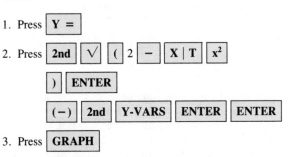

3. Press $\boxed{\textbf{GRAPH}}$

This is the level curve corresponding to $z = 2$.

The second level curve is outside the first, but to picture in the viewing screen the relative extremum of the function we want the level curves to hone in on the point in the xy-plane where the relative extremum occurs. This means that we must choose values of z that produce smaller circles. We know in this example that the function has a relative minimum at $(0,0,0)$, so the intersection of the xy-plane, $z = 0$, and the function is merely this point. Thus we will get smaller and smaller circles as we choose positive values of z that approach 0. Graph the level curves for this function using the method above for $z = 0.5, 0.2, 0.1, 0.05,$ and 0.01.

Example 2

Consider a more difficult function, $f(x,y) = x^3 - 3x^2 + y^2$. Let us first try to locate the relative extrema graphically and then use calculus to verify our guess. Graph several level curves to try and locate the relative extrema. For this function first try the level curve corresponding to $z = 0$ and graph $0 = x^3 - 3x^2 + y^2$. Solve for y and get

$$y = \pm\sqrt{-x^3 + 3x^2}$$

Next graph $y = \sqrt{-x^3 + 3x^2}$ and $y = -\sqrt{-x^3 + 3x^2}$.

Casio

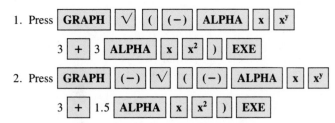

TI-81

1. Press $\boxed{\text{Y} =}$

2. Press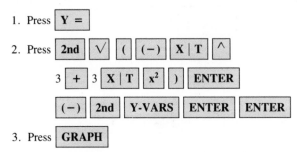

3. Press $\boxed{\textbf{GRAPH}}$

Try several additional values until you see concentric closed curves that seem to be honing in on a single point. From calculus we determine the function has a relative minimum at $(2,0,-4)$, so the level curves that illustrate this behavior of the function are those close to and greater than $z = -4$. What is the behavior of the function at the point $(0,0,0)$? How can you tell from the calculator that the function has a saddle point?

Further Explorations

One way to simulate a three-dimensional coordinate system with a graphics calculator is to sketch the graph of $y = x$ in the standard viewing screen and let it represent the x-axis, then let the x-axis in the screen represent the y-axis of the three-dimensional coordinate system and let the original y-axis represent the z-axis of the three-dimensional system. In problems 1 to 4 use this depiction of a three-dimensional coordinate system to visualize the given points.

1. $(0,1,2)$
2. $(1,1,2)$
3. $(1,1,-2)$
4. $(2,3,3)$

Trigonometric Functions

9

Sydney's underground railway system was built using trigonometry. (H. Armstrong Roberts)

Chapter Overview

The study of circles, angles, and triangles dates back to the dawn of civilization. Trigonometric functions are defined in terms of these figures. Today they are used to describe the workings of compact disc players, telephones, radios, and television sets because sound is composed of waves that are modeled by periodic functions, which are functions that repeat their values periodically.

Section 9–1 introduces the radian measure of a circle in terms of the more familiar degree measure. The two fundamental trigonometric functions, sine and cosine, are defined in Section 9–2, where their graphs are drawn and relationships between them are developed. Section 9–3 examines the derivative and integral of the sine and cosine functions. The calculus is then extended in Section 9–4 to the other four trigonometric functions: tangent, cotangent, secant, and cosecant.

CASE STUDY PREVIEW

In the early 1980s the Melbourne, Australia, government started an extensive, controversial plan to build a complex underground system of tunnels. Part of the tunnel project would serve as a railway and another part would carry sewerage. The tunnel-boring operations revealed a lack of knowledge of the forces that were generated when cutting through diverse rock formations. Mining engineers in charge of the project contacted the University of Melbourne to start a research project to design rock-boring tools with the proper shaft. Because such machines use rotating cutter heads, trigonometric functions must be used to model the operation of the shaft. The Case Study shows how several design problems were overcome.

9–1 Radian Measure

The practice of measuring angles by degrees dates back to the ancient Babylonians around the third century B.C. One full revolution is divided into 360 equal parts, each representing 1 degree. A degree is further divided into minutes and seconds. Although this form of angular measurement is still common, degree measure is being widely replaced by radian measure. In mathematics it is especially convenient to use radians because degrees are geometric quantities whereas radians are numbers. This allows us to perform operations on radians that would be impossible to do with degrees. Other branches of science are using radian measure more, and the international metric system has officially adopted radian measure (see Fig. 1). In this section we study radian measure and its connection to degree measure.

Radian Measure of Angles

The circumference of the unit circle is 2π units. Since one full revolution of the terminal side traverses the circumference of the circle, there are 2π radians in one full revolution of the circle. Hence the angular measurement of 2π radians equals

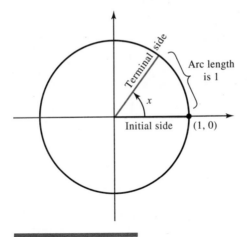

FIGURE 1 x is an angle with a measure of 1 radian.

360°. We write

$$2\pi \text{ radians } = 360° \qquad\qquad (1)$$

Similarly, one-half of a revolution, or 180°, is π radians. Expressed as an equation this is

$$\frac{1}{2} 2\pi \text{ radians } = 180°$$

$$\pi \text{ radians } = 180°$$

A right angle, or 90°, is one-quarter of a revolution, so

$$\frac{1}{4} 2\pi \text{ radians } = 90°$$

$$\frac{\pi}{2} \text{ radians } = 90°$$

Figure 2 shows these two familiar angles.

These particular instances suggest the general relationship between angular measurement in radians and degrees. From equation (1) we can write

$$1° = \frac{2\pi}{360} \text{ radians } = \frac{\pi}{180} \text{ radians}$$

This shows that to convert from degrees to radians, multiply the number of degrees by $\pi/180$.

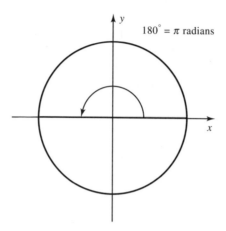

$180° = \pi$ radians

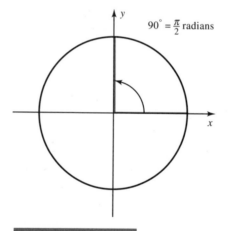

$90° = \frac{\pi}{2}$ radians

FIGURE 2

Let r be the measurement of an angle in radians and d be its measurement in degrees. Then to convert from $d°$ to r radians, use the formula

$$r = d° = d \cdot \frac{\pi}{180} \text{ radians}$$

To convert from r radians to $d°$, use the formula

$$d° = r = r \cdot \frac{180}{\pi} \text{ degrees}$$

From now on we will assume that an angle is measured in radians unless otherwise stated. Examples 1 and 2 illustrate these formulas.

EXAMPLE 1

Problem

Convert the angles from degrees to radians: (a) 60°, (b) 30°, (c) 135°.

Solution Use the first formula in the display.

(a) $60° = 60 \cdot \dfrac{\pi}{180}$ radians $= \dfrac{\pi}{3}$ radians

(b) $30° = 30 \cdot \dfrac{\pi}{180}$ radians $= \dfrac{\pi}{6}$ radians

(c) $135° = 135 \cdot \dfrac{\pi}{180}$ radians $= \dfrac{3\pi}{4}$ radians

These angles are shown in Figure 3.

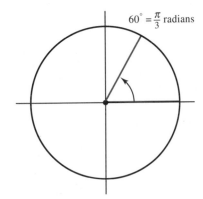

$60° = \frac{\pi}{3}$ radians

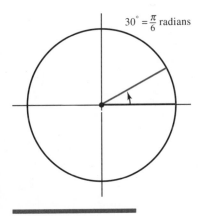

$30° = \frac{\pi}{6}$ radians

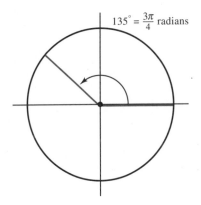

$135° = \frac{3\pi}{4}$ radians

FIGURE 3

EXAMPLE 2

Problem

Convert the angles from radians to degrees: (a) $\dfrac{\pi}{4}$, (b) $\dfrac{2\pi}{3}$, (c) $\dfrac{3\pi}{2}$.

Solution Use the second formula in the display.

(a) $\dfrac{\pi}{4} = \dfrac{\pi}{4} \cdot \dfrac{180}{\pi}$ degrees $= 45°$

(b) $\dfrac{2\pi}{3} = \dfrac{2\pi}{3} \cdot \dfrac{180}{\pi}$ degrees $= 120°$

(c) $\dfrac{3\pi}{2} = \dfrac{3\pi}{2} \cdot \dfrac{180}{\pi}$ degrees $= 270°$

These angles are shown in Figure 4.

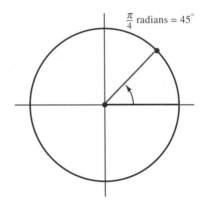

$\frac{\pi}{4}$ radians = 45°

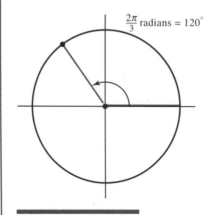

$\frac{2\pi}{3}$ radians = 120°

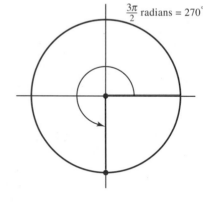

$\frac{3\pi}{2}$ radians = 270°

FIGURE 4

Extending Radian Measure

Often applications require more than one revolution. For example, a combination lock usually requires more than one revolution to open, and a lathe needs many revolutions to form a piece of wood or metal. If t is the radian measure of an angle, then the geometric interpretation of t is the distance traversed on the unit circle by the point P, starting from the point $(1, 0)$ and proceeding in a counterclockwise direction. (See Figure 5.) The radius of the unit circle turns through $t/2\pi$ revolutions. For $t > 2\pi$ the angle goes through more than one revolution. To illustrate:

$$t = 4\pi \text{ radians} \quad \text{corresponds to} \quad \frac{4\pi}{2\pi} = 2 \text{ revolutions}$$

$$t = 6\pi \text{ radians} \quad \text{corresponds to} \quad \frac{6\pi}{2\pi} = 3 \text{ revolutions}$$

$$t = 7\pi \text{ radians} \quad \text{corresponds to} \quad \frac{7\pi}{2\pi} = 3.5 \text{ revolutions}$$

Example 3 shows the connection between degree and radian measure for angles of more than one revolution.

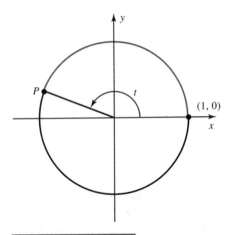

FIGURE 5 Radian measure t is the length traversed by P on the circumference of the unit circle.

EXAMPLE 3

Problem

Convert the angles from radians to degrees and express the angle in terms of the number of revolutions that it makes: (a) $\dfrac{9\pi}{4}$, (b) $\dfrac{10\pi}{3}$, (c) $\dfrac{9\pi}{2}$.

Solution Use the second formula in the display.

(a) $\dfrac{9\pi}{4} = \dfrac{9\pi}{4} \cdot \dfrac{180}{\pi}$ degrees $= 405°$

The angle makes $(9\pi/4)/(2\pi) = 9/8$ revolutions.

(b) $\dfrac{10\pi}{3} = \dfrac{10\pi}{3} \cdot \dfrac{180}{\pi}$ degrees $= 600°$

The angle makes $(10\pi/3)/(2\pi) = 5/3$ revolutions.

(c) $\dfrac{9\pi}{2} = \dfrac{9\pi}{2} \cdot \dfrac{180}{\pi}$ degrees $= 810°$

The angle makes $(9\pi/2)/(2\pi) = 9/4$ revolutions. (See Figure 6.)

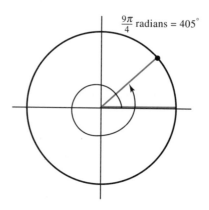

$\frac{9\pi}{4}$ radians $= 405°$

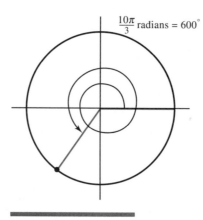

$\frac{10\pi}{3}$ radians $= 600°$

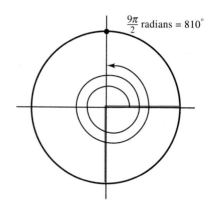

$\frac{9\pi}{2}$ radians $= 810°$

FIGURE 6

It is also important in many applications to distinguish between clockwise and counterclockwise motion. If the terminal side of the angle is rotated in the clockwise direction, we say the distance traversed on the unit circle is negative. Hence the radian measure is negative if the rotation is in the clockwise direction. For instance, if the angle is $-\pi/4$ then the terminal side is in the fourth quadrant. The angle it makes with the initial side is 45°, but since it is rotated in the clockwise direction, the angle is negative. Example 4 illustrates several negative angles.

EXAMPLE 4

Problem

Convert the angles from radians to degrees and express the angle in terms of the revolutions that it makes: (a) $-\dfrac{\pi}{6}$, (b) $-\dfrac{4\pi}{3}$, (c) $-\dfrac{9\pi}{2}$.

Solution (a) $-\dfrac{\pi}{6} = -\dfrac{\pi}{6} \cdot \dfrac{180}{\pi}$ degrees $= -30°$

The angle makes $(\pi/6)/(2\pi) = 1/12$ revolution in the clockwise direction.

(b) $-\dfrac{4\pi}{3} = -\dfrac{4\pi}{3} \cdot \dfrac{180}{\pi}$ degrees $= -240°$

The angle makes $(4\pi/3)/(2\pi) = 2/3$ revolution in the clockwise direction.

(c) $-\dfrac{9\pi}{2} = -\dfrac{9\pi}{2} \cdot \dfrac{180}{\pi}$ degrees $= -810°$

The angle makes $(9\pi/2)/(2\pi) = 9/4$ revolutions in the clockwise direction. (See Figure 7.)

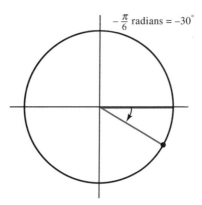

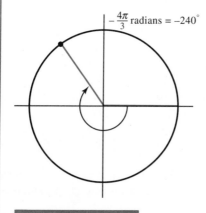

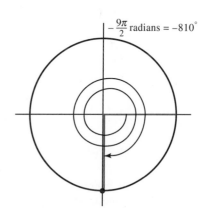

FIGURE 7

Angular Speed

Consider a wheel rotating around its center at a constant rate. As a point on the circumference travels a certain distance it sweeps an angle x of the circle. The **angular speed** w is the rate at which the angle x changes per unit of time. We express ω in terms of x and time t by

$$\omega = \frac{x}{t}$$

Angular speed is usually measured in radians per second (rad/s). For example, if a wheel is rotating at a rate of 1 revolution per second, then its angular speed is 2π rad/s because

$$\omega = \frac{2\pi}{1} = 2\pi$$

E X A M P L E   5

The Case Study mentions a shaft of a rock-cutting machine that rotates at 90 revolutions per minute. Find the angular speed of the shaft.

Solution The rate of 90 rpm corresponds to $90/60 = 3/2$ revolutions per second. Since the angular speed of 1 revolution per second is 2π rad/s, the angular speed of the shaft is

$$\omega = \frac{3}{2} \cdot 2\pi = 3\pi \text{ rad/s}$$

EXERCISE SET 9–1

In Problems 1 to 6 convert the angles from degrees to radians.

1. $90°$ 2. $45°$ 3. $210°$

4. $150°$ 5. $270°$ 6. $330°$

In Problems 7 to 12 convert the angles from radians to degrees.

7. $\dfrac{\pi}{6}$ 8. $\dfrac{5\pi}{6}$ 9. $\dfrac{5\pi}{2}$

10. $\dfrac{3\pi}{4}$ 11. $\dfrac{7\pi}{4}$ 12. $\dfrac{7\pi}{3}$

In Problems 13 to 18 convert the angles from radians to degrees and express the angle in terms of the number of revolutions that it makes.

13. $\dfrac{11\pi}{4}$ 14. $\dfrac{7\pi}{3}$ 15. $\dfrac{7\pi}{2}$

16. $\dfrac{17\pi}{4}$ 17. $\dfrac{14\pi}{3}$ 18. $\dfrac{25\pi}{6}$

In Problems 19 to 24 convert the angles from radians to degrees and express the angle in terms of the number of revolutions that it makes.

19. $-\dfrac{5\pi}{6}$ 20. $-\dfrac{5\pi}{3}$ 21. $-\dfrac{11\pi}{2}$

22. $-\dfrac{5\pi}{4}$ 23. $-\dfrac{14\pi}{3}$ 24. $-\dfrac{29\pi}{6}$

In Problems 25 to 28 find the radian measure of the angle whose measure is given in degrees.

25. $390°$ 26. $540°$

27. $1080°$ 28. $1110°$

In Problems 29 to 32 find the radian measure of the angle that makes the given number of revolutions in a clockwise direction.

29. two complete revolutions

30. four complete revolutions

31. two and one-third revolutions

32. three and one-sixth revolutions

In Problems 33 to 36 find the radian measure of the angle that makes the given number of revolutions in a counterclockwise direction.

33. 2 complete revolutions

34. 4.5 revolutions

35. 3.25 revolutions

36. 5.75 revolutions

37. Turning a combination lock two and one-sixth revolutions clockwise corresponds to an angle of what radian measure?

38. Find the radian measure of the angle that the hour hand of a clock defines as it advances from 3 P.M. to the given time: (a) 6 P.M., (b) 9 P.M., (c) 6 A.M.

39. The shaft of a rock-cutting machine rotates at 60 rpm. Find the angular speed of the shaft.

40. The shaft of a rock-cutting machine rotates at 100 rpm. Find the angular speed of the shaft.

Referenced Exercise Set 9–1

1. The National Safety Council has developed a method for determining safe approach speeds for vehicles at intersections where the view is obstructed.* The method uses a sight, or visibility, triangle whose legs are the perpendicular paths of two vehicles approaching the intersection at right angles. The "sight line" is the line that connects the eyes of the drivers of the two vehicles. The obstruction may be a building or trees or something similar. If the sight line passes through the obstruction, the drivers cannot see each other. The angles made by the sight line and the vehicle paths are critical in computing safe approach speeds. Two common angles made by sight lines are 60° and 30°, respectively. Convert these angles to radians.

2. The transmission of sound waves can be visualized by picturing the propagation of waves on the surface of water. For example, when a tuning fork is struck, air molecules vibrate between the prongs, and this motion causes a wave to move away from the fork in much the same way a pebble dropped in an pond causes a wave on water. Trigonometry is used to model this phenomenon.† The stage that a particular air molecule has reached in its vibration is called its "phase," which is measured in radians. The phase p is related to the angular speed x of the particle by $p = tx + b$. Find p in radians when $t = .2$ seconds, $x = 50$ rad/s, and b is a one-quarter revolution.

3. Beech Aircraft, with headquarters in Wichita, Kansas, is a large producer of general aviation aircraft. In the late 1970s a customer requested that the Beech Super King Air, a twin turboprop aircraft, be modified to increase flying time from 9 hours to 11 hours.‡ This required an increase in fuel capacity of about 100 gallons. Beech decided to install new fuel tanks inside the wing tips. In order to avoid increasing wing stress too much, they decided to position the new tanks so they would drain by gravity into the main tanks. This avoided using heavy pumps. The angle of declination of the tanks was important. Several prototypes were built with angles of (a) 2°, (b) 3°, and (c) 3.5°. Convert these angles to radians.

*Louis J. Pignataro, *Traffic Engineering: Theory and Practice*, Englewood Cliffs, NJ, Prentice-Hall, 1973, pp. 338–339.

†Robert Erickson, *Sound Structure in Music*, Berkeley, CA, University of California Press, 1975, pp. 1–60.

‡R. D. Marwill, "Tip Tanks for the Beech Super King Air," *Engineering Case Library*, American Society for Engineering Education, ECL 246, 1982, pp. 1–41.

9–2 The Sine and the Cosine

In many applications the values of a variable quantity repeat themselves at regular intervals. For example, in ecology the size of a particular population increases and decreases periodically. Trigonometric functions are used to model this type of behavior. In this section we examine the sine and cosine functions. In Section 10–4 the other four trigonometric functions are defined in terms of these two functions.

There are two common definitions of the sine and cosine. One depends on right triangles, the other on the unit circle. Each definition has its advantage. The right triangle definition makes it easier to compute the functional values of certain common angles. The circular definition allows us to extend the domain of the functions. We start with the right triangle definition and study how to compute specific functional values. Then we incorporate this definition with the more general approach.

Right Triangle Definition

Let the number x represent the radian measure of an angle and picture the angle in the right triangle in Figure 1. The trigonometric functions are ratios of one side of the triangle to another. The sine of an angle x, denoted **sin x,** and the cosine of an angle x, denoted **cos x,** are defined in the display.

DEFINITION

$$\sin x = \frac{\text{length of opposite side}}{\text{length of hypotenuse}}$$

$$\cos x = \frac{\text{length of adjacent side}}{\text{length of hypotenuse}}$$

For most values of x the functional values of sin x and cos x are difficult to compute. A calculator with sin and cos keys is a great help. The values can also be taken from a table of trigonometric functions, such as the one in the Appendix. There are a few values of x, however, whose functional values can be derived from the 30-60-90 degree and the 45-45-90 degree triangles shown in Figure 2. The Pythagorean Theorem can be used to calculate the lengths of the sides of these triangles. Then the definition is used to calculate the sine and cosine of the angles.

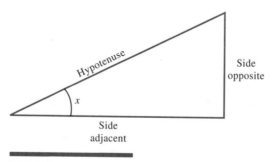

FIGURE 1

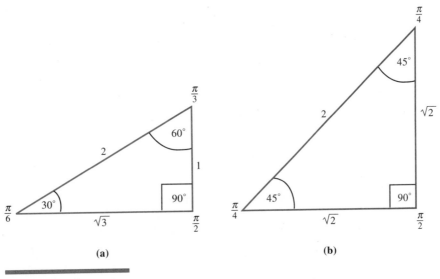

FIGURE 2

For instance, in the 30-60-90 triangle the 30° angle has a radian measure of $\pi/6$. The length of the side opposite this angle is 1, while the length of the hypotenuse is 2. By the definition sin $\pi/6 = 1/2$. In Example 1 we compute several additional values from these triangles.

EXAMPLE 1

Problem

Use the triangles in Figure 2 to compute (a) cos $\pi/6$, (b) sin $\pi/3$, (c) cos $\pi/3$, (d) sin $\pi/4$, (e) cos $\pi/4$.

Solution (a) To compute cos $\pi/6$, use the 30-60-90 degree triangle, where $\pi/6$ corresponds to 30°. The length of the side adjacent to this angle is $\sqrt{3}$, while the length of the hypotenuse is 2. By definition, cos $\pi/6 = \sqrt{3}/2$.
(b) To compute sin $\pi/3$, use the 30-60-90 triangle where $\pi/3$ corresponds to 60°. Then sin $\pi/3 = \sqrt{3}/2$.
(c) Using the 30-60-90 triangle, we get cos $\pi/3 = 1/2$.
(d) To compute sin $\pi/4$ use the 45-45-90 triangle. Then sin $\pi/4 = \sqrt{2}/2$;
(e) Using the 45-45-90 triangle shown in Figure 2b, cos $\pi/4 = \sqrt{2}/2$.

If the lengths of the sides of a right triangle are known, then the sine and cosine of the angles can be computed without knowing the measure of the angle. Example 2 illustrates this.

EXAMPLE 2

Consider the right triangle in Figure 3 whose sides are of lengths 3, 4, and 5, respectively.

Problem

Compute (a) sin x and (b) cos x.

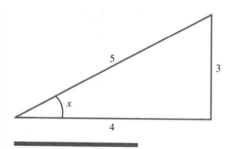

FIGURE 3

Solution (a) To compute sin x, note that the length of the opposite side is 3 while the length of the hypotenuse is 5. By definition, sin $x = 3/5$.
(b) To compute cos x, note that the length of the adjacent side is 4, while the length of the hypotenuse is 5. By definition, cos $x = 4/5$.

The Unit Circle Definition

A major obstacle to defining sin x and cos x in terms of lengths of sides of a triangle is that these functions are then defined only for angles between 0 and $\pi/2$ radians. To extend these definitions, we view them in a slightly different way by placing the angle in standard position in a unit circle. Then the angle is a central angle whose terminal side intersects the unit circle at a point P. Drop a perpendicular from P to the horizontal axis. This forms a right triangle whose hypotenuse is 1 since it is a radius of the unit circle. See Figure 4. The length of the side opposite the angle is the second coordinate of P. The length of the side adjacent to the angle is the first coordinate of P. Thus the coordinates of P are (cos x, sin x).

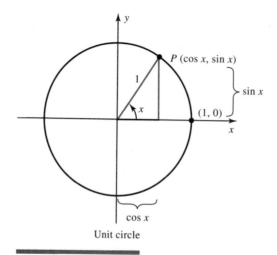

FIGURE 4

This view of sin x and cos x agrees with the right triangle definition for x between 0 and $\pi/2$. The advantage is that we do not have to restrict x to this interval.

UNIT CIRCLE DEFINITION OF SINE AND COSINE

Let x be an angle in standard position with a terminal side that intersects the unit circle at a point P. The first coordinate of P is cos x and the second coordinate is sin x. (See Figure 5.)

Example 3 shows how to compute sine and cosine for some angles that are not between 0 and $\pi/2$. The computation relies on Example 1.

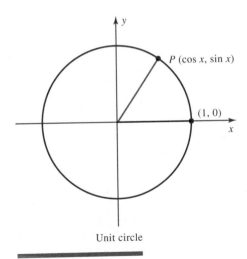

Unit circle

FIGURE 5

EXAMPLE 3

Problem

Use the triangles in Figure 6 to compute (a) $\cos(-\pi/6)$, (b) $\sin(-\pi/6)$, (c) $\sin(2\pi/3)$, (d) $\sin(13\pi/4)$.

Solution (a) To compute $\cos(-\pi/6)$, note that the terminal side of the angle $-\pi/6$ is in the fourth quadrant. To find the coordinates of P, notice that the perpendicular from P to the x-axis forms a triangle that is congruent to the 30-60-90 triangle in Figure 2. Thus the lengths of the sides are the same, except that the y-coordinate is negative. Therefore P has the coordinates $(\sqrt{3}/2, -1/2)$. From the definition $\cos(-\pi/6) = \sqrt{3}/2$.

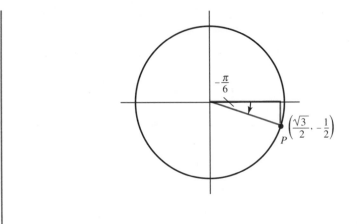

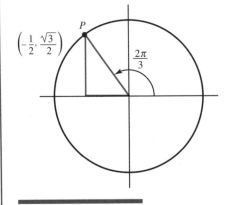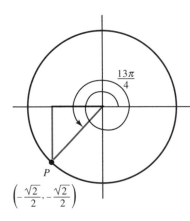

FIGURE 6

(b) To compute $\sin(-\pi/6)$, use the same point P in (a). Then $\sin(-\pi/6) = -1/2$.

(c) To compute $\sin(2\pi/3)$, use the 30-60-90 triangle formed in the second quadrant. The coordinates of P are $(-1/2, \sqrt{3}/2)$. Therefore $\sin(2\pi/3) = \sqrt{3}/2$.

(d) To compute $\sin(13\pi/4)$, use the 45-45-90 triangle formed in the third quadrant. The coordinates of P are $(-\sqrt{2}/2, -\sqrt{2}/2)$. Therefore $\sin(13\pi/4) = -\sqrt{2}/2$.

Table 1 gives values of $\sin x$ and $\cos x$ for angles x that are multiples of $\pi/6$ and $\pi/4$ up to π. Each value can be verified by calculations similar to those in Example 3. For other angles use a calculator or Table 5 in the Appendix. Table 1 also gives the values of three quadrantal angles, which are angles whose terminal sides lie on an axis. For example, for $x = 0$, the point P, which is the intersection of the terminal side and the unit circle, is $(1, 0)$. From Table 1, $\sin 0 = 0$ and $\cos 0 = 1$.

Table 1

x	0	$\dfrac{\pi}{6}$	$\dfrac{\pi}{4}$	$\dfrac{\pi}{3}$	$\dfrac{\pi}{2}$	$\dfrac{2\pi}{3}$	$\dfrac{3\pi}{4}$	$\dfrac{5\pi}{6}$	π
$\sin x$	0	$\dfrac{1}{2}$	$\dfrac{\sqrt{2}}{2}$	$\dfrac{\sqrt{3}}{2}$	1	$\dfrac{\sqrt{3}}{2}$	$\dfrac{\sqrt{2}}{2}$	$\dfrac{1}{2}$	0
$\cos x$	1	$\dfrac{\sqrt{3}}{2}$	$\dfrac{\sqrt{2}}{2}$	$\dfrac{1}{2}$	0	$-\dfrac{1}{2}$	$-\dfrac{\sqrt{2}}{2}$	$-\dfrac{\sqrt{3}}{2}$	-1

Properties of Sine and Cosine

There are several important relationships between sin x and cos x. For every value x the point (cos x, sin x) is the intersection of the terminal side and the unit circle. The equation of the unit circle is $X^2 + Y^2 = 1$. This means that every point on the unit circle satisfies this equation. Therefore, replacing X with cos x and Y with sin x yields

$$\cos^2 x + \sin^2 x = 1$$

This relationship holds for every value of x. For this reason it is called an **identity.** The notation used for raising a trigonometric function to a power is illustrated in this identity. We write $\cos^2 x$ to represent $(\cos x)^2$. We now present several additional identities.

If x is any angle, then $x + 2\pi$ represents the angle x plus one revolution, while $x + 2n\pi$ represents x plus n revolutions. Each of these angles has the same point of intersection of its terminal side with the unit circle, so that the sines of the angles are equal and the cosines of the angles are equal. This is also true of $x - 2n\pi$. This yields the following result.

$$\sin(x \pm 2n\pi) = \sin x \qquad \cos(x \pm 2n\pi) = \cos x$$

Example 3 showed that $\cos(-\pi/6) = \cos \pi/6$ because the point P on the unit circle corresponding to $\pi/6$ has the same first coordinate as the point corresponding to $-\pi/6$. This is true for all angles, so $\cos(-x) = \cos x$. Example 3 also showed that $\sin(-\pi/6) = -\sin \pi/6$ because the second coordinates of the corresponding points P are negatives of each other, so $\sin(-x) = -\sin x$. See Figure 7.

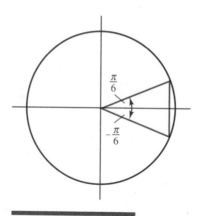

FIGURE 7

$$\sin(-x) = -\sin x \qquad \cos(-x) = \cos x$$

Another identity that is true for all numbers x and y is

$$\sin(x + y) = \sin x \cos y + \cos x \sin y$$

This is called the "sine of a sum of two angles" formula. Its proof can be found in most trigonometry textbooks. Since it holds for any two angles, we can substitute $x = \pi/2$ and $y = -x$ to derive another important identity.

$$\sin\left(\frac{\pi}{2} - x\right) = \sin\frac{\pi}{2}\cos(-x) + \cos\frac{\pi}{2}\sin(-x)$$
$$= (1)\cos(-x) + (0)\sin(-x)$$
$$= \cos(-x) = \cos x$$

Similarly, we can show that $\cos(\pi/2 - x) = \sin x$.

$$\sin\left(\frac{\pi}{2} - x\right) = \cos x \qquad \cos\left(\frac{\pi}{2} - x\right) = \sin x$$

Other identities include

$$\sin(x - y) = \sin x \cos y - \cos x \sin y$$
$$\sin 2x = 2 \sin x \cos x$$
$$\cos 2x = \cos^2 x - \sin^2 x$$
$$\sin^2 x = (1/2)(1 - \cos 2x)$$
$$\cos^2 x = (1/2)(1 + \cos 2x)$$

Graphs of Sine and Cosine

Six x and cos x measure quantities that repeat their values over certain periods. Their graphs display this periodicity property. To sketch their graphs, refer back to Table 1. First, plot points on the graph of each function for values of x that range from 0 to π. Then extend the table to include values of x up to 2π, and plot those points. Next, connect the points with a smooth curve.

The graph of sin x is given in Figure 8. Because the graph repeats itself every 2π units, we say that the sine function is periodic with period 2π. The graph of the cosine function, shown in Figure 9, can be drawn in a similar manner. This function is also periodic with period 2π.

The graph of the sine function occurs naturally in many settings. Trigonometry is especially useful in music. Figure 10 pictures the vibrating string on a

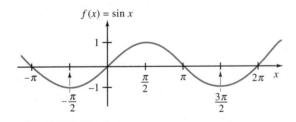

FIGURE 8

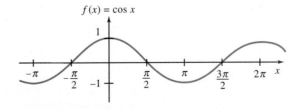

FIGURE 9

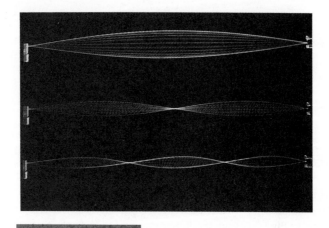

FIGURE 10

violin.* It shows the string displacement on the vertical axis versus time on the horizontal axis. As we will see in the next section, there is a close connection between trigonometry and sound.

*Thanks to Dr. Robert Weinberg and Mr. David Nairns, Temple University, Department of Physics.

EXERCISE SET 9–2

In Problems 1 to 4 use the triangles in Figure 2 to compute the given value.

1. $2 \cos \dfrac{\pi}{6}$ 2. $3 \sin \dfrac{\pi}{3}$

3. $1 + \cos \dfrac{\pi}{3}$ 4. $4 - \sin \dfrac{\pi}{4}$

In Problems 5 to 8 the right triangle in the accompanying figure has sides of lengths 5, 12, and 13, respectively. Compute the given value.

5. $\cos a$ 6. $\sin a$

7. $\sin b$ 8. $\cos b$

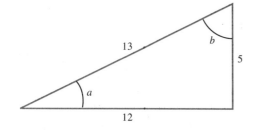

In Problems 9 to 20 compute the functional value.

9. $\cos\left(-\dfrac{\pi}{3}\right)$ 10. $\sin\left(-\dfrac{\pi}{3}\right)$

11. $\sin \dfrac{4\pi}{3}$ 12. $\cos \dfrac{4\pi}{3}$

13. $\sin\left(-\dfrac{4\pi}{3}\right)$ 14. $\cos\left(-\dfrac{4\pi}{3}\right)$

15. $\cos \dfrac{5\pi}{4}$ 16. $\sin\left(-\dfrac{5\pi}{4}\right)$

17. $\sin \dfrac{15\pi}{4}$ 18. $\cos \dfrac{15\pi}{4}$

19. $\sin\left(-\dfrac{17\pi}{6}\right)$ 20. $\cos \dfrac{19\pi}{6}$

In Problems 21 to 24 use a calculator or the table in the Appendix to find the functional values.

21. $\sin .5$ 22. $\cos .6$

23. $\sin .1$ 24. $\cos 1/10$

In Problems 25 to 28 find a number x such that $0 \le x \le \pi/2$ and x satisfies the given property.

25. $\sin x = \cos \dfrac{3\pi}{2}$ 26. $\sin x = \sin\left(-\dfrac{3\pi}{2}\right)$

27. $\cos x = \sin 3\pi$ 28. $\sin x = \sin 5\pi$

In Problems 29 to 32 sketch the graph of the function.

29. $f(x) = 2 \sin x$ 30. $f(x) = 3 \sin x$

31. $f(x) = 1 + 2 \cos x$

32. $f(x) = -2 + 3 \sin x$

In Problems 33 to 36 sketch the graph of the function.

33. $f(x) = \sin 2x$ 34. $f(x) = 2 \sin 3x$

35. $f(x) = 1 - 2 \cos 2x$

36. $f(x) = -2 + 3 \sin 4x$

37. Why are $\sin x$ and $\cos x$ called "periodic" functions?

38. Find the largest and the smallest values of $f(x)$ and the value of x where they occur for the following functions: (a) $f(x) = \sin 2x$, (b) $f(x) = 2 \sin x$, (c) $f(x) = 3 + 2 \cos 4x$.

39. Consider the formula for the movement of a typical air molecule of a simple sound

$$y(x) = 0.002 \cos(440\pi x)$$

Find the amplitude which is the largest value the function takes on.

In Problems 40 to 43 match the graphs in Figures a to d to the function that is a solution of the given differential equation.

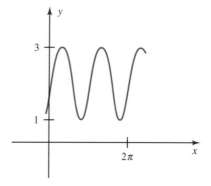

(b)

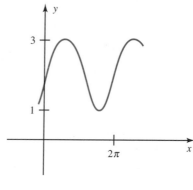

(c)

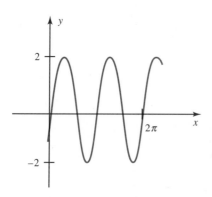

(a)

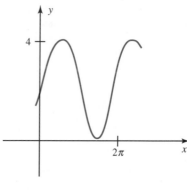

(d)

40. $y = \sin x + 2$ 42. $y = 2 \sin x + 2$

41. $y = \sin 2x + 2$ 43. $y = 2 \sin 2x$

Referenced Exercise Set 9–2

1. Kudzu is a weedy vine that forms a dense canopy over bare ground. It has been used as cattle fodder, for erosion control, and as a means to reclaim mine spoils. It may eventually be used as a source of sugar, and as fiber for clothing, wallpaper, and paper. In an article studying the growth patterns of kudzu in central Maryland, Forseth and Teramura measured how the sun affected the leaf orientation by computing the cosine of incidence, defined as the cosine of the angle between the leaf lamina and a line perpendicular to the sun's direct beam.* The cosine of incidence directly affects the amount of irradiation that reaches the ground. For one type of kudzu leaf, the cosine of incidence was .9, whereas it was .8 for another type. Assuming the angles of incidence were between 0 and $\pi/2$ radians, find the angles corresponding to these values.

2. In his book showing the influence of mathematics in western culture, Morris Kline gives the following formula of the sound of a violin, which relates the position $y(t)$ at time t of a particular air molecule vibrating because of the violin strings:†

   $$y(t) = .06 \sin 18t + .02 \sin 36t + .01 \sin 54t$$

   For each term, find the amplitude, which is the largest value that the function takes on.

3. In an article on how to make violins, Itokawa and Kumagai use the following formula to calculate the conductivity $C(x)$ of the f-holes of a violin for the angle x between the f-holes.‡ The f-holes of a violin amplify the vibration of the strings. Conductivity measures the speed of the sound wave through the f-holes.

   $$C(x) = \frac{\pi}{2.9\sqrt{\cos x}}$$

For a given violin, $x = 84°$. Compute $C(x)$ for this value of x.

Cumulative Exercise Set 9–2

1. Convert the angle $-750°$ to radians.

2. Convert the angle $\frac{8}{3}\pi$ to degrees.

3. Find the measure in (a) radians and (b) degrees of the angle that makes $\frac{4}{3}$ revolutions in a counterclockwise direction.

4. What angle does the hour hand of a clock define as it advances from 8 P.M. on one day until noon the next day in (a) radians and (b) degrees?

5. Find each value without using a calculator or a table.
   (a) $1 + \cos \pi$ (b) $2 \sin(\frac{\pi}{2})$

6. Use a calculator or Table 5 in the Appendix to find the given value.
   (a) $\sin 75°$ (b) $\cos(\frac{\pi}{6})$

7. Compute the values of (a) $1 - \sin a$ and (b) $2 \cos b$ from the accompanying 5-12-13 right triangle.

8. Sketch the graph of the function $f(x) = (\frac{1}{2})\cos x$.

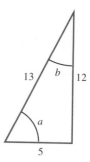

*Irwin N. Forseth and Alan H. Teramura, "Kudzu Leaf Energy Budget and Calculated Transpiration: At the Influence of Leaflet Orientation," *Ecology,* Vol. 67, 1987, pp. 564–570.

†Morris Kline, *Mathematics in Western Culture,* London, Oxford University Press, 1953, pp. 287–303.

‡Hideo Itokawa and Chihiro Kumagai, "On the Study of Violin and Its Making," *Musical Acoustics: Piano and Wood Instruments,* Earle Kent (ed.), Stroudsburg, PA, Dowden, Hutchinson and Ross, 1977, pp. 55–84.

9–3 **Differentiation and Integration**

The graphs of the sine and cosine (Figures 8 and 9 in Section 9–2) show the close connection between the two functions. The graph of the sine is identical to the graph of the cosine when the sine curve is moved to the left $\pi/2$ units. This is because of the identity $\sin(\pi/2 - x) = \cos x$. In this section we will see another similarity between these functions when we calculate their derivatives. The rules governing their derivatives are straightforward, but the proofs of the rules are not. The proofs are given at the end of the section.

The Derivative of the Sine

To get an intuitive grasp of the derivative of the sine, superimpose the graphs of the sine and the cosine as in Figure 1. Then identify the slope of the tangent line at several points of the sine curve. For instance, the slope at $x = 0$ is 1; the slope at $x = \pi/2$ is 0; the slope at $x = \pi$ is -1. In each case the slope of the sine function at $x = a$ is equal to $\cos a$. This is true for all x, meaning that the derivative of $\sin x$ is $\cos x$ for all x. This is recorded in the following rule.

$$D_x \sin x = \cos x$$

A more rigorous verification of this rule is given at the end of this section.

If u is a function of x, we can apply the chain rule to get the sine version of the chain rule.

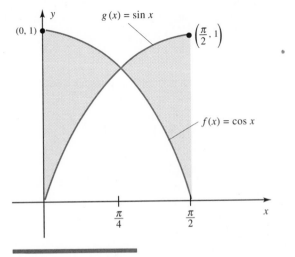

FIGURE 1

If u is a function of x, then

$$D_x \sin u = \cos u \cdot D_x u$$

Example 1 illustrates how to apply the sine version of the chain rule. It should be noted that $\sin x^2 \neq (\sin x)^2$ since $(\sin x)^2$ is written $\sin^2 x$. To see this, let x be a specific value, say $x = \dfrac{\pi}{4}$, and compute each expression.

EXAMPLE 1

Problem

Compute the derivative of the functions: (a) $f(x) = \sin 5x$, (b) $f(x) = \sin x^2$, (c) $f(x) = \sin(x^3 - 5x)$.

Solution (a) Let $u = 5x$. Then $D_x u = D_x 5x = 5$. Thus

$$f'(x) = D_x \sin 5x = (\cos 5x)5 = 5 \cos 5x$$

(b) Let $u = x^2$. Then $D_x u = D_x x^2 = 2x$. Thus

$$f'(x) = D_x \sin x^2 = (\cos x^2)2x = 2x \cos x^2$$

(c) Let $u = x^3 - 5x$. Then $D_x u = D_x(x^3 - 5x) = 3x^2 - 5$. Thus

$$f'(x) = D_x \sin(x^3 - 5x) = [\cos(x^3 - 5x)](3x^2 - 5)$$
$$= (3x^2 - 5) \cos(x^3 - 5x)$$

The Derivative of the Cosine

Once the derivative of $\sin x$ is known the derivative of $\cos x$ can be derived by using the chain rule and the two trigonometric identities

$$\sin\left(\frac{\pi}{2} - x\right) = \cos x \qquad \text{and} \qquad \cos\left(\frac{\pi}{2} - x\right) = \sin x$$

Start with the first identity and take its derivative, using the sine version of the chain rule with $u = \pi/2 - x$.

$$D_x \cos x = D_x \sin\left(\frac{\pi}{2} - x\right)$$

$$= \cos\left(\frac{\pi}{2} - x\right) D_x \left(\frac{\pi}{2} - x\right)$$

$$= \cos\left(\frac{\pi}{2} - x\right)(-1)$$

$$= -\cos\left(\frac{\pi}{2} - x\right)$$

$$= -\sin x$$

The last conclusion is drawn from the second trigonometric identity above. This result is recorded in the display together with the cosine version of the chain rule.

$$D_x \cos x = -\sin x$$

If u is a function of x, then

$$D_x \cos u = -\sin u \cdot D_x u$$

Example 2 illustrates how to apply the cosine version of the chain rule.

EXAMPLE 2

Problem

Compute the derivative of the functions: (a) $f(x) = x^2 \cos 5x$, (b) $f(x) = (5 + 3 \cos x^2)^3$.

Solution (a) Use the product rule and the cosine version of the chain rule with $u = 5x$. Then $D_x u = D_x 5x = 5$. Thus

$$f'(x) = D_x(x^2 \cos 5x) = x^2 D_x \cos 5x + (\cos 5x)D_x x^2$$
$$= x^2(-\sin 5x)5 + (\cos 5x)2x$$
$$= -5x^2 \sin 5x + 2x \cos 5x$$

(b) Two versions of the chain rule must be used. First $f(x)$ is of the form $g(x)^n$, where $g(x) = 5 + 3 \cos x^2$. Thus the extended power rule is used to find $f'(x)$. That is, $f'(x) = ng(x)^{n-1}g'(x)$. To compute $g'(x)$, we must use the cosine version of the chain rule with $u = x^2$. This results in

$$f'(x) = ng(x)^{n-1}g'(x)$$
$$= 3(5 + 3 \cos x^2)^2 D_x(5 + 3 \cos x^2)$$
$$= 3(5 + 3 \cos x^2)^2(-3 \sin x^2)(D_x x^2)$$
$$= 3(5 + 3 \cos x^2)^2(-3 \sin x^2)(2x)$$
$$= -18x(5 + 3 \cos x^2)^2(\sin x^2)$$

Integration

After practicing the differentiation formulas on several functions involving sine and cosine, it becomes easy to remember that $D(\sin x) = \cos x$ and $D(\cos x) = -\sin x$. We are now going to develop integration formulas for sine and cosine. You might find that the similarity between the differentiation formulas and the integration formulas will cause some difficulty at first, but the integration formulas too will become much easier to remember after you have worked several exercises involving them.

Recall that $\int f(x) \, dx = F(x) + C$ if $F'(x) = f(x)$. By setting $f(x) = \sin x$

we get the first formula in the accompanying display. The second formula is obtained by setting $f(x) = \cos x$.

$$\int \sin x \, dx = -\cos x + C$$

$$\int \cos x \, dx = \sin x + C$$

These formulas can be used with integration techniques to evaluate a variety of integrals involving the sine and cosine functions. Example 3 uses substitution to compute two such integrals.

EXAMPLE 3

Problem

Compute the integrals: (a) $\int 5 \cos 5x \, dx$, (b) $\int x \sin x^2 \, dx$.

Solution (a) Use the substitution $u = 5x$. Then $du = 5 \, dx$. Thus

$$\int 5 \cos 5x \, dx = \int (\cos 5x)(5 \, dx)$$

$$= \int \cos u \, du$$

$$= \sin u + C$$

$$= \sin 5x + C$$

(b) Use the substitution $u = x^2$. Then $du = 2x \, dx$ and $(1/2)du = x \, dx$. Thus

$$\int x \sin x^2 \, dx = \int (\sin x^2)(x \, dx)$$

$$= \int (\sin u) \frac{1}{2} \, du = \frac{1}{2} \int \sin u \, du$$

$$= -\frac{1}{2} \cos u + C$$

$$= -\frac{1}{2} \cos x^2 + C$$

The next example requires a more complicated substitution. It also shows how to compute a definite integral involving trigonometric functions.

EXAMPLE 4

Problem

Compute the definite integral.

$$\int_0^{\pi/2} 4 \sin^3 x \cos x \, dx$$

Solution First find the indefinite integral

$$\int 4 \sin^3 x \cos x \, dx$$

Use the substitution $u = \sin x$. Then $du = \cos x \, dx$. Thus

$$\int 4 \sin^3 x \cos x \, dx = \int 4u^3 \, du = u^4 + C = \sin^4 x + C$$

The definite integral is then evaluated as follows:

$$\int_0^{\pi/2} 4 \sin^3 x \cos x \, dx = \sin^4 x \, \Big|_0^{\pi/2}$$

$$= \sin^4 \left(\frac{\pi}{2} \right) - \sin^4(0)$$

$$= 1^4 - 0^4 = 1$$

The next example shows how to find areas of regions bounded by trigonometric functions.

EXAMPLE 5

Problem

Find the area of the region bounded by the graphs of $f(x) = \sin x$ and $g(x) = \cos x$ from $x = 0$ to $x = \frac{\pi}{2}$. See Figure 1.

Solution To calculate the area of a region bounded by two curves, we need to determine the intervals where $f(x) \geq g(x)$. In those intervals we integrate $f(x) - g(x)$. Otherwise we integrate $g(x) - f(x)$. First determine where the graphs intersect by setting $f(x) = g(x)$. This yields

$$\sin x = \cos x \qquad \text{for} \qquad 0 \leq x \leq \frac{\pi}{2}$$

From Figure 1 the only value of x satisfying these properties is $x = \pi/4$. Also, from Figure 1 we have

$$\cos x \geq \sin x \qquad \text{for} \qquad 0 \leq x \leq \frac{\pi}{4}$$

$$\sin x \geq \cos x \qquad \text{for} \qquad \frac{\pi}{4} \leq x \leq \frac{\pi}{2}$$

To compute the area, two integrals are required:

$$\int_0^{\pi/4} (\cos x - \sin x)\, dx + \int_{\pi/4}^{\pi/2} (\sin x - \cos x)\, dx$$

$$= (\sin x + \cos x)\Big|_0^{\pi/4} + (-\cos x - \sin x)\Big|_{\pi/4}^{\pi/2}$$

$$= \left[\frac{\sqrt{2}}{2} + \frac{\sqrt{2}}{2} - (0 + 1) \right] + \left[-0 - 1 - \left(-\frac{\sqrt{2}}{2} - \frac{\sqrt{2}}{2} \right) \right]$$

$$= 2\sqrt{2} - 2$$

Verification of the Derivative of the Sine

To verify that the derivative of the sine is the cosine, we need the following result:

$$\lim_{h \to 0} \frac{\sin h}{h} = 1 \tag{1}$$

An intuitive geometric argument of this result utilizes the graph in Figure 2. The slope of the tangent line of $y = \sin x$ at $x = 0$ appears to be 1. The secant lines that pass through $(0, 0)$ and $(h, \sin h)$ have slope $(\sin h)/h$. Since these slopes approximate the slope of the tangent line, they must approach 1. This means that the limit of $(\sin h)/h$ as h approaches 0 is 1.

Equation (1) can also be verified by using a calculator. Choose values of h that approach 0 and compute successive values of $(\sin h)/h$. Table 1 has several such calculations. In a similar way, the following result can be verified:

$$\lim_{h \to 0} \left[\frac{\cos h - 1}{h} \right] = 0 \tag{2}$$

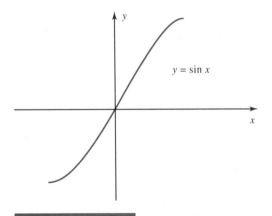

FIGURE 2 The tangent line at $x = 0$ appears to have slope 1.

Table 1

h	.1	.01	.001	.0001	.00001	$-.001$	$-.0001$
$\sin h$	.09983	.01000	.00100	.00010	.00001	$-.00100$	$-.00010$
$\dfrac{\sin h}{h}$	.99833	.99998	.99999	.99999	.99999	.99999	.99999

We can verify the formula for the derivative of the sine function by using equations (1) and (2) along with the formula for the sine of a sum.

$$\sin(x + h) = \sin x \cos h + \cos x \sin h$$

From the definition of the derivative we have

$$f'(x) = \lim_{h \to 0} \frac{f(x + h) - f(x)}{h}$$

$$= \lim_{h \to 0} \left[\frac{\sin(x + h) - \sin x}{h} \right]$$

$$= \lim_{h \to 0} \left[\frac{\sin x \cos h + \cos x \sin h - \sin x}{h} \right]$$

Factor $\sin x$ from the first and third terms and split the fraction into two fractions. This yields

$$f'(x) = \lim_{h \to 0} \left[\sin x \, \frac{(\cos h - 1)}{h} + \cos x \, \frac{(\sin h)}{h} \right]$$

Use the fact that the limit of a sum is the sum of the limits. Also, since $\sin x$ and $\cos x$ are not affected as h approaches 0 they can be passed by the limit sign.

$$f'(x) = \sin x \left[\lim_{h \to 0} \frac{(\cos h - 1)}{h} \right] + \cos x \left[\lim_{h \to 0} \frac{(\sin h)}{h} \right]$$

Apply equations (1) and (2).

$$f'(x) = \sin x \cdot (0) + \cos x \cdot (1)$$

$$= \cos x$$

This verifies that $D_x \sin x = \cos x$.

Periodic Motion

Periodic motion is a natural phenomenon that occurs frequently in various settings. Simple examples are the oscillating motion of a pendulum and the rise and fall of the population of an animal species. To see how trigonometric functions model periodic motion, consider a vertical spring of natural length L. If a body of mass m is attached to the end of the spring, the spring is stretched to a new length, $L + 1$, as shown in Figure 3. If the mass is pulled beyond its equilibrium (resting) position, then the spring will exert a force F to restore itself to the resting position. This force is proportional to the distance pulled y and is in the opposite direction. With k a positive constant of proportionality, this can be expressed as

$$F = -ky$$

By Newton's second law of motion the sum of the forces acting on the mass is given by $F = ma$, where a is the acceleration of the particle. By definition the acceleration is the second derivative of y: $a = D_x^2 y = y''(x)$. This yields the differential equation

$$my''(x) = -ky \tag{3}$$

The solution of this equation can be shown to be

$$y(x) = A \cos(ax + b)$$

for constants A, a, and b, with $a = \sqrt{k/m}$. To see that this equation is a solution of the differential equation, compute y''.

$$y'(x) = -Aa \sin(ax + b) \qquad y''(x) = -Aa^2 \cos(ax + b)$$

Substituting this result into equation (3) yields

$$my''(x) = -mAa^2 \cos(ax + b) = -mA \left(\frac{k}{m}\right) \cos(ax + b)$$

$$= -kA \cos(ax + b)$$

$$= -ky(x)$$

Let us apply this concept of periodic motion to sound. The variety of sounds produced by musical instruments, voices, radios, and television sets is too great to study in one model. But a simple sound, that of a tuning fork, can be modeled by an elementary trigonometric function. When a prong of a tuning fork is struck, both prongs oscillate inward and outward very rapidly. This motion is similar to a stretched spring that moves according to a restoring force. The movement of the prongs causes the air around them to move back and forth, which in turn causes pressure to build. This results in waves, called sound waves, spreading from the fork. The phenomenon is similar to moving a stick rapidly in a pool of water. The

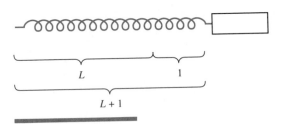

FIGURE 3

motion of the stick causes waves to spread through the pool. Just as the wave in the water is made up of the oscillating movement of individual water molecules, so too does a sound wave consist of the oscillating movement of individual air molecules.

The motion of a single molecule of air can be modeled after the motion of a stretched string. Its motion is given by the formula

$$y(x) = A \cos(2\pi ax + b)$$

The variable x is the time (measured in seconds) that the molecule is in motion and y is the **displacement,** or the molecule's distance (measured in inches) from its rest position. The constant A is the **amplitude,** which is the maximum displacement, and the constant a is the frequency, or the number of oscillations that the molecule makes per second. A typical formula for a simple sound is

$$y(x) = 0.001 \cos(2\pi 440x)$$
$$= 0.001 \cos(880\pi x)$$

This means that an air molecule oscillating in accordance with this formula moves back and forth about its rest position 440 times per second, and the farthest point it reaches from its rest position is 0.001 inch.

More complex sounds are made up of oscillating air molecules that move according to formulas such as

$$y(x) = A \cos(ax + b) + B \sin(cx + d)$$

For example, in the article "On the Study of Violin and Its Making," Itokawa and Kumagai consider the frequency characteristics of the violin.* The tone quality at particular pitches is studied and various graphs are generated. The data are approximated by different combinations of sine and cosine functions. For instance, one particular violin at the "Wolfe-note" pitch exhibited the following behavior:

$$y(x) = .003 \cos(450\pi x) + .013 \sin(410\pi x)$$

*Hideo Itokawa and Chihiro Kumagai, "On the Study of Violin and Its Making," *Musical Accoustics: Piano and Wood Instruments,* Earle Kent (ed.), Stroudsburg, PA, Dowden, Hutchinson and Ross, 1977.

EXERCISE SET 9–3

In Problems 1 to 6 compute the derivative of the function.

1. $f(x) = \sin 6x$

2. $f(x) = 4 \sin 6x$

3. $f(x) = \sin x^{-4}$

4. $f(x) = -2 \sin x^{-4}$

5. $f(x) = \sin(2x^4 - 5x^2)$

6. $f(x) = 3 \sin(x^{-1} - x^{1/2})$

In Problems 7 to 10 compute the derivative of the function.

7. $f(x) = x^2 \cos 6x$

8. $f(x) = 4x^3 \cos(-6x)$

9. $f(x) = (x + \cos 5x)^{-2}$

10. $f(x) = (x^2 + 2 \cos 3x)^{1/2}$

In Problems 11 to 16 compute the integral.

11. $\int 9 \cos 3x \, dx$

12. $\int 4 \cos(2x - 1) \, dx$

13. $\int x^2 \sin x^3 \, dx$

14. $\int x^{-2} \cos x^{-1} \, dx$

15. $\int 5 \cos^4 x \sin x \, dx$

16. $\int \dfrac{\cos x}{\sin x} \, dx$

In Problems 17 and 18 compute the definite integral.

17. $\int_0^{\pi/6} \sin x \cos^3 x \, dx$ 18. $\int_0^{\pi/4} \cos x \sin^2 x \, dx$

In Problems 19 and 20 find the area of the region bounded by the graphs of functions.

19. $f(x) = \sin x$ and $g(x) = \cos x$ from
$$x = -\frac{\pi}{4} \text{ to } x = \frac{\pi}{2}.$$

20. $f(x) = \sin x$ and $g(x) = \cos x$ from
$$x = -\frac{\pi}{4} \text{ to } x = \pi.$$

In Problems 21 to 26 find $f'(x)$.

21. $f(x) = \sin(e^x) + e^{\sin x}$

22. $f(x) = \ln(\sin x) + \sin(\ln x)$

23. $f(x) = (\sin 3x + e^{3x})^{1/3}$

24. $f(x) = \sin(\cos 3x)$ 25. $f(x) = \cos^2 x^3$

26. $f(x) = x \sin^2(4x + 1)$

In Problems 27 to 30 evaluate the integrals.

27. $\int e^x \cos(e^x) \, dx$ 28. $\int \frac{\sin(\ln x)}{x} \, dx$

29. $\int (\cos x)e^{\sin x} \, dx$ 30. $\int \frac{\sin 3x}{\cos 3x} \, dx$

In Problems 31 and 32 find $f'(x)$.

31. $f(x) = \sin(e^{3x}) + e^{\sin 3x}$

32. $f(x) = \sin^2(\cos 3x)$

33. Evaluate $\int \sin^3 x \, dx$

34. Find $f''(x)$ for $f(x) = \sin x \cos x$.

35. Compute $f'(x)$ where $f(x) = \sin^2 x + \cos^2 x$. Give a reason why $f'(x)$ is such a simple function.

36. A water trough is to be constructed from three metal sheets each 1 meter wide and 5 meters long plus end panels in the shape of trapezoids. Find the angle at which the long panels should be joined so as to provide a trough of maximum volume.

37. Use a calculator to verify that
$$\lim_{h \to 0} \left[\frac{\cos h - 1}{h} \right] = 0$$

Referenced Exercise Set 9–3

1. In describing how sound waves travel, Beranek shows sound pressure $p(x)$ when the air molecule is at position x satisfies the differential equation*
$$p''(x) = -w^2 p(x)$$
where w is a constant. Verify that a solution of this differential equation is $p(x) = \cos wt$.

2. When the read-write mechanism of one of IBM's first large mainframe computers developed failure problems, engineers had to solve the problem quickly or IBM would lose the trust and credibility of its customers. In an article describing how IBM uncovered and solved the problem, Piziali shows that the difficulty was in the failure of a part called the pickup-head link.† This is the device that picks up information stored magnetically and moves the information to another position in the computer; it might, for instance, be moved to be processed by the printer. The pickup-head link was failing because it was put under too much stress over long periods as it came into contact with the drum on which the information was stored. Upon contact, the link was deflected, or bent. The deflection $D(x)$ depended on the amount of pressure x, also referred to as shock, measured in pounds per square inch. The formula for $D(x)$ was
$$D(x) = \frac{.125}{\cos .04x^{1/2}}$$
The rate of change of $D(x)$ as x increased was critical. Compute $D'(x)$.

Cumulative Exercise Set 9–3

1. Convert the angles from radians to degrees and express the angle in terms of the number of revolutions that it makes.
   (a) $11\pi/2$
   (b) $8\pi/3$
   (c) $-11\pi/4$

2. Find the radian measure of the angle whose measure is given in degrees.
   (a) $300°$
   (b) $750°$
   (c) $-1080°$

*Leo L. Beranek, *Noise Reduction,* New York, McGraw-Hill, 1980, pp. 20–25.
†R. Piziali, "The Pickup-head Link Failure," *Engineering Case Library,* Washington, DC, American Society for Engineering Education, No. 186, 1972, pp. 1–16.

3. Find the radian measure of the angle that makes the given amount of revolutions in a clockwise direction.
   (a) two and one-third revolutions
   (b) four and one-sixth revolutions
   (c) three and seven-eighths revolutions

4. The shaft of a rock-cutting machine rotates at 80 revolutions per minute. Find the angular speed of the shaft.

5. Compute the functional value.
   (a) $2 + 3 \sin 4\pi/3$
   (b) $2 \cos 4\pi/3 + 3 \sin (-4\pi/3)$

6. Use Table 5 in the Appendix to find the functional values.
   (a) $\sin .1$
   (b) $\cos \frac{1}{5}$

7. Find a number x such that $0 < x < \frac{\pi}{2}$ and x satisfies $\sin x = \cos (-3\pi/2)$.

8. Sketch the graph of the function $f(x) = 1 - 2 \sin x$.

9. Compute the derivative of each function.
   (a) $f(x) = \sin 3x + 2 \cos 4x$
   (b) $f(x) = \sin x^{-2}$
   (c) $f(x) = x^2 \cos(-2x)$

10. Compute the integral.
    (a) $\displaystyle\int 9 \sin(3x) \, dx$
    (b) $\displaystyle\int x^2 \cos x^3 \, dx$

11. Compute the definite integral.
    $$\int_0^{\pi/3} \sin x \cos^4 x \, dx$$

12. Evaluate the integral.
    (a) $\displaystyle\int e^{2x} \sin(e^{2x}) \, dx$
    (b) $\displaystyle\int \frac{\cos(\ln x) \, dx}{x}$

9–4   The Other Trigonometric Functions

There are four trigonometric functions in addition to the sine and cosine. In this section we define them and compute their derivatives. Of the remaining four functions the tangent is the most important, in part because each of the other three functions, the cosecant, secant, and cotangent, is the reciprocal of the sine, cosine, and tangent, respectively. For this reason most of our attention will be focused on the tangent.

Definitions

The product and quotient of the sine and cosine functions appear so often that some of them are given specific names for easy reference. Together with their abbreviations, these functions are the **tangent** function, expressed as tan x; the **cotangent** function, or cot x; the **secant** function, or sec x; and the **cosecant** function, or csc x. They are defined as follows:

$$\tan x = \frac{\sin x}{\cos x} \qquad \cot x = \frac{\cos x}{\sin x}$$

$$\sec x = \frac{1}{\cos x} \qquad \csc x = \frac{1}{\sin x}$$

Each function is defined for all values of x except where the denominator equals 0. For instance, cot x is defined for all x except where $\sin x = 0$. That is, cot x is not defined for $x = 0$, π, 2π, and, in general, all $x = n\pi$, where n is an integer.

A number of identities involving the trigonometric functions are studied in a course of trigonometry. In this section we cover only one. Several additional identities are mentioned in the exercises. To derive the identity, start with the identity derived in the previous section.

$$\sin^2 x + \cos^2 x = 1$$

Divide each term in this equation by $\cos^2 x$.

$$\frac{\sin^2 x}{\cos^2 x} + \frac{\cos^2 x}{\cos^2 x} = \frac{1}{\cos^2 x}$$

The first term equals $\tan^2 x$, which is standard notation for $(\tan x)^2$. The second term equals 1, and the third term equals $\sec^2 x$. This yields

$$\tan^2 x + 1 = \sec^2 x$$

The right triangle interpretation of the **tangent** is derived from the definitions of the sine and cosine. Recall that the sine is the ratio of the length of the opposite side and the length of the hypotenuse. The cosine is the ratio of the length of the adjacent side and the length of the hypotenuse. From Figure 1 we get

$$\tan x = \frac{\sin x}{\cos x} = \frac{(\text{opposite side})/(\text{hypotenuse})}{(\text{adjacent side})/(\text{hypotenuse})}$$

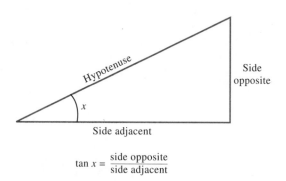

$$\tan x = \frac{\text{side opposite}}{\text{side adjacent}}$$

FIGURE 1

Therefore when x is restricted to $0 < x < \pi/2$, we have the following formula:

$$\tan x = \frac{\text{opposite side}}{\text{adjacent side}}$$

Example 1 illustrates how to evaluate the tangent function at some values of x.

EXAMPLE 1

Problem

For the angle x in Figure 2 find (a) tan x, (b) cot x, (c) sec x, (d) csc x.

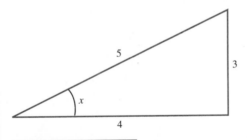

FIGURE 2

Solution (a) Using the formula with the length of the opposite side 3 units and the length of the adjacent side 4 units, tan $x = 3/4$.
(b) Since cot $x = (\cos x)/(\sin x)$, we see that cot $x = 1/\tan x$. Thus cot $x = 1/(3/4) = 4/3$.
(c) By definition sec $x = 1/\cos x$. Since cos $x = 4/5$, sec $x = 5/4$.
(d) By definition csc $x = 1/\sin x$. Since sin $x = 3/5$, csc $x = 5/3$.

The Graph of the Tangent

The tangent is defined for all x where cos $x \neq 0$. That is, tan x is not defined for $x = \pi/2 + 2n\pi$ for all integers n. At these values tan x has a vertical asymptote because the denominator approaches 0 and the numerator approaches 1 as x approaches any of these values. The tangent is a periodic function with period π. The graph of tan x is given in Figure 3.

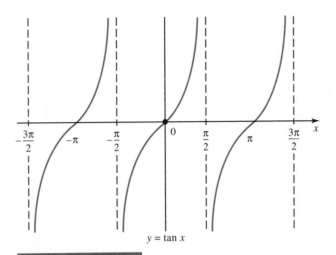

$y = \tan x$

FIGURE 3

The Derivative of the Tangent

The formula for the derivative of tan x can be derived from its definition by using the quotient rule.

$$D_x \tan x = D_x \left[\frac{\sin x}{\cos x} \right]$$

$$= \frac{(\cos x)(\cos x) - (\sin x)(-\sin x)}{(\cos x)^2}$$

$$= \frac{\cos^2 x + \sin^2 x}{\cos^2 x}$$

$$= \frac{1}{\cos^2 x} = \sec^2 x$$

Thus the derivative of tan x is $\sec^2 x$. This fact and the tangent version of the chain rule are recorded in the next display.

$$D_x \tan x = \sec^2 x$$

If u is a function of x, then

$$D_x \tan u = \sec^2 u \, D_x u$$

EXAMPLE 2

Problem

Compute the derivative of the functions: (a) $f(x) = \tan 5x$, (b) $f(x) = \tan x^2$, (c) $f(x) = \tan(x^3 - 5x)$.

Solution (a) Let $u = 5x$. Then $D_x u = D_x 5x = 5$. Thus

$$f'(x) = D_x \tan 5x = (\sec^2 5x)5 = 5 \sec^2 5x$$

(b) Let $u = x^2$. Then $D_x u = D_x x^2 = 2x$. Thus

$$f'(x) = D_x \tan x^2 = (\sec^2 x^2)2x = 2x \sec^2 x^2$$

(c) Let $u = x^3 - 5x$. Then $D_x u = D_x(x^3 - 5x) = 3x^2 - 5$. Thus

$$f'(x) = D_x \tan(x^3 - 5x)$$

$$= [\sec^2(x^3 - 5x)](3x^2 - 5)$$

$$= (3x^2 - 5) \sec^2(x^3 - 5x)$$

The next example shows how to use the derivative of the tangent with other differentiation formulas.

EXAMPLE 3

Problem

Compute the derivative of the functions: (a) $f(x) = \tan^3 5x$, (b) $f(x) = (1 - \tan x^2)^{-4}$.

Solution (a) Use the extended power rule with $u = \tan 5x$ and $n = 3$. The derivative of $\tan 5x$ was computed in Example 2. Thus

$$f'(x) = D_x \tan^3 5x = 3 (\tan^2 5x)(D_x \tan 5x)$$
$$= 3(\tan^2 5x)(5 \sec^2 5x)$$
$$= 15(\tan^2 5x)(\sec^2 5x)$$

(b) Use the extended power rule with $u = 1 - \tan x^2$ and $n = -4$. The derivative of $\tan x^2$ was computed in Example 2. Thus

$$f'(x) = D_x(1 + \tan x^2)^{-4}$$
$$= -4(1 + \tan x^2)^{-5}D_x(\tan x^2)$$
$$= -4(1 + \tan x^2)^{-5}(2x \sec^2 x^2)$$
$$= -8x(1 + \tan x^2)^{-5}(\sec^2 x^2)$$

Derivatives of the Other Functions

The derivatives of the other trigonometric functions can also be derived by the quotient rule. The exercises will ask for their derivations. The formulas are

$$D_x \cot x = -\csc^2 x$$

$$D_x \sec x = \sec x \tan x$$

$$D_x \csc x = -\csc x \cot x$$

Incorporating these formulas into the chain rule yields the following, where u is a function of x.

$$D_x \cot u = -\csc^2 u \, D_x u$$

$$D_x \sec u = \sec u \tan u \, D_x u$$

$$D_x \csc u = -\csc u \cot u \, D_x u$$

The next example illustrates how to apply these formulas.

EXAMPLE 4

Problem

Compute the derivative of the functions: (a) $f(x) = \cot 5x$, (b) $f(x) = \cot^4 5x$, (c) $f(x) = \sec 2x + \csc x^2$.

Solution (a) Use the formula for $D_x \cot u$ with $u = 5x$. Then $D_x u = 5$. Thus

$$f'(x) = D_x \cot 5x = -\csc^2 5x \, D_x 5x$$
$$= -5 \csc^2 5x$$

(b) Use the extended power rule with $u = \cot 5x$ and $n = 4$. The derivative of $\cot 5x$ was computed in part (a). Thus

$$f'(x) = D_x \cot^4 5x = 4(\cot^3 5x)(D_x \cot 5x)$$

$$= 4(\cot^3 5x)(-5 \csc^2 5x)$$

$$= -20(\cot^3 5x)(\csc^2 5x)$$

(c) Use the derivative of a sum rule.

$$f'(x) = D_x(\sec 2x + \csc x^2)$$

$$= D_x \sec 2x + D_x \csc x^2$$

$$= (\sec 2x \tan 2x)D_x 2x + (-\csc x^2 \cot x^2)D_x x^2$$

$$= 2 \sec 2x \tan 2x - 2x \csc x^2 \cot x^2$$

EXERCISE SET 9–4

In Problems 1 to 4 find the given functional value of angle a where a is the angle in the accompanying figure.

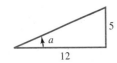

12

1. $\tan a$ 2. $\cot a$

3. $\sec a$ 4. $\csc a$

In Problems 5 to 12 compute the derivative of the function.

5. $f(x) = \tan(5x + 1)$ 6. $f(x) = \tan(x^2 + 5x)$

7. $f(x) = \tan(x^3 - 5x)$

8. $f(x) = \tan(x^{-3} - e^x)$

9. $f(x) = \tan^5 4x$ 10. $f(x) = \tan^{1/5} 5x$

11. $f(x) = (3x + \tan x^2)^{1/2}$

12. $f(x) = (\sin x + \tan 2x)^{-2}$

In Problems 13 to 15 derive the formula for the derivative of the given trigonometric function.

13. $f(x) = \cot x$ 14. $f(x) = \sec x$

15. $f(x) = \csc x$

In Problems 16 to 30 compute the derivative of the function.

16. $f(x) = \cot(5x + 3)$ 17. $f(x) = \sec(x^2 + 3x)$

18. $f(x) = \csc(x^2 - 5)$

19. $f(x) = \cot(x^{-3} - e^x)$

20. $f(x) = \sec^3 2x$ 21. $f(x) = \csc^{1/2} 3x$

22. $f(x) = (3x + \cot x^3)^{-2}$

23. $f(x) = \ln \sec x$ 24. $f(x) = e^{\csc x}$

25. $f(x) = e^{\tan 6x}$

26. $f(x) = \ln \cot(4x - 1)$

27. $f(x) = \cot e^{3x}$

28. $f(x) = \tan \ln(3x + 1)$

29. $f(x) = \tan 2x \cot 3x$

30. $f(x) = \sec 4x \tan(4x + 1)$

31. Express the differentiation formulas in this section as antidifferentiation formulas.

In Problems 32 to 38 evaluate the integral $\int f(x)\, dx$.

32. $f(x) = \sec^2 3x$

33. $f(x) = \sec 4x \tan 4x$

34. $f(x) = x \sec^2 x^2$ 35. $f(x) = x^2 \csc^2 x^3$

36. $f(x) = \sec^2 x \tan x$

37. $f(x) = \csc^2 x \cot^2 x$

38. $f(x) = \tan 2x$

Referenced Exercise Set 9–4

1. Sound propagation in water has many applications, from sonar to fishing techniques. In an article studying the individual sound characteristics of the Mediterra-

nean Sea, Leroy measured the relationship between the range $R(x)$ of a ray of sound when emitted into the sea at an angle x with the surface.* The relationship was found to be

$$R(x) = a \tan x + b \tan x^{1/2}$$

where a and b are constants depending on the time of year and the position on the sea. The rate of change of $R(x)$ is an important measure. Compute $R'(x)$.

2. There are many ways to describe how musical sounds are transmitted. In one such model of the sound of a piano, the impact of the hammer on a string causing it to vibrate is regarded as an electrical transmission.† The imparting of energy to the string by the hammer blow is analogous to connecting an inductance across a transmission line between its terminals. The impedance $Z(x)$ is a function of length of the string x given by the relationship

$$Z(x) = a \cot(2\pi bx)$$

where a and b are constants. Compute $Z'(x)$.

Cumulative Exercise Set 9–4

1. Convert the angle $300°$ to radians.

2. How many revolutions does the hour hand on a clock make if the hour hand makes an angle of $\frac{5}{2}\pi$ with 12 o'clock?

3. Use a calculator or Table 5 in the Appendix to find the value of (a) $1 + \cos 20°$ and (b) $\sqrt{2} \sin(\frac{\pi}{4})$.

4. Sketch the graph of the function $f(x) = 1 - \frac{1}{2}\sin x$.

In Problems 5 to 8 compute the derivative of the function.

5. $f(x) = \frac{x}{2} \sin \pi x$

6. $f(x) = \sin 2x \cos \frac{x}{2}$

7. $f(x) = \sqrt{\tan 5x}$

8. $f(x) = \dfrac{2}{\sin 2x}$

In Problems 9 to 11 compute the integral.

9. $\displaystyle\int \sin \pi x \, dx$

10. $\displaystyle\int \sin 2x \cos 2x \, dx$

11. $\displaystyle\int \sec^2 x \, dx$

12. Find the area of the region between the curve $y = \cos x$ and the x-axis from $x = -\pi$ to $x = \pi$.

*C. C. Leroy, ''Sound Propagation in the Mediterranean Sea,'' *Underwater Accoustics*, Vol. 2, V. M. Albers (ed.), New York, Plenum Press, 1977, pp. 203–234.

†N. H. Fletcher, ''A Vibrating String as an Electrical Transmission,'' *Musical Accoustics: Piano and Wind Instruments*, Earle L. Kent (ed.), Stroudsburg, PA, Dowden, Hutchinson and Ross, 1977, pp. 31–39.

CASE STUDY   Melbourne's Underground Railway: Shaft Design*

All too often students equate mathematics with a magic wand—if you have a problem, fetch a math formula and ''presto!''—out pops the solution. Real-world applications could not be further from this scenario. The development of most models involves trial-and-error techniques, mixed with a large dose of frustration. This Case Study will give you a glimpse of what problem solving in the real world is like.

Melbourne's Tunnel System   In the late 1970s the city government of Melbourne, Australia, decided to construct a large tunnel system beneath the city. One part of the system would consist of an underground railway and another

*Adapted from W. P. Lewis, ''A Case Study in Shaft Design,'' *Engineering Case Library*, Washington, DC, American Society for Engineering Education, No. ECL 240, 1984.

would carry sewerage to a new treatment plant. The tunnel-boring operations revealed a lack of knowledge about the forces that are generated when cutting through different types of rock strata. The escalating costs and incessant delays were the cause of politicans' heated rhetoric and taxpayers' lament. A research program on rock-cutting had to be started immediately to design a special machine for the tunnel-boring task.

To keep expenditures in line with projected budgets, tunneling rates through the various types of rock strata had to be predicted accurately. An important variable was the wear of the cutting tools, because there would be great stress on them. Accurate predictions of wear would enable researchers to plan efficiently for maintenance shutdowns. Exacerbating the problem was the fact that relevant information for local operating conditions was nonexistent. So an experimental test machine would have to be designed and tested. A young master's degree student, Gray Bailey, was given the task of designing and constructing the experimental machine.

Bailey had confidence in most aspects of his design, but he wanted an expert opinion on the major design feature, the shaft that operates the cutting head. Since most of the stress of the cutting process would be on the shaft, the force on the shaft would have to be accurately determined. Bailey hired a University of Melbourne researcher, Ian McPherson, to help in the shaft design. The key decision was the diameter of the shaft. If it was too small, it would not wear well. But too large a shaft would decrease the efficiency of the motor and adversely affect the tunneling rate as well as the wear on the motor.

In their first meeting McPherson asked Bailey for as many details of the project as he could relate. Bailey described the type of machine they were considering. "We have an electric motor available," said Bailey, "and based on research the maximum tangential force on the tool during cutting is most unlikely to exceed 50 kN (kiloNewtons). If you allow another 5% margin on this, I am sure your calculations would be very safe."

Realizing that clients are frequently apt to forget or overlook relevant pieces of information, McPherson often extended the meeting with questions such as "Do you mind if I ask once again if that is all you can tell me?"

After several of these prompts, Bailey assured him, "Now you know as much as I do!" They parted, and McPherson started the design. He felt confident he could solve the problem. Figures 1 and 2 show McPherson's schematic of the shaft layout. As the shaft turns, the cutting edges of the tool bore into the rock and essentially scrape it away. This puts tangential force F_t on the cutting edge of the tool. To calculate the maximum torque F, he used the formula

$$F = F_t r$$

where r is the radius of the tool. Since $r = 1.09$ and Bailey said to take F_t as $50 + .05(50) = 52.5$, McPherson calculated the maximum torque to be

$$F = 52.5(1.09) = 57.2$$

As he was about to turn his attention to the force on the shaft, McPherson realized he did not know the direction of rotation with respect to the motor.

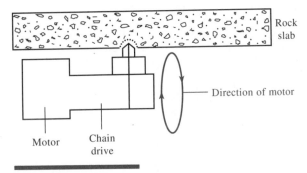

FIGURE 1

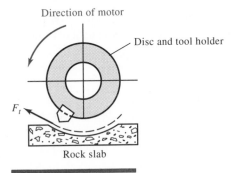

FIGURE 2

Should the rotation be through a negative (clockwise) angle of rotation or a positive (counterclockwise) angle? A phone call to Gray Bailey determined that Bailey had given no thought to the position of the drive motor. McPherson decided to base his design on a positive angle of rotation because it represented a more adverse load on the motor. If the client decided on the alternate design, there would be less load, so the design would still have ample support.

Next McPherson had to determine the force directly on the shaft. There were several points on the shaft that were subject to stress. He started with the force on the sprocket. Since F is the maximum torque, the force on the shaft would be the sum of the vertical components of the force F at each point of the circumference. Suppose the cutting edge of the tool makes an angle x with the vertical axis. He represented the tangential force per unit of length of circumference by f. Assume f is constant over the entire circumference. Let dF represent the force on a small part of the sprocket that causes the tool to rotate. If this small part of the sprocket on the circumference subtends an angle dx, then it has length $r\,dx$. Then $dF = fr\,dx$. Represent the vertical component of the force by F_v and the horizontal component by F_h. The force on the shaft at the sprocket is F_v. Then

$$dF_v = fr\,dx\,\sin x \qquad \text{and} \qquad F_h = fr\,dx\,\cos x$$

The sum of the individual forces dF is F. Since the force acts when the sprockets are engaged, it acts on the shaft from $x = 0$ to $x = \pi$, that is, from the point where the edge is facing down to where it is vertical as it rotates clockwise. This sum is given by

$$F_v = \int dF_v = \int_0^\pi fr \sin x \, dx = fr \int_0^\pi \sin x \, dx = fr(\cos x) \Big|_0^\pi = 2fr$$

$$F_h = \int dF_h = \int_0^\pi fr \cos x \, dx = fr \int_0^\pi \cos x \, dx = fr(\sin x) \Big|_0^\pi = 0$$

The maximum torque is $F = \pi rf \approx 57.2$. Solving this equation for f yields

$$f = \frac{57.2}{\pi r}$$

The first integral above states that the force on the shaft is $F_v = 2fr$. Substituting the preceding equation into this equation yields

$$F_v = 2fr = 2r \left(\frac{57.2}{\pi r} \right) = \frac{114.4}{\pi} \approx 36.4$$

This is the value of the force on the shaft at the sprocket.

McPherson then isolated three additional points on the shaft where there would be significant forces. He computed these forces in a similar manner. The equation for the shaft diameter was reduced to

$$d = 8.58n^{1/3} \text{ cm}$$

where n is a safety factor. He was unsure what value of n to use because the local conditions of rock strata were unknown. He decided to graph d as a function of n (Figure 3). From the graph he felt that a diameter of 14 cm would be reasonable because it would give a safety factor of just greater than 4.

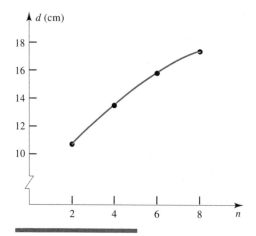

FIGURE 3

McPherson called a meeting with Bailey and presented his design, including a shaft diameter of 14 cm. Ever cautious, McPherson kept asking himself, even during this "final" conversation, "What else could go wrong?" Since the shaft cutting edge scrapes along the rock, the shaft will tend to bend in the direction opposite from the scraping. This is called deflection.

"I just realized I forgot to ask you to specify the maximum allowable deflection," said McPherson.

Bailey was jolted. "Will the shaft deflect?"

"Of course it will! You've got high loads and a long bearing span. I feel deflection will be significant."

"Will it? Oh well, I'll leave it to you. You're the expert."

McPherson muttered, "There's more to this shaft design problem than meets the eye."

Back to the drawing board. The calculations had to be reworked with deflection parameters included. McPherson's final calculation showed that to account for the maximum allowable deflection, d would have to be greater than 16.7. To allow a little margin of additional safety, he recommended a shaft of 18 cm. Bailey, however, felt that this would make the shaft too heavy. McPherson analyzed whether the design could be amended with support for the shaft. This led to a radical new design, which not only supported the shaft but also essentially eliminated the torque on the shaft, thus making it a much more efficient and reliable machine.

Bailey's experimental machine was built with this new design and it worked well. Construction began soon after tests showed the machine would run efficiently. If you ever visit Melbourne, take a ride on the underground railway, built in part by the McPherson–Bailey tunnel-boring machine.

Case Study Exercises

In Problems 1 to 4 compute the maximum torque on the shaft for the given values of r and F_t.

1. $r = 1.1$, $F_t = 52$
2. $r = 1.5$, $F_t = 55$
3. $r = 1.25$, $F_t = 51.5$
4. $r = 1.15$, $F_t = 50.5$

In Problems 5 to 8 compute the vertical component F_v of F for the given values of r and F_t.

5. $r = 1.1$, $F_t = 52$
6. $r = 1.5$, $F_t = 55$
7. $r = 2.5$, $F_t = 50$
8. $r = 2.15$, $F_t = 50.5$

CHAPTER REVIEW

Key Terms

9–1 Radian Measure
Radian   Angular Speed

9–2 The Sine and Cosine Functions
Sine   Identity
Cosine

9–3 Differentiation and Integration
Periodic Motion   Amplitude
Displacement

9–4 The Other Trigonometric Functions
Tangent   Secant
Cotangent   Cosecant

Summary of Important Concepts

$$r \text{ (radians)} = \frac{\pi}{180} d \text{ (degrees)}$$

$$\sin^2 + \cos^2 x = 1$$

$$\sin(x + 2n\pi) = \sin x$$

$$\cos(x + 2n\pi) = \cos x$$

$$\sin(-x) = -\sin x$$

$$\cos(-x) = \cos x$$

$$\sin(x + y) = \sin x \cos y + \cos x \sin y$$

$$\sin\left(\frac{\pi}{2} - x\right) = \cos x$$

$$\cos\left(\frac{\pi}{2} - x\right) = \sin x$$

$$D_x \sin u = \cos u \, D_x u$$

$$D_x \cos u = -\sin u \, D_x u$$

$$D_x \tan u = \sec^2 u \, D_x u$$

$$D_x \cot u = -\csc^2 u \, D_x u$$

$$D_x \sec u = \sec u \tan u \, D_x u$$

$$D_x \csc u = \csc u \cot u \, D_x u$$

REVIEW PROBLEMS

1. Convert the angles from degrees to radians:
   (a) 135°, (b) 210°.

2. Convert the angles from radians to degrees:
   (a) $\dfrac{8\pi}{3}$, (b) $\dfrac{11\pi}{6}$.

3. Convert the angles from radians to degrees and express the angle in terms of the revolutions that it makes: (a) $\dfrac{11\pi}{3}$, (b) $\dfrac{25\pi}{6}$.

4. Find the radian measure of the angle that makes two and one-third revolutions in a clockwise direction.

5. Turning a combination lock two and one-sixth revolutions corresponds to an angle of what radian measure?

In Problems 6 to 9 find the functional value.

6. $\cos\left(-\dfrac{4\pi}{3}\right)$

7. $\sin\dfrac{15\pi}{4}$

8. $\sin .7$

9. $\cos\dfrac{1}{10}$

10. Find a number x such that $0 \le x \le \dfrac{\pi}{2}$ and $\sin x = \sin 5\pi$.

In Problems 11 and 12 compute the derivative of the function.

11. $f(x) = 4\sin(x^{-2} - x^{1/3})$

12. $f(x) = (x + 2\cos 3x)^{1/4}$

In Problems 13 and 14 compute the integral.

13. $\displaystyle\int x^3 \sin x^4\, dx$

14. $\displaystyle\int \dfrac{\sin x}{\cos^2 x}\, dx$

15. Find $f'(x)$: $f(x) = \ln(\cos x) + \sin(\ln x)$.

In Problems 16 to 20 compute the derivative of the function.

16. $f(x) = \tan(x^2 + 5x)$

17. $f(x) = (\sec x + \tan 2x)^{-2}$

18. $f(x) = \sec^4 2x$ 19. $f(x) = e^{\csc x}$

20. $f(x) = \sec 2x \tan(3x + 2)$

GRAPHICS CALCULATOR EXPLORATIONS

Graphs of functions involving sin x are very common throughout mathematics and its applications. Starting with the function $y = \sin x$ you can use your graphics calculator to develop a facility for understanding more complicated trigonometric functions, such as $y = x \sin x$ and $y = x^2 \sin x$.

One of the most important properties of the sine function is its oscillatory behavior. The graph of $y = \sin x$ oscillates between the upper boundary $y = 1$ and the lower boundary $y = -1$. The function $y = 2\sin x$ oscillates between the bounds $y = 2$ above and $y = -2$ below. What is the behavior of the graph of $y = 0.5\sin x$? Graph these three functions in the same viewing screen, making sure the calculator is in radian mode and that the range is set to adequately see the oscillatory behavior of each function.

Casio

1. Press [GRAPH] [SIN] [EXE]

2. Press [GRAPH] 2 [SIN] [ALPHA] [x] [EXE]

3. Press [GRAPH] .5 [SIN] [ALPHA] [x] [EXE]

TI-81

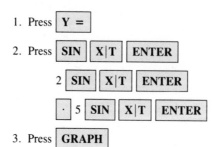

1. Press Y =

2. Press SIN X|T ENTER

2 SIN X|T ENTER

. 5 SIN X|T ENTER

3. Press GRAPH

In Example 1 we investigate the behavior of functions of the form $y = a \sin x$ for a constant a.

Example 1

What is the behavior of $y = 5 \sin x$? It has a similar graph to the graphs in the calculator display above. Thus the boundaries between which it oscillates are $y = 5$ and $y = -5$. In the same viewing screen graph $y = 5 \sin x$ and its two boundaries, $y = 5$ and $y = -5$.

Casio

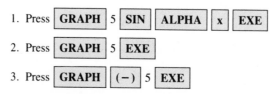

1. Press GRAPH 5 SIN ALPHA x EXE

2. Press GRAPH 5 EXE

3. Press GRAPH (−) 5 EXE

TI-81

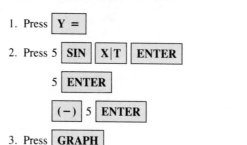

1. Press Y =

2. Press 5 SIN X|T ENTER

5 ENTER

(−) 5 ENTER

3. Press GRAPH

In general, the graph of $y = a \sin x$ has the two boundaries $y = a$ and $y = -a$.

Duplicate the above calculations with a few additional functions to investigate their behavior. In the following examples we will build on them. Example 2 compares their graphs with a more complicated function.

Example 2

What is the behavior of the graph of $y = x \sin x$? Since the function is not a constant times $\sin x$ you would expect its graph to be different from those in the preceding display, but it is important to see how similar its behavior is to the function $y = a \sin x$. The function $y = x \sin x$ also has an oscillatory behavior, but its bounds are not horizontal lines, they are the two lines $y = x$ and $y = -x$. Try to picture what the viewing screen will look like with these three graphs in it, then graph them and determine how close your guess was.

Casio

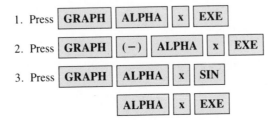

TI-81

1. Press $\boxed{Y =}$

2. Press $\boxed{X|T}$ $\boxed{ENTER}$

$\boxed{(-)}$ $\boxed{X|T}$ $\boxed{ENTER}$

$\boxed{X|T}$ $\boxed{SIN}$ $\boxed{X|T}$ $\boxed{ENTER}$

3. Press $\boxed{GRAPH}$

  Can you give an algebraic explanation why the graph of $y = x \sin x$ has the bounds $y = x$ and $y = -x$?

Example 3

What is the behavior of the graph of the function $y = (x + 1) \sin x$? Its graph is similar to $y = x \sin x$ except that it oscillates between the boundaries $y = x + 1$ and $y = -x - 1$. Graph these three functions in the same viewing screen.

Casio

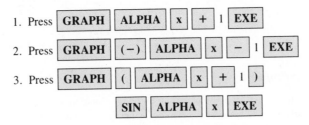

TI-81

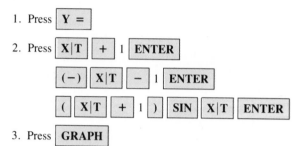

1. Press $\boxed{Y =}$

2. Press $\boxed{X|T}$ $\boxed{+}$ 1 $\boxed{\text{ENTER}}$

   $\boxed{(-)}$ $\boxed{X|T}$ $\boxed{-}$ 1 $\boxed{\text{ENTER}}$

   $\boxed{(}$ $\boxed{X|T}$ $\boxed{+}$ 1 $\boxed{)}$ $\boxed{\text{SIN}}$ $\boxed{X|T}$ $\boxed{\text{ENTER}}$

3. Press $\boxed{\text{GRAPH}}$

In the Further Explorations we ask you to predict the behavior of the graphs of $y = (ax + b) \sin x$. They have an oscillatory behavior. What are their boundaries?

Example 4

What is the behavior of the graph of the function $y = x^2 \sin x$? Its graph is similar to $y = x \sin x$ except that it oscillates between the boundaries $y = x^2$ and $y = -x^2$. Graph these three functions in the same viewing screen.

Casio

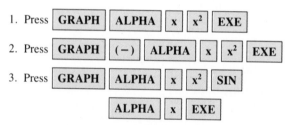

1. Press $\boxed{\text{GRAPH}}$ $\boxed{\text{ALPHA}}$ $\boxed{x}$ $\boxed{x^2}$ $\boxed{\text{EXE}}$

2. Press $\boxed{\text{GRAPH}}$ $\boxed{(-)}$ $\boxed{\text{ALPHA}}$ $\boxed{x}$ $\boxed{x^2}$ $\boxed{\text{EXE}}$

3. Press $\boxed{\text{GRAPH}}$ $\boxed{\text{ALPHA}}$ $\boxed{x}$ $\boxed{x^2}$ $\boxed{\text{SIN}}$

   $\boxed{\text{ALPHA}}$ $\boxed{x}$ $\boxed{\text{EXE}}$

TI-81

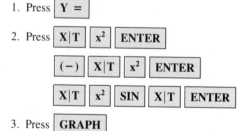

1. Press $\boxed{Y =}$

2. Press $\boxed{X|T}$ $\boxed{x^2}$ $\boxed{\text{ENTER}}$

   $\boxed{(-)}$ $\boxed{X|T}$ $\boxed{x^2}$ $\boxed{\text{ENTER}}$

   $\boxed{X|T}$ $\boxed{x^2}$ $\boxed{\text{SIN}}$ $\boxed{X|T}$ $\boxed{\text{ENTER}}$

3. Press $\boxed{\text{GRAPH}}$

In general, what is the behavior of $y = f(x) \sin x$? Example 5 gives one more illustration with a function whose graph is a bit more unusual.

Example 5

What is the behavior of the graph of the function $y = (\cos x) \sin x$? It oscillates between the boundaries $y = \cos x$ and $y = -\cos x$. Graph these three functions in the same viewing screen.

Casio

1. Press $\boxed{\textbf{GRAPH}}$ $\boxed{\textbf{COS}}$ $\boxed{\textbf{EXE}}$

2. Press $\boxed{\textbf{GRAPH}}$ $\boxed{\textbf{(−)}}$ $\boxed{\textbf{COS}}$ $\boxed{\textbf{ALPHA}}$ $\boxed{\textbf{x}}$ $\boxed{\textbf{EXE}}$

3. Press $\boxed{\textbf{GRAPH}}$ $\boxed{\textbf{COS}}$ $\boxed{\textbf{ALPHA}}$ $\boxed{\textbf{x}}$ $\boxed{\textbf{SIN}}$

   $\boxed{\textbf{ALPHA}}$ $\boxed{\textbf{x}}$ $\boxed{\textbf{EXE}}$

TI-81

1. Press $\boxed{\textbf{Y =}}$

2. Press $\boxed{\textbf{COS}}$ $\boxed{\textbf{X|T}}$ $\boxed{\textbf{ENTER}}$

   $\boxed{\textbf{(−)}}$ $\boxed{\textbf{COS}}$ $\boxed{\textbf{X|T}}$ $\boxed{\textbf{ENTER}}$

   $\boxed{\textbf{COS}}$ $\boxed{\textbf{X|T}}$ $\boxed{\textbf{SIN}}$ $\boxed{\textbf{X|T}}$ $\boxed{\textbf{ENTER}}$

3. Press $\boxed{\textbf{GRAPH}}$

Further Explorations

In problems 1 to 12 graph the given function and determine the bounds of the graph.

1. $y = 10 \sin x$
2. $y = (x - 1) \sin x$
3. $y = (2x + 1) \sin x$
4. $y = 2x^2 \sin x$
5. $y = x^3 \sin x$
6. $y = (x^2 + 1) \sin x$
7. $y = 5 \cos x$
8. $y = x \cos x$
9. $y = x^2 \cos x$
10. $y = (\cos x + 1) \sin x$
11. $y = (x - 1) \sin x$
12. $y = (2x + 1) \sin x$
13. Describe the graph of $y = (ax + b) \sin x$.
14. Describe the graph of $y = (ax + b) \cos x$.

Appendix A: The Graphing Calculator

The advent of inexpensive handheld graphics calculators and programmable calculators is revolutionizing the way we teach and learn mathematics. It is advantageous for every student to learn how to use these extremely powerful and convenient machines. Their potential is enormous.

There is a section of Graphics Calculator Explorations at the end of each chapter that illustrates some geometric aspect of that chapter. Each Exploration includes several problems that invite you to use these calculators to explore other ideas in that chapter more fully.

This appendix is designed to get you started. It is by no means meant to replace your calculator's manual. Once you become familiar with your calculator you will have fun designing your own experiments, and once you know how to accomplish a few simple tasks the manual will not seem as intimidating.

There are several outstanding calculators on the market. We illustrate those by Casio and Texas Instruments (the TI-81.) As the technology improves and costs plummet, more manufacturers will join the market. Each has its own advantages and we do not recommend one over the others. Although all graphics calculators accomplish the same tasks, they have slightly different keyboards and capabilities. Even if your calculator is different from the Casio or the TI-81, the examples should be self-explanatory, and a brief perusal of your manual should enable you to translate the actual key sequences to fit your calculator.

One of the most powerful and compelling capabilities of these calculators is their ability to sketch almost instantaneously the graph of any function. The calculators have "built-in" functions that you can graph by simply pressing the appropriate keys. The remainder of this appendix will demonstrate, separately for the Casio and TI-81, how to sketch the graphs of such functions.

Casio

To sketch the graph $y = x^2$ press the following:

| Graph | x^2 | EXE |

To sketch the graph $y = x^{-1}$ press the following:

| Graph | x^{-1} | EXE |

It is important to keep in mind that whenever sketching the graph of the function you must set the range on the x-axis and the range on the y-axis. When sketching the graph of one of the built-in functions the calculator uses preset ranges. Only the part of the graph within these ranges appears on the screen. If you want to see any other part of the graph, or wish to magnify the graph, you must overwrite the preset ranges.

This is not a new idea. When graphing a function on a piece of paper, you determine the ranges. For example suppose you wanted to sketch the graph of the following system of linear equations:

$$3x + y = 9$$
$$2x + y = 8$$

You would most likely find the intercepts first; the x-intercepts are 3 and 4 and the y-intercepts are 9 and 8. You would then determine how large you want the graph to be on your paper and then plot the largest intercept, 9, just inside the range you decided to have for the graph. In other words, a little thought and computation goes into the decision of the magnitude of the range of your graph on a piece of paper. The same type of care must be taken when sketching graphs with the calculator.

To set the ranges on the axes, press the | Range | key and the following appears:

Range
    Xmin: -5
       max:5
       scl:2
    Ymin: -10
       max:5
       scl:2

Xmin represents the leftmost point on the x-axis, Xmax the rightmost point on the x-axis, and Ymin and Ymax the lowermost and uppermost points on the y-axis. "scl" represents the scale on each axis.

The position of the minus sign, $-$, before 5 in Xmin will be blinking. The position that is blinking denotes the location of the cursor. The next character you press will appear where the cursor is located. To change the position of the cursor, use the arrow keys | $\Rightarrow$ | | $\Leftarrow$ | | $\Uparrow$ | | $\Downarrow$ |. To change the values of the ranges, move the cursor to the appropriate value and press the number to which you wish to change it. When you want to enter the appropriate number, press the execute key | EXE |. The cursor will automatically move to the next value. If you do not want to change a value, just press | EXE |. Now press | G$\longleftrightarrow$T |, which shifts you to the graphing mode (G) of the calculator from the text mode (T) that you were in. The screen will show the coordinate system as you defined it.

If you are satisfied and want to graph a function, press the clear key $\boxed{\text{AC}}$.

Suppose you want to graph the two preceding linear equations:

$$3x + y = 9$$
$$2x + y = 8$$

The first step is to press $\boxed{\text{Graph}}$. On the screen appears

Graph Y = _

You must alter the equations so that they fit the form to which the calculator is set. The equations become

$$y = 9 - 3x$$
$$y = 8 - 2x$$

The variable x is entered by pressing two keys, the red $\boxed{\text{ALPHA}}$ key and the $\boxed{+}$ key. Every key of the calculator has three definitions and the $\boxed{\text{SHIFT}}$ and $\boxed{\text{ALPHA}}$ keys give you the alternate definitions. Thus "x" appears in red under the $\boxed{+}$ key. The key sequence that sketches the graphs of the first linear equation is

$\boxed{\text{Graph}}$ $\boxed{8}$ $\boxed{-}$ $\boxed{2}$ $\boxed{\text{ALPHA X}}$ $\boxed{\text{EXE}}$

Appearing on the screen before you press $\boxed{\text{EXE}}$ is

Graph Y = 8 − 2X

When you press $\boxed{\text{EXE}}$ the graph of the function appears. To graph the second function, press $\boxed{\text{AC}}$ or $\boxed{\text{G}\longleftrightarrow\text{T}}$ and enter the following key sequence:

$\boxed{\text{Graph}}$ $\boxed{9}$ $\boxed{-}$ $\boxed{3}$ $\boxed{\text{ALPHA X}}$ $\boxed{\text{EXE}}$

Both graphs appear on the screen at once. Suppose you wanted to find the point of intersection graphically; you can use the very powerful "zoom-in" feature of the calculator. Of course, it is possible to magnify the graph by changing the ranges, but the zoom-in feature allows you to magnify the graph almost instantaneously. Use the $\boxed{\text{SHIFT}}$ $\boxed{*}$ key sequence to zoom. To find the point of intersection, first press $\boxed{\text{SHIFT}}$ $\boxed{\text{Trace}}$ to access the cursor. It will be blinking and the coordinates of the point in the coordinate system where it is blinking appear on the screen when you press $\boxed{\text{SHIFT}}$ $\boxed{\text{X}\longleftrightarrow\text{Y}}$. Use the arrow keys to move the cursor to the point of intersection. Read the coordinates by pressing $\boxed{\text{SHIFT}}$ $\boxed{\text{X}\longleftrightarrow\text{Y}}$. To get better accuracy, zoom-in and move the cursor to the point of intersection. You can zoom-in as many times as you want, each time obtaining better accuracy.

TI-81

To sketch the graph of $y = x^2$ press:

$\boxed{\text{Y}=}$ $\boxed{\text{X}|\text{T}}$ $\boxed{x^2}$ $\boxed{\text{ENTER}}$ $\boxed{\text{GRAPH}}$

To sketch the graph $y = x^{-1}$ press the following:

$\boxed{\text{Y}=}$ $\boxed{\text{X}|\text{T}}$ $\boxed{x^{-1}}$ $\boxed{\text{ENTER}}$ $\boxed{\text{GRAPH}}$

When sketching the graph of one of the built-in functions the calculator uses preset ranges. Sometimes when sketching the graph of other functions you must set the range on the x-axis and the range on the y-axis. Only the part of the graph within these ranges appears on the screen. If you want to see any other part of the graph, or to magnify the graph, you must overwrite the preset ranges by pressing $\boxed{\text{RANGE}}$ and entering the desired values.

This is not a new idea. When graphing a function on a piece of paper, you determine the ranges. For example, suppose you wanted to sketch the graph of the following system of linear equations:

$$3x + y = 9$$
$$2x + y = 8$$

You would most likely find the intercepts first; the x-intercepts are 3 and 4 and the y-intercepts are 9 and 8. You would then determine how large you want the graph to be on your paper and then plot the largest intercept, 9, just inside the range you decided to have for the graph. In other words, a little thought and computation goes into the decision of the magnitude of the range of your graph on a piece of paper. The same type of care must be taken when sketching graphs with the calculator.

To set the ranges on the axes, press the $\boxed{\text{RANGE}}$ key and the following appears:

RANGE
    Xmin: -10
    Xmax:10
    Xscl:1
    Ymin: -10
    Ymax:10
    Yscl:1
    Xres:1

Xmin represents the leftmost point on the x-axis, Xmax the rightmost point on the x-axis, and Ymin and Ymax the lowermost and uppermost points on the y-axis. "scl" represents the scale on each axis, while "Xres" represents the resolution.

The position of the minus sign on Xmin: -10 will be blinking. The position that is blinking denotes the location of the cursor. The next character you press will appear where the cursor is located. To change the position of the cursor use the arrow keys $\boxed{\Rightarrow}$ $\boxed{\Leftarrow}$ $\boxed{\Uparrow}$ $\boxed{\Downarrow}$. To change the values of the ranges, move the cursor to the appropriate value and press the number to which you wish to change it. After entering the appropriate number, press $\boxed{\text{ENTER}}$. The cursor will automatically move to the next value.

Pressing $\boxed{\text{GRAPH}}$ is all that is required for the TI-81 to draw the graph of the function within the ranges of values you defined. To graph the system of linear equations

$$3x + y = 9$$
$$2x + y = 8$$

you must first solve each one for y:

$$y = 9 - 3x$$
$$y = 8 - 2x$$

Then press $\boxed{\text{Y}=}$. The variable x is entered by pressing $\boxed{\text{X}|\text{T}}$. The key sequence that defines the equation $y = 9 - 3x$ is

$$9 \boxed{-} 3 \boxed{\text{X}|\text{T}} \boxed{\text{ENTER}}$$

The TI-81 names this function Y_1. It is possible to work with three other functions simultaneously. After entering the first function the cursor moves to Y_2. The key sequence for entering $y = 8 - 2x$ is

$$8 \boxed{-} 2 \boxed{\text{X}|\text{T}} \boxed{\text{ENTER}}$$

Now press $\boxed{\text{GRAPH}}$ to draw the functions. The TI-81 draws Y_1 first and Y_2 second.

The "zoom" feature can be used to find the point of intersection graphically. Of course you can magnify the graph by changing the ranges, but the zoom-in feature allows you to magnify the graph almost instantaneously. First, use the arrow keys to position the cursor as close to the point of intersection as possible. The coordinates at the bottom of the screen provide a first approximation. Press the $\boxed{\text{TRACE}}$ key, and use the left arrow key and the right arrow key to position the cursor as close as possible to the point of intersection. Then press $\boxed{\text{ZOOM}}$, which provides a menu of seven items. Press 2 | ENTER |, then use the "trace" feature to obtain a better approximation to the point of intersection. By repeating this process one or two more times you will see (literally) that the point of intersection occurs at $x = 1$, $y = 6$. That these are the correct values can be checked algebraically by substitution into the equations.

Appendix B:
Tables

Table 1 Exponentials and Their Reciprocals

x	e^x	e^{-x}
10	22026.46580	0.00005
9	8103.08393	0.00012
8	2980.95799	0.00034
7	1096.63316	0.00091
6	403.42879	0.00248
5	148.41316	0.00674
4	54.59815	0.01832
3	20.08554	0.04979
2	7.38906	0.13534
1.9	6.68589	0.14957
1.8	6.04965	0.16530
1.7	5.47395	0.18268
1.6	4.95303	0.20190
1.5	4.48169	0.22313
1.4	4.05520	0.24660
1.3	3.66930	0.27253
1.2	3.32012	0.30119
1.1	3.00417	0.33287
1.0	2.71828	0.36788
0.9	2.45960	0.40657
0.8	2.22554	0.44933
0.7	2.01375	0.49659
0.6	1.82212	0.54881
0.5	1.64872	0.60653
0.4	1.49183	0.67032
0.3	1.34986	0.74082
0.2	1.22140	0.81873
0.1	1.10517	0.90484
0	1.00000	

Table 2 Natural Logarithms

x	ln x	x	ln x	x	ln x	x	ln x
0.05	−2.99573	2	0.69315	25	3.21888	100	4.60517
0.10	−2.30259	3	1.09861	30	3.40120	150	5.01064
0.15	−1.89712	4	1.38629	35	3.55535	200	5.29832
0.20	−1.60944	5	1.60944	40	3.68888	250	5.52146
0.25	−1.38629	6	1.79176	45	3.80666	300	5.70378
0.30	−1.20397	7	1.94591	50	3.91202	350	5.85793
0.35	−1.04982	8	2.07944	55	4.00733	400	5.99146
0.40	−0.91629	9	2.19722	60	4.09434	450	6.10925
0.45	−0.79851	10	2.30259	65	4.17439	500	6.21461
0.50	−0.69315	11	2.39790	70	4.24850	550	6.30992
0.55	−0.59784	12	2.48491	75	4.31749	600	6.39693
0.60	−0.51083	13	2.56495	80	4.38203	650	6.47697
0.65	−0.43078	14	2.63906	85	4.44265	700	6.55108
0.70	−0.35667	15	2.70805	90	4.49981	750	6.62007
0.75	−0.28768	16	2.77259	95	4.55388	800	6.68461
0.80	−0.22314	17	2.83321			850	6.74524
0.85	−0.16252	18	2.89037			900	6.80239
0.90	−0.10536	19	2.94444			950	6.85646
0.95	−0.05129	20	2.99572			1000	6.90776
1.00	0.00000						

Table 3 Basic Integration Formulas

1. $\displaystyle\int e^x \, dx = e^x + C$

2. $\displaystyle\int \frac{1}{x} \, dx = \ln |x| + C$

3. $\displaystyle\int a^x \, dx = \frac{a^x}{\ln a} + C, a > 0 \quad (a \neq 1)$

4. $\displaystyle\int \frac{1}{\sqrt{a^2 + x^2}} \, dx = \ln |x + \sqrt{x^2 + a^2}| + C$

5. $\displaystyle\int \frac{1}{\sqrt{x^2 - a^2}} \, dx = \ln |x + \sqrt{x^2 - a^2}| + C$

6. $\displaystyle\int \frac{1}{a^2 - x^2} \, dx = \frac{1}{2a} \ln \left| \frac{a + x}{a - x} \right| + C \quad (x^2 < a^2)$

7. $\displaystyle\int \frac{1}{x^2 - a^2} \, dx = -\frac{1}{2a} \ln \left| \frac{x + a}{x - a} \right| + C \quad (a^2 < x^2)$

8. $\displaystyle\int \frac{1}{x\sqrt{a^2 - x^2}} \, dx = -\frac{1}{a} \ln \left| \frac{a + \sqrt{a^2 - x^2}}{x} \right| + C \quad (0 < x < a)$

9. $\displaystyle\int x^n \ln x \, dx = x^{n+1} \left[\frac{\ln x}{n + 1} - \frac{1}{(n + 1)^2} \right] + C \quad (n \neq -1)$

10. $\displaystyle\int \frac{1}{x\sqrt{a^2 + x^2}} \, dx = -\frac{1}{a} \ln \left| \frac{a + \sqrt{a^2 + x^2}}{x} \right| + C$

11. $\displaystyle\int \ln |x| \, dx = x(\ln |x| - 1) + C$

12. $\displaystyle\int \frac{x}{ax + b} \, dx = \frac{x}{a} - \frac{b}{a^2} \ln |ax + b| + C$

13. $\displaystyle\int \frac{x}{(ax + b)^2} \, dx = \frac{b}{a^2(ax + b)} + \frac{1}{a^2} \ln |ax + b| + C$

14. $\displaystyle\int \frac{1}{x(ax + b)} \, dx = \frac{1}{b} \ln \left| \frac{x}{ax + b} \right| + C$

15. $\displaystyle\int \frac{1}{x(ax + b)^2} \, dx = \frac{1}{b(ax + b)} + \frac{1}{b^2} \ln \left| \frac{x}{ax + b} \right| + C$

16. $\displaystyle\int \sqrt{x^2 \pm a^2} \, dx = \frac{1}{2}x \sqrt{x^2 \pm a^2} + \frac{a^2}{2} \ln |x + \sqrt{x^2 \pm a^2}| + C$

17. $\displaystyle\int \frac{\sqrt{a^2 \pm x^2}}{x} \, dx = \sqrt{a^2 \pm x^2} - a \ln \left| \frac{a + \sqrt{a^2 \pm x^2}}{x} \right| + C$

Table 3 Basic Integration Formulas (*continued*)

18. $\displaystyle\int x^n e^{ax}\ dx = \frac{1}{a} x^n e^{ax} - \frac{n}{a} \int x^{n-1} e^{ax}\ dx + C$

19. $\displaystyle\int \sin^2 u\ du = \frac{u}{2} - \frac{1}{4} \sin 2u + C$

20. $\displaystyle\int \sin^3 u\ du = -\frac{1}{3} \cos u(\sin^2 u + 2) + C$

21. $\displaystyle\int \cos^2 u\ du = \frac{1}{2} u + \frac{1}{4} \sin 2u + C$

22. $\displaystyle\int \cos^3 u\ du = \frac{1}{3} \sin u\ (\cos^2 u + 2) + C$

23. $\displaystyle\int \frac{du}{1 \pm \sin u} = \mp \tan \left(\frac{\pi}{4} \mp \frac{u}{2} \right) + C$

24. $\displaystyle\int \frac{du}{1 + \cos u} = \tan \left(\frac{u}{2} \right) + C$

25. $\displaystyle\int \frac{du}{1 - \cos u} = -\cot \left(\frac{u}{2} \right) + C$

Table 4 Trigonometric Functions

Deg	Rad	Sin	Cos	Tan	Deg	Rad	Sin	Cos	Tan
0	0.0000	0.0000	1.0000	0.0000	45	0.7854	0.7071	0.7071	1.0000
1	0.0175	0.0175	0.9998	0.0175	46	0.8029	0.7193	0.6947	1.0355
2	0.0349	0.0349	0.9994	0.0349	47	0.8203	0.7314	0.6820	1.0724
3	0.0524	0.0523	0.9986	0.0524	48	0.8378	0.7431	0.6691	1.1106
4	0.0698	0.0698	0.9976	0.0699	49	0.8552	0.7547	0.6561	1.1504
5	0.0873	0.0872	0.9962	0.0875	50	0.8727	0.7660	0.6428	1.1918
6	0.1047	0.1045	0.9945	0.1051	51	0.8901	0.7771	0.6293	1.2349
7	0.1222	0.1219	0.9925	0.1228	52	0.9076	0.7880	0.6157	1.2799
8	0.1396	0.1392	0.9903	0.1405	53	0.9250	0.7986	0.6018	1.3270
9	0.1571	0.1564	0.9877	0.1584	54	0.9425	0.8090	0.5878	1.3764
10	0.1745	0.1736	0.9848	0.1763	55	0.9599	0.8192	0.5736	1.4281
11	0.1920	0.1908	0.9816	0.1944	56	0.9774	0.8290	0.5592	1.4826
12	0.2094	0.2079	0.9781	0.2126	57	0.9948	0.8387	0.5446	1.5399
13	0.2269	0.2250	0.9744	0.2309	58	1.0123	0.8480	0.5299	1.6003
14	0.2443	0.2419	0.9703	0.2493	59	1.0297	0.8572	0.5150	1.6643
15	0.2618	0.2588	0.9659	0.2679	60	1.0472	0.8660	0.5000	1.7321
16	0.2793	0.2756	0.9613	0.2867	61	1.0647	0.8746	0.4848	1.8040
17	0.2967	0.2924	0.9563	0.3057	62	1.0821	0.8829	0.4695	1.8807
18	0.3142	0.3090	0.9511	0.3249	63	1.0996	0.8910	0.4540	1.9626
19	0.3316	0.3256	0.9455	0.3443	64	1.1170	0.8988	0.4384	2.0503
20	0.3491	0.3420	0.9397	0.3640	65	1.1345	0.9063	0.4226	2.1445
21	0.3665	0.3584	0.9336	0.3839	66	1.1519	0.9135	0.4067	2.2460
22	0.3840	0.3746	0.9272	0.4040	67	1.1694	0.9205	0.3907	2.3559
23	0.4014	0.3907	0.9205	0.4245	68	1.1868	0.9272	0.3746	2.4751
24	0.4189	0.4067	0.9135	0.4452	69	1.2043	0.9336	0.3584	2.6051
25	0.4363	0.4226	0.9063	0.4663	70	1.2217	0.9397	0.3420	2.7475
26	0.4538	0.4384	0.8988	0.4877	71	1.2392	0.9455	0.3256	2.9042
27	0.4712	0.4540	0.8910	0.5095	72	1.2566	0.9511	0.3090	3.0777
28	0.4887	0.4695	0.8829	0.5317	73	1.2741	0.9563	0.2924	3.2709
29	0.5061	0.4848	0.8746	0.5543	74	1.2915	0.9613	0.2756	3.4874
30	0.5236	0.5000	0.8660	0.5774	75	1.3090	0.9659	0.2588	3.7321
31	0.5411	0.5150	0.8572	0.6009	76	1.3265	0.9703	0.2419	4.0108
32	0.5585	0.5299	0.8480	0.6249	77	1.3439	0.9744	0.2250	4.3315
33	0.5760	0.5446	0.8387	0.6494	78	1.3614	0.9781	0.2079	4.7046
34	0.5934	0.5592	0.8290	0.6745	79	1.3788	0.9816	0.1908	5.1446
35	0.6109	0.5736	0.8192	0.7002	80	1.3963	0.9848	0.1736	5.6713
36	0.6283	0.5878	0.8090	0.7265	81	1.4137	0.9877	0.1564	6.3138
37	0.6458	0.6018	0.7986	0.7536	82	1.4312	0.9903	0.1392	7.1154
38	0.6632	0.6157	0.7880	0.7813	83	1.4486	0.9925	0.1219	8.1442
39	0.6807	0.6293	0.7771	0.8098	84	1.4661	0.9945	0.1045	9.5144
40	0.6981	0.6428	0.7660	0.8391	85	1.4835	0.9962	0.0872	11.4301
41	0.7156	0.6561	0.7547	0.8693	86	1.5010	0.9976	0.0698	14.3007
42	0.7330	0.6691	0.7431	0.9004	87	1.5184	0.9986	0.0523	19.0811
43	0.7505	0.6820	0.7314	0.9325	88	1.5359	0.9994	0.0349	28.6363
44	0.7679	0.6947	0.7193	0.9657	89	1.5533	0.9998	0.0175	57.2900
					90	1.5708	1.0000	0.0000	—

Answers to Odd-Numbered Exercises

CHAPTER 1

Exercise Set 1–1

1.

x	−2	−1	0	1	2	3
y	−7	−5	−3	−1	1	3
point	(−2, −7)	(−1, −5)	(0, −3)	(1, −1)	(2, 1)	(3, 3)

3.

x	−2	−1	0	1	2	3
y	6	5	4	3	2	1
point	(−2, 6)	(−1, 5)	(0, 4)	(1, 3)	(2, 2)	(3, 1)

5.

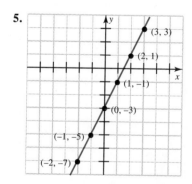

7.

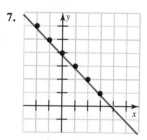

9. $2x + y = 6$ $y = 6 - 2x$

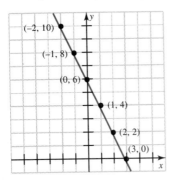

11. $x - y = 3$

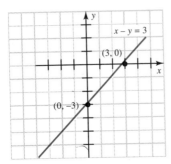

13. $3x - y = -1$

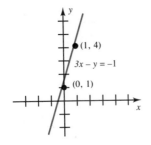

15.

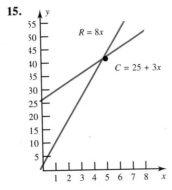

17.

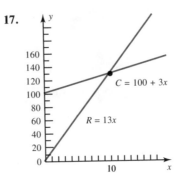

19. (a) $\$58,100$
    (b) 10%

21.

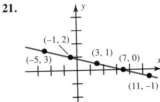

23.

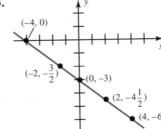

25.

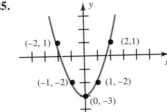

27.

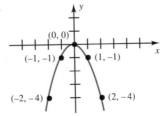

29.

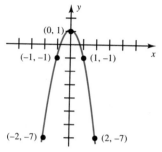

31.

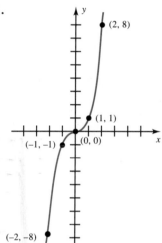

33.

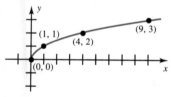

35.

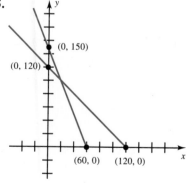

37.

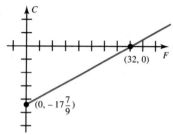

39. 37° Celsius
41. $50,000

43. (a)

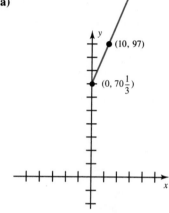

(b) 169 cm
(c) 43 cm
45. This is the break-even point.
47. More than 125 items
49. 66,706
51. Figure (a)

Referenced Exercise Set 1–1

1. (a) 72.4%
   (b) 1978
3. (a) 2 hours
   (b) 3 hours
   (c) 1 hour, 20 minutes

5. (a) $T = 0.03x$
   (b) $T = 0.05x$

Exercise Set 1–2

1.

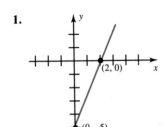

3.

5. $m = 2$
7. $m = -\frac{3}{2}$
9. $m = 2$

11.

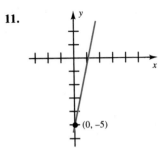

13.

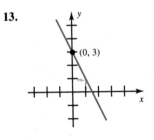

15. Not parallel
17. (a) Undefined **(b)** $m = 0$
19. $y = -7x + 15$

21.

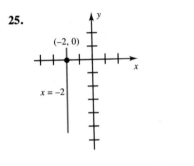

$y = 2x + 7$

23.

$y = -3$

$(-5, -3)$

25.

$(-2, 0)$

$x = -2$

27.

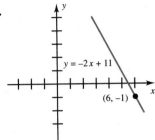

29.

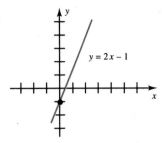

31.

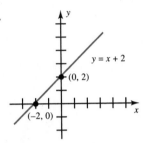

33.

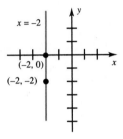

35. (a) Positive   **(b)** Zero   **(c)** Negative

37. (a) $y = -\frac{3}{2}x + 6$ **(b)** $m = -\frac{3}{2}$ **(c)** y-intercept $= 6$

39. (a) $y = -\frac{5}{2}x + 5$ **(b)** $y = -\frac{5}{2}x - 5$

41. $y = -4x + 1$

43. (a)

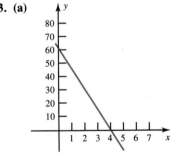

  (b) $15,000$

  (c) The y-intercept represents the worth of the work station when it was purchased.

45. (a) $y = 48643.3x + 17558072$

  (b) $18,044,505$

47. The population of Iowa decreased from 1980 to 1990.

49. Figure d

51. Figure a

53. No, for example, $xy = 1$.

Referenced Exercise Set 1–2

1. (a) $y = .5x + 7.5$

(b)

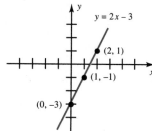

3. (a) Dead: $y = (1/50)x + 15$
(b) Average: $y = (1/100)x + 10$
(c) Live: $y = (1/100)x + 20/7$

5. (a) During the first few minutes of running, the percentage of energy supplied to the body by carbohydrates increases.
(b) After a few minutes of running, the percentage of energy supplied to the body by carbohydrates decreases.

Cumulative Exercise Set 1–2

1.

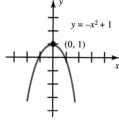

5. $y = -4x$
7. $2x - 4y = -20$
9. $27,000$

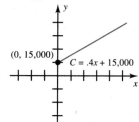

3.

Exercise Set 1–3

1. $1, -3, -4$ **3.** $1, 3, \frac{9}{4}$
5. (a) $2a - 3$ **(b)** $2x + 3$ **(c)** $2x + 2h - 3$
   (d) 2

7. (a) $-a^2 + a - 3$ **(b)** $-x^2 - 5x - 3$
   (c) $-x^2 - 2xh - h^2 + x + h + 3$
   (d) $-2x + 1 - h$

9. 69,888,000   **11.** $x \geq -1$

13. $x \leq \frac{1}{2}$

15.

x	-2	-1	0	1	2
$f(x)$	-5	-6	-5	-2	3

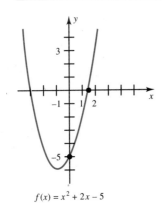

$f(x) = x^2 + 2x - 5$

17.

x	-2	-1	0	1	2
$f(x)$	18	4	-4	-6	-2

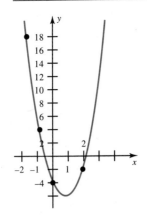

19. (a) Not a function   (b) Function   (c) Function

21. $2h$ **23.** $-2xh - h^2 + h$

25. -1 **27.** $-4x - 2h - 4$

29. $x \neq 2$ **31.** $-1 \leq x \leq 1$

33. $x \geq 1$ or $x \leq -1$

35. $x \neq 0$ and $x \neq 2$

37. $x \geq 3$ but $x \neq 4$

39. (a) $\{-2, 2\}$ (b) none   (c) $\{-0.5, 0.5\}$

41. Function

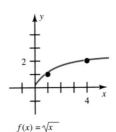

$f(x) = \sqrt{x}$

Referenced Exercise Set 1–3

1. (a) yes, $d = 6t^{2/3}$ or $d = 6\sqrt[3]{t^2}$ (b) 49.92 miles   (c) 0.19 hour   or   11 minutes, 33 seconds

3. yes

5. (a) $f(x) = 0.3x - 564$ (b) 33   (c) 1977

Cumulative Exercise Set 1–3

1. (a)

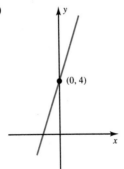

(0, 4)

(b)

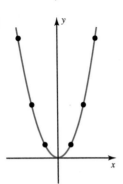

(c) (4, 16) and (−1, 1)

3. $y = -2x - 6$

5.

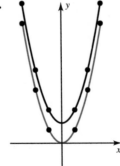

7. −1
9. \$4,789,000,000

Exercise Set 1–4

1.

x	0	1	2	3
y	0	5	4	9

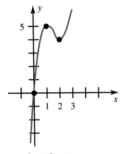

$f(x) = 2x^3 - 9x^2 + 12x$

3.

x	0	1	2	−1
y	1	2	1	10

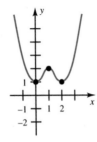

$f(x) = x^4 - 4x^3 + 4x^2 + 1$

5. 3   **7.** The line $x = 5$
9. The lines $x = 3$ and $x = -3$

11.

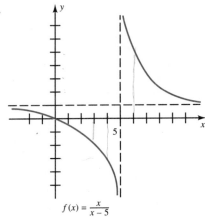

$$f(x) = \frac{x}{x-5}$$

13.

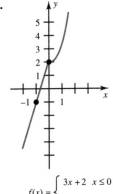

$$f(x) = \begin{cases} 3x+2 & x \le 0 \\ x^2+2 & x > 0 \end{cases}$$

15.

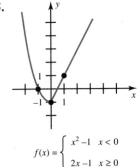

$$f(x) = \begin{cases} x^2-1 & x < 0 \\ 2x-1 & x \ge 0 \end{cases}$$

17. $f(0) = 2$ **19.** 30%

21. (a) $x > 0$
   (b) Four items should be produced to minimize cost.

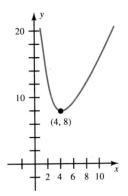

23.

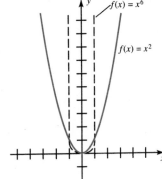

$f(x) = x^6$

$f(x) = x^2$

25. 2
27. The lines $x = 3$ and $x = -3$

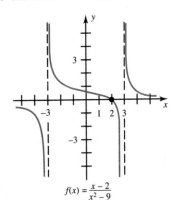

$$f(x) = \frac{x-2}{x^2-9}$$

29. The line $x = 0$

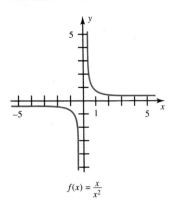

$$f(x) = \frac{x}{x^2}$$

31. The lines $x = 0$, $x = 1$, and $x = -1$

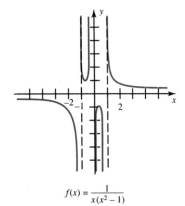

$$f(x) = \frac{1}{x(x^2 - 1)}$$

33. The lines $x = -\frac{3}{2}$ and $x = 1$

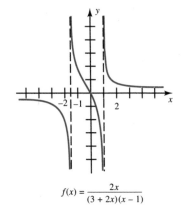

$$f(x) = \frac{2x}{(3 + 2x)(x - 1)}$$

35.

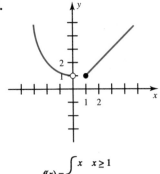

$$f(x) = \begin{cases} x & x \geq 1 \\ x^2 + 1 & x \leq 0 \end{cases}$$

37.

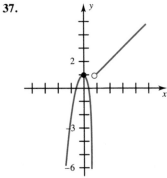

$$f(x) = \begin{cases} 1 - 2x - 5x^2 & x \leq 1 \\ x & x > 1 \end{cases}$$

39.

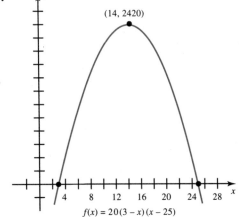

(14, 2420)

$$f(x) = 20(3 - x)(x - 25)$$

Maximum profit occurs when 14 items are produced.

41. Figure a
43. Figure c

45.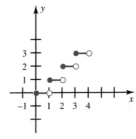

$$f(x) = \text{INT}(x) \text{ for } x \geq 0$$

Referenced Exercise Set 1–4

1. 40%

3. (a)

 $$f(x) = \begin{cases} .15x & \text{if } 0 \leq x \leq 29{,}300 \\ .27(x - 29{,}300) + 4350 & \text{if } x > 29{,}300 \end{cases}$$

(b)

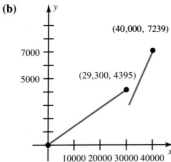

(c) 4395   **(d)** 4350.27

(e) Tax on income of $29,301 is $44.73 less than tax on income of $29,300.

5. (a) 16 minutes and 5 seconds

(b) 16 hours, 16 minutes and 1 second

Cumulative Exercise Set 1–4

1.

3. $y = x + 1$
5. $x \geq 4$

7. $\dfrac{h(2x + h)}{h} = 2x + h,\ h \neq 0$

9.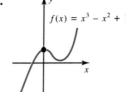

11.

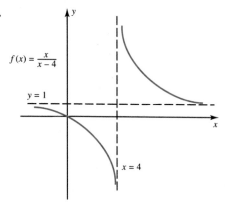

$f(x) = \dfrac{x}{x - 4}$

$y = 1$

$x = 4$

Exercise Set 1–5

1. (a) $(f + g)(x) = \dfrac{x + 2}{x - 1} + \dfrac{x}{x + 2}$

$\qquad = \dfrac{2x^2 + 3x + 4}{(x - 1)(x + 2)}$

(b) $(f \cdot g)(x) = \dfrac{(x + 2)}{(x - 1)} \cdot \dfrac{x}{(x + 2)} = \dfrac{x}{x - 1}$

(c) $(f/g)(x) = \dfrac{(x + 2)^2}{x(x - 1)}$

3. (a) $(f + g)(x) = \dfrac{x}{x + 1} + \dfrac{1}{x} = \dfrac{x^2 + x + 1}{x(x + 1)}$

(b) $(f \cdot g)(x) = \dfrac{1}{x + 1}$

(c) $(f/g)(x) = \dfrac{x^2}{x + 1}$

5. (a) All real numbers except $x = 1$ and $x = -2$
(b) All real numbers except $x = -1$ and $x = 0$

7. (a) All real numbers except $x = 1$, $x = -2$, and $x = 0$
(b) All real numbers except $x = -1$ and $x = 0$

9. (a) $\tfrac{9}{2}$ **(b)** 2   **(c)** 8

11. (a) $\tfrac{3}{2}$ **(b)** -1 **(c)** -4

13. (a) $(f \circ g)(x) = \sqrt{\dfrac{x}{1 - x}}$

(b) $(g \circ f)(x) = \dfrac{\sqrt{x}}{1 - \sqrt{x}}$

15. (a) $(f \circ g)(x) = \dfrac{2(x - 2)}{(x + 2)}$

(b) $(g \circ f)(x) = \dfrac{x - 1}{x + 1}$

17. $g(x) = \sqrt{x}$, $k(x) = x^3 + 2x - 5$,
$f(x) = (g \circ k)(x)$

19. $g(x) = \dfrac{1}{x^2} + x^3$, $k(x) = x - 2$,
$f(x) = (g \circ k)(x)$

21. $Q(t) = -0.005t^4 + 0.16t^3 - 3.28t^2 + 32t + 20$

23. (a) $(f + g + h)(x) = \dfrac{2x^3 - x - 10}{(x - 1)(x + 2)(x - 2)}$

(b) $(f \cdot g \cdot h)(x) = \dfrac{x}{(x - 1)(x - 2)}$

25. (a) $(f + g + h)(x) = \dfrac{2x^2 - x + 3}{(x + 3)(x - 3)}$

(b) $(f \cdot g \cdot h)(x) = \dfrac{x(x - 2)}{(x + 3)^2(x - 3)}$

27. 0

29. $\sqrt{x^2} = |x|$

31. $\dfrac{x + 1}{2x + 1}$

33. No
$(f \circ g)(1) = \tfrac{2}{3}$
$(g \circ f)(1) = 3$

35. $g(x) = \sqrt[3]{x^2}$, $k(x) = x - 1$, $f(x) = (g \circ k)(x)$

37. $\dfrac{2a^2 - 3a + 3}{a^2 - 1}$

39. $\dfrac{1}{a + 1} + \sqrt{a}$

41. No
$(f \cdot g)(x) = \dfrac{x}{x - 1}$
$(f \circ g)(x) = \dfrac{3x + 4}{-2}$

43. (a) All $x \geq 0$ except $x = 200$ **(b)** $0 \leq x \leq 100$
(c) 10 hours   **(d)** 40 hours   **(e)** 66.7%

45. (a)

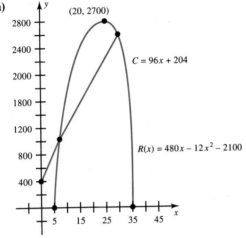

$C = 96x + 204$

$R(x) = 480x - 12x^2 - 2100$

(b) $8 \le x \le 24$

47. (a) $\dfrac{1}{\sqrt{x + 1}}$ **(b)** $\dfrac{1}{\sqrt{x + 1}}$ **(c)** $\dfrac{1}{\sqrt{x + 1}}$

Referenced Exercise Set 1–5

1. PEX

3. (a) $x = 7$
   (b) $x = 16$ and $x = 17$ (in a 24-hour clock)
   (c) $x = 18$

Cumulative Exercise Set 1–5

1. $y = (\frac{5}{3})x$

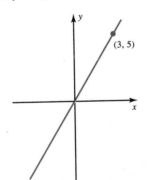

3. (a) The vertical line $x = 2$
   (b) There is no y-intercept.
5. (a) -8
   (b) $-2x^2 - 4xh - 2h^2$
   (c) $h(-4x - 2h)$
   (d) $-4x - 2h$

7. (a)

x	-2	-1	0	1	2
$f(x)$	16	9	8	7	0

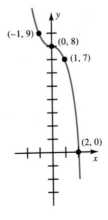

(b) 3

9.

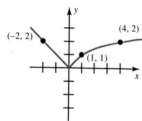

11. 2

Exercise Set 1–6

1. $x = -4$, $x = 2$

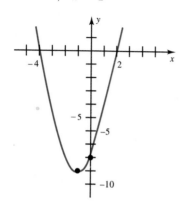

3. $x = -3$

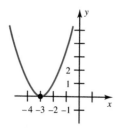

5. No solution

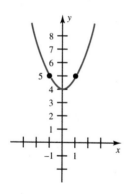

7. Domain: real numbers except $x = \frac{3}{2}$, -1
   x-intercept: $x = 2$
9. Domain: real numbers except $x = \frac{3}{4}$, -2
   x-intercept: none
11. -2, $\frac{3}{2}$, 2
13. $x = 2$, $x = -2$
15. $x = 4$, $x = -3$
17. $(1, 9)$ and $(8, 16)$
19. $(2, 12)$ and $(7, 7)$
21. $x = 2$

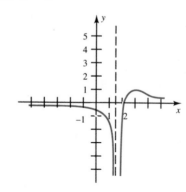

23. $x = -1$

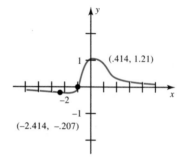

25. Domain: all real numbers except $x = 0$ and $x = 2$
  x-intercept: none
27. Domain: all real numbers except $x = 0$ and $x = 2$
  x-intercept: none
29. $(-1, 1)$ and $(2, 4)$
31. $(0, 0)$ and $(1, 1)$

33. $(3, 13)$ and $(-3, 13)$
35. 2 seconds and 5 seconds after the object is thrown
37. 3.5 seconds after the object is thrown
39. 8 units and 24 units
41. $(2, 5)$ and $(4, 8)$; 2
43. $(2, 35)$ and $(8, 75)$; 30

Referenced Exercise Set 1–6

1. $x = 1$
3. (a) Americans have been eating more poultry than pork since 1982.
  (b) Americans have been eating more poultry than beef since 1987.

Cumulative Exercise Set 1–6

1. $y = -x - 2$

3. $f(x) = x^2 + 2x - 1$

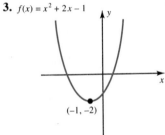

$(-1, -2)$

5. $f + g = x^3 + x^2 + x - 1$
  $f \cdot g = (x^2 + x)(x^3 - 1) = x^5 + x^4 - x^2 - x$
7. $f(x) = h(g(x))$ where $h(x) = x^4$ and $g(x) = x^3 + 2$
9. The domain is all x except $x = 2$; the x-intercept is $x = 0$.
11. $(1, 7)$ and $(6, 12)$

Case Study Exercises

1. 0.58252679
5. No

3. 0.5826
7. 25.16 pounds

11. (a) 1052.05 **(b)** 1035.30
13. Sheridan Shuttles

9.

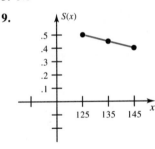

Review Exercises

1. $-9, -10, -9, -6, 6$

3. $2xh + h^2 - 4h$

5. All real numbers except $x = 1$

7. Yes

9. The lines $x = 2$, $x = 3$, and $x = -3$.

11. (a) $\dfrac{2x^2 + 3x + 4}{(x - 1)(x + 2)}$ **(b)** $\dfrac{x}{x - 1}$ **(c)** $\dfrac{(x + 2)^2}{x(x - 1)}$

13. (a) $\frac{9}{2}$ **(b)** 2   **(c)** 8

15. $g(x) = x - \sqrt{x}$, $k(x) = x + 2$, $f(x) = (g \circ k)(x)$

17. No x-intercepts

19. $x = 1, 2, -2$

21. Domain: all real numbers except -2 and 1 x-intercept: $x = -1$

CHAPTER 2

Exercise Set 2–1

1. (a) 2   **(b)** 2

3. (a) -4 **(b)** -4

5. (a) 1   **(b)** 0.01

7. (a) 1   **(b)** 0.001

9. 2

11. 8

13. $y = 2x - 6$

15. $y = 8x + 18$

17. $-4x$

19. $2x$

21. (a) 4.1   **(b)** 4.01   **(c)** 4.001

23. (a) 8.2   **(b)** 8.02   **(c)** 8.002

25. (a) 0.24846   **(b)** 0.24984   **(c)** 0.24998

27. 4

29. $y = 4x - 1$

31. $4x - 4$

33. $3x^2 + 1$

35. $-32t + 192$

37. $-32t + 128$

39. 0

41. 0

43. $2ax + b$

Referenced Exercise Set 2–1

1. (a) 26.025   **(b)** 36.85

3. (a) 2.05%   **(b)** -0.1%

5. 20%

7. (a) Positive   **(b)** Reagan

Exercise Set 2–2

1. 3

5. No limit

9. No limit

13. 0

17. 2

21. No limit

25. $\frac{1}{6}$

29. -4

3. 6

7. No limit

11. 3

15. 0

19. No limit

23. $\frac{1}{2}$

27. 0

31. $\frac{1}{3}$

33. 3

37. 0

41. 4

45. 18

49. 0

53. 2

57. 0

61. $90°$

35. 0

39. 0

43. $\frac{1}{4}$

47. 8

51. 2

55. 3

59. No limit

63. 200 miles

Referenced Exercise Set 2–2

1.

x	.05	.06	.07	.08
$f(x)$	.73	.86	.93	.96

$$\lim_{x \to .08} f(x) = .96$$

3. (a) 1   **(b)** 1

Cumulative Exercise Set 2–2

1. -3.99
3. $y = -4x - 4$

5. 0
7. 0

Exercise Set 2–3

1. $2x$

3. $\dfrac{1}{\sqrt{2x}}$

5. $-\dfrac{1}{2\sqrt{x}}$

7. $-\dfrac{1}{x^2}$

9. $-\dfrac{1}{x^2}$

11. $y = -2x + 6$

13. $y = -x - 6$
15. (a) 6 seconds   **(b)** 576 feet

17. (a) $\dfrac{35}{16} = 2.1875$ seconds   **(b)** $\dfrac{1225}{16} = 76.5625$ feet

19. 7

21. 6

23. $-\dfrac{1}{2x\sqrt{x}}$

25. $-\dfrac{2}{x^3}$

27. $y = -\frac{1}{4}x + 1$

29. $y = \frac{1}{4}x + \frac{3}{4}$

31. $y = -5$
33. (a) 4 seconds   **(b)** 320 feet
35. 12 seconds
37. $2ax + b$
39. (a) Yes   **(b)** Yes
41. (a) Yes   **(b)** No
43. (a) No   **(b)** No

Referenced Exercise Set 2–3

1. (a) 15%   **(b)** 28%   **(c)** 33%   **(d)** 28%
3. (a) 6,500 rpm   **(b)** 450 bhp
5. (a) The derivative   **(b)** March and June   **(c)** November

Cumulative Exercise Set 2–3

1. 4
3. 4.1, 4.01
5. 0

7. Does not exist
9. $t = 2$ sec
11. $-3x^{-4}$

Exercise Set 2–4

1. Continuous
3. Not continuous
5. Not continuous

7. Continuous
9. Continuous
11. Continuous

13. Not continuous
15. $x = 0$ and $x = 1$
17. $x = 4$ and $x = -4$
19. $x = 2$ and $x = -4$
21. $x = 0$
23. $x = 0$ and $x = 3$
25. Differentiable
27. Not differentiable
29. Not differentiable
31. Not differentiable
33. Differentiable
35. Not differentiable
37. Not differentiable
39. Not differentiable
41. Yes
43. Not continuous
45. Continuous everywhere
Differentiable everywhere except $x = 0$

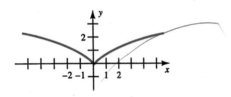

47. Continuous everywhere except $x = 1$
Differentiable everywhere except $x = 1$

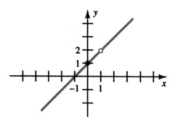

49. Continous at $x = 2$, differentiable at $x = 2$
51. Continuous at $x = 2$, not differentiable at $x = 2$
53. Not continuous at $x = 2$, not differentiable at $x = 2$

Cumulative Exercise Set 2–4

1. -1
3. $-\frac{1}{3}$
5. $6x$

7. $y = -6x - 3$
9. Continuous and differentiable
11. $x = 0$

Review Exercises

1. (a) 3   (b) 2.1
5. 2

3. $2x - 1$
7. $\frac{4}{3}$

9. 0
13. $y = 5x - 23$

11. $2x - 3$

CHAPTER 3

Exercise Set 3–1

1. $3x^2$
5. $-15x^{-4} + 5$
7. $\frac{2}{3}x^{-2/3} + \frac{3}{2}x^{-1/2} - 4$
9. $-\frac{2}{3}x^{-4/3} - \frac{7}{4}x^{-5/4}$

3. $8x^3$

11. $0.3x^{-0.7} + 1.2x^{-0.6}$

13. $-\dfrac{3}{x^4} + 12x^3$

15. $-\dfrac{1.5}{x^{2.5}} - \dfrac{1}{x^{5/4}} + 9.6x^{1.4}$

17. $7x^{2.5} - \dfrac{5.6}{x^{2.4}} - \dfrac{5.2}{x^{2.3}}$

19. $\dfrac{0.125}{x^{0.5}} - \dfrac{0.192}{x^{1.4}}$

21. $\dfrac{1}{\sqrt{x}} - 2$

23. $\dfrac{5}{2\sqrt{x}} + \dfrac{1}{\sqrt{x^3}}$

25. $-\dfrac{9}{\sqrt{3}} x^{-3/2} - \dfrac{2}{3\sqrt[3]{4}} x^{-4/3}$

27. $y = 4x - 1$ **29.** $y = 2x + 1$

31. (a) $v = -32t + 32,\ a = -32$
    (b) $t = 1$ second
    (c) 16 feet

33. 12   **35.** 10

Referenced Exercise Set 3–1

3. $-0.048x + 0.575$

Exercise Set 3–2

1. $4x^3(x + 5) + x^4 = 5x^4 + 20x^3$
3. $6x^2(3x - 7) + (3)(2x^3) = 24x^3 - 42x^2$
5. $(-3x^{-4})(3x^2 - 17) + (6x)(x^{-3} + 5)$
           $= -3x^{-2} + 51x^{-4} + 30x$
7. $(\frac{1}{3}x^{-2/3} + 1)(3x^{1/2} - 4x) + (\frac{3}{2}x^{-1/2} - 4)(x^{1/3} + x)$
9. $(-\frac{1}{3}x^{-4/3} - \frac{7}{4}x^{-5/4})(4x^2 - 5) + (8x)(x^{-1/3} + 7x^{-1/4})$
11. $4(2x + 1)$

13. $6x^2(x^3 + 1)$ **15.** $\dfrac{1}{(2x + 1)^2}$

17. $\dfrac{x^3 - 10}{3x^3}$ **19.** $y = -3x + 12$

21. $y = 2$

23. $L = \dfrac{0.5x^2 + 3}{0.5x^2 + 5};\quad L' = \dfrac{2x}{(0.5x^2 + 5)^2}$

25. $2x(x^3 + 15)(x^2 - 17) + 3x^4(x^2 - 17)$
    $+ 2x^3(x^3 + 15) = 7x^6 - 85x^4 + 60x^3 - 510x$
27. $-x^{-2}(x^{1/3} + 2)(x^2 - 3)$
    $+ \frac{1}{3}x^{-5/3}(x^2 - 3) + 2(x^{1/3} + 2)$

29. $\dfrac{x^4 - 21x^2 - 70x}{(x^2 - 7)^2}$

31. $\dfrac{-x^4 - 10x^2 + 3}{x^2(x^2 - 1)^2}$

33. $C'(x) = 23 - 138x^2$

35. $-\dfrac{100}{x^2} - 0.4$

Referenced Exercise Set 3–2

1. $q'(x) = 0.26x(x - 10) + 0.13(x - 10)^2$

3. $C'(x) = \dfrac{-7.12(13.7 - 2x)}{(13.7x - x^2)^2}$

Cumulative Exercise Set 3–2

1. $-6x^{-3} - 4x^3$
3. $y = -x - 2$
5. $36x^2(x^{-5} - 9) - 5x^{-6}(12x^3 + 1)$
        $= -5x^{-6} - 24x^{-3} - 324x^2$

7. $2x(x - 5)(x^2 + 7) + x^2(x^2 + 7) + x^2(x - 5)(2x)$
        $= 5x^4 - 20x^3 + 21x^2 - 70x$

Exercise Set 3–3

1. $15(5x + 2)^2$ **3.** $9x^2(x^3 + 4)^2$

5. $3(3x^2 + 6x)(x^3 + 3x^2)^2 = 9x(x + 2)(x^3 + 3x^2)^2$

7. $\frac{1}{2}x^{1/2}(x^{3/2} + 4)^{-2/3}$

9. $-6(\frac{1}{2}x^{-1/2} - 2x^{-2})(x^{1/2} + 2x^{-1})^{-7}$

11. $(5x^2 + 2)^2 + 20x^2(5x^2 + 2)$

$$= (5x^2 + 2)(25x^2 + 2)$$

13. $2x(x^3 + 2x)^2 + 2x^2(x^3 + 2x)(3x^2 + 2)$

$$= 8x^3(x^2 + 1)(x^2 + 2)$$

15. $\dfrac{2x(x^3 - x)^3 - 3x^2(x^3 - x)^2(3x^2 - 1)}{(x^3 - x)^6}$

$$= \dfrac{x^2(-7x^2 + 1)}{(x^3 - x)^4}$$

17. $y = 8$ **19.** $25y + 7x = 12$

21. $0.16[5 - 0.2(t - 6)^2](t - 6)$

23. $L = \dfrac{x^2 + 5}{x^2 + 10};\quad L' = \dfrac{10x}{(x^2 + 10)^2}$

25. $2x^3(x^3 + 2x^2)^3(7x + 10)$

27. $x^{-1/2}(x^{3/2} + 4)^{-3} - 9x(x^{3/2} + 4)^{-4}$

29. $x(x^2 + 1)(x^3 + 1)^2(13x^3 + 9x + 4)$

31. $\dfrac{x(2x^3 + x + 3)}{(x^2 + 1)^{1/2}(x^3 + 3)^{2/3}}$

33. $\dfrac{-11x^{3/2} + 26x^{-5/2}}{(x^2 + 2x^{-2})^4}$

35. $\dfrac{-x^2 + 1}{(x^2 + 1)^2}$

37. $\dfrac{-x^2 + 4x + 5}{(x^2 + 5)^2}$

39. $-1000(200 - x)$

Referenced Exercise Set 3–3

1. $p(x) = 0.007(16 - 0.41x^{-1/2})^2$

$\qquad\qquad - 0.33(16 - 0.41x^{-1/2}) + 100$

$p'(x) = -0.02173x^{-3/2} - 0.00118x^{-2}$

3. (c) The domain is $\{-1\}$.

Cumulative Exercise Set 3–3

1. $6x^2 + \dfrac{1}{2\sqrt{x}} + \dfrac{1}{x^2}$

3. $2x - 4$

5. $6x(x^2 - 1)^2 - 3x^2 - 1$

7. $y = 3$

9. $y = -12x + 32$

11. (a) $(x^3 - x - 2)(3x^2 - 1) +$

$\qquad\qquad (x^3 - x - 2)(3x^2 - 1)$

(b) $2(x^3 - x - 2)(3x^2 - 1)$

(c) Yes

Exercise Set 3–4

1. 2 **3.** $24x^2$

5. $\dfrac{8}{x^3} = 8x^{-3}$ **7.** $\dfrac{12}{x^5} = 12x^{-5}$

9. $12x^2 - 2$

11. $\frac{8}{9}x^{-7/3} + \frac{35}{16}x^{-9/4}$

13. $-0.21x^{-1.7} - 0.72x^{-1.6}$

15. 26

17. $\frac{3}{8}$ **19.** $2 + 12t$

21. $-\frac{1}{4}g^{-3/2}$

23. (a) $t = 1$ **(b)** $a(t)$ is never 0

25. (a) $t = \frac{1}{3}$ and $t = 1$ **(b)** $t = \frac{2}{3}$

27. $a(t) = -32$

29. $1.8 - 0.018x$

31. $0.96x - 195.2$

33. $12x^2 + 4$

35. $\dfrac{24x^2 - 4}{(2x^2 + 1)^3}$

37. $\dfrac{1}{(x^2 + 1)^{3/2}}$ **39.** $0, 2$

41. $1, -1$

43. $f^{(3)}(x) = 24x;\quad f^{(4)}(x) = 24$

45. $f^{(4)}(x) = 48x^{-5} + 120x^{-6};$

$\qquad f^{(3)}(x) = -12x^{-4} - 24x^{-5}$

47. $f^{(n)}(x) = (-1)^n(n + 1)!\, x^{-(n+2)}$

Referenced Exercise Set 3–4

1. 0.0625

Cumulative Exercise Set 3–4

1. $6x^{-3} - 0.4x^{-0.6} + 5$
3. $6x^2 - 4x^3 - 10x^4 + 6x^5$
5. $3(3x^4 + x^5)^2(12x^3 + 5x^4)$

7. $21x^2(1 - 3x)^2 - 6(7x^3 + 2)(1 - 3x)$
9. $12x^2 - 18x^{-4}$
11. $a(t) = -32$

Case Study Exercises

1. 81.11   **3.** 84.17

Review Exercises

1. $8x + 3x^2$
3. $-20x^{-6} - \frac{1}{3}x^{-2/3}$

5. $5x + 2y = 5$ **7.** $\dfrac{x^4 + 6x^2}{(x^2 + 2)^2}$

9. $\dfrac{\frac{1}{3}(x^{-2/3})(x^{4/5} + 2) - (\frac{4}{5}x^{-1/5})(x^{1/3})}{(x^{4/5} + 2)^2}$

11. $4(2x^3 + x)^3(6x^2 + 1)$

13. $6x + 9(3x^2 + 4)(x^3 + 4x)^2$
15. $x(x^2 + 1)(x^3 + 1)^2(13x^3 + 9x + 4)$
17. $y = 24x - 8$

19. $\dfrac{6x(7x^3 - 8)}{(x^3 + 4)^4}$

21. $\dfrac{(-1)^n(n + 2)!\ x^{-(n+3)}}{2}$

CHAPTER 4

Exercise Set 4–1

1. Increasing on $0 < x < \infty$
Decreasing on $-\infty < x < 0$
Horizontal tangent at $x = 0$

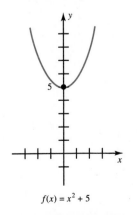

$f(x) = x^2 + 5$

3. Increasing on $-\infty < x < 0$
Decreasing on $0 < x < \infty$
Horizontal tangent at $x = 0$

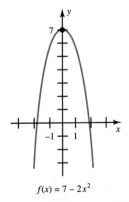

$f(x) = 7 - 2x^2$

5. Increasing on $0 < x < \infty$
Decreasing on $-\infty < x < 0$
Horizontal tangent at $x = 0$

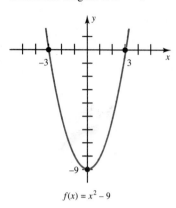

$f(x) = x^2 - 9$

7. Increasing on $2 < x < \infty$
Decreasing on $-\infty < x < 2$
Horizontal tangent at $x = 2$

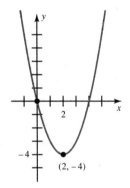

$f(x) = x(x - 4)$

9. Increasing on $-2 < x < \infty$
Decreasing on $-\infty < x < -2$
Horizontal tangent at $x = -2$

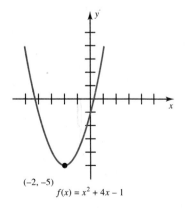

$f(x) = x^2 + 4x - 1$

11. Increasing on $-\infty < -3$ and $1 < x < \infty$
Decreasing on $-3 < x < 1$
Horizontal tangents at $x = -3$ and $x = 1$

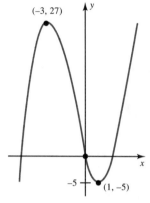

$f(x) = x^3 + 3x^2 - 9x$

13. Increasing on $-\infty < x < -1$ and $3 < x < \infty$
Decreasing on $-1 < x < 3$
Horizontal tangents at $x = -1$ and $x = 3$

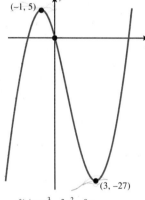

$f(x) = x^3 - 3x^2 - 9x$

15. Increasing on $-1 < x < 3$
Decreasing on $-\infty < x < -1$ and $3 < x < \infty$
Horizontal tangents at $x = -1$ and $x = 3$

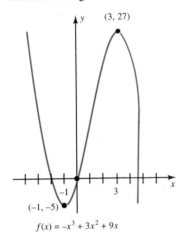

$(3, 27)$

-1

3

$(-1, -5)$

$f(x) = -x^3 + 3x^2 + 9x$

17. Increasing on $-\infty < x < -3$ and $3 < x < \infty$,
Decreasing on $-3 < x < 0$ and $0 < x < 3$
Horizontal tangents at $x = -3$ and $x = 3$

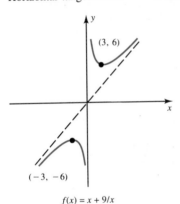

$(3, 6)$

$(-3, -6)$

$f(x) = x + 9/x$

19. Increasing on $-\infty < x < -2\sqrt{3}$
and $2\sqrt{3} < x < \infty$
Decreasing on $-2\sqrt{3} < x < 0$ and $0 < x < 2\sqrt{3}$
Horizontal tangents at $x = -2\sqrt{3}$
and $x = 2\sqrt{3}$

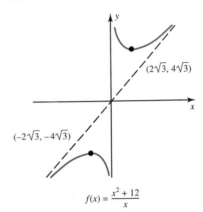

$(2\sqrt{3}, 4\sqrt{3})$

$(-2\sqrt{3}, -4\sqrt{3})$

$f(x) = \dfrac{x^2 + 12}{x}$

21. Increasing on $-\infty < x < -\dfrac{2\sqrt{3}}{3}$

and $\dfrac{2\sqrt{3}}{3} < x < \infty$

Decreasing on $-\dfrac{2\sqrt{3}}{3} < x < \dfrac{2\sqrt{3}}{3}$

Horizontal tangents at $x = -\dfrac{2\sqrt{3}}{3}$ and $x = \dfrac{2\sqrt{3}}{3}$

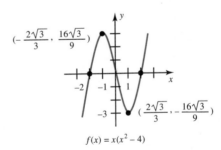

$\left(-\dfrac{2\sqrt{3}}{3}, \dfrac{16\sqrt{3}}{9}\right)$

-2 -1 1

-3 $\left(\dfrac{2\sqrt{3}}{3}, -\dfrac{16\sqrt{3}}{9}\right)$

$f(x) = x(x^2 - 4)$

23. Increasing on $-\infty < x < 0$ and $\frac{4}{3} < x < \infty$
Decreasing on $0 < x < \frac{4}{3}$
Horizontal tangents at $x = 0$ and $x = \frac{4}{3}$

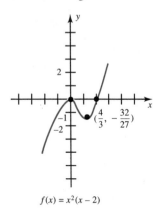

$$f(x) = x^2(x - 2)$$

25. Increasing on $-\infty < x < -1$ and $-1 < x < \infty$
Never decreasing
No horizontal tangents

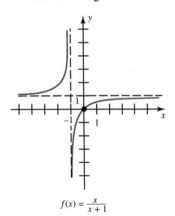

$$f(x) = \frac{x}{x + 1}$$

27. Increasing on $-\infty < x < -1$ and $-1 < x < \infty$
Never decreasing
No horizontal tangents

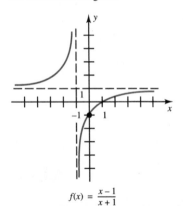

$$f(x) = \frac{x - 1}{x + 1}$$

29. Increasing on $0 < x < 6$ and $8 < x < 10$
Decreasing on $6 < x < 8$

31.

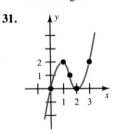

33.

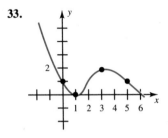

35. Increasing on $-\infty < x < 0$ and $4 < x < \infty$
Decreasing on $0 < x < 2$ and $2 < x < 4$
Horizontal tangents at $x = 0$ and $x = 4$

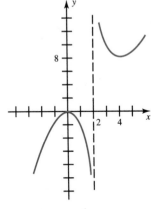

$$f(x) = \frac{x^2}{x - 2}$$

37. Increasing on $-\sqrt{2} < x < 0$ and $\sqrt{2} < x < \infty$
Decreasing on $-\infty < x < -\sqrt{2}$ and $0 < x < \sqrt{2}$
Horizontal tangents at $x = -\sqrt{2}$, $x = 0$, and
$x = \sqrt{2}$

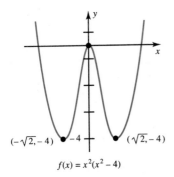

$(-\sqrt{2}, -4)$ $(\sqrt{2}, -4)$

$f(x) = x^2(x^2 - 4)$

39. Increasing on $-\dfrac{\sqrt{10}}{2} < x < 0$

and $\dfrac{\sqrt{10}}{2} < x < \infty$

Decreasing on $-\infty < x < -\dfrac{\sqrt{10}}{2}$

and $0 < x < \dfrac{\sqrt{10}}{2}$

Horizontal tangents at $x = -\dfrac{\sqrt{10}}{2}$, $x = 0$,

and $x = \dfrac{\sqrt{10}}{2}$

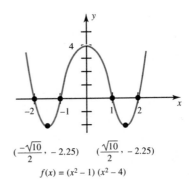

$(\dfrac{-\sqrt{10}}{2}, -2.25)$ $(\dfrac{\sqrt{10}}{2}, -2.25)$

$f(x) = (x^2 - 1)(x^2 - 4)$

41. Increasing on $-1 < x < 2$ and $5 < x < \infty$
Decreasing on $-\infty < x < -1$ and $2 < x < 5$
Horizontal tangents at $x = -1$, $x = 2$, and $x = 5$

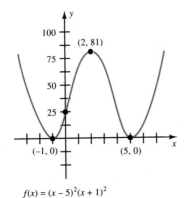

$f(x) = (x - 5)^2(x + 1)^2$

43. (a) Increasing in $(-1, 0)$; decreasing in $(0, 3)$
    (b) Positive in $(-1, 0)$; negative in $(0, 3)$
45. (a) Increasing in $(-2, -1)$ and $(2, 3)$; decreasing in $(-1, 1)$, $(1, 2)$, and $(3, 4)$
    (b) Positive in $(-2, -1)$ and $(2, 3)$; negative in $(-1, 1)$, $(1, 2)$, and $(3, 4)$

Referenced Exercise Set 4–1

1. Slope of tangent line is negative for $x < 1$.
Slope of tangent line is positive for $x > 1$.

3. (a) Increasing on $(-\infty, 0)$ and $(30, \infty)$
Decreasing on $(0, 30)$

(b) It represents a percentage, and the percentage should go down.

(c) $f(20) = 25.93$

(d) When $x = 0$, $f(0) = 100$ and when $x = 30$, $f(30) = 0$. Since $f(x)$ represents a percentage, it must vary between 100% and 0%, so x must be in the interval $[0, 30]$; also, no shipments were made with more than 30% of the eggs below A quality.

Exercise Set 4–2

1. Relative minimum at $(0, -4)$

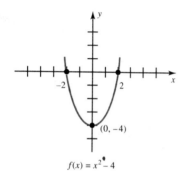

$f(x) = x^2 - 4$

3. Relative maximum at $(0, 2)$

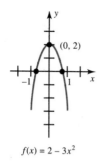

$f(x) = 2 - 3x^2$

5. Relative minimum at $(2, -4)$

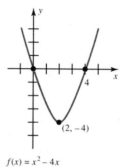

$f(x) = x^2 - 4x$

7. Relative maximum at $(-\frac{2}{3}, \frac{4}{27})$
Relative minimum at $(0, 0)$

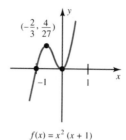

$f(x) = x^2 (x + 1)$

9. Relative maximum at $(0, -1)$
Relative minimum at $(4, -33)$

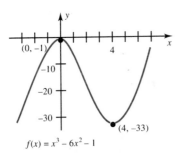

$f(x) = x^3 - 6x^2 - 1$

11. Relative maximum at $(-3, 27)$
Relative minimum at $(1, -5)$

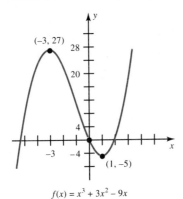

$f(x) = x^3 + 3x^2 - 9x$

13. Relative maximum at $(-1, 5)$
Relative minimum at $(3, -27)$

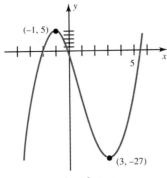

$f(x) = x^3 - 3x^2 - 9x$

15. Relative maximum at $(0, 0)$
Relative minimum at $(-1, -1)$ and $(1, -1)$

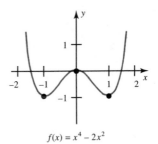

$f(x) = x^4 - 2x^2$

17. Relative maximum at $(2, 16)$ and $(-2, 16)$
Relative minimum at $(0, 0)$

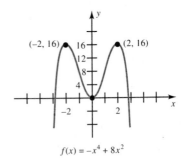

$f(x) = -x^4 + 8x^2$

19. Relative maximum at $(-3, -6)$
Relative minimum at $(3, 6)$

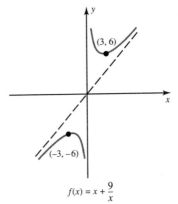

$f(x) = x + \dfrac{9}{x}$

21. Relative maximum at $(-2\sqrt{3}, -4\sqrt{3})$
Relative minimum at $(2\sqrt{3}, 4\sqrt{3})$

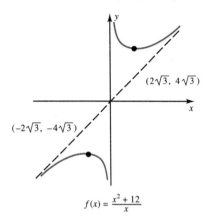

$(2\sqrt{3}, 4\sqrt{3})$

$(-2\sqrt{3}, -4\sqrt{3})$

$f(x) = \dfrac{x^2 + 12}{x}$

23. No relative maximum
Relative minimum at $(3, -27)$

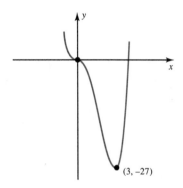

$(3, -27)$

$f(x) = x^3(x - 4)$

25. Relative maximum at $(0, 0)$
Relative minimum at $(-1, -1)$ and $(1, -1)$

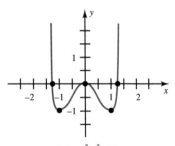

$f(x) = x^2(x^2 - 2)$

27. Relative minimum at $(0, 0)$
No relative maximum

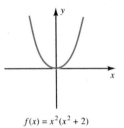

$f(x) = x^2(x^2 + 2)$

29. No relative maximum or minimum

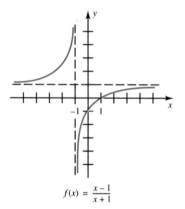

$f(x) = \dfrac{x - 1}{x + 1}$

31. Relative minimum at $(3, 2)$
No relative maximum

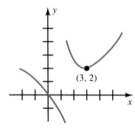

$(3, 2)$

33. Relative maximum at $(5, 4)$
Relative minimum at $(-1, 1)$

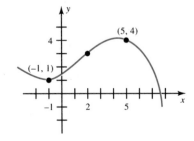

$(5, 4)$

$(-1, 1)$

35. $x = 4$

37. Relative maximum at $(-3, 162)$
Relative minimum at $(3, -162)$

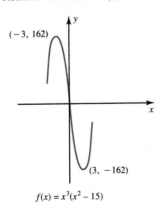

$$f(x) = x^3(x^2 - 15)$$

39. Relative maximum at $(0, 4)$
Relative minimum at $\left(-\dfrac{\sqrt{10}}{2}, -2.25\right)$
and $\left(\dfrac{\sqrt{10}}{2}, -2.25\right)$

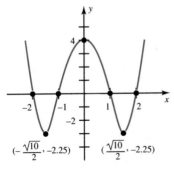

$$f(x) = (x^2 - 1)(x^2 - 4)$$

41. Relative maximum at $(2, 81)$
Relative minimum at $(-1, 0)$ and $(5, 0)$

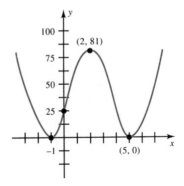

$$f(x) = (x - 5)^2 (x + 1)^2$$

43. Relative minimum at $(5, 0)$
No relative maximum

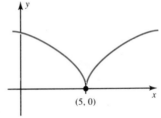

$$f(x) = (x - 5)^{2/3}$$

Referenced Exercise Set 4–2

1. 13,243 hours

3. $x = 1$ and $x = -1$ are excluded

Cumulative Exercise Set 4–2

1. (a) $(-\infty, 0)$
   (b) $(0, 4)$

   (c)
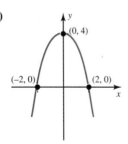

3. (a) $(-\infty, 0)$ and $(0, \infty)$
   (b) None

   (c)
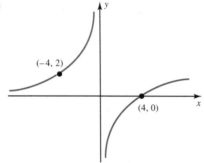

5.
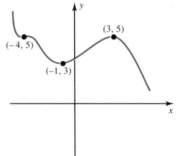

7. Relative minimum at $x = 0$
   Relative maxima at $x = 2$ and $x = -2$

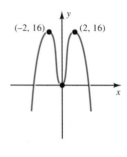

Exercise Set 4–3

1. Concave down: $(-2, 0)$
   Concave up: $(0, 2)$
   Point of inflection: $(0, 0)$
3. Concave down: $(-1, 2)$
   Concave up: $(-2, -1)$
   Point of inflection: $x = -1$
5. Concave down: $(-1, 0)$
   Concave up: $(-2, -1)$ and $(0, 2)$
   Point of inflection: $x = -1, x = 0$
7. Concave up: $(-\infty, \infty)$
   Concave down: nowhere
   No point of inflection
9. Concave up: $(2, \infty)$
   Concave down: $(-\infty, 2)$
   Point of inflection: $x = 2$
11. Concave up: $(-1, \infty)$
   Concave down: $(-\infty, -1)$
   Point of inflection: $x = -1$

13. Relative maximum at $(-1, 5)$
   Relative minimum at $(3, -27)$
15. Relative maximum at $(0, 0)$
   Relative minimum at $(-1, -1)$ and $(1, -1)$
17. Relative maximum at $(2, 16)$ and $(-2, 16)$
   Relative minimum at $(0, 0)$

19.
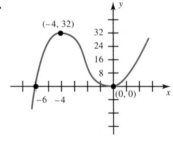

$f(x) = x^3 + 6x^2$

21.

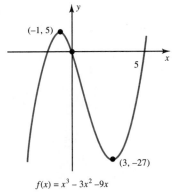

$(-1, 5)$

5

$(3, -27)$

$f(x) = x^3 - 3x^2 - 9x$

23.

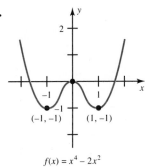

2

-1

1

$(-1, -1)$

$(1, -1)$

$f(x) = x^4 - 2x^2$

25.

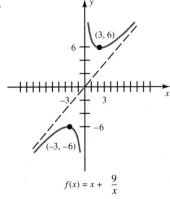

$(2, 96)$

$f(x) = 64x - 2x^4$

27.

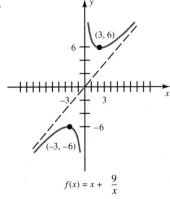

$(3, 6)$

6

-3

3

-6

$(-3, -6)$

$f(x) = x + \dfrac{9}{x}$

29.

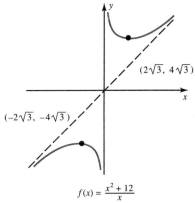

$(2\sqrt{3}, \ 4\sqrt{3})$

$(-2\sqrt{3}, \ -4\sqrt{3})$

$f(x) = \dfrac{x^2 + 12}{x}$

31.

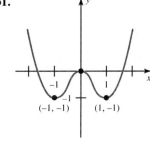

-1

1

$(-1, -1)$

$(1, -1)$

$f(x) = x^2(x^2 - 2)$

33.

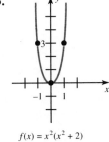

3

-1

1

$f(x) = x^2(x^2 + 2)$

35.

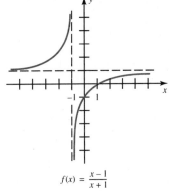

-1

1

$f(x) = \dfrac{x - 1}{x + 1}$

37.

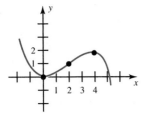

39.

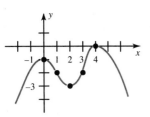

41.

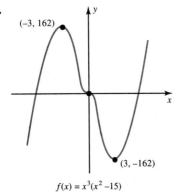

$f(x) = x^3(x^2 - 15)$

43.

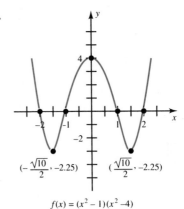

$f(x) = (x^2 - 1)(x^2 - 4)$

45.

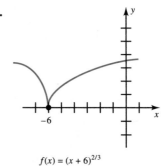

$f(x) = (x + 6)^{2/3}$

47. (a) Increasing in $(-2, -1)$ and $(1, 2)$; decreasing in $(-1, 1)$
   (b) Positive in $(-2, -1)$ and $(1, 2)$; negative in $(-1, 1)$
   (c) Positive in $(0, 2)$; negative in $(-2, 0)$

49. (a) Increasing in $(-2, 0)$ and $(0, 2)$; decreasing in $(-3, -2)$
   (b) Positive in $(-2, 0)$ and $(0, 2)$; negative in $(-3, -2)$
   (c) Positive in $(-3, -\frac{1}{2})$ and $(0, 1)$; negative in $(-\frac{1}{2}, 0)$ and $(1, 2)$

Referenced Exercise Set 4–3

1. 1982
3. 205.83

Cumulative Exercise Set 4–3

1. Increasing on $6 < x < \infty$
Decreasing on $-\infty < x < 6$
Horizontal tangent at $x = 6$

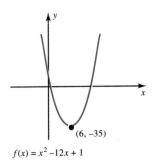

$f(x) = x^2 - 12x + 1$

3. Increasing on $-\infty < x < -4$ and on $4 < x < \infty$
Decreasing on $-4 < x < 0$ and on $0 < x < 4$
Horizontal tangents at $x = -4$ and $x = 4$

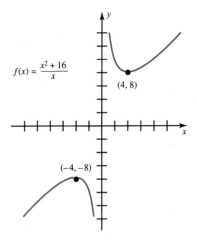

$f(x) = \dfrac{x^2 + 16}{x}$

5. Relative maximum at $x = -1$
Relative minimum at $x = 0$

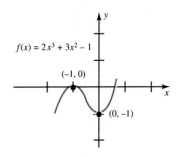

$f(x) = 2x^3 + 3x^2 - 1$

7. Relative maximum at $x = 2$
Relative minimum at $x = -2$

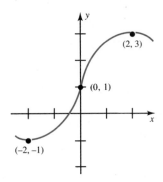

9. Relative maximum at $x = -2$
Relative minimum at $x = 3$

11.

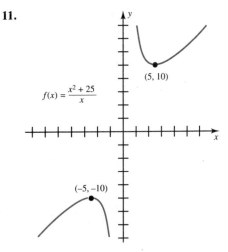

$f(x) = \dfrac{x^2 + 25}{x}$

Exercise Set 4–4

1. Absolute maximum at $f(4) = 25$
   Absolute minimum at $f(-1) = 0$
3. Absolute maximum at $f(-2) = 9$
   Absolute minimum at $f(6) = -55$
5. Absolute maximum at $f(2) = 21$
   Absolute minimum at $f(0) = 1$
7. Absolute maximum at $f(-2) = 5$
   Absolute minimum at $f(-3) = 1$
9. Absolute maximum at $f(-2) = 9$ and $f(2) = 9$
   Absolute minimum at $f(1) = 0$ and $f(-1) = 0$
11. Absolute maximum at $f(4) = 12$
    Absolute minimum at $f(-1) = -18$
13. Absolute maximum at $f(-1) = 64$
    Absolute minimum at $f(1) = 0$ and $f(3) = 0$
15. 256 feet; 4 seconds
17. 10 ft × 20 ft

19. 200 and 100
21. 2000 ft × 1000 ft
23. 6.23 miles from the point he wishes to reach
25. Base: 4 inches × 4 inches
    Side: 2 inches high
27. Base: 3.175 inches × 3.175 inches
    Side: 3.175 inches
29. Length: 28 inches
    Width: 14 inches
31. $\dfrac{6}{1 + \dfrac{\pi}{4}}$ feet = 3.36 feet

33. 6.32 × 9.49
37. Base: 1.88 inches × 1.88 inches
    Height: 2.82 inches

Referenced Exercise Set 4–4

1. (1.98, 17.96)

3. (28, 1.0828)

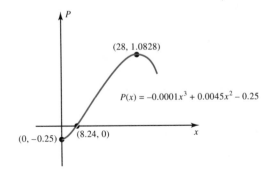

$P(x) = -0.0001x^3 + 0.0045x^2 - 0.25$

Cumulative Exercise Set 4–4

1. (a) $(-\infty, 0)$ and $(\frac{8}{3}, \infty)$
   (b) $x = 0$ and $x = \frac{8}{3}$

   (c)

   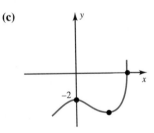

3.

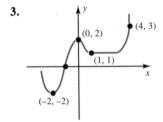

5.

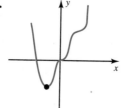

9. Relative maximum at $x = -\frac{3}{5}$
Relative minimum at $x = 1$

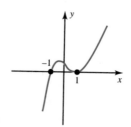

7. Relative maximum at $x = -\sqrt[4]{\frac{1}{5}}$
Relative minimum at $x = \sqrt[4]{\frac{1}{5}}$

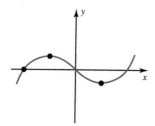

11. $\frac{1}{2}$

Exercise Set 4–5

1. $y' = -\dfrac{x}{3y}$

3. $y' = \dfrac{-y - 2xy}{x + x^2}$

5. $y' = \dfrac{3x^2 + y^3}{-3xy^2 + 4y^3}$

7. $y' = \dfrac{-2x - y^3}{3xy^2 + y^{-2}}$

9. $-\dfrac{2}{3}$

11. 0

13. $D_t y = -\dfrac{2x}{3y^2};\ D_t x = -\dfrac{4}{3}$

15. $D_t y = \dfrac{2x - 3}{2y};\ D_t x = -1$

17. 4.5 mph

19. $\dfrac{dh}{dt} = \dfrac{20}{81\pi}\ \dfrac{\text{cm}}{\text{min}} = 0.0786\ \dfrac{\text{cm}}{\text{min}}$

21. 4 ft/sec

23. $\frac{9}{7}$ ft/sec

25. 0.245 cm/sec

27. $\frac{5}{8}$ ft/sec

29. 12 ft/sec

31. 235.75 mph

33. -1.8 m³/sec

Referenced Exercise Set 4–5

1. 6.58 inches $\times$ 3.42 inches

Cumulative Exercise Set 4–5

1. Increasing on $-\infty < x < 0$ and on $4 < x < \infty$
Decreasing on $0 < x < 4$
Horizontal tangents at $x = 0$ and $x = 4$

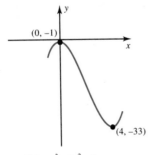

$f(x) = x^3 - 6x^2 - 1$

3.

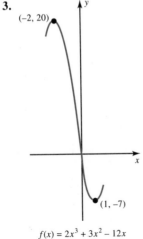

$f(x) = 2x^3 + 3x^2 - 12x$

5.

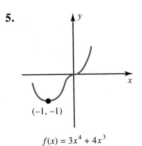

$f(x) = 3x^4 + 4x^3$

7. Absolute minimum $f(-3) = -80$
Absolute maximum $f(0) = 1$

9. Set $x = \dfrac{16 - \sqrt{76}}{6}$;

the box is x by $6 - 2x$ by $10 - 2x$

11. $x' = \dfrac{4(100)^{\frac{3}{4}}}{11}$

Case Study Exercises

1. 29.487   **3.** 25.496
5. 5:00 P.M. because it is rush hour

Review Exercises

1. Increasing on $0 < x < \infty$
Decreasing on $-\infty < x < 0$
Horizontal tangent at $x = 0$

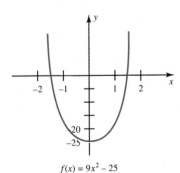

$f(x) = 9x^2 - 25$

3. Increasing on $-\infty < x < 3$ and $3 < x < \infty$
No horizontal tangent

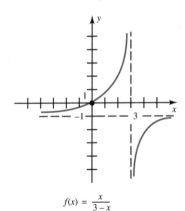

$f(x) = \dfrac{x}{3-x}$

5. Relative minimum at $(-3, -19)$

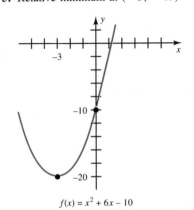

$f(x) = x^2 + 6x - 10$

7. Relative maximum at $(\sqrt{2}, 4)$ and $(-\sqrt{2}, 4)$
Relative minimum at $(0, 0)$

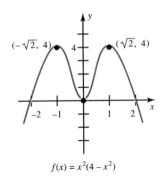

$f(x) = x^2(4 - x^2)$

9. Relative minimum at $(1, -1)$
11. Concave up: $0 < x < 1$
Concave down: $-\infty < x < 0$ and $1 < x < \infty$

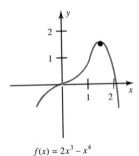

$f(x) = 2x^3 - x^4$

13. Absolute maximum of 4 at $x = 0$
Absolute minimum of -16 at $x = -2$
15. Absolute maximum of 6 at $x = 8$
Absolute minimum of -0.3849 at $x = 0.19245$
17. $y' = \dfrac{-2x - y}{x + 2y}$ **19.** $\frac{1}{24}$ m/min

CHAPTER 5

Exercise Set 5–1

1.

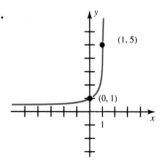

$f(x) = 5^x$

3.

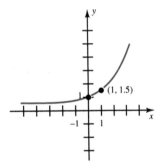

$f(x) = 1.5^x$

5.

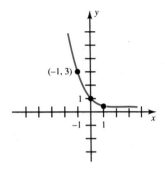

$f(x) = (\frac{1}{3})^x$

7.

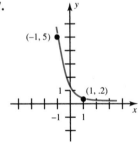

$f(x) = .2^x$

9. 5   **11.** 4   **13.** 2

15.

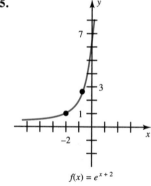

$f(x) = e^{x+2}$

17.

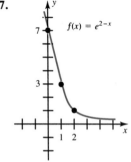

$f(x) = e^{2-x}$

19. (a) 59.6   **(b)** 60
21. (a) 27.8   **(b)** 28

23.

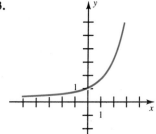

$$f(x) = 1.1^x$$

25.

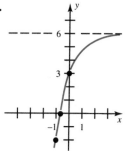

$$f(x) = 6 - 3e^{-x}$$

27.

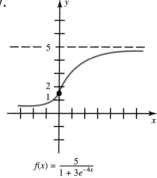

$$f(x) = \frac{5}{1 + 3e^{-4x}}$$

29.

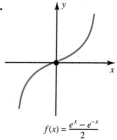

$$f(x) = \frac{e^x - e^{-x}}{2}$$

31. 3
33. 0
35.

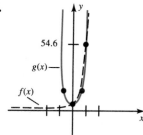

37. 81
39. **(a)** 16°F   **(b)** 10°F

41. **(a)**

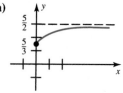

(b) Increasing   **(c)** Level off and approach 2.5 million

43. No

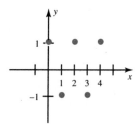

45. **(a)** $e^3 \approx 20.1$ **(b)** $e^4 \approx 54.6$

Referenced Exercise Set 5–1

1. $U(G) \approx 0.808 < 0.817$
risk-averse

3. **(a)** $t = 3$, $P(t) = 0.039203$
(b) $t = 8$, $P(t) = 0.5722409$
(c) $t = 16$, $P(t) = 0.997199$

Case Study Exercises

1. 22 pounds
3. **(a)** 1925 calories   **(b)** 4 pounds
5. 7.9 pounds

7. Let w be your present weight.
(a) $w(t) = (w - 10) + 10e^{-0.005t}$
(b) $(w - 4)$ pounds

Exercise Set 5–2

1.

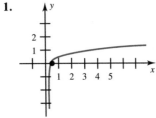

$f(x) = \log_{10} 2x$

3.

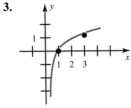

$f(x) = \log_3 x$

5.

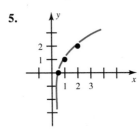

$f(x) = \ln (3x)$

7.

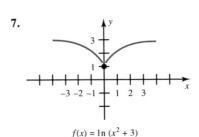

$f(x) = \ln (x^2 + 3)$

9. $\ln 3 + 4 \ln x$
11. $\ln 3 + \ln x + 2 \ln y - \ln(x - y)$
13. **(a)** $3x$ **(b)** $5 - x$
15. **(a)** 5   **(b)** 0.5
17. 11.9 years   **19.** 3.8 years

21.

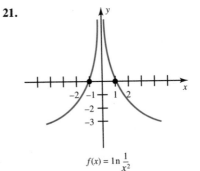

$f(x) = \ln \dfrac{1}{x^2}$

23.

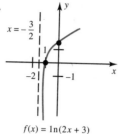

$f(x) = \ln(2x + 3)$

25.

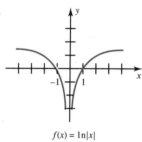

$f(x) = \ln|x|$

27. $x \neq 0$
29. $x \neq 0$
31. $\frac{1}{2}[\ln x + \ln y]$
33. $-\frac{1}{2}[3 \ln x + \ln y]$
35. $\ln(e^x) = x$
37. $\dfrac{\ln 3}{2} \approx 0.5493$ **39.** $-\ln 20$
41. $e^{0.2}$ **43.** $\dfrac{2}{\ln 3}$
45. (a) 1   (b) 4
47. (a) $\frac{1}{3}$ (b) $\frac{81}{64}$
49. (a) error   (b) error
51. about 11 minutes
53. $\ln 2/\ln 1.1 \approx 7.3$ years
55. $\ln 4/\ln 1.05 \approx 28.4$ years
57. $t = \dfrac{\ln 3}{\ln 1.08} \approx 14.27$ years
59. $t = \dfrac{\ln(\frac{1}{2})}{-0.005} \approx 138.6$ days

Referenced Exercise Set 5–2

1. (a) 0.8 hour   (b) 0.7 hour   (c) 0.48 hour
    (d) 0.23 hour
3. (a) .95 hour   (b) .92 hour   (c) .84 hour
    (d) .71 hour

5. $R(0.2) = 0.16$
7. The year 2022

Cumulative Exercise Set 5–2

1.

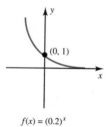

(0, 1)

$f(x) = (0.2)^x$

3. $x = -1$
5. $2 \ln |x| + 3 \ln |y| - \ln|x - y|$
7. $x = e^{-1/6}$

Case Study Exercises

1. $M = 3.2,$ $S \approx 2.56$
3. \$0.50 to \$1.00
5. \$22.63   **7.** \$128

Exercise Set 5–3

1. $\dfrac{1}{x}$

3. $\dfrac{1}{x}$

5. $\dfrac{1}{x + 5}$

7. $\dfrac{2x}{x^2 + 3}$

9. $\dfrac{8x}{1 + 4x^2}$

11. $\dfrac{6x^2 - \dfrac{1}{x^2}}{2x^3 + \dfrac{1}{x}} = \dfrac{6x^4 - 1}{2x^5 + x}$

13. $1 + \ln 4x$

15. $-\dfrac{3 \ln(x^2 - 1)}{x^4} + \dfrac{2}{x^2(x^2 - 1)}$

17. $\dfrac{\dfrac{(2x - 5)}{x} - 2 \ln 3x}{(2x - 5)^2} = \dfrac{(2x - 5) - 2x \ln 3x}{x(2x - 5)^2}$

19. $\dfrac{-5}{3 - 5x} = \dfrac{5}{5x - 3}$

21. $\dfrac{-\frac{5}{2}x^{3/2}}{4 - x^{5/2}}$

23. $1 + \ln |7x|$

25. Minimum at $x = e^{-1}$

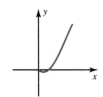

27. Maximum at $x = e$

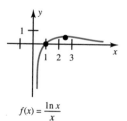

$f(x) = \dfrac{\ln x}{x}$

29. Inelastic in $[5, 8)$, elastic in $(8, 15]$

31. $2(x + \ln x)\left(1 + \dfrac{1}{x}\right)$

33. $\dfrac{2x \ln x - x}{(\ln x)^2}$

35. $\dfrac{2x + 1/x}{x^2 + \ln x} = \dfrac{2x^2 + 1}{x(x^2 + \ln x)}$

37. $\dfrac{10}{x} \cdot \dfrac{1}{\ln 10} = \dfrac{10}{x \ln 10}$

41. $f''(x) = \dfrac{(-1)^{n+1}(n - 1)!}{x^n}$

43. $f(2) \approx 1.39$ but $g(2) \approx 0.48$; $f'(2) = 1$ but $g'(2) = \ln 2 \approx 0.69$

Referenced Exercise Set 5–3

1. $A = me$, $A'(1) = e$

Cumulative Exercise Set 5–3

1.

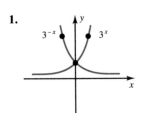

3. (a) $x = -2$ **(b)** $x = 6$ **(c)** $x = 2$

5.

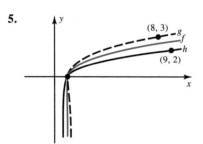

7. (a) $x = 3$ **(b)** $x = 1$

9. $\dfrac{-2x}{4 - x^2}$

11. $(1 - \ln x)^2 - 2(1 - \ln x)$ or $-(1 - \ln x)(1 + \ln x)$

Exercise Set 5–4

1. $6e^{6x}$

3. $3e^{3x + 5}$

5. $28e^{7x + 5}$

7. $2xe^{x^2 + 3}$

9. $16xe^{1 + 4x^2}$

11. $(x^{-2} - 6x^2)e^{2x^3 + x^{-1}}$

13. $(4x + 1)e^{4x}$

15. $(2x^{-2} - 3x^{-4})e^{x^2 - 1}$

17. $\dfrac{(6x - 17)e^{3x}}{(2x - 5)^2}$

19. $5e^{5x} - 5e^{-5x}$

21. $\dfrac{e^{-x}}{(2 + e^{-x})^2}$

23. $\dfrac{2e^{-2x} - e^{x}}{(1 + e^{x} + e^{-2x})^2}$

25.

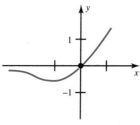

$f(x) = xe^x$

27.

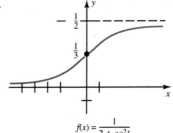

$f(x) = \dfrac{1}{2 + e^{-2x}}$

29. $\dfrac{1 + 4e^{4x}}{x + e^{4x}}$

31. $2(\ln x + e^{5x})\left(\dfrac{1}{x} + 5e^{5x}\right)$

33. $\dfrac{(2x - 1)e^{x^2} + 2x - x^2}{e^x}$

35. $\left(\dfrac{1}{x} + 3x^2 \ln x\right)e^{x^3}$

37. $f^n(x) = 2^n e^{2x}$

39. Figure c

41. Figure d

Referenced Exercise Set 5–4

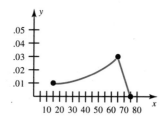

3. 1.98 months

Cumulative Exercise Set 5–4

1.

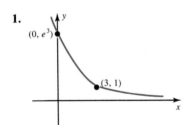

$f(x) = e^{3-x}$

3. $\frac{2}{3}$

5. $2x \ln(1 - x^3) - 3x^4/(1 - x^3)$

7. $(x - \ln x)^2/x + 2(\ln x)(x - \ln x)(1 - 1/x)$

9. $-10xe^{2-5x^2}$

11. $\dfrac{10e^{-5x}}{(3 + e^{-5x})^2}$

Review Exercises

1.

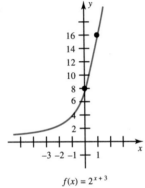

$f(x) = 2^{x+3}$

3.

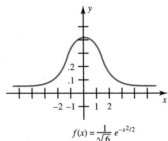

$f(x) = \dfrac{1}{\sqrt{6}} e^{-x^2/2}$

5.

$f(x) = \ln(4 - x^2)$

7. $-\ln(2.1) \approx -0.74$

9. $e^2 \approx 7.4$

11. $\dfrac{60[\ln(1 + 3x)^4]^4}{1 + 3x}$

13. $\dfrac{1}{2x}$

15. $x(x + 2)e^x + e^{-x}$

17. Absolute minimum at $x = 0$
No points of inflection

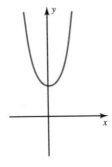

CHAPTER 6

Exercise Set 6–1

1. $\frac{3}{2}x^2 + 5x + C$
3. $\frac{2}{5}x^5 - 7x + C$
5. $-6x^{-1} + \frac{5}{2}x^2 + C$
7. $\frac{4}{3}x^{3/2} + \frac{9}{4}x^{4/3} - \frac{5}{2}x^2 + C$
9. $\frac{9}{2}x^{2/3} + \frac{28}{3}x^{3/4} - 5x + C$
11. $\frac{x^{1.2}}{1.2} + \frac{3x^{2.4}}{2.4} + C$
13. $-x^{-1} - \frac{5}{3}x^{-3} + C$
15. $-\frac{1}{6x^6} + \frac{16x^{3/4}}{3} + \frac{16x^{5/4}}{5} + C$
17. $\frac{3e^{2x}}{2} + 5\ln|x| + C$
19. $\frac{1}{4}x^4 - 2e^{-x} + \frac{2}{3}x^{3/2} - 5\ln|x| + C$

21. $\frac{10}{3}x^{3/2} + \frac{1}{2}e^{-4x} + \ln|x| + C$
23. $18\ln|x| + \frac{3}{\sqrt[3]{4}}x^{2/3} - \frac{1}{7}e^{7x} + C$
25. $y = \frac{1}{3}x^3 - 2x + \frac{2}{3}$
27. $y = 4\ln|x| - x^2 + 2 - 4\ln 2$
29. $\frac{2}{3}x^3 - 7\ln|x| + C$ **31.** $6x + \frac{5}{4}x^4 + C$
33. $\frac{4}{9}x^9 - \frac{14}{3}x^6 + \frac{49}{3}x^3 + C$
35. $S(t) = \frac{t^3}{3} + t^2$
37. 3257 years
39. $k = -6.93 \times 10^{-4}$
41. $k = -0.0693$

Referenced Exercise Set 6–1

1. 6,969 years   **3.** $m' = -0.068m$

Exercise Set 6–2

1. $3.125 = \frac{25}{8}$ **3.** $2.71875 = \frac{174}{64}$
5. $11.375 = \frac{91}{8}$

7.
n	2	4	8	10
area	8	7	6.5	6.4

9.
n	2	4	8	10
area	7	5.75	5.1875	5.08

11.
n	2	4	8	10
area	14	11.5	10.375	10.16

13. 24   **15.** 20
17. 6   **19.** 2
21. 10.5   **23.** 6
25. $\frac{64}{3}$ **27.** $\frac{140}{3}$
29. $\frac{5}{6}$ **31.** 1.0986

Referenced Exercise Set 6–2

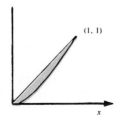

(1, 1)

x

Cumulative Exercise Set 6–2

1. $2x^2 + (6/x) + C$

3. $4e^x - 3(\ln |x|) + C$

5. $s(t) = 2 \cdot \ln(t^2 + 1) + 4$

7. 1.94

Exercise Set 6–3

1. $\frac{13}{2}$

3. $\frac{4}{3}$

5. $-\frac{5}{6}$

7. $-\dfrac{476}{3}$

9. $\dfrac{481}{32}$

11. 0

13. 12.5

15. $\frac{64}{3}$

17. 1

19. $\frac{1}{2}$

21. 25

23. 20

25. 2.08

27. 71.5068

29. -17.2047

31. 5

33. $\frac{59}{12}$

35. 0.5754

37. -1

39. -1

Referenced Exercise Set 6–3

1. 1,000

3. 19.84

Cumulative Exercise Set 6–3

1. $-3x^{-1} - x^5/5 + 5x + C$

3. $3e^{2x}/2 - 5 \ln |x| + C$

5. 30

7. 42

9. $\frac{23}{4}$

11. 24

Exercise Set 6–4

1. $\frac{14}{3}$

3. $\frac{20}{3}$

5. 10

7. $\frac{1}{6}$

9. $\frac{1}{6}$

11. $\frac{1}{6}$

13. $\frac{1}{6}$

15. $\frac{1}{6}$

17. $\frac{32}{3}$

19. $\frac{32}{3}$

21. $\frac{1}{12}$

23. 12

25. 13

27. 13

29. 16

31. C.S. $= \frac{32}{3}$

P.S. $= \frac{40}{3}$

33. C.S. $= \frac{76}{3}$

P.S. $= \frac{76}{3}$

35. $1 - \ln 2 \cong 0.3069$

37. 8.107

39. 6.08

Referenced Exercise Set 6–4

1. C.S. $= 3166.67$

3. C.S. $= \frac{7}{32}$, P.S. $= \frac{7}{192}$

Cumulative Exercise Set 6–4

1. $(x^2/6) + C$

3. $f(x) = (\ln x) + 5$

5. (a) $\frac{29}{4}$ (b) $\frac{22}{3}$

7. $\frac{64}{5}$

9. 8

11. $\frac{1}{2}$

Case Study Exercises

1. 4.004   **3.** 13.2

5. 0.36

Review Exercises

1. $\frac{25}{4}x^{4/5} - 5x^{2/5} - \frac{3}{2}x^2 + C$

3. $y = -\dfrac{1}{x} - \dfrac{x^2}{2} + \dfrac{3}{2}$

5. 8

9. 637.519

7. 344.16

11. $\frac{1}{6}$

CHAPTER 7

Exercise Set 7–1

1. $\dfrac{(x^2 + 5)^4}{4} + C$ **3.** $\dfrac{(x^4 - 7)^3}{3} + C$

5. $\dfrac{(3x^2 + 5)^4}{24} + C$ **7.** $\dfrac{(3x^4 - 5)^4}{48} + C$

9. $\dfrac{(3x^6 - 5)^{5/2}}{45} + C$ **11.** $\dfrac{-(x^4 - 7)^{-2}}{4} + C$

13. $\dfrac{(6x^{-2} + 5)^{-1}}{12} + C$ **15.** $\dfrac{(x^2 + 5x)^{-2}}{-2} + C$

17. $\dfrac{(x^4 - 4x)^{3/2}}{6} + C$

19. $\frac{1}{2}e^{2x} + \frac{5}{4}e^{4x} + C$

21. $\frac{1}{4}e^{x^4} - 2e^{-x} + C$

23. $\frac{2}{3} \ln |3x + 1| + C$

25. $\frac{1}{3} \ln |x^3 - 1| + C$

27. $\frac{7}{6}$

29. 0

31. $\frac{1}{3}(10^{3/2} - 1) = 10.2076$

33. $\dfrac{19}{480}$ **35.** $\dfrac{(e^x + 1)^2}{2} + C$

37. $\dfrac{(\ln x)^2}{2} + C$

39. $\frac{1}{4}x^4 - \frac{2}{3}x^3 + \frac{1}{2}x^2 + C$

41. $\dfrac{-2}{375}(1 - 5x)^{3/2}(15x + 2) + C$

43. $\frac{3}{16}(x^2 - 1)^{8/3} + \frac{3}{10}(x^2 - 1)^{5/3} + C$

45. $\frac{1}{4}[\ln(1 + x^2)]^2 + C$

Referenced Exercise Set 7–1

1. $0.5(x + 1) + 0.3 \ln(x + 1) + C$

Exercise Set 7–2

1. $2xe^x - 2e^x + C$

3. $\frac{1}{3}xe^{3x} - \frac{1}{9}e^{3x} + C$

5. $\dfrac{2}{15}x(3x + 5)^{5/2} - \dfrac{4}{315}(3x + 5)^{7/2} + C$

7. $-\frac{3}{10}x(5x - 2)^{-2} - \frac{3}{50}(5x - 2)^{-1} + C$

9. $\frac{1}{2}x^2 \ln x - \frac{1}{4}x^2 + C$

11. $\frac{1}{3}x^3 \ln x - \frac{1}{9}x^3 + C$

13. $x^2e^x - 2xe^x + 2e^x + C$

15. $-\dfrac{1}{2}(2x + 5)(x + 5)^{-2} - (x + 5)^{-1} + C$

17. $\dfrac{2}{3}x^2(x - 4)^{3/2} - \dfrac{8}{15}(x)(x - 4)^{5/2}$

$+ \dfrac{16}{105}(x - 4)^{7/2} + C$

19. $\frac{1}{3}x^2(1 - 3x)^{-1} - \frac{2}{27}(1 - 3x) + \frac{2}{27} \ln |1 - 3x| + C$

21. $\frac{1}{4}e^2 + \frac{1}{4}$

23. $\frac{5\sqrt{5}}{3} + \frac{9}{5} \approx 5.5268$

25. $\frac{8}{81}$

27. $\frac{2}{3} x^2(x + 1)^{3/2} - \frac{8}{15} x(x + 1)^{5/2}$

$$+ \frac{16}{105} (x + 1)^{7/2} + C$$

29. $\frac{2}{3} x^3(x + 1)^{3/2} - \frac{4}{5} x^2(x + 1)^{5/2}$

$$+ \frac{16}{35} x(x + 1)^{7/2} - \frac{32}{315} (x + 1)^{9/2} + C$$

31. $\frac{1}{3}x^2(x^2 + 1)^{3/2} - \frac{2}{15}(x^2 + 1)^{5/2} + C$

33. $\frac{1}{2}x^2e^{x^2} - \frac{1}{2}e^{x^2} + C$

35. $\frac{2942}{105} \approx 28.0190$

37. $\frac{x^2}{2} \ln x - \frac{x^2}{4} + x \ln x - x + C$

39. $-\frac{\ln x}{x} - \frac{1}{x} + C$

Referenced Exercise Set 7–2

1. 0.3093

Cumulative Exercise Set 7–2

1. $\frac{(2x^4 + 3)^6}{48} + C$

3. $\frac{3(x^4 - 4x^2)^{5/3}}{20} + C$

5. $3xe^{2x}/2 - 3e^{2x}/4 + C$

7. $\frac{x^6 \ln x}{6} - \frac{x^6}{36} + C$

Exercise Set 7–3

1. $\frac{1}{2(3x + 2)} + \frac{1}{4} \ln \left| \frac{x}{3x + 2} \right| + C$

3. $-\frac{1}{4(3x - 4)} + \frac{1}{16} \ln \left| \frac{x}{3x - 4} \right| + C$

5. $\frac{1}{6} \ln \left| \frac{1 + 3x}{1 - 3x} \right| + C$

7. $\frac{1}{24} \ln \left| \frac{4 + 3x}{4 - 3x} \right| + C$

9. $\frac{1}{24} \ln \left| \frac{3 + x}{3 - x} \right| + C$

11. $\frac{1}{2}[\frac{1}{2}x^2\sqrt{x^4 + 9} + \frac{9}{2} \ln |x^2 + \sqrt{x^4 + 9}|] + C$

13. $\frac{1}{4}[x^2\sqrt{4x^4 + 9} + \frac{9}{2} \ln |2x^2 + \sqrt{4x^4 + 9}|] + C$

15. $\frac{1}{2}[x^2\sqrt{4x^4 - 25} + \frac{25}{2} \ln |2x^2 + \sqrt{4x^4 - 25}|] + C$

17. $-\frac{1}{5}(\ln \frac{3}{4} - \ln 2) = 0.196$

19. $\frac{1}{4}(\ln \frac{1}{4} - \ln \frac{1}{5}) - \frac{3}{80} \approx 0.018$

21. $\frac{\sqrt{2}}{4} + \frac{1}{4} \ln |1 + \sqrt{2}|$

23. $\frac{1}{3}x^3e^{3x} - \frac{1}{3}x^2e^{3x} + \frac{2}{9}xe^{3x} - \frac{2}{27}e^{3x} + C$

25. $\frac{2}{25(5x + 2)} + \frac{1}{25} \ln |5x + 2| + C$

27. $-\frac{1}{4} \ln \left| \frac{x}{x - 4} \right| + C$

29. $-\ln \left| \frac{1 + \sqrt{1 - 9x^2}}{3x} \right| + C$

31. $\frac{1}{6} \ln |3x^2 + \sqrt{16 + 9x^4}| + C$

33. $-\frac{1}{3} \ln \left| \frac{3 + \sqrt{9 + 4x^2}}{2x} \right| + C$

35. $\frac{1}{36} \ln \left| \frac{3 + 2x^3}{3 - 2x^3} \right| + C$

37. $\frac{x^6}{2} \left[\frac{\ln x^2}{3} - \frac{1}{9} \right] + C$

Referenced Exercise Set 7–3

1. $\dfrac{-t^2}{2} - 10t - 100 \ln |t - 10| + C$

Cumulative Exercise Set 7–3

1. $(\frac{2}{15})(1 + 5x)^{3/2} + C$

3. $9x - 2x^3 + (\frac{1}{5})x^5 + C$

5. $-xe^{-x} - e^{-x} + C$

7. $\frac{1}{2} \ln(2x - 2) + \frac{1}{2}x + C$

9. $\frac{3}{8}$

11. $\frac{8}{3}$

Exercise Set 7–4

1. 9.5

3. 73

5. 22.5

7. 333

9. 23.88

11. 0.8863

13. 1.683

15. 1.644

17. 0.9436

19. 0.6766

21. 2.72

23. 62.48

25. 38.5861

27. 11.5345

29. 2.748

31. 58.489

33. $\frac{73}{45}$

35. Simpson's Rule: 1.62
Trapezoidal Rule: 1.68
Fundamental Theorem: 1.61

37. Simpson's Rule: 1.609
Trapezoidal Rule: 1.614
Fundamental Theorem: 1.609

Referenced Exercise Set 7–4

1. 913.5

3. 413.02

Cumulative Exercise Set 7–4

1. $\dfrac{(x^6 + 6x)^{4/3}}{8} + C$

3. $\dfrac{4x^{5/4} \ln x}{5} - \dfrac{16x^{5/4}}{25} + C$

5. $\dfrac{1}{3(2x + 3)} + \dfrac{1}{9} \ln \left| \dfrac{x}{2x + 3} \right| + C$

7. $\dfrac{x^2}{4}(25 + x^4)^{1/2} + \dfrac{25}{4} \ln (x^2 + (25 + x^4)^{1/2}) + C$

9. 22

11. 0.886

Case Study Exercises

1. 1.7%

3. 2.48%

5. 3.2% (from Exercise 4)

Review Exercises

1. $\dfrac{(x^2 + x)^{-2}}{-2} + C$ **3.** $-\frac{15}{8}$

5. $-\frac{3}{4}(1 - x)^{4/3} + \frac{6}{7}(1 - x)^{7/3} - \frac{3}{10}(1 - x)^{10/3} + C$

7. $-\dfrac{1}{7(5x - 7)} + \dfrac{1}{49}\ln\left|\dfrac{x}{5x - 7}\right| + C$

9. $\dfrac{1}{6\sqrt{14}}\ln\left|\dfrac{\sqrt{14} + 3x}{\sqrt{14} - 3x}\right| + C$

11. 0.8862

CHAPTER 8

Exercise Set 8–1

1. 1   **3.** -9

5. 2   **7.** 63

9. All ordered pairs for which $x \neq 2$

11. All ordered pairs for which $x > -1$

13.

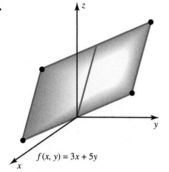

$f(x, y) = 3x + 5y$

15.

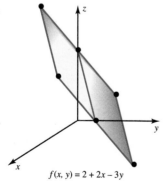

$f(x, y) = 2 + 2x - 3y$

17.

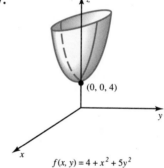

$(0, 0, 4)$

$f(x, y) = 4 + x^2 + 5y^2$

19.

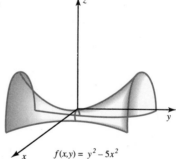

$f(x,y) = y^2 - 5x^2$

21.

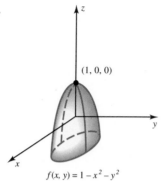

$(1, 0, 0)$

$f(x, y) = 1 - x^2 - y^2$

23.

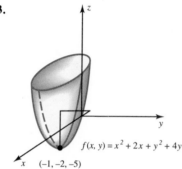

$f(x, y) = x^2 + 2x + y^2 + 4y$

$(-1, -2, -5)$

25.

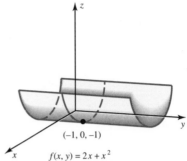

$(-1, 0, -1)$

$f(x, y) = 2x + x^2$

27. No

29. A vertical line will cut the surface in no more than one point.

31. $P = 40x + 50y + 60z$

33. Figure a

35. Figure d

Referenced Exercise Set 8–1

1. Tuesday, \$142,108

3. $y \approx 0.001$

Exercise Set 8–2

1. $f_x(x, y) = 6x$
  $f_y(x, y) = \frac{5}{2}y^{-1/2}$

3. $f_x(x, y) = y^2 - 10x$
  $f_y(x, y) = 2xy$

5. $f_x(x, y) = y^2 - 6xy$
  $f_y(x, y) = 2xy - 3x^2$

7. $f_x(x, y) = -2x^{-3} + y^{-2} - 2xy^3$
  $f_y(x, y) = -2xy^{-3} - 3x^2y^2 + 4$

9. $f_x(x, y) = e^{2y} - \dfrac{2}{2x + 3y}$

  $f_y(x, y) = 2xe^{2y} - \dfrac{3}{2x + 3y}$

11. $f_x(x, y) = 8(2x + 3y)^3$
  $f_y(x, y) = 12(2x + 3y)^3$

13. $f_x(x, y) = -2(2 + 4y)(2x + 4xy)^{-3}$
  $f_y(x, y) = -8x(2x + 4xy)^{-3}$

15. $f_x(1, 0) = 0$
  $f_y(2, 1) = 20.5$

17. $f_x(1, 0) = 1$
  $f_y(2, 1) = -18 + e$

19. $f_{xx}(x, y) = 6$
  $f_{xy}(x, y) = 0$

  $f_{yy}(x, y) = \dfrac{30}{y^4}$

21. $f_{xx}(x, y) = 6y$
  $f_{xy}(x, y) = 6x + 10e^{2y}$
  $f_{yy}(x, y) = 20xe^{2y}$

23. $f_{xx} = 2y^2e^x + xy^2e^x + 4e^{2x}$
  $f_{yy} = 2xe^x$
  $f_{xy} = 2ye^x + 2xye^x$

25. $x = 1$ and $y = 1$

27. $f_{xxx} = 24x - 6$
  $f_{xxy} = 0$
  $f_{xyy} = 0$
29. $f_{xxx} = 0$
  $f_{xxy} = 2$
  $f_{xyy} = -4e^{2y}$

31. $f_x(x, y, z) = \lim_{h \to 0} \dfrac{f(x + h, y, z) - f(x, y, z)}{h}$

33. $f_x = -6x^{-3} + \frac{1}{3}x^{-2/3}z - 2xz^3 + yz$
  $f_y = xz$

Referenced Exercise Set 8–2

1. $\dfrac{\partial C}{\partial Q} = 1.87Q^{0.1}F^{1.5}L^{0.36}P^{0.5}$

  $\dfrac{\partial C}{\partial F} = 2.55Q^{1.1}F^{0.5}L^{0.36}P^{0.5}$

  $\dfrac{\partial C}{\partial L} = 0.612Q^{1.1}F^{1.5}L^{-0.64}P^{0.5}$

  $\dfrac{\partial C}{\partial P} = 0.85Q^{1.1}F^{1.5}L^{0.36}P^{-0.5}$

3. (a) $0.07D^{-0.44}A^{0.06}$ **(b)** $0.0075D^{0.56}A^{-0.94}$

Cumulative Exercise Set 8–2

1. (a) 99 **(b)** -101
3. The plane through the triangle traced in the accompanying figure.

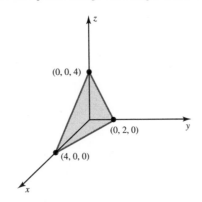

5. $f_x = \dfrac{2x}{y}, f_y = -\dfrac{x^2}{y^2}$

7. $f_x(0, 1) = 2, f_y(0, 1) = 1$

Exercise Set 8–3

1. Relative minimum at $(0, 0)$
3. No
5. No   **7.** $(-3, -4)$
9. $(3, -3)$
11. $(0, +1)$ and $(0, -1)$

13. Saddle point at $(-2, 0)$
  No relative maximum or minimum
15. Relative minimum at $(4, 0)$
  Saddle point at $(4, -2)$

17. Relative minimum at $(0, -1)$
Saddle point at $(0, 1)$
19. Relative minimum at $(4, 3)$
21. Saddle point at $(-3, 1)$
23. Relative minimum at $(2, 2)$
Saddle point at $(0, 0)$
25. Saddle point at $(0, 1)$
Saddle point at $(0, -1)$

27. Saddle point at $(4, 2)$ and $(0, -2)$
Relative minimum at $(4, -2)$ and relative maximum at $(0, 2)$
29. Saddle point at $(4, -2.225)$
Relative minimum at $(4, 0.225)$
31. $(12, 2)$ is a relative minimum.
33. $l \times w \times h = 4\sqrt[3]{2} \times 4\sqrt[3]{2} \times 2\sqrt[3]{2}$

Referenced Exercise Set 8–3

1. $x = \dfrac{a}{2b},\ y = 0$

Cumulative Exercise Set 8–3

1.

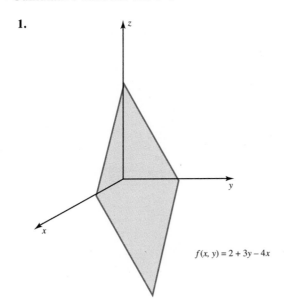

$f(x, y) = 2 + 3y - 4x$

3.

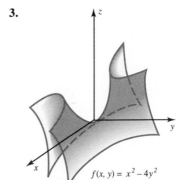

$f(x, y) = x^2 - 4y^2$

5. $f_x(x, y) = 3x^2 - 4xy$
$f_y(x, y) = -2x^2 + 4y^3$
7. $f_x(x, y) = 12x^2(x^3 + 5y)^3$
$f_y(x, y) = 20(x^3 + 5y)^3$
9. Saddle point at $(0, -2)$
11. Saddle point at $(6, -9)$

Exercise Set 8–4

1. $(\frac{1}{2}, \frac{1}{2})$ **3.** $(-1, \frac{1}{2})$
5. $(-1, 0)$ and $(-\frac{5}{2}, \frac{9}{4})$
7. $(1.63, 0.185)$ and $(0.2, 0.9)$
9. $(19.26, 4.63)$ and $(7.4, -1.3)$
11. $(\frac{1}{4}, \frac{1}{2})$
13. $(-\frac{1}{2}, \frac{2}{3})$ **15.** $(\frac{1}{3}, \frac{2}{3}, \frac{2}{3})$
17. $(100, 100, 100)$

19. The numbers are 10, 10, 10.
21. $(25, 2500)$
23. $(200, 200, 100)$
25. $\sqrt{10}/2$ at $(\frac{3}{2}, \frac{1}{2})$
27. $(\frac{1}{2}, 0)$, $(\frac{1}{6}, -\frac{2}{3})$
29. $(\frac{1}{4}, \frac{1}{6}, 1)$ **31.** $(2, 0, 0)$, $(\frac{6}{15}, \frac{16}{15}, \frac{8}{15})$
33. $(1, 1, 1)$ **35.** $x = y = 2, z = 4$

37. $x = y = 14$, $z = 28$

39. (a) $C(x, y) = 1000 + 100x + 50y$

   (b) $x = y = 2.5$

41. $P(x, y) = 160x + 80y - 20x^2 - 4y^2 - 8xy - 15$; $x = 2.5$, $y = 7.5$

43. $6 \times 6 \times 6$ inches

Referenced Exercise Set 8–4

1. $\left(\sqrt{\dfrac{27a}{118}}, \sqrt{\dfrac{32a}{177}} \right)$

Cumulative Exercise Set 8–4

1. (a) 0   (b) $\{(x, y) | y \neq 0\}$

3. $f_x = \dfrac{-y^2}{(x - y)^2}$

   $f_y = \dfrac{x^2}{(x - y)^2}$

5. No

7. Relative minimum at $(4, 1)$

9. $x = 9$, $y = 7$

11. $x = 3$, $y = 6$, $z = 6$

Exercise Set 8–5

1. $\dfrac{7x^2 y^6}{2} + \dfrac{3x^4 y^{-2}}{2} - 5x + C(y)$

3. $xy^8 - 4x^{-4}y^5 - 3xy - \frac{1}{2}y^2 + C(x)$

5. $31y + 2y^3 \ln 2 - 5$

7. $10x^4 + \dfrac{8}{x} - 10$

9. -2 **11.** 674.44

13. $\frac{32}{3}$ **15.** 3

17. $\dfrac{2x}{9}(x^2 + 3y)^{3/2} + C(x)$

19. $\ln |3x^2 + 2xy| + C(y)$

21. $\frac{1}{3}(x^2 + 3y)^{3/2} + C(y)$

23. $\dfrac{2x}{9}(x^2 + 15)^{3/2} - \dfrac{2x}{9}(x^2 + 12)^{3/2}$

25. $\ln |75 + 10y| - \ln |27 + 6y|$

27. $\frac{1}{3}(16 + 3y)^{3/2} - \frac{1}{3}(3y)^{3/2}$

29. $(\ln 2)(\ln 3)$

31. 3.05

33. $\frac{1}{2}(e^2 - e - 1 + \dfrac{1}{e})$

35. 256

37. (a) $\frac{832}{9}$ (b) $\frac{832}{9}$

39. $(\frac{1}{3})e^6 - e^2 + (\frac{2}{3})$

43. $\displaystyle\int_0^1 \int_0^{2 - x^2} dy\, dx$ and $\displaystyle\int_0^2 \int_0^{\sqrt{2 - y}} dx\, dy$

45. $\displaystyle\int_{-1}^0 \int_{x^2}^{2 + x} dy\, dx + \int_0^1 \int_{x^2}^{2 - x} dy\, dx$

   and $\displaystyle\int_0^1 \int_{-\sqrt{y}}^{\sqrt{y}} dx\, dy + \int_1^2 \int_{y - 2}^{2 - y} dx\, dy$

Referenced Exercise Set 8–5

1. 95.9 k

Cumulative Exercise Set 8–5

1.

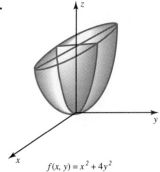

$f(x, y) = x^2 + 4y^2$

3. $f_x(x, y) = 2x - 6x^2y$
$f_y(x, y) = -2x^3 + 2y^{-2/3}$
5. Saddle point at $(0, 6)$
7. $(\frac{1}{6}, \frac{1}{4})$
9. 1
11. -4.062

Exercise Set 8–6

1. 1/15
3. 1/9
5. 424/21
7. 2/3
9. 13/70
11. 28/5
13. 12
15. $\frac{1}{2}(e^7 - e^6 - e^3 + e^2)$
17. 16/3

19. 36
21. 128/9
23. ln 3
25. $(4\sqrt{3}/5) + 1$
27. $\ln(17)/4$
29. 8/3
31. $e^3 - 2e + (1/e)$
33. 54
35. 1

Cumulative Exercise Set 8–6

1. (a) 2 (b) Not defined (c) $\{(x, y)|x \neq -y\}$
3. (a) $f_x = 12x^3 - 4y^3$, $f_y = -12xy^2 - 2$
  (b) $f_{xy} = f_{yx} = -12y^2$, $f_{xx} = 36x^2$, $f_{yy} = -24xy$
5. $x = 9$, $y = 7$

7. 0
9. $(e/2) - 1$
11. 80/3

Case Study Exercises

1. $x = 111.11$
  $y = 37$
3. $x = 122.22$
  $y = 43.46$
5. $x = 133.33$
  $y = 27.78$

Case Study Exercises

1. 10
3. 104
5. 32 cubic units
7. 36,000

9. No, because $b^2 = 2a^2$ and $a^3 = 3b^3$ cannot be satisfied simultaneously.
11. 9000

Review Exercises

1. -1

3.

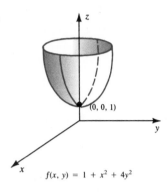

$f(x, y) = 1 + x^2 + 4y^2$

5. $f_x(x, y) = y^3 - 8(2x + 3y)^3$
   $f_y(x, y) = 3xy^2 - 12(2x + 3y)^3$

7. $f_{xx} = -\frac{18}{25}x^{-8/5} + 2y^{-2}$
   $f_{yy} = 6x^2y^{-4}$
   $f_{xy} = f_{yx} = -4xy^{-3}$

9. Critical point $(\frac{10}{11}, \frac{5}{11})$

11. Relative minimum at $(4, -2)$
   Relative maximum at $(0, 2)$
   Saddle points at $(0, -2)$ and $(4, 2)$

13. $(\frac{5}{6}, \frac{5}{2})$

15. The numbers are 15, 15, and 15.

17. $7x^2y^6 + \dfrac{3x^4y^{-2}}{2} - xy + \dfrac{x^2}{2} - x + C(y)$

19. -2 **21.** $\frac{2}{3}$

CHAPTER 9

Exercise Set 9–1

1. $\dfrac{\pi}{2}$ **3.** $\dfrac{7\pi}{6}$

5. $\dfrac{3\pi}{2}$ **7.** $30°$

9. $450°$ **11.** $315°$

13. $495°$, $\frac{11}{8}$ revolutions

15. $630°$, $\frac{7}{4}$ revolutions

17. $840°$, $\frac{7}{3}$ revolutions

19. $-150°$, $\frac{5}{12}$ revolutions in the clockwise direction

21. $-990°$, $\frac{11}{4}$ revolutions in the clockwise direction

23. $-840°$, $\frac{7}{3}$ revolutions in the clockwise direction

25. $\dfrac{13\pi}{6}$ **27.** 6π

29. 4π **31.** $\dfrac{14\pi}{3}$

33. 4π **35.** $\dfrac{13\pi}{2}$

37. $\dfrac{13\pi}{3}$ **39.** 2π rad/sec

Referenced Exercise Set 9–1

1. $\dfrac{\pi}{3}$ radians, $\dfrac{\pi}{6}$ radians

3. (a) $\dfrac{\pi}{90}$ radians (b) $\dfrac{\pi}{60}$ radians (c) $\dfrac{7\pi}{360}$ radians

Exercise Set 9–2

1. $\sqrt{3}$ **3.** $\frac{3}{2}$

5. $\frac{12}{13}$ **7.** $\frac{12}{13}$

9. $\frac{1}{2}$ **11.** $-\dfrac{\sqrt{3}}{2}$

13. $\dfrac{\sqrt{3}}{2}$

15. $-\dfrac{\sqrt{2}}{2}$

17. $-\dfrac{\sqrt{2}}{2}$

19. $-\frac{1}{2}$

21. 0.4794

23. 0.0998

25. 0

27. $\dfrac{\pi}{2}$

29.

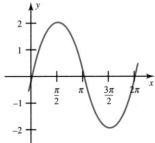

$f(x) = 2 \sin x$

31.

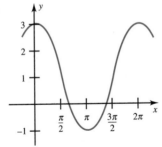

$f(x) = 1 + 2 \cos x$

33.

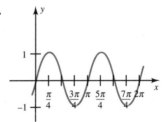

$f(x) = \sin 2x$

35.

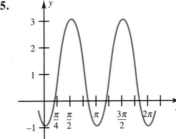

$f(x) = 1 - 2 \cos 2x$

37. Values of $f(x)$ repeat at specified intervals.
39. Amplitude $= 0.002$
41. Figure b
43. Figure a

Referenced Exercise Set 9–2

1. 0.4510, 0.6435

3. 3.3507

Cumulative Exercise Set 9–2

1. $-\frac{25}{6}\pi$
3. (a) $\frac{8}{3}\pi$ **(b)** $480°$

5. (a) 0 **(b)** 2
7. (a) $\frac{1}{13}$ **(b)** $\frac{24}{13}$

Exercise Set 9–3

1. $f'(x) = 6 \cos 6x$
3. $f'(x) = -4x^{-5} \cos x^{-4}$

5. $f'(x) = (8x^3 - 10x)\cos(2x^4 - 5x^2)$
7. $f'(x) = 2x \cos 6x - 6x^2 \sin 6x$

9. $f'(x) = -2(x + \cos 5x)^{-3}(1 - 5 \sin 5x)$

11. $3 \sin 3x + C$ **13.** $-\dfrac{1}{3} \cos x^3 + C$

15. $-\cos^5 x + C$ **17.** $\dfrac{7}{64}$

19. $2\sqrt{2} - 1 \approx 1.828$

21. $f'(x) = e^x \cos e^x + (\cos x)(e^{\sin x})$

23. $f'(x) = (\cos 3x + e^{3x})(\sin 3x + e^{3x})^{-2/3}$

25. $f'(x) = -6x^2 \sin x^3 \cos x^3$

27. $\sin(e^x) + C$ **29.** $e^{\sin x} + C$

31. $f'(x) = 3e^{3x} \cos(e^{3x}) + 3e^{\sin 3x} (\cos 3x)$

33. $-\cos x + \dfrac{\cos^3 x}{3} + C$

35. $0, f(x) = 1$

Cumulative Exercise Set 9–3

1. (a) $990°$, $2\frac{3}{4}$ revolutions
    (b) $480°$, $1\frac{1}{3}$ revolutions
    (c) $-495°$, $1\frac{3}{8}$ revolutions in the clockwise direction

3. (a) $\dfrac{14\pi}{3}$ (b) $\dfrac{25\pi}{3}$ (c) $\dfrac{31\pi}{4}$

5. (a) $2 - \dfrac{3\sqrt{3}}{2}$ (b) $-1 - \dfrac{3\sqrt{3}}{2}$

7. 0

9. (a) $3 \cos 3x - 8 \sin 4x$
    (b) $-x^{-3} \sin x^{-2}$
    (c) $2x \cos(-2x) + 2x^2 \sin(-2x)$

11. $\frac{31}{160}$

Exercise Set 9–4

1. $\dfrac{5}{12}$ **3.** $\dfrac{13}{12}$

5. $f'(x) = 5 \sec^2(5x + 1)$

7. $f'(x) = (3x^2 - 5)\sec^2(x^3 - 5x)$

9. $f'(x) = 20 \tan^4 4x \sec^2 4x$

11. $f'(x) = \dfrac{1}{2} (3 + 2x \sec^2 x^2)(3x + \tan x^2)^{-1/2}$

17. $f'(x) = (2x + 3)\sec(x^2 + 3x)\tan(x^2 + 3x)$

19. $f'(x) = -(3x^{-4} + e^x)\csc^2(x^{-3} - e^x)$

21. $f'(x) = -\dfrac{3}{2} \csc^{1/2}(3x)\cot(3x)$

23. $f'(x) = \tan x$

25. $f'(x) = 6 \sec^2(6x)e^{\tan 6x}$

27. $f'(x) = -3e^{3x} \csc^2(e^{3x})$

29. $f'(x) = 2 \sec^2 2x \cot 3x - 3 \csc^2 3x \tan 2x$

31. $\displaystyle\int \sec^2 x \, dx = \tan x + C$

$\displaystyle\int \csc^2 x \, dx = -\cot x + C$

$\displaystyle\int \sec x \tan x \, dx = \sec x + C$

$\displaystyle\int \csc x \cot x \, dx = -\csc x + C$

33. $\dfrac{\sec 4x}{4} + C$ **35.** $-\dfrac{\cot x^3}{3} + C$

37. $-\dfrac{\cot^3 x}{3} + C$

Referenced Exercise Set 9–4

1. $a \sec^2 x + \dfrac{b \sec^2 x^{1/2}}{2x^{1/2}}$

Cumulative Exercise Set 9–4

1. $\dfrac{5\pi}{3}$

3. (a) 1.9397 (b) 1

5. $\dfrac{\pi}{2} x \cos \pi x + \dfrac{1}{2} \sin \pi x$

7. $\dfrac{5 \sec^2 5x}{2\sqrt{\tan 5x}}$

9. $-\pi \cos \pi x + C$

11. $\tan x + C$

Case Study Exercises

1. 57.2

3. 64.375

5. 36.43

7. 79.6

Chapter Review

1. (a) $\dfrac{3\pi}{4}$ (b) $\dfrac{7\pi}{6}$

3. (a) 660°, $\dfrac{11}{6}$ revolutions

  (b) 750°, $\dfrac{25}{12}$ revolutions

5. $\dfrac{13\pi}{3}$

7. $-\dfrac{\sqrt{2}}{2} \approx -0.707$

9. 0.9950

11. $f'(x) = 4\left(-2x^{-3} - \dfrac{1}{3}x^{-2/3}\right) \cos(x^{-2} - x^{1/3})$

13. $f'(x) = \dfrac{-\cos x^4}{4} + C$

15. $-\tan x + \dfrac{\cos(\ln x)}{x}$

17. $-2(\sec x \tan x + 2 \sec^2 2x)(\sec x + \tan 2x)^{-3}$

19. $-\csc x \cot x (e^{\csc x})$

Index